Joint Function Preservation

Alberto Gobbi · John G. Lane
Umile Giuseppe Longo · Ignacio Dallo
Editors

Joint Function Preservation

A Focus on the Osteochondral Unit

Editors
Alberto Gobbi
O.A.S.I. Bioresearch Foundation
Gobbi NPO
Milan
Italy

John G. Lane
Musculoskeletal and Joint Research
Foundation
San Diego, CA
USA

Umile Giuseppe Longo
Orthopaedics and Traumatology Unit
University Campus Biomedico
Roma
Italy

Ignacio Dallo
OASI Bioresearch Foundation
Gobbi NPO
Milan
Italy

ISBN 978-3-030-82960-5   ISBN 978-3-030-82958-2   (eBook)
https://doi.org/10.1007/978-3-030-82958-2

This Springer imprint is published by the registered company Springer Nature Switzerland AG
The registered company address is: Gewerbestrasse 11, 6330 Cham, Switzerland

There is a trend among book authors to thank every famous person who has inspired them and of course I would like to do the same as without their inspiration I would not be where I am today. But during 2020 the world faced the nightmare of Covid and we were all forced to spend more time at home with our families, and for me Isakos is a big family. Despite this challenging period, together with my friends we have found the inspiration and the energy to write this book.

—Alberto Gobbi and John G. Lane

I dedicate my first book as a co-editor to my parents, who have always been with me and helped me achieve my dreams. To my three children Nicolás, Antonia, and Federico, especially to my wife Lucrecia for their unconditional love and understanding of my many hours of work.

—Ignacio Dallo

Foreword

Before I proceed, I would like to thank Alberto Gobbi and his co-editors for giving me the privilege and honor of introducing this book.

This book delves into the knowledge of articular wellness with the help of recognized experts in the musculoskeletal system.

Joint function may be a well-known concept, but there is still a long way to go to have a deeper understanding of biologics, articular homeostasis, synovia, and the cross-talking concept between the cartilage and its subjacent tissue when the knee function is compromised.

"The concept that subchondral bone can act as an effective shock absorber is not new. It was established by Dr. Physick in 1827 that a ball made of cancellous bone absorbs the energy applied to it, as contrasted with an ivory ball." However, one realizes that in spite of this idea from way back, the importance of the subchondral bone has suffered neglect for decades. This book aims to give the subchondral bone the attention it deserves especially during the process of osteoarthritis.

When subchondral bone is altered, there can be a variety of reasons such as inflammation, edema, avascular necrosis, and other factors. In the same manner, different evaluation and diagnostic systems are available that are fundamental for an accurate assessment.

At the present time, prosthetic treatment continues to be the first treatment option. However, biologic treatments are gaining momentum, and new therapies are appearing which broaden our horizons to improving quality of life in a population with a longer life expectancy. Regeneration is the ultimate goal, and to achieve this, scaffold cells and signaling proteins are imperative, all of this under genetic control.

The characteristics of each articulation have not been forgotten by the editors. The shoulder, elbow, wrist, hip, knee, and ankle joints have all been given due focus. Of course, we cannot forget rehabilitation, which plays a fundamental role in the restoration of function and is also tackled in the book.

To the reader, I hope this introduction transmits the importance and relevance of this work to update and increment our current knowledge of the subchondral bone, the subjacent tissue to cartilage.

Again, I would like to express my gratitude to Alberto and the co-writers of this publication.

Ramon Cugat
Orthopaedic Surgical Department
Hospital Quiron Barcelona
Barcelona, Spain

Introduction to Joint Function Preservation

"But out of limitations comes creativity"—Debbi Allen

Joint osteoarthritis (OA) is one of the largest economic burdens that we face in the modern-day world. This global epidemic is in part due to the "baby boomer" generation that are now suffering from aging-related disease and conditions, such as joint OA, that significantly reduce quality of life and "healthspan." The limited capacity to self-regenerate essential structures within the joint is the number one reason joint preservation is so challenging following traumatic injury or progressive wear-and-tear. Due to this inherent regenerative *limitation*, the precedence of developing effective preservative treatment modalities has been set forth by the World Health Organization to reduce the burden of joint OA and improve overall health during the aging process. As a result, there have been tremendous strides made in the development of novel joint preservation strategies to maintain the structural components and restore biomechanical function. The challenge, as with most biologically and biomechanically complex systems, is to establish a basis through scientific evidence to support these preservation strategies.

Like many others in the scientific community that have risen to this challenge, my laboratory over the last 25 years has been dedicated to improving and enhancing musculoskeletal tissue repair after injury, disease, and aging. Although it has been speculated for numerous years that the high regenerative potential of adult stem cells is due to their multipotentiality, current findings appear to indicate that very few donor cells actually differentiate and participate in the regeneration of these injured musculoskeletal tissues; rather, the vast majority of cells reconstituting the regenerated tissues are host-derived. This concept is further supported by results that have shown that interrupting paracrine signaling (i.e., by blocking VEGF and angiogenesis) of implanted stem cells decreases regeneration and repair capacity in injured, well-vascularized tissues, such as skeletal muscle and bone [1]. Indeed, several blood vessel cells, immune and inflammatory cells, circulating progenitor and resident cells play a role in the regeneration and repair processes at the site of injury; however, the identity of the host cells involved in the repair processes following stem cell transplantation remains unclear. My laboratory and others have also demonstrated that blood vessel walls harbor pericytes which are at the origin of mesenchymal stem cells (MSCs) (1). We therefore believe that promotion of angiogenesis accelerate musculoskeletal tissue repair by creating a supply of adult stem cells in the regenerating area [1]. Unlike most musculoskeletal tissues that are well-vascularized and require angiogenesis for repair, articular cartilage is a hypovascular tissue that requires reduced angiogenic activity for successful adult stem cell-mediated repair in articular cartilage [2, 3]. In the case of articular cartilage repair with adult stem cells, the implanted stem cells also exert a paracrine effect on the local microenvironment, which has a beneficial effect on articular cartilage repair. The host cells participating in the articular cartilage repair process are likely derived from the bone marrow, synovial cells, and other joint derived cells [2, 3]. *To this end, it is clear that the interplay between stem cells and the host micro-milieu act as a unit and highlights the importance of preserving the "niche" while developing regenerative therapies.*

While many stem cell therapies for OA are under investigation, none are currently FDA- approved for modifying the course of the disease. Of the many adult stem cell treatments available for the treatment of OA, bone marrow stem cells (BMSCs) in bone marrow concentrate (BMC) are the most clinically translatable and already in clinical use since they can be harvested using a minimally invasive approach and do not require in vitro expansion. There is, however, significant potential for improving the therapeutic efficacy of BMSC/BMC treatment for OA. The number of senescent cells in BMC increases with age, and these cells release pro-inflammatory cytokines/chemokines, proteases, as well as other senescence-associated secretory phenotypes (SASP) that can impair stem cell function and likely contribute to the development and progression of OA [4]. Compounds (senolytic agents) have been identified that specifically kill senescent cells, abrogating systemic SASP factors and leading to improved outcomes for a variety of musculoskeletal disorders including joint OA [5]. Our group has recently demonstrated that blocking fibrosis with oral losartan (a TGF-β1 inhibitor) can improve cartilage repair by promoting regeneration of hyaline cartilage while reducing the amount of fibrocartilage [6]. *Thus, it is believed that combining senolytic and/or anti-fibrotic agents with orthobiologic therapies (i.e., BMC, BMSCs, and platelet-rich plasma [PRP]) will lead to significantly better outcomes than stem cell treatment alone, representing a new treatment paradigm for joint preservation.*

It is generally understood that not all degenerative and age-related orthopedic disorders can be treated with biological therapies alone; sometimes surgical intervention is necessary to repair damaged structures or address biomechanical abnormalities that create destructive loading in the joint. These combinatorial approaches, incorporating cell-based and biological therapies along with FDA-approved medications (i.e., losartan, anti-angiogenic drugs, senolytic agents) to block or remove certain deleterious factors and cells, respectively, can restore the homeostatic functions of the joint after surgical intervention, facilitating recovery and tissue regeneration. These new therapeutic solutions (in isolation or combined with surgical procedures) as well as current concepts on the osteochondral pathology and characterization are well described in this volume.

Dr. Alberto Gobbi, a dominant figure in the orthopedic field, brought together knowledgeable scientists and clinicians to assemble the *Joint Function Preservation* volume to contribute their professional insight and review contemporary solutions on managing and improving joint preservation and function. Dr. Gobbi is a pioneer in the development of new biological therapies also known as "OrthoBiologics" for degenerative disease like OA. This volume introduces several biologically and surgically based clinical solutions to preserve the local niche, improve the regenerative potential of stem cells as well as biomechanical function of the joint's osteochondral unit. This volume also places emphasis on the fundamental biology of the osteochondral unit and the joint's milieu to improve our understanding of its dynamic function and healing responses in an injured state. This information is important to highlight when discussing joint preservation solutions to delay the progression of OA and re-establish the biological niche. The preservation of the osteochondral unit and joint function is the focal point of this volume and the basis of many orthopedic practices.

Johnny Huard
Director of Center for Regenerative Sports Medicine,
Steadman Philippon Research Institute, Scientific
Director of Proof Point Biologics,
The Steadman Clinic
Vail, CO, USA
Department of Clinical & Biomedical Science,
College of Veterinary Medicine
Colorado State University
Fort Collins, CO, USA

References

1. Huard J. Stem cells, blood vessels, and angiogenesis as major determinants for musculoskeletal tissue repair. J Orthop Res. 2019;37(6):1212–20.
2. Kubo S, Cooper GM, Matsumoto T, Phillippi JA, Corsi KA, Usas A, et al. Blocking vascular endothelial growth factor with soluble Flt-1 improves the chondrogenic potential of mouse skeletal muscle-derived stem cells. Arthritis Rheum. 2009;60(1):155–65.
3. Matsumoto T, Cooper GM, Gharaibeh B, Meszaros LB, Li G, Usas A, et al. Cartilage repair in a rat model of osteoarthritis through intraarticular transplantation of muscle-derived stem cells expressing bone morphogenetic protein 4 and soluble Flt-1. Arthritis Rheum. 2009;60(5):1390–405.
4. Xu M, et al. Transplanted senescent cells induce an osteoarthritis-like condition in mice. J Gerontol A Biol Sci Med Sci. 2017;72(6):780–5.
5. Zhu Y, et al. New agents that target senescent cells: the flavone, fisetin, and the BCL-XL inhibitors, A1331852 and A1155463. Aging (Albany NY). 2017;9(3):955–63.
6. Utsunomiya H, Gao X, Deng Z, Cheng H, Nakama G, Scibetta AC, et al. Biologically regulated marrow stimulation by blocking tgf-beta1 with losartan oral administration results in hyaline-like cartilage repair: a rabbit osteochondral defect model. Am J Sports Med. 2020;48(4):974–84.

Preface

The osteochondral unit is composed of hyaline cartilage connected through a zone of calcified cartilage to the subchondral cortical bone. It transfers load-bearing weight over the joint to allow normal joint articulation and movement and must withstand a combination of compressive, tensile, and shear stresses. The dynamic relationship between cartilage and bone is crucial for joint health and integrity. Damage and degeneration of the osteochondral unit can severely limit patients' quality of life, impacting joint function and leading to several disabling diseases, such as osteoarthritis. Joint pain is often the primary symptom of osteochondral unit disease. Early detection and diagnosis are important for the appropriate treatment. This book is a state-of-the-art guide on osteochondral unit and analyzes a fascinating area of medicine that will continue to grow.

The molecular biology and mechanical properties of the osteochondral unit are vital aspects of many surgical disciplines.

Basic research into the complexities of cartilage/bone crosstalk and the in vivo development of the osteochondral unit is fundamental to the improvement of joint therapies and tissue restoration.

This book provides a comprehensive review of some of the most important scientific and clinically relevant topics in biology, biomechanics, mechanisms of healing, and surgical strategies.

It is hoped that this book will become a valuable resource for clinical understanding, postgraduate research, and resident training.

The book is written by distinguished contributors who have graciously provided new techniques and experimental findings in their respective fields.

Milan, Italy
San Diego, CA, USA
Roma, Italy
Milan, Italy

Alberto Gobbi
John G. Lane
Umile Giuseppe Longo
Ignacio Dallo

Contents

Joint Function and Dysfunction

1

Abigail L. Campbell, Mathew J. Hamula,
and Bert R. Mandelbaum

1.1 Introduction

Articular cartilage is a vital component of an intricate system that constitutes the knee. The purpose of cartilage is to provide a low friction surface for motion as well as a cushion on which to transmit forces efficiently and effectively. It lacks access to either abundant nutrients or progenitor cells rendering it vulnerable to injury and with little capacity to mount a regenerative response. Partial-thickness defects generally do not involve injury to the vasculature; however, chondroprogenitor cells in marrow and blood cannot enter the damaged region. Therefore, these defects have a limited healing potential and typically progress. On the other hand, full-thickness lesions that penetrate the subchondral bone have a higher likelihood of intrinsic repair though typically will go on to heal with fibrocartilage with inferior mechanical properties to native articular cartilage [1]. Understanding and treating dysfunction of the osteochondral unit of the knee requires a fundamental knowledge of physiology and pathophysiology.

Chondropenia, literally meaning "deficiency of cartilage,", describes dysfunction of the osteochondral unit that ranges from mild structural abnormalities to full-blown osteoarthritis. The thickness and volume of articular cartilage follows a paradigm somewhat analogous to Wolff's Law, in that form and mass follow function in bone remodeling. Cartilage demonstrates a directly proportional change in thickness that has a linear dose–response correlation with repetitive loading activities. If the integrity of the functional weight-bearing unit is lost, either through acute injury or chronic microtrauma in the high-impact athlete, a chondropenic response is initiated that can include loss of articular cartilage volume and stiffness, elevation of contact pressures, and development or progression of articular cartilage defects.

Age, obesity, overuse, hormonal factors such as menopause, and trauma are the main risk factors for the onset of chondropenia [2, 3]. The chondropenic cascade leading to chondral lesions and joint degeneration can also be exacerbated by the presence of additional pathology such as ligamentous instability, malalignment, and meniscal injury [4].

Abnormal mechanical stress increases not only nitric oxide (NO) production, but also matrix metalloproteinase (MMP) activity [5]. Abnormal joint forces also disturb chondrocyte metabolism via surface mechanoreceptors such as integrins, which stimulate pro-inflammatory cytokine activity and synthesis [3, 6].

As athletes experience higher rates of knee injury as well as repetitive and abnormal loading

A. L. Campbell · M. J. Hamula
B. R. Mandelbaum (✉)
Cedars Sinai, Kerlan Jobe Orthopedic Institute,
Los Angeles, CA, USA
e-mail: b.mandelbaum@smog-ortho.net

© ISAKOS 2022
A. Gobbi et al. (eds.), *Joint Function Preservation*, https://doi.org/10.1007/978-3-030-82958-2_1

of the joint, they are therefore at greater risk to develop chondropenia [7]. Athletes are in fact at significant risk for symptomatic degenerative joint disease relatively early on in their lives [8–12]. Acute injury has a significant effect on cartilage, and long-term follow-up studies reveal that articular cartilage defects in athletes show a direct link between chondral damage and the development of osteoarthritis [11].

Cartilage injuries of the knee are ubiquitous and affect over one third of athletes compared to less than one fifth of the general population [8]. These injuries can cause significant morbidity and are frequently career-ending. Acute chondral injuries occur in 9–60% of anterior cruciate ligament (ACL) ruptures and over 90% of patellar dislocations [8, 13]. Articular cartilage defects of the femoral condyles have been observed in up to 50% of athletes undergoing anterior cruciate ligament (ACL) reconstruction with an increased propensity in female athletes [7, 14].

Focal cartilage defects have been reported in 60–67% of individuals undergoing knee arthroscopy [15, 16]. Even when treated with state-of-

the-art surgical modalities, it is often difficult to return to previous levels of performance. Cartilage injury can portend a poor prognosis even in healthy athletes. A 2018 ESSKA study of 31 high-level athletes undergoing matrix-associated cartilage transplantation reported that at 10-year follow-up, only 58% of patients returned to pre-injury level of sport [17].

Micro- or macro-trauma creates a catabolic environment for cartilage: inflammatory cytokine production of interleukins-1β, -6, and -8, tumor necrosis factor-α, MMPs 1, 3, 13, and nitric oxide (NO) disrupt the biochemical homeostasis, decreasing collagen formation and increasing degradation [3, 18–20].

The aforementioned factors of repetitive loading, hormonal influence, abnormal loading, and altered mechanics contribute to deleterious effects on the osteochondral unit. The clinical results of these changes manifests in "falling off" the dose–response chondropenic curve proposed by the senior author [21]. Specifically, performance level or response is decreased as a function of dose (activity), as cartilage volume is lost (Fig. 1.1).

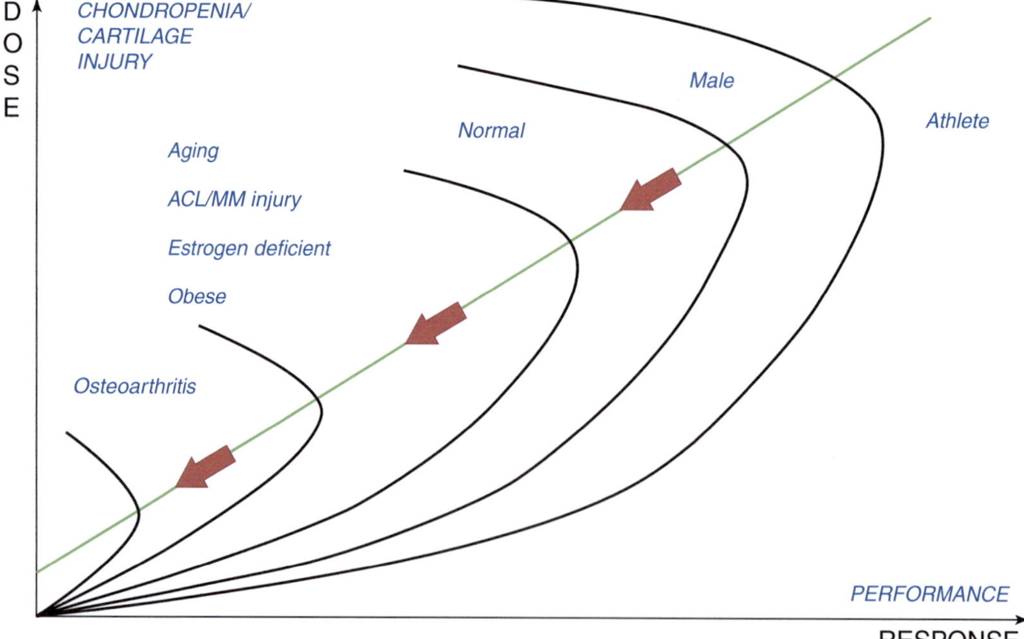

Fig. 1.1 "Dose–response" Chondropenic curve. Correlation model between performance levels (response) and activity performed (dose) as function of joint degeneration

Cartilage injuries have the potential to limit patients' livelihood and athletes' future in their respective occupations, even when addressed operatively. It is therefore imperative for the managing physician to maximize their armamentarium of conservative treatments. This chapter will discuss evaluation and management of dysfunction of the osteochondral unit, with a focus on the active patient. Operative techniques will not be addressed in depth: the focus will be patient care from presentation to postoperative issues. Management strategies will be discussed in the context of the chondral management paradigm: chondroprotection, chondrofacilitation, and resurfacing including an algorithm for recommended care.

1.2 Clinical Evaluation and Classification

Clinical evaluation begins with a thorough history and physical examination. Care should be taken to elicit any history of trauma, either recent or remote, swelling, instability, or mechanical symptoms. Medical history is also relevant: medications, hormonal abnormalities, and systemic diseases can affect the function of the knee. The physical examination should specifically include evaluation for the presence of swelling, effusion, pain to palpation, catching, locking, and special tests to evaluate for concomitant pathology. Range of motion is important and noting any pain with mid-range, terminal flexion, or terminal extension.

Imaging is a crucial adjunct in assessing patients with chondral disease. Plain radiographs are able to evaluate for osteochondral defects, loose bodies, joint space narrowing, alignment, and patellofemoral anatomy. Advanced imaging in the form of magnetic resonance imaging (MRI) is the current standard of diagnostic imaging affording great detail of chondral lesions and underlying bony involvement. Despite advances in MRI technology, chondral lesions may still remain undetected until arthroscopy. One potential application of the Nanoscope (Arthrex, Naples, FL) is to assist with diagnosis in cases where the MRI is not sensitive enough to pick up a lesion. Patient selection is important, however, since it can be difficult to tolerate in the office setting without sedation or pain medication.

The purpose of any classification system is three-fold: distinguish subtle differences in pathology by capturing relevant factors, facilitate communication between clinicians, and guide management. There are several classification systems today including the Outerbridge, Bauer and Shariaree, and cartilage severity score (CSS). Our preferred method is the CSS which provides a scoring system out of 100 including all of the articular surfaces of the knee as well as meniscal integrity. We have found that it is helpful in conveying to patients the severity of cartilage injury whether focal or global. There is also a comprehensive method developed by the International Cartilage Repair Society (ICRS). This score accounts for nine variables: etiology, defect thickness, lesion size, degree of containment, location, ligamentous integrity, meniscal integrity, alignment, and relevant factors in the patient history.

1.3 Indications for Non-operative Management

With the recent advances in cartilage restoration, it may seem trivial to discuss the non-operative management of chondral lesions. However, there are substantial advances in treatment modalities that avoid invasive procedures and significant recovery time and rehabilitation. Additionally, with surgical management there is no guarantee of return to pre-injury levels of function. First, it is important to discuss the indications and contra-indications for non-operative management.

The indications for non-operative management are essentially patients with no significant relative or absolute contra-indications. Patients can consider non-operative treatment of symptomatic cartilage lesions in the absence of any significant red flag symptoms such as mechanical symptoms of locking or catching secondary to a loose body or concurrent reparable meniscal tear. Those with partial-thickness or full-thickness cartilage lesions can consider an initial trial of non-operative management as long as the risks and benefits are discussed thoroughly. Relative contra-indications of non-operative management include concomitant ligamentous or meniscal injury that may predispose the knee to more rapid degeneration. Any significant osteochondral or chondral loose body is an absolute contra-indication to non-operative management and should undergo arthroscopic loose body removal. Furthermore, there is a role for non-operative treatments of patients who may at some point benefit from surgical intervention and for postoperative patients to optimize outcomes and prevent revision surgery.

1.4 Chondroprotection, Chondrofacilitation, and Resurfacing: A Framework for Management

When considering management of osteochondral unit dysfunction in the active patient, it is helpful to have a framework that captures the nuances of pathophysiology and provides guidance for treatment options. Murray et al. [22] outlined in a previous paper three general categories to address chondral pathology:

1. Chondroprotection: strategies that aim to prevent loss of existing cartilage.
2. Chondrofacilitation: strategies that seek to facilitate intrinsic repair of damaged articular cartilage.
3. Resurfacing: improvements in chondral surface function are sought through replacement rather than intrinsic repair of cartilage defects with hyaline cartilage. These include autologous chondrocyte implantation (ACI) in all of its current permutations, autograft and allograft transplantation, and synthetics including scaffolds that fill the defect.

This chapter is not intended to delineate operative techniques but will focus on patient management from presentation to postoperative care. There is a significant cohort of patients that require either chondroprotection or chondrofacilitation postoperatively after a resurfacing procedure. Broadly speaking, we will discuss three groups of patients: non-operative treatment entirely, patients that will go on to need cartilage repair, postoperative patients from a cartilage repair or resurfacing that benefit from chondrofacilitation and chondroprotection in order to maximize outcomes and prevent the need for revision surgery.

1.5 Chondroprotection

The aim of chondroprotection is to promote cartilage homeostasis and prevent the chondropenic cascade that can ultimately lead to loss of structural integrity. As such there are numerous treatment recommendations with varying degrees of supporting evidence. These methods can be characterized as dynamic modifications or pharmacological interventions.

1.5.1 Prevention

The goal is to address any modifiable risk factors with the best protocols to date. Injury prevention programs such as the FIFA 11+ are recommended to reduce risk of intra-articular knee injury, particularly in the female athlete [23].

1.5.2 Acute Injury: Aspiration

In the presence of acute injury with hemarthrosis present, for example, ACL rupture, cartilage is exposed to myriad pro-inflammatory molecules [24]. In the setting of acute injury, aspiration of the knee is recommended to remove the pro-inflammatory mediators in the acutely injured knee. This may mitigate the catabolic effects discussed in this chapter's Introduction.

1.5.3 Weight Loss/Exercise

Joint function is an interplay between motion and the forces that act on it. However, there are limits to modifications that we can recommend as clinicians that have overwhelming supporting evidence. For early osteoarthritis (OA), for example, there is evidence to support lower extremity muscle strengthening for pain and offloading effects [25–28]. Weight loss can reduce peak loads in the knee joint and abductor moment at the knee by a scale of 2.2 kg decrease in peak load for every 1 kg of weight loss [29].

In addition to the weight loss benefits discussed previously, exercise is recommended for knee cartilage disease by the Osteoarthritis Research Society International and the American College of Rheumatology [30, 31]. A 2020 randomized trial published in the New England Journal of Medicine found physical therapy superior to glucocorticoid injection for knee osteoarthritis at 1 year, with those receiving therapy having less pain and functional disability (WOMAC) than those who received glucocorticoid injection [32]. Exercise programs in patients with exacerbations of knee osteoarthritis have been shown to improve symptoms with a relatively low rate of poor effects [33, 34]. Favorable inflammatory biomarker profiles were found with exercise programs in randomized studies [35]. Exercise may have an epigenetic effect as well. MicroRNA–target interactions have been implicated in cartilage disease as well as muscle homeostasis related to exercise [36].

Blood flow restriction therapy is being utilized for various orthopedic applications, and there is some early evidence that it may improve pain while minimizing joint stress in knee osteoarthritis [37, 38]. Exercise is therefore recommended as a staple of first-line management for cartilage disease of the knee. Regarding the use of bracing, there is no level I evidence to support its effect and all available studies are equivocal [39].

1.5.4 Supplements

Glucosamine is a monosaccharide that in vitro has been shown to increase chondrocyte aggre-can production and decrease inflammatory and degradative mediators [40–42]. Chondroitin sulfate is a structural component of cartilage that adds compression strength to the cartilage matrix. Animal studies have demonstrated a chondroprotective effect by anti-inflammatory and anti-degradative effects, as well as stimulation of hyaluronic acid and proteoglycans [43–45].

There are dozens of studies assessing chondroitin sulfate and glucosamine supplementation for the use in cartilage disease of the knee. Examining oral supplementation in humans, a meta-analysis and systematic review of all randomized studies was conducted in 2018 reported that the use of either glucosamine or chondroitin sulfate significantly improved visual analog scale (VAS) pain scores, but did not have this effect when combined and did not affect Western Ontario and McMaster Universities Osteoarthritis Index (WOMAC) score [46]. However, two randomized studies reported reduction on joint space narrowing with chondroitin sulfate [47, 48]. Based on available evidence, chondroitin sulfate supplementation may improve symptoms and mitigate progression of cartilage degeneration in the knee.

Curcumin, a compound found in turmeric, has been studied for use in the knee for its potential anti-inflammatory effect. In animal studies, curcumin administration has a chondroprotective rather than chondrofacilitative action, leading to an increase in the number of chondrocytes and collagen content but not increasing cartilage thickness [49, 50]. However, despite its promising results in recent animal studies, there is little evidence in clinical outcomes with human use. It has been shown to be safe for use in humans for the indication of knee chondral disease [51].

1.5.5 Estrogen

Estrogen plays a well-understood role in the modulation of bone density. Its effect on cartilage has only been recently elucidated. Animal studies have demonstrated that estrogen inhibits degradation of cartilage's extracellular matrix, and that estrogen therapy can reduce the degree of cartilage degeneration [52, 53]. A large cohort study in humans identified post-menopausal status as an independent risk factor for cartilage degeneration [54]. Certain estrogen receptors have been implicated in cartilage catabolism by upregulating matrix metalloproteinases [55, 56]. Due to this relationship, female patients in peri- or postmenopausal age groups experiencing knee pain due to cartilage disease should be referred to an endocrinologist or women's health specialist for hormonal evaluation. Developing a relationship with a local physician in this specialty is highly recommended to optimize patient care.

1.5.6 Steroid

Steroid injections are frequently performed in the knee. While the short-term improvement in pain has been established for use in the knee [57], there is evidence that extended use may have deleterious effects on articular cartilage [58, 59]. While there is concern for possible catabolic effect on cartilage, there is also evidence that intra-articular steroid injections in the knee may have an anabolic effect [60]. We recommend intra-articular steroid injection for use during a flare of knee pain in the absence of acute injury, and one should not fear intermittent use as this treatment can be very effective for acute pain. However, the treating provider should keep in mind that a steroid injection is not a solution for an osteochondral unit injury or dysfunction in the knee.

1.5.7 Future Directions in Chondroprotection

The positive effects of exercise continue to be elucidated as well as supplementation that may be related to diet. Whole body health including diet and exercise will likely become a focus of both preventative and treatment approaches for cartilage injury and disease. As there are no simple and infallible invasive solutions to cartilage injury, prevention in the context of overall health and wellness is likely to become the focus of early management, thereby providing cartilage care before treatment becomes necessary.

1.6 Chondrofacilitation

Once structural damage has occurred, the goal is to facilitate intrinsic repair by creating a harmony between the innate biology and the local articular cartilage milieu. The goals of injectable therapies are to deliver essential growth factors or temper inflammation in order to promote the regeneration or healing response of functional hyaline cartilage. These elements may be used as sole non-operative techniques or as adjuncts to surgical techniques. The focus of this section will be to discuss them in the three groups of patients previously outlined.

1.6.1 Hyaluronic Acid

Hyaluronic acid (HA) is a major component of synovial fluid that has anti-inflammatory effects and may stimulate proteoglycan production. Initially developed as an avian-derived product, most HA is now produced by biological fermentation. HA has multiple functions in the native knee: lubrication, load absorption, fluid homeostasis, and analgesia [61]. Its mechanism of action in cartilage disease specifically comprises proteoglycan and glycosaminoglycan synthesis, anti-inflammatory effect, mechanical lubrication, and analgesia [62]. HA can be utilized as a multiple-injection series or one injection only, based on molecular weight and concentration.

There are myriad products available today including high molecular weight and extended-release. Both molecular weight and HA concentration can influence HA's efficacy, which should be taken into consideration when reading literature on this subject. Animal studies show promising data in its chondrofacilitative effects [50, 63, 64], including a benefit in early administration after acute cartilage injury [65]. Human studies examining intra-articular HA have been widely published, with positive clinical benefits in randomized trials [66, 67]. Of three randomized trials comparing HA and placebo that assessed structural changes on knee MRI, two trials reported no difference in joint space width loss between HA and placebo [68, 69], while one found significantly less joint space loss in both medial and lateral compartments [70]. Clinically, HA has been shown to delay total knee arthroplasty [62, 71].

For these reasons, HA is a valuable asset to the provider managing knee pain due to cartilage acute injury or chondropenia. In our clinic, we often administer HA with steroid in the first of a three-injection series. The addition of steroid to this first injection has anecdotally improved patients' pain faster and allowed earlier return to activities. HA can also be combined with PRP, though evidence behind combination therapy is currently limited. This combination will be discussed further in this chapter.

1.6.2 Platelet-Rich Plasma

Platelet-rich plasma (PRP) in its current iteration has been demonstrated to be safe and contains significant concentrations of autologous growth factors and proteins that may augment intrinsic repair [72]. The current definition includes quantitative criteria, specifically requiring PRP to contain more than one million platelets per milliliter (mL) of serum as this critical concentration shows the most promise in terms of stimulating a healing response [73, 74]. The other factor in PRP formulations is the white blood cell concentration, with leukocyte-rich PRP (LR-PRP) and leukocyte-poor PRP (LP-PRP). While the use of PRP to treat cartilage injuries has rapidly expanded over the last decade, there remains a sparsity of evidence for use in isolated setting in the treatment of chondral lesions. Lui et al. [75] conducted a study showing superior cartilage healing with intra-articular injections of PRP compared to HA in a rabbit model with 5 mm focal chondral defects. Further animal studies on autologous conditioned plasma and platelet-enriched fibrin scaffolds have shown similar superior results [76, 77]. Additionally, combining PRP with HA has been shown to increase the release of growth factors [78].

There is limited clinical evidence to support the use of PRP in vivo for chondral lesions and OA. In some head-to-head comparisons, hyaluronic acid injections seem to outperform PRP alone in terms of pain relief [79–82]. Other studies, including recent meta-analyses and randomized controlled trials, have overall shown more consistent evidence for LP-PRP for intra-articular use in the treatment of chondral lesions and OA compared with placebo and hyaluronic acid [80, 83–86]. In general, LP-PRP likely produces less of an inflammatory response than LR-PRP within the intra-articular environment which may ultimately prove more therapeutically beneficial.

Further studies on standardized formulations are needed to make definitive recommendations on isolated PRP for the non-operative treatment of chondral lesions. However, PRP has been reported to improve cartilage regeneration when used alongside microfracture and osteochondral allograft implantation. In a mouse model, LR-PRP injection was compared to saline injection in femoral condylar focal cartilage defects and found increased cartilage regeneration and collagen II in the repair tissue in the PRP group. This suggests that there is a role for PRP at least as an adjunct, particularly in patients who may at some point benefit from a cartilage restoration procedure or following a surgery in order to enhance chondro-facilitation. A recent study by Everhart et al. [87] demonstrated an improved healing rate in meniscal repairs with the use of PRP at the time of surgery although there was no difference when a concomitant anterior cruciate ligament (ACL) reconstruction was performed. For now, there is a growing body of evidence that PRP is helpful in conjunction with surgical procedures and can facilitate intrinsic repair of cartilage lesions. There is still not enough evidence to recommend its isolated use on focal chondral lesions. However, it may provide a useful temporizing measure for an athlete's mid-season as a non-surgical treatment option prior to an arthroscopic debridement or cartilage restoration procedure.

1.6.3 Bone Marrow Aspirate Concentrate

Bone marrow aspirate concentrate (BMAC) has gained popularity and widespread use as it is relatively easy to harvest and one of the few treatment options acceptable under the US Food and Drug Administration (FDA) guidelines [88]. It can be used to give growth factors to the injury site, such as vascular endothelial growth factor, platelet-derived growth factor, transforming growth factor-beta, and bone morphogenic proteins. This is in addition to the mesenchymal stem cells (MSCs) present in the concentrate. BMAC shows a lot of potential, particularly in the treatment of osteochondral lesions of the tibial plateau where the use of osteochondral allograft is limited by size, shape, or location. There are several studies on the use of BMAC in chondral lesions [89–94], the vast majority with promising results. In general, there were more favorable results when BMAC was used with a scaffold. Given that some studies were inconclusive or showed negative results with BMAC alone, there is currently limited use for BMAC in isolation for the treatment of chondral lesions. However, in conjunction with a scaffold, including even HA, there is some promising data showing improvement in function. BMAC has been reported as a valuable augment to microfracture, matrix-associated chondrocyte implantation, and osteochondral allograft implantation. It has also improved cartilage repair compared with microfacture in an animal model [95]. At this time, BMAC is a valuable addition to our armamentarium when combined with scaffolds. Its role in the non-operative paradigm is confined to intra-articular injection combined with HA in patients who can tolerate the harvest in the clinic setting.

1.6.4 Cellular-Based Therapies

Cellular-based therapies are an attractive option in cartilage restoration. It is important to be cognizant of nomenclature when it comes to this heterogeneous group of therapeutic agents. Stem cells are defined as undifferentiated progenitor cells that are capable of proliferation, regeneration, self-maintenance, and replication [96]. Mesenchymal stem cells (MSCs) are of particular interest in the treatment of chondral lesion due to their accessibility and greater homogeneity in cell division [97]. The Mesenchymal and Tissue Stem Cell Committee of the International Society for Cellular Therapy in 2006 defined the minimal criteria for a human cell to be classified as an MSC: (1) the ability to adhere to plastic when maintained in standard culture conditions; (2) expression of CD105, CD73, and CD90; (3) the lack of expression of CD45, CD34, CD14, or CD11b, CD79alpha or CD19, and HLA-DR surface molecules; and (4) the ability to differentiate into osteoblasts, adipocytes, and chondroblasts in vitro [4]. Chang et al. also postulated that MSCs also have an anti-inflammatory effect based on preclinical trials in small mammals [96]. The two most popular options due to ease of collection are adipose-derived and bone marrow-derived MSCs.

Adipose-derived stem cells (ASC) are relatively easy to harvest and result in a high yield of stem cells [98]. They have been shown to differentiate into chondrocytes in vitro and in vivo [99]. Intra-articular injections of ASCs have been reported to improve patient-reported outcomes for knee osteoarthritis as well as increase cartilage volume [100]. ASCs have been found to induce chondrocyte proliferation and extracellular matrix production as well [101]. This promising therapy still has limited clinical studies however.

Bone marrow-derived MSCs (BMSCs) are even more appealing due to their ease of collection. Sites of extraction include the iliac crest, tibia, or femur. One issue is that yield is typically low and the stem cells must be isolated and expanded in cell culture prior to utilization and this process can take up to 3 weeks. There are several animal models showing the positive effect of MSCs when combined with a matrix or scaffold [102, 103] as well as intra-articular injection of MSCs [104]. Gobbi et al. reported BMAC use in combination with a collagen scaffold and found 80% filling of defects and improved clinical outcomes at 3 years [105]. Although it seems highly promising, there is still a sparsity of literature showing clinical efficacy in humans. Chahla et al. [106] conducted a systematic review of studies evaluating the intra-articular injection of cell-based therapies in the knee. Only six studies were included, several of which were level III designs. While no major adverse events were reported, the improvement was modest and the quality of evidence was poor. Better studies are needed to definitively say that cellular-based therapies are recommended for the non-operative management of chondral lesions.

1.6.5 Osseous Involvement

Chondropenia results from a dose–response repetitive injury that leads to loss of articular cartilage volume. Once chondral lesions and osseous changes begin to occur the pathogenesis of osteoarthritis is well under way. Lesions can either extend through the full-thickness of the cartilage and involve the bone, or simply be accompanied by changes in the subchondral bone. Some of the structural changes that have been observed in the subchondral bone in severe osteoarthritis include bone marrow lesions, loss of mineralization, and progressive replacement of the marrow with fibroneurovascular mesenchymal tissue [107–109]. There is a growing interest in understanding and addressing both the osseous and chondral components of joint degeneration. Bone marrow lesions in osteoarthritis represent a late finding in degenerative joint disease and have been treated with various medications aimed at preventing bone resorption or promoting bone regeneration with varying degrees of success in clinical studies [110–115]. While no studies exist looking at the effect of bracing on bone marrow lesions in the tibiofemoral joint, a randomized controlled trial showed decrease in bone marrow lesion volume with 6 weeks of a pull-on patella sleeve in the patellofemoral joint [116].

There has been some recent investigation into combining intraosseous infiltration of injectable therapies combined with intra-articular to allow infiltration into the cartilage from both internal and external pathways, thereby treating the entire osteochondral unit. Early clinical results of combined intra-articular and intraosseous PRP therapy are promising [117, 118], but long-term data is not yet available. In the presence of subchondral bone edema, this may provide an effective solution to address the inflammatory pathways related to pain and edema. The goal will be to intervene in this process early on and alter the natural history of joint degeneration before the onset of osteoarthritis.

1.6.6 Future Directions in Chondrofacilitation

The goal of facilitating intrinsic cartilage repair without surgical intervention is an ambitious one. As we continue to improve our understanding of the chondropenic cascade and catabolic process of joint degeneration, there will be more potential opportunities for therapeutic interventions. An example of this is Wnt signaling, which has been established as an important factor in the pathogenesis of osteoarthritis. It contributes to differentiation of osteoblasts and chondrocytes, as well as the production of catabolic proteases. A relatively novel Wnt pathway inhibitor, small-molecule 04690 (SM04690) has been shown in a rodent model to induce the differentiation of functional chondrocytes and increase cartilage thickness and cartilage regeneration [119]. Additionally, Deshmukh et al. showed protection from cartilage catabolic activity. This novel therapeutic agent is currently undergoing phase 2B trials and has already demonstrated safety in human applications [120]. It is an exciting prospect to be able to stimulate chondral genesis, in addition to chondrofacilitation and chondroprotection.

There may not be a single therapy that provides effective treatment of cartilage lesions in the making. However, given the complexity of cartilage homeostasis, and by extension chondral pathology, it is more likely the answer will be some combination of therapies. The more immediate future may focus on combining the healing pro-inflammatory effects of PRP or mesenchymal stem cells of BMAC with a scaffold such as HA in a way that could target the chondral lesion effectively. As our understanding of the current modalities improves, we may be on the precipice of a transformation in our non-operative approach to cartilage lesions. Additionally, chondroprotection of cartilage restoration or resurfacing procedures is of paramount importance.

1.7 Chondrorestoration and Resurfacing

While this is not an operative technique guide, we will briefly discuss operative strategies for articular cartilage injury in the athlete. As chondrofacilitative strategies seek to support and augment the body's ongoing attempts to produce hyaline cartilage from the site of injury, chondrorestoration and resurfacing approaches originate from within the lesion itself through transplantation (allogenic or autologous) or implantation of autologous chondrocytes. The literature is influenced by the fact that most studies use different techniques, outcome measures, and differing lengths of follow-up precluding definitive comparison. As such, current AAOS, OARSI, and NICE guidelines conclude that there is no clear superiority for any specific technique and recommend that treatment strategies should be based on individual patient factors. We will outline the key chondrorestorative and resurfacing options, their indications, and available results. The goal is restoration and resurfacing is creating a surface of hyaline cartilage. The addition of biologic augmentation is often indicated, as described in the previous section.

1.7.1 Microfracture

For lesions less than 2 cm^2 that do not have underlying osseous defects, microfracture can be performed. Perforation of subchondral bone generates conduits to the vascularized bone marrow allowing migration of marrow cells and intrinsic repair. The main drawback is the limited durability of the new articular surface, which is predominantly fibrocartilage. While short-term outcomes are good [121], the long-term data behind microfracture has been disappointing [122, 123]. The focus has shifted to augmenting and optimizing microfracture rather than performing it as a stand-alone procedure. As larger microfracture holes have been associated with bony impaction and walling off of marrow, nanofractures have been described using thinner awls (1 mm) that protrude to a controlled depth of 9 mm. Preservation of trabecular architecture with this technique has been confirmed using high-definition CT [124]. Concomitant use of PRP or BMAC may improve outcomes over microfracture alone as well [95, 125].

1.7.2 Osteochondral Autograft Implantation

Indications for osteochondral autograft implantation are for osteochondral lesions <2 cm^2. Osteochondral implantation replaces mature hyaline cartilage with autograft tissue including a segment of underlying bone. There are several commercially available systems. Defects have been successfully addressed in young athletes although long-term results in this population are still unclear. In a 17-year prospective multicenter study, good to excellent results were reported in 91% of femoral, 86% of tibial, and 74% of patellofemoral mosaicplasty in athletes [126]. A prospective, randomized study reported significant superiority of osteochondral transfer over microfracture at 3 years [127]. Limitations include the potential for incongruity and graft height mismatch that can result in early wear [128].

1.7.3 Osteochondral Allograft Implantation

Indications for osteochondral allograft implantation are for >2 cm^2 full-thickness chondral defects with or without osseous defect or AVN. Osteochondral allograft transfer (OALT) procedures avoid the challenges of matching chondral thickness, geometry, and donor-site morbidity that limit autologous transfer procedures. Several studies have shown that transplanted bone is well-incorporated by the host with good articular cartilage function. A 91% success rate was reported at 5 years, 85% at 7.5 years, and 75% at 10 years with femoral and patellofemoral allografting, with overall better outcomes on condyles than the patellofemoral joint similar to other cartilage techniques [129, 130]. Although osteochondral allograft transplantation has better durability than microfracture, there remains a long-term decrease in graft survival [131]. While concern is present in using osteochondral implantation techniques following microfracture, a recent study found similar outcomes, satisfaction, and reoperation rates for both autologous chondrocyte implantation and osteochondral allograft transplantation following failed microfracture [132].

Concerns about graft sterility, rejection, access to allograft, and cost are limiting factors. As with autologous osteochondral plugs, graft subsidence, lack of integration, and peripheral chondrocyte death may occur. If graft incorporation occurs, however, good to excellent outcomes are generally achieved with accelerated return to sport [92].

1.7.4 Autologous Chondrocyte Implantation

Indications for autologous chondrocyte implantation (ACI) are focal lesions of 1–10 cm^2, or failed microfracture. Contra-indications include >8 mm depth of bone loss, kissing lesions, osteoarthritis, and inflammatory arthritis. ACI is a two-step procedure: first, harvesting chondrocytes from a healthy non-weight-bearing portion of the knee, second, implantation of culture-expanded autologous chondrocytes under a periosteal flap (first-generation ACI), a collagen membrane (second-generation ACI), or onto a membrane carrier or porous scaffold prior to implantation (third-generation ACI, MACI).

Good to excellent results have been reported in 85–92% of patients at 2 years, with femoral condyle lesions generally producing better results than defects in the patellofemoral joint [133]. Sustained improvements seen in large, symptomatic, full-thickness lesions of the distal femur treated with ACI have been reported in the majority of patients at up to 10 years [134]. A recent long-term study reported increased stiffness of the repair tissue in the first 5 years following ACI, most rapidly in the first 2 years, with final stiffness similar to hyaline cartilage [135]. When performed in elite athletes, ACI resulted in a successful return to sport extending to 5 years and beyond [136, 137]. The main disadvantage of ACI is the time required for tissue maturation,

therefore extending ultimate return-to-sport time; however, promising short-term data has recently been reported for single-stage procedures [138].

1.7.5 Rehabilitation and Return to Sport

Rehabilitation aims to return the patient to sport, to prevent subsequent or further reinjury, and to minimize the risk of cartilage degeneration. An individualized approach should be taken, and it should be recognized that not all athletes will return to pre-injury levels of function after cartilage surgery.

Rehabilitation must be adapted to the type of chondrorestorative or resurfacing procedure performed and each athlete's sport-specific demands. Additional procedures performed must also be taken into consideration. We utilize a stepwise approach consisting of an initial protection and joint activation phase, a progressive joint loading and functional restoration phase, and finally an activity restoration phase. The length of rehabilitation is not time-based, and rather depends on the athlete's performance within each stage. A key benefit of osteochondral grafting is that early weight-bearing can be tolerated. This is not the same with ACI/MACI or microfracture, where the repair construct has to be given time to mature and incorporate. Combined procedures (ACL reconstruction, tibial osteotomy, meniscal procedures) do not adversely affect the return-to-sport rate following cartilage repair although rehabilitation can be modified addressing the concomitant procedure [139].

Prospective studies have shown that 33–96% of athletes return to sport after ACI, with 60–80% returning to the same level. Average return-to-sport time following ACI is 18–25 months [140]. Return to competition has been reported in 59–66% of athletes after microfracture, with 57% returning to their pre-operative level of performance at 8–17 months [121, 141]. Sporting return has been reported in 91–93% of athletes after osteochondral transfer at mean 6.5–7 months [127]. There has been higher return-to-sport rate reported for osteochondral autograft implantation compared to microfracture [142].

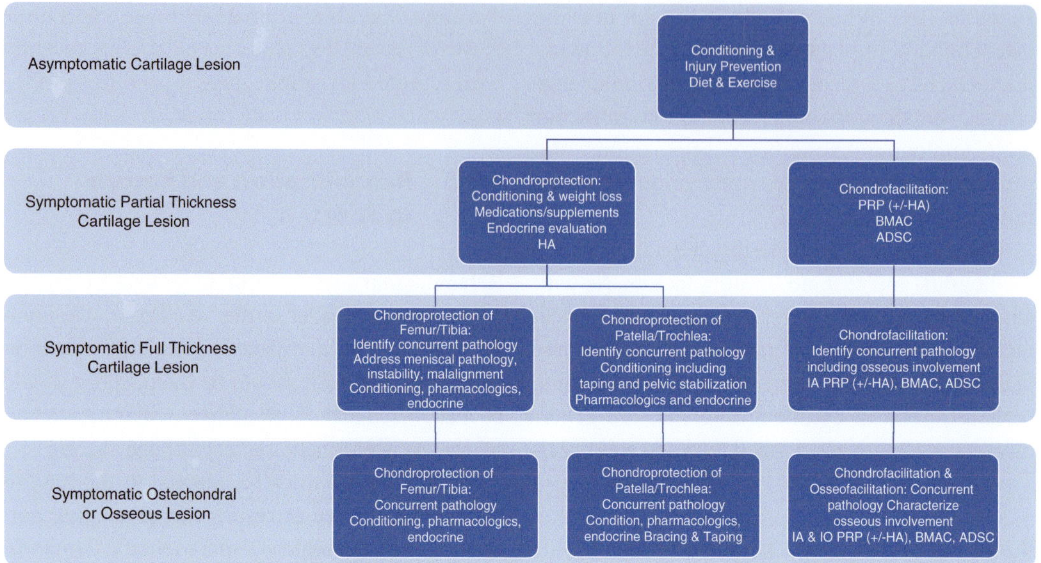

Fig. 1.2 A non-operative treatment algorithm for the management of cartilage lesions based on chondroprotection and chondrofacilitation in chondral, osteochondral, and osseous lesions. *HA* hyaluronic acid, *PRP* platelet-rich plasma, *BMAC* bone marrow aspirate concentrate, *ADSC* adipose-derived stem cells, *IA* intra-articular, *IO* intraosseous

Eighty-eight percentage of athletes returned to partial activity and 79% returned to full activity after knee osteochondral allograft transplantation at average 9.6 months [143]. Regardless of the technique used, the time to return to sport is longer for younger and more competitive athletes [144]. The absence of prior surgery, higher pre-injury level of sport, and shorter pre-operative duration of symptoms correlate with higher return-to-sport rates [145].

1.8 Treatment Algorithm

We offer our current treatment algorithm focusing on full spectrum management of osteochondral unit dysfunction in the athlete, based on the principles of chondroprotection, chondrofacilitation, and resurfacing. Asymptomatic lesions, so long as there are no absolute indications for surgical management, should be monitored and treated with conditioning, minimizing high-impact joint loading when possible, and injury prevention protocols. Diet and exercise can also play a pivotal role in maintaining functionality.

Once cartilage lesions become symptomatic, first-line treatment should include a comprehensive analysis and discussion of dietary and exercise programs. This may include supplementation as discussed in the Chondroprotection section of this chapter. Chondroprotective measures include conditioning, weight loss, medications, supplements, and endocrine evaluation. Chondroprotection also involves identifying concurrent pathology such as meniscal tears, instability, and malalignment and potentially third-line treatment of surgical management.

Second-line modalities can be broadly categorized as chondroprotective or chondrofacilitative. Chondrofacilitation should be individualized to the patient and pathology. Non-operative management is outlined in Fig. 1.2.

Third-line treatment comprises surgical management of osteochondral unit dysfunction, accompanied by appropriate targeted rehabilitation, incorporating the chondroprotective and chondrofacilitative elements described in these sections. In cases with structural injury, surgical management is indicated as outlined in Fig. 1.3.

Fig. 1.3 A treatment algorithm for the management of articular cartilage defects in athletes based on protection of existing cartilage, chondrofacilitation, and chondrorestoration/resurfacing. *ACI* autologous chondrocyte implantation, *PF* patellofemoral, *OCD* osteochondral defect, *AVN* avascular necrosis. *Athletes undergoing tibiofemoral realignment osteotomy should be counseled on the poor prognosis of competitive sporting return

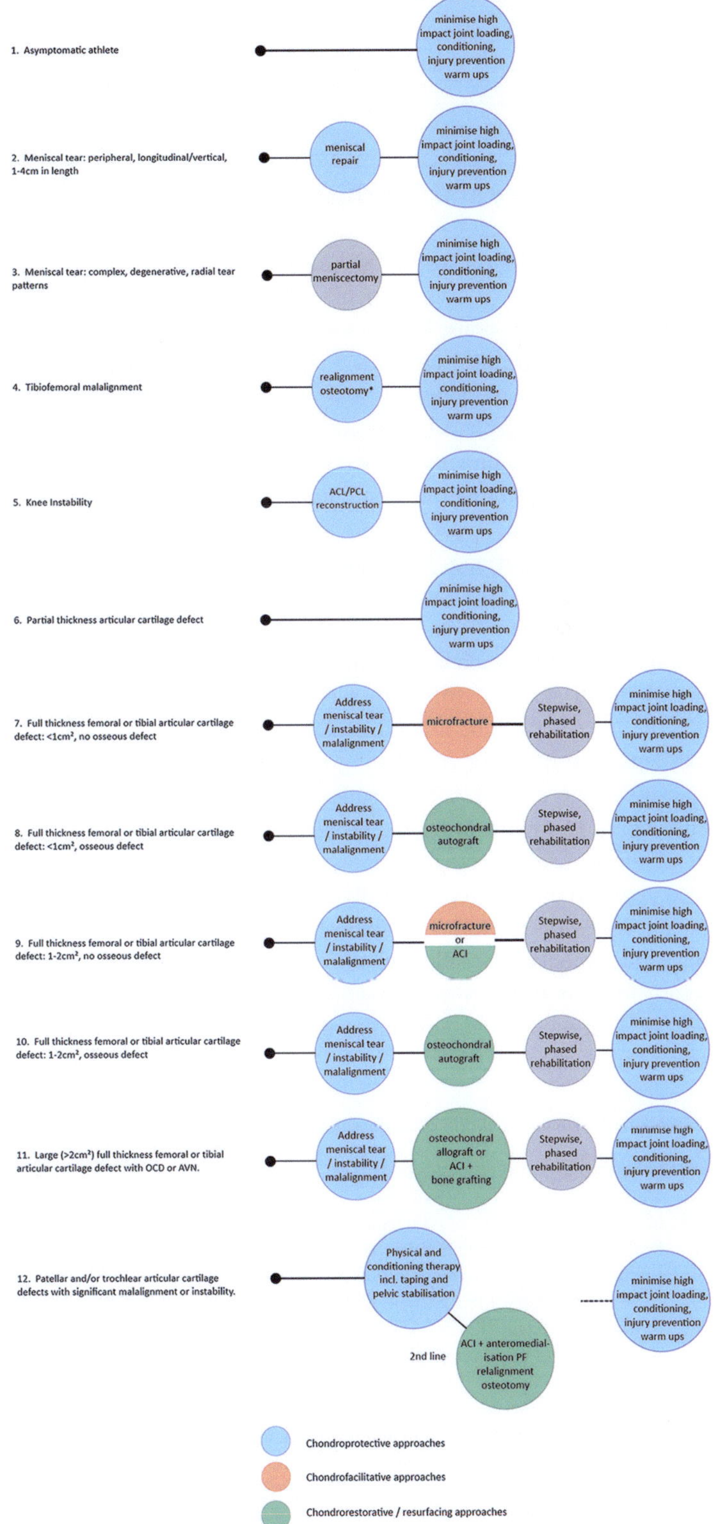

As our understanding and therapeutic techniques continue to evolve, this algorithm will expand significantly. From Murray et al. [146] (Mandelbaum KSSTA).

1.9 Summary and Conclusion

A multitude of non-operative modalities exist for the prevention of chondropenia and treatment of cartilage lesions. It is an exciting prospect as orthopedic surgeons and other practitioners become more critical of current surgical solutions for cartilage lesions or seek to help patients who previously would not have had any worthwhile treatment options. The goal is an ambitious one to prevent chondropenia and protect chondral surfaces by stimulating regeneration of native functional hyaline cartilage using growth factors and anti-inflammatory therapies. Surgical techniques aimed at restoring chondral surfaces still play a crucial role, and the focus should be to facilitate and protect cartilage restoration or resurfacing procedures. Currently, there is no single satisfactory all-encompassing treatment for the broad spectrum of chondral lesions. Therefore, an individualized approach is required that fully involves the patient in the discussion. The aims are to maximize the potential for athletes and patients to return to their full sporting or working activities, prevent reinjury, and minimize the progression of joint degeneration.

References

1. Murray IR, Corselli M, Petrigliano FA, Soo C, Peault B. Recent insights into the identity of mesenchymal stem cells: implications for orthopaedic applications. Bone Joint J. 2014;96-B(3):291–8.
2. Aydin AT, Ozenci M, Gür S. Chondropenia: early-stage degenerative disease. Acta Orthop Traumatol Turc. 2007;
3. Speziali A, Delcogliano M, Tei M, Placella G, Chillemi M, Tiribuzi R, et al. Chondropenia: current concept review. Musculoskelet Surg. 2015;
4. Dominici M, Le Blanc K, Mueller I, Slaper-Cortenbach I, Marini F, Krause D, et al. Minimal criteria for defining multipotent mesenchymal stromal cells. The International Society for Cellular Therapy position statement. Cytotherapy. 2006;8(4):315–7.
5. Fermor B, Brice Weinberg J, Pisetsky DS, Misukonis MA, Banes AJ, Guilak F. The effects of static and intermittent compression on nitric oxide production in articular cartilage explants. J Orthop Res. 2001;19(4):729–37.
6. Pulai JI, Chen H, Im H-J, Kumar S, Hanning C, Hegde PS, et al. NF-κB mediates the stimulation of cytokine and chemokine expression by human articular chondrocytes in response to fibronectin fragments. J Immunol. 2005;174(9):5781–8.
7. Arendt E, Dick R. Knee injury patterns among men and women in collegiate basketball and soccer: NCAA data and review of literature. Am J Sports Med. 1995;23(6):694–701.
8. Flanigan DC, Harris JD, Trinh TQ, Siston RA, Brophy RH. Prevalence of chondral defects in athletes' knees: a systematic review. Med Sci Sports Exerc. 2010;42(10):1795–801.
9. Walczak BE, McCulloch PC, Kang RW, Zelazny A, Tedeschi F, Cole BJ. Abnormal findings on knee magnetic resonance imaging in asymptomatic NBA players. J Knee Surg. 2008;21(1):27–33.
10. Roos H, Lindberg H, Gärdsell P, Lohmander LS, Wingstrand H. The prevalence of gonarthrosis and its relation to meniscectomy in former soccer players. Am J Sports Med. 1994;22(2):219–22.
11. Messner K, Maletius W. The long-term prognosis for severe damage to weight-bearing cartilage in the knee: a 14-year clinical and radiographic follow-up in 28 young athletes. Acta Orthop Scand. 1996;67(2):165–8.
12. Brophy RH, Rodeo SA, Barnes RP, Powell JW, Warren RF. Knee articular cartilage injuries in the National Football League: epidemiology and treatment approach by team physicians. J Knee Surg. 2009;22(4):331–8.
13. Brophy RH, Zeltser D, Wright RW, Flanigan D. Anterior cruciate ligament reconstruction and concomitant articular cartilage injury: incidence and treatment. Arthroscopy. 2010;26(1):112–20.
14. Piasecki DP, Spindler KP, Warren TA, Andrish JT, Parker RD. Intraarticular injuries associated with anterior cruciate ligament tear: findings at ligament reconstruction in high school and recreational athletes. An analysis of sex-based differences. Am J Sports Med. 2003;31(4):601–5.
15. Widuchowski W, Widuchowski J, Trzaska T. Articular cartilage defects: study of 25,124 knee arthroscopies. Knee. 2007;14(3):177–82.
16. Curl WW, Krome J, Gordon ES, Rushing J, Smith BP, Poehling GG. Cartilage injuries: a review of 31,516 knee arthroscopies. Arthroscopy. 1997;13(4):456–60.
17. Zaffagnini S, Vannini F, Di Martino A, Andriolo L, Sessa A, Perdisa F, et al. Low rate of return to pre-injury sport level in athletes after cartilage sur-

gery: a 10-year follow-up study. Knee Surg Sports Traumatol Arthrosc. 2019;27(8):2502–10.

18. Krasnokutsky S, Samuels J, Abramson SB. Osteoarthritis in 2007. Bull NYU Hosp Jt Dis. 2007;65(3):222–8.

19. Cameron M, Buchgraber A, Passler H, Vogt M, Thonar E, Fu F, et al. The natural history of the anterior cruciate ligament-deficient knee. Changes in synovial fluid cytokine and keratan sulfate concentrations. Am J Sports Med. 1997;25(6):751–4.

20. Stefan Lohmander L, Roos H, Dahlberg L, Hoerrner LA, Lark MW. Temporal patterns of stromelysin-1, tissue inhibitor, and proteoglycan fragments in human knee joint fluid after injury to the cruciate ligament or meniscus. J Orthop Res. 1994;12(1):21–8.

21. McAdams TR, Mithoefer K, Scopp JM, Mandelbaum BR. Articular cartilage injury in athletes. Cartilage. 2010;1(3):165–79.

22. Murray IR, Benke MT, Mandelbaum BR. Management of knee articular cartilage injuries in athletes: chondroprotection, chondrofacilitation, and resurfacing. Knee Surg Sports Traumatol Arthrosc. 2016;24(5):1617–26.

23. Steffen K, Emery CA, Romiti M, Kang J, Bizzini M, Dvorak J, et al. High adherence to a neuromuscular injury prevention programme (FIFA 11+) improves functional balance and reduces injury risk in Canadian youth female football players: a cluster randomised trial. Br J Sports Med. 2013;47(12):794–802.

24. Swärd P, Frobell R, Englund M, Roos H, Struglics A. Cartilage and bone markers and inflammatory cytokines are increased in synovial fluid in the acute phase of knee injury (hemarthrosis)—a cross-sectional analysis. Osteoarthr Cartil. 2012;20(11):1302–8.

25. Berger MJ, McKenzie CA, Chess DG, Goela A, Doherty TJ. Quadriceps neuromuscular function and self-reported functional ability in knee osteoarthritis. J Appl Physiol. 2012;113(2):255–62.

26. Chang A, Hayes K, Dunlop D, Song J, Hurwitz D, Cahue S, et al. Hip abduction moment and protection against medial tibiofemoral osteoarthritis progression. Arthritis Rheum. 2005;52(11):3515–9.

27. Fransen M, McConnell S. Land-based exercise for osteoarthritis of the knee: a metaanalysis of randomized controlled trials. J Rheumatol. 2009;36(6):1109–17.

28. Segal NA, Glass NA, Torner J, Yang M, Felson DT, Sharma L, et al. Quadriceps weakness predicts risk for knee joint space narrowing in women in the MOST cohort. Osteoarthr Cartil. 2010;18(6):769–75.

29. Aaboe J, Bliddal H, Messier SP, Alkjaer T, Henriksen M. Effects of an intensive weight loss program on knee joint loading in obese adults with knee osteoarthritis. Osteoarthr Cartil. 2011;19(7):822–8.

30. Kolasinski SL, Neogi T, Hochberg MC, Oatis C, Guyatt G, Block J, et al. American College of Rheumatology/Arthritis Foundation guideline for the Management of Osteoarthritis of the hand, hip, and knee. Arthritis Rheumatol. 2020;72(2):220–33.

31. Bannuru RR, Osani MC, Vaysbrot EE, Arden NK, Bennell K, Bierma-Zeinstra SMA, et al. OARSI guidelines for the non-surgical management of knee, hip, and polyarticular osteoarthritis. Osteoarthr Cartil. 2019;27(11):1578–89.

32. Deyle GD, Allen CS, Allison SC, Gill NW, Hando BR, Petersen EJ, et al. Physical therapy versus glucocorticoid injection for osteoarthritis of the knee. N Engl J Med. 2020;382(15):1420–9.

33. Bartholdy C, Klokker L, Bandak E, Bliddal H, Henriksen M. A standardized "rescue" exercise program for symptomatic flare-up of knee osteoarthritis: description and safety considerations. 2016;46(11):942–6.

34. Lee K-H, Lim J-W, Park Y-G, Ha Y-C. Erratum to: vitamin D deficiency is highly concomitant but not strong risk factor for mortality in patients aged 50 year and older with hip fracture. J Bone Metab. Korea (South). 2016;23:49.

35. Bricca A, Juhl CB, Steultjens M, Wirth W, Roos EM. Impact of exercise on articular cartilage in people at risk of, or with established, knee osteoarthritis: a systematic review of randomised controlled trials. Br J Sports Med. 2019;53(15):940–7.

36. Shorter E, Sannicandro AJ, Poulet B, Goljanek-Whysall K. Skeletal muscle wasting and its relationship with osteoarthritis: a mini-review of mechanisms and current interventions. Curr Rheumatol Rep. 2019;21(8):40.

37. Ferraz RB, Gualano B, Rodrigues R, Kurimori CO, Fuller R, Lima FR, et al. Benefits of resistance training with blood flow restriction in knee osteoarthritis. Med Sci Sports Exerc. 2018;50(5):897–905.

38. Minniti MC, Statkevich AP, Kelly RL, Rigsby VP, Exline MM, Rhon DI, et al. The safety of blood flow restriction training as a therapeutic intervention for patients with musculoskeletal disorders: a systematic review. Am J Sports Med. 2020;48(7):1773–85.

39. Duivenvoorden T, Brouwer RW, van Raaij TM, Verhagen AP, Verhaar JA, Bierma-Zeinstra SM. Braces and orthoses for treating osteoarthritis of the knee. Cochrane Database Syst Rev. 2015;(3):CD004020.

40. Azuma K, Osaki T, Tsuka T, Imagawa T, Okamoto Y, Takamori Y, et al. Effects of oral glucosamine hydrochloride administration on plasma free amino acid concentrations in dogs. Mar Drugs. 2011;9(5):712–8.

41. Imagawa K, de Andres MC, Hashimoto K, Pitt D, Itoi E, Goldring MB, et al. The epigenetic effect of glucosamine and a nuclear factor-kappa B (NF-kB) inhibitor on primary human chondrocytes—implications for osteoarthritis. Biochem Biophys Res Commun. 2011;405(3):362–7.

42. Dalirfardouei R, Karimi G, Jamialahmadi K. Molecular mechanisms and biomedical applications of glucosamine as a potential multifunctional therapeutic agent. Life Sci. 2016;152:21–9.

43. Hui JH, Chan SW, Li J, Goh JC, Li L, Ren XF, et al. Intra-articular delivery of chondroitin sulfate for the treatment of joint defects in rabbit model. J Mol Histol. 2007;38(5):483–9.

44. Monfort J, Pelletier JP, Garcia-Giralt N, Martel-Pelletier J. Biochemical basis of the effect of chondroitin sulphate on osteoarthritis articular tissues. Ann Rheum Dis. 2008;67(6):735–40.

45. Ronca F, Palmieri L, Panicucci P, Ronca G. Anti-inflammatory activity of chondroitin sulfate. Osteoarthr Cartil. 1998;6 Suppl A:14–21.

46. Simental-Mendia M, Sanchez-Garcia A, Vilchez-Cavazos F, Acosta-Olivo CA, Pena-Martinez VM, Simental-Mendia LE. Effect of glucosamine and chondroitin sulfate in symptomatic knee osteoarthritis: a systematic review and meta-analysis of randomized placebo-controlled trials. Rheumatol Int. 2018;38(8):1413–28.

47. Kahan A, Uebelhart D, De Vathaire F, Delmas PD, Reginster JY. Long-term effects of chondroitins 4 and 6 sulfate on knee osteoarthritis: the study on osteoarthritis progression prevention, a two-year, randomized, double-blind, placebo-controlled trial. Arthritis Rheum. 2009;60(2):524–33.

48. Hochberg MC, Zhan M, Langenberg P. The rate of decline of joint space width in patients with osteoarthritis of the knee: a systematic review and meta-analysis of randomized placebo-controlled trials of chondroitin sulfate. Curr Med Res Opin. 2008;24(11):3029–35.

49. Nicoliche T, Maldonado DC, Faber J, Silva M. Evaluation of the articular cartilage in the knees of rats with induced arthritis treated with curcumin. PLoS One. 2020;15(3):e0230228.

50. Zhang Z, Leong DJ, Xu L, He Z, Wang A, Navati M, et al. Curcumin slows osteoarthritis progression and relieves osteoarthritis-associated pain symptoms in a post-traumatic osteoarthritis mouse model. Arthritis Res Ther. 2016;18(1):128.

51. Shep D, Khanwelkar C, Gade P, Karad S. Safety and efficacy of curcumin versus diclofenac in knee osteoarthritis: a randomized open-label parallel-arm study. Trials. 2019;20(1):214.

52. Xu X, Li X, Liang Y, Ou Y, Huang J, Xiong J, et al. Estrogen modulates cartilage and subchondral bone remodeling in an ovariectomized rat model of postmenopausal osteoarthritis. Med Sci Monit. 2019;25:3146–53.

53. Xu K, Sha Y, Wang S, Chi Q, Liu Y, Wang C, et al. Effects of bakuchiol on chondrocyte proliferation via the PI3K-Akt and ERK1/2 pathways mediated by the estrogen receptor for promotion of the regeneration of knee articular cartilage defects. Cell Prolif. 2019;52(5):e12666.

54. Lou C, Xiang G, Weng Q, Chen Z, Chen D, Wang Q, et al. Menopause is associated with articular cartilage degeneration: a clinical study of knee joint in 860 women. Menopause. 2016;23(11):1239–46.

55. Son YO, Park S, Kwak JS, Won Y, Choi WS, Rhee J, et al. Estrogen-related receptor gamma causes osteoarthritis by upregulating extracellular matrix-degrading enzymes. Nat Commun. 2017;8(1):2133.

56. Liang Y, Duan L, Xiong J, Zhu W, Liu Q, Wang D, et al. E2 regulates MMP-13 via targeting miR-140 in IL-1beta-induced extracellular matrix degradation in human chondrocytes. Arthritis Res Ther. 2016;18(1):105.

57. Arroll B, Goodyear-Smith F. Corticosteroid injections for osteoarthritis of the knee: meta-analysis. BMJ. 2004 Apr;328(7444):869.

58. McAlindon TE, LaValley MP, Harvey WF, Price LL, Driban JB, Zhang M, et al. Effect of intra-articular triamcinolone vs saline on knee cartilage volume and pain in patients with knee osteoarthritis: a randomized clinical trial. JAMA. 2017;317(19):1967–75.

59. Raynauld J-P, Buckland-Wright C, Ward R, Choquette D, Haraoui B, Martel-Pelletier J, et al. Safety and efficacy of long-term intraarticular steroid injections in osteoarthritis of the knee: a randomized, double-blind, placebo-controlled trial. Arthritis Rheum. 2003;48(2):370–7.

60. Klocke R, Levasseur K, Kitas GD, Smith JP, Hirsch G. Cartilage turnover and intra-articular corticosteroid injections in knee osteoarthritis. Rheumatol Int. 2018;38(3):455–9.

61. Adams ME, Lussier AJ, Peyron JG. A risk-benefit assessment of injections of hyaluronan and its derivatives in the treatment of osteoarthritis of the knee. Drug Saf. 2000;23(2):115–30.

62. Altman R, Lim S, Steen RG, Dasa V. Hyaluronic acid injections are associated with delay of Total knee replacement surgery in patients with knee osteoarthritis: evidence from a large U.S. health claims database. PLoS One. 2015;10(12):e0145776.

63. Pashuck TD, Kuroki K, Cook CR, Stoker AM, Cook JL. Hyaluronic acid versus saline intra-articular injections for amelioration of chronic knee osteoarthritis: a canine model. J Orthop Res. 2016;34(10):1772–9.

64. Barreto RB, Sadigursky D, de Rezende MU, Hernandez AJ. Effect of hyaluronic acid on chondrocyte apoptosis. Acta Ortop Bras. 2015;23(2):90–3.

65. Kaplan LD, Lu Y, Snitzer J, Nemke B, Hao Z, Biro S, et al. The effect of early hyaluronic acid delivery on the development of an acute articular cartilage lesion in a sheep model. Am J Sports Med. 2009;37(12):2323–7.

66. Petterson SC, Plancher KD. Single intra-articular injection of lightly cross-linked hyaluronic acid reduces knee pain in symptomatic knee osteoarthritis: a multicenter, double-blind, randomized, placebo-controlled trial. Knee Surg Sports Traumatol Arthrosc. 2019;27(6):1992–2002.

67. Petrella RJ, Petrella M. A prospective, randomized, double-blind, placebo controlled study to evaluate the efficacy of intraarticular hyaluronic acid for osteoarthritis of the knee. J Rheumatol. 2006;33(5):951–6.

68. Pham T, Le Henanff A, Ravaud P, Dieppe P, Paolozzi L, Dougados M. Evaluation of the symptomatic and

structural efficacy of a new hyaluronic acid compound, NRD101, in comparison with diacerein and placebo in a 1 year randomised controlled study in symptomatic knee osteoarthritis. Ann Rheum Dis. 2004;63(12):1611–7.

69. Listrat V, Ayral X, Patarnello F, Bonvarlet JP, Simonnet J, Amor B, et al. Arthroscopic evaluation of potential structure modifying activity of hyaluronan (Hyalgan) in osteoarthritis of the knee. Osteoarthr Cartil. 1997;5(3):153–60.

70. Wang Y, Hall S, Hanna F, Wluka AE, Grant G, Marks P, et al. Effects of Hylan G-F 20 supplementation on cartilage preservation detected by magnetic resonance imaging in osteoarthritis of the knee: a two-year single-blind clinical trial. BMC Musculoskelet Disord. 2011;12:195.

71. Dasa V, Lim S, Heeckt P. Real-world evidence for safety and effectiveness of repeated courses of hyaluronic acid injections on the time to knee replacement surgery. Am J Orthop (Belle Mead NJ). 2018;47(7)

72. Foster TE, Puskas BL, Mandelbaum BR, Gerhardt MB, Rodeo SA. Platelet-rich plasma: from basic science to clinical applications. Am J Sports Med. 2009;37(11):2259–72.

73. Dhillon RS, Schwarz EM, Maloney MD. Platelet-rich plasma therapy—future or trend? Arthritis Res Ther. 2012;14(4):219.

74. Rughetti A, Giusti I, D'Ascenzo S, Leocata P, Carta G, Pavan A, et al. Platelet gel-released supernatant modulates the angiogenic capability of human endothelial cells. Blood Transfus. 2008;6(1):12–7.

75. Liu J, Song W, Yuan T, Xu Z, Jia W, Zhang C. A comparison between platelet-rich plasma (PRP) and hyaluronate acid on the healing of cartilage defects. PLoS One. 2014;9(5):e97293.

76. Milano G, Deriu L, Sanna Passino E, Masala G, Saccomanno MF, Postacchini R, et al. The effect of autologous conditioned plasma on the treatment of focal chondral defects of the knee. An experimental study. Int J Immunopathol Pharmacol. 2011;24(1 Suppl 2):117–24.

77. Goodrich LR, Chen AC, Werpy NM, Williams AA, Kisiday JD, Su AW, et al. Addition of mesenchymal stem cells to autologous platelet-enhanced fibrin scaffolds in chondral defects: does it enhance repair? J Bone Joint Surg Am. 2016;98(1):23–34.

78. Iio K, Furukawa K-I, Tsuda E, Yamamoto Y, Maeda S, Naraoka T, et al. Hyaluronic acid induces the release of growth factors from platelet-rich plasma. Asia Pac J Sports Med Arthrosc Rehabil Technol. 2016;4:27–32.

79. Kon E, Mandelbaum B, Buda R, Filardo G, Delcogliano M, Timoncini A, et al. Platelet-rich plasma intra-articular injection versus hyaluronic acid viscosupplementation as treatments for cartilage pathology: from early degeneration to osteoarthritis. Arthroscopy. 2011;27(11):1490–501.

80. Sanchez M, Fiz N, Azofra J, Usabiaga J, Aduriz Recalde E, Garcia Gutierrez A, et al. A randomized clinical trial evaluating plasma rich in growth factors (PRGF-Endoret) versus hyaluronic acid in the short-term treatment of symptomatic knee osteoarthritis. Arthroscopy. 2012;28(8):1070–8.

81. Filardo G, Kon E, Di Martino A, Di Matteo B, Merli ML, Cenacchi A, et al. Platelet-rich plasma vs hyaluronic acid to treat knee degenerative pathology: study design and preliminary results of a randomized controlled trial. BMC Musculoskelet Disord. 2012;13:229.

82. Filardo G, Di Matteo B, Di Martino A, Merli ML, Cenacchi A, Fornasari P, et al. Platelet-rich plasma intra-articular knee injections show no superiority versus viscosupplementation: a randomized controlled trial. Am J Sports Med. 2015 Jul;43(7):1575–82.

83. Riboh JC, Saltzman BM, Yanke AB, Fortier L, Cole BJ. Effect of leukocyte concentration on the efficacy of platelet-rich plasma in the treatment of knee osteoarthritis. Am J Sports Med. 2016;44(3):792–800.

84. Patel S, Dhillon MS, Aggarwal S, Marwaha N, Jain A. Treatment with platelet-rich plasma is more effective than placebo for knee osteoarthritis: a prospective, double-blind, randomized trial. Am J Sports Med. 2013;41(2):356–64.

85. Cerza F, Carni S, Carcangiu A, Di Vavo I, Schiavilla V, Pecora A, et al. Comparison between hyaluronic acid and platelet-rich plasma, intra-articular infiltration in the treatment of gonarthrosis. Am J Sports Med. 2012;40(12):2822–7.

86. Cole BJ, Karas V, Hussey K, Pilz K, Fortier LA. Hyaluronic acid versus platelet-rich plasma. Am J Sports Med. 2017;45(2):339–46.

87. Everhart JS, Cavendish PA, Eikenberry A, Magnussen RA, Kaeding CC, Flanigan DC. Platelet-rich plasma reduces failure risk for isolated meniscal repairs but provides no benefit for meniscal repairs with anterior cruciate ligament reconstruction. Am J Sports Med. 2019;47(8):1789–96.

88. Gobbi A, Chaurasia S, Karnatzikos G, Nakamura N. Matrix-induced autologous chondrocyte implantation versus multipotent stem cells for the treatment of large patellofemoral chondral lesions: a nonrandomized prospective trial. Cartilage. 2015;6(2):82–97.

89. Gobbi A, Whyte GP. One-stage cartilage repair using a hyaluronic acid-based scaffold with activated bone marrow-derived mesenchymal stem cells compared with microfracture: five-year follow-up. Am J Sports Med. 2016;44(11):2846–54.

90. Enea D, Cecconi S, Calcagno S, Busilacchi A, Manzotti S, Gigante A. One-step cartilage repair in the knee: collagen-covered microfracture and autologous bone marrow concentrate. A pilot study. Knee. 2015;22(1):30–5.

91. Krych AJ, Nawabi DH, Farshad-Amacker NA, Jones KJ, Maak TG, Potter HG, et al. Bone marrow con-

centrate improves early cartilage phase maturation of a scaffold plug in the knee: a comparative magnetic resonance imaging analysis to platelet-rich plasma and control. Am J Sports Med. 2016;44(1):91–8.

92. Krych AJ, Pareek A, King AH, Johnson NR, Stuart MJ, Williams RJ 3rd. Return to sport after the surgical management of articular cartilage lesions in the knee: a meta-analysis. Knee Surg Sports Traumatol Arthrosc. 2017;25(10):3186–96.

93. Skowronski J, Skowronski R, Rutka M. Large cartilage lesions of the knee treated with bone marrow concentrate and collagen membrane—results. Ortop Traumatol Rehabil. 2013;15(1):69–76.

94. Skowronski J, Rutka M. Osteochondral lesions of the knee reconstructed with mesenchymal stem cells—results. Ortop Traumatol Rehabil. 2013;15(3):195–204.

95. Fortier LA, Potter HG, Rickey EJ, Schnabel LV, Foo LF, Chong LR, et al. Concentrated bone marrow aspirate improves full-thickness cartilage repair compared with microfracture in the equine model. J Bone Joint Surg Am. 2010;92(10):1927–37.

96. Potten CS, Loeffler M. Stem cells: attributes, cycles, spirals, pitfalls and uncertainties. Lessons for and from the crypt. Development. 1990;110(4):1001–20.

97. Chang Y-H, Liu H-W, Wu K-C, Ding D-C. Mesenchymal stem cells and their clinical applications in osteoarthritis. Cell Transplant. 2016;25(5):937–50.

98. Ruetze M, Richter W. Adipose-derived stromal cells for osteoarticular repair: trophic function versus stem cell activity. Expert Rev Mol Med. 2014;16:e9.

99. Wu L, Cai X, Zhang S, Karperien M, Lin Y. Regeneration of articular cartilage by adipose tissue derived mesenchymal stem cells: perspectives from stem cell biology and molecular medicine. J Cell Physiol. 2013;228(5):938–44.

100. Jo CH, Lee YG, Shin WH, Kim H, Chai JW, Jeong EC, et al. Intra-articular injection of mesenchymal stem cells for the treatment of osteoarthritis of the knee: a proof-of-concept clinical trial. Stem Cells. 2014;32(5):1254–66.

101. Bosetti M, Borrone A, Follenzi A, Messaggio F, Tremolada C, Cannas M. Human lipoaspirate as autologous injectable active scaffold for one-step repair of cartilage defects. Cell Transplant. 2016;25(6):1043–56.

102. Jang K-M, Lee J-H, Park CM, Song H-R, Wang JH. Xenotransplantation of human mesenchymal stem cells for repair of osteochondral defects in rabbits using osteochondral biphasic composite constructs. Knee Surg Sports Traumatol Arthrosc. 2014;22(6):1434–44.

103. Jung M, Kaszap B, Redohl A, Steck E, Breusch S, Richter W, et al. Enhanced early tissue regeneration after matrix-assisted autologous mesenchymal stem cell transplantation in full thickness chondral defects in a minipig model. Cell Transplant. 2009;18(8):923–32.

104. Nam HY, Karunanithi P, Loo WC, Naveen S, Chen H, Hussin P, et al. The effects of staged intra-articular injection of cultured autologous mesenchymal stromal cells on the repair of damaged cartilage: a pilot study in caprine model. Arthritis Res Ther. 2013;15(5):R129.

105. Gobbi A, Karnatzikos G, Sankineani SR. One-step surgery with multipotent stem cells for the treatment of large full-thickness chondral defects of the knee. Am J Sports Med. 2014;42(3):648–57.

106. Chahla J, Piuzzi NS, Mitchell JJ, Dean CS, Pascual-Garrido C, LaPrade RF, et al. Intra-articular cellular therapy for osteoarthritis and focal cartilage defects of the knee: a systematic review of the literature and study quality analysis. J Bone Joint Surg Am. 2016;98(18):1511–21.

107. Campbell TM, Churchman SM, Gomez A, McGonagle D, Conaghan PG, Ponchel F, et al. Mesenchymal stem cell alterations in bone marrow lesions in patients with hip osteoarthritis. Arthritis Rheumatol. 2016;68(7):1648–59.

108. Cox LGE, van Donkelaar CC, van Rietbergen B, Emans PJ, Ito K. Alterations to the subchondral bone architecture during osteoarthritis: bone adaptation vs endochondral bone formation. Osteoarthr Cartil. 2013;21(2):331–8.

109. Suri S, Walsh DA. Osteochondral alterations in osteoarthritis. Bone. 2012;51(2):204–11.

110. Klement MR, Sharkey PF. The significance of osteoarthritis-associated bone marrow lesions in the knee. J Am Acad Orthop Surg. 2019;27(20):752–9.

111. Pelletier J-P, Roubille C, Raynauld J-P, Abram F, Dorais M, Delorme P, et al. Disease-modifying effect of strontium ranelate in a subset of patients from the phase III knee osteoarthritis study SEKOIA using quantitative MRI: reduction in bone marrow lesions protects against cartilage loss. Ann Rheum Dis. 2015;74(2):422–9.

112. Varenna M, Zucchi F, Failoni S, Becciolini A, Berruto M. Intravenous neridronate in the treatment of acute painful knee osteoarthritis: a randomized controlled study. Rheumatology (Oxford). 2015;54(10):1826–32.

113. Laslett LL, Doré DA, Quinn SJ, Boon P, Ryan E, Winzenberg TM, et al. Zoledronic acid reduces knee pain and bone marrow lesions over 1 year: a randomised controlled trial. Ann Rheum Dis. 2012;71(8):1322–8.

114. Jäger M, Tillmann FP, Thornhill TS, Mahmoudi M, Blondin D, Hetzel GR, et al. Rationale for prostaglandin I2 in bone marrow oedema—from theory to application. Arthritis Res Ther. 2008;10(5):R120.

115. Mayerhoefer ME, Kramer J, Breitenseher MJ, Norden C, Vakil-Adli A, Hofmann S, et al. Short-term outcome of painful bone marrow oedema of the knee following oral treatment with iloprost or tramadol: results of an exploratory phase II study of 41 patients. Rheumatology (Oxford). 2007;46(9):1460–5.

116. Callaghan MJ, Parkes MJ, Hutchinson CE, Gait AD, Forsythe LM, Marjanovic EJ, et al. A randomised trial of a brace for patellofemoral osteoarthritis targeting knee pain and bone marrow lesions. Ann Rheum Dis. 2015;74(6):1164–70.

117. Su K, Bai Y, Wang J, Zhang H, Liu H, Ma S. Comparison of hyaluronic acid and PRP intra-articular injection with combined intra-articular and intraosseous PRP injections to treat patients with knee osteoarthritis. Clin Rheumatol. 2018;37(5):1341–50.

118. Sánchez M, Delgado D, Pompei O, Pérez JC, Sánchez P, Garate A, et al. Treating severe knee osteoarthritis with combination of intra-osseous and intra-articular infiltrations of platelet-rich plasma: an observational study. Cartilage. 2019;10(2):245–53.

119. Deshmukh V, Hu H, Barroga C, Bossard C, Kc S, Dellamary L, et al. A small-molecule inhibitor of the Wnt pathway (SM04690) as a potential disease modifying agent for the treatment of osteoarthritis of the knee. Osteoarthr Cartil. 2018;26(1):18–27.

120. Yazici Y, McAlindon TE, Fleischmann R, Gibofsky A, Lane NE, Kivitz AJ, et al. A novel Wnt pathway inhibitor, SM04690, for the treatment of moderate to severe osteoarthritis of the knee: results of a 24-week, randomized, controlled, phase 1 study. Osteoarthr Cartil. 2017;25(10):1598–606.

121. Mithoefer K, Mcadams T, Williams RJ, Kreuz PC, Mandelbaum BR. Clinical efficacy of the microfracture technique for articular cartilage repair in the knee: an evidence-based systematic analysis. Am J Sports Med. 2009;37(10):2053–63.

122. Gobbi A, Nunag P, Malinowski K. Treatment of full thickness chondral lesions of the knee with microfracture in a group of athletes. Knee Surg Sports Traumatol Arthrosc. 2005;13(3):213–21.

123. Zamborsky R, Danisovic L. Surgical techniques for knee cartilage repair: an updated large-scale systematic review and network meta-analysis of randomized controlled trials. Arthroscopy. 2020;36(3):845–58.

124. Zedde P, Cudoni S, Giachetti G, Manunta ML, Masala G, Brunetti A, et al. Subchondral bone remodeling: comparing nanofracture with microfracture. An ovine in vivo study. Joints. 2016;

125. Milano G, Sanna Passino E, Deriu L, Careddu G, Manunta L, Manunta A, et al. The effect of platelet rich plasma combined with microfractures on the treatment of chondral defects: an experimental study in a sheep model. Osteoarthr Cartil. 2010;18(7):971–80.

126. Hangody L, Dobos J, Baló E, Pánics G, Hangody LR, Berkes I. Clinical experiences with autologous osteochondral mosaicplasty in an athletic population: a 17-year prospective Multicenter study. Am J Sports Med. 2010;38(6):1125–33.

127. Gudas R, Kalesinskas RJ, Kimtys V, Stankevičius E, Toliušis V, Bernotavičius G, et al. A prospective randomized clinical study of mosaic osteochondral autologous transplantation versus microfracture for the treatment of osteochondral defects in the knee joint in young athletes. Arthroscopy. 2005;21(9):1066–75.

128. Koh JL, Wirsing K, Lautenschlager E, Zhang LO. The effect of graft height mismatch on contact pressure following osteochondral grafting: a biomechanical study. Am J Sports Med. 2004;32(2):317–20.

129. Bugbee WD. Fresh osteochondral allografts. J Knee Surg. 2002;15(3):191–5.

130. Ghazavi MT, Pritzker KP, Davis AM, Gross AE. Fresh osteochondral allografts for post-traumatic osteochondral defects of the knee. J Bone Joint Surg Br. 1997;79(6):1008–13.

131. Gross AE, Shasha N, Aubin P. Long-term followup of the use of fresh osteochondral allografts for post-traumatic knee defects. Clin Orthop Relat Res. 2005;(435):79–87.

132. Riff AJ, Huddleston HP, Cole BJ, Yanke AB. Autologous chondrocyte implantation and osteochondral allograft transplantation render comparable outcomes in the setting of failed marrow stimulation. Am J Sports Med. 2020;48(4):861–70.

133. Peterson L, Brittberg M, Kiviranta I, Åkerlund EL, Lindahl A. Autologous chondrocyte transplantation: biomechanics and long-term durability. Am J Sports Med. 2002;30(1):2–12.

134. Moseley JB, Anderson AF, Browne JE, Mandelbaum BR, Micheli LJ, Fu F, et al. Long-term durability of autologous chondrocyte implantation: a multicenter, observational study in US patients. Am J Sports Med. 2010;38(2):238–46.

135. Paatela T, Vasara A, Nurmi H, Kautiainen H, Jurvelin JS, Kiviranta I. Biomechanical changes of repair tissue after autologous chondrocyte implantation at long-term follow-up. Cartilage. 2020;

136. Mithöfer K, Peterson L, Mandelbaum BR, Minas T. Articular cartilage repair in soccer players with autologous chondrocyte transplantation: functional outcome and return to competition. Am J Sports Med. 2005;33(11):1639–46.

137. Mithoefer K, Peterson L, Saris DBF, Mandelbaum BR. Evolution and current role of autologous chondrocyte implantation for treatment of articular cartilage defects in the football (soccer) player. Cartilage. 2012;3(1 Suppl):31S–6S.

138. Słynarski K, de Jong WC, Snow M, Hendriks JAA, Wilson CE, Verdonk P. Single-stage autologous chondrocyte-based treatment for the repair of knee cartilage lesions: two-year follow-up of a prospective single-arm multicenter study. Am J Sports Med. 2020;48(6):1327–37.

139. Steadman JR, Miller BS, Karas SG, Schlegel TF, Briggs KK, Hawkins RJ. The microfracture technique in the treatment of full-thickness chondral lesions of the knee in National Football League players. J Knee Surg. 2003;16(2):83–6.

140. Mithoefer K, Hambly K, Della Villa S, Silvers H, Mandelbaum BR. Return to sports participation after articular cartilage repair in the knee: scientific evidence. Am J Sports Med. 2009;37(Suppl 1):167S–76S.

141. Mithoefer K, Williams RJ, Warren RF, Wickiewicz TL, Marx RG. High-impact athletics after knee articular cartilage repair: a prospective evaluation of the microfracture technique. Am J Sports Med. 2006;34(9):1413–8.

142. Krych AJ, Harnly HW, Rodeo SA, Williams RJ. Activity levels are higher after osteochondral autograft transfer mosaicplasty than after microfracture for articular cartilage defects of the knee: a retrospective comparative study. J Bone Joint Surg Am. 2012;94(11):971–8.

143. Krych AJ, Robertson CM, Williams RJ. Return to athletic activity after osteochondral allograft transplantation in the knee. Am J Sports Med. 2012;40(5):1053–9.

144. Harris JD, Brophy RH, Siston RA, Flanigan DC. Treatment of chondral defects in the athlete's knee. Arthroscopy. 2010;26(6):841–52.

145. Mithoefer K, Hambly K, Logerstedt D, Ricci M, Silvers H, Della Villa S. Current concepts for rehabilitation and return to sport after knee articular cartilage repair in the athlete. J Orthop Sports Phys Ther. 2012;42(3):254–73.

146. Murray IR, Benke MT, Mandelbaum BR. Management of knee articular cartilage injuries in athletes: chondroprotection, chondrofacilitation, and resurfacing. Knee Surg Sports Traumatol Arthrosc. 2016;24:1617–26. https://doi.org/10.1007/s00167-015-3509-8.

Overview of Orthobiologics and Joint Function

Ignacio Dallo, Rachel M. Frank, Hannah Bradsell, Nicolas S. Piuzzi, and Alberto Gobbi

2.1 Introduction

The *Bio* is not new but it is now a key component in the practice of modern orthopaedics and sports medicine; nowadays, there is a growing interest on the use of biologic treatments which incorporate tissue engineering strategies: cells, scaffolds, and signaling molecules [1, 2]. Furthermore, patients are currently seeking "stem cells" or regenerative medicine treatments in response to both unmet treatment needs and marketing efforts that are often outpacing clinical evidence and regulatory control [3–5]. With the advent of injectable therapies using endogenous growth factors and cells inserted directly into the tissue to potentially facilitate healing, decrease inflammation, and subsequently provoke an analgesic effect after an injury or illness. Injections have the advantage of being "minimally invasive" with relatively low risk of complications. Commonly used biological approaches include platelet-rich plasma (PRP), bone marrow aspirate concentrate (BMAC), adipose tissue, and allogenic amniotic fluid. These injectable treatments may contribute to a regenerative microenvironment with the potential to improve healing rates and function in patients with musculoskeletal problems; however, to date, only symptomatic improvements have been reported, and structure modifying treatments are lacking. The American Academy of Orthopaedic Surgeons (AAOS) defined these biological as substances that can be found naturally in the body that aid in injury healing [6]. The term bio-orthopaedic includes all the biological treatment options for different orthopaedic conditions [7].

The application of biological therapies has the potential to facilitate the healing mechanism of tissues with limited healing potential and vascularity such as tendons, cartilage, meniscus, and ligaments. However, in order to understand and advance the field of biologic treatments it is important to understand the potential and limitations of the different components in tissue engineering approaches. The term "stem cell" has been overused based on the consensus of the expert's opinion [8]. Although stem and progenitor cells contribute to homeostasis, remodeling, and repair of tissues, there currently is no available stem cell treatment in orthopaedics. Therefore, it is recommended that the use of minimally manipulated cell products and tissue-derived culture-expanded cells be referred to as "cell therapy," to allow a better representation to the nature of these treatments. Basic science

I. Dallo (✉) · A. Gobbi
O.A.S.I. Bioresearch Foundation, Milan, Italy
e-mail: info@drignaciodallo.com.ar;
gobbi@cartilagedoctor.it

R. M. Frank · H. Bradsell
University of Colorado School of Medicine,
Aurora, CO, USA

N. S. Piuzzi
Cleveland Clinic, Cleveland, OH, USA

© ISAKOS 2022
A. Gobbi et al. (eds.), *Joint Function Preservation*, https://doi.org/10.1007/978-3-030-82958-2_2

research has provided proof of the concept that some cell therapy approaches (e.g., bone marrow-derived mesenchymal stem cells (MSCs)) may downregulate inflammation and produce an analgesic effect. Nonetheless, despite studies show promising clinical results for cartilage injuries and ligaments tears, conclusive clinical evidence is still missing [9]. As such, there is a continued need for high-quality basic science and clinical investigation into the safety and efficacy of biological therapies including cell-based therapies. It is recommended that physicians and institutions offering biologic therapies establish patient registries for surveillance and quality assessments. Several clinical trials are currently being performed evaluating these noninvasive therapies despite limited understanding of the underlying pathologic basis of the disease and without a complete characterization of their components. Additional studies are needed to identify optimal formulations and in defining the ideal dose and timing of orthobiologics for various orthopaedic conditions [10].

2.2 Bio-Orthopaedics and Orthobiologics Treatments

Orthobiologic nomenclature can be somewhat confusing and, in many cases, even misleading depending on the product being discussed. Lack of consensus on how to describe orthobiologics has led to challenges in interpreting the literature and comparing one orthobiologic product to another. Bio-Orthopaedics is the modern branch of orthopaedics that include all orthobiologic treatments aimed to enhance the biologic response of connective tissues in an effort to optimize the repair process and improve clinical outcomes [11]. Accurate nomenclature and reporting of currently available orthobiologics will facilitate advancement in the field. Different tissue sources have been used to obtain orthobiologic therapies including blood, bone marrow, and fat among others. For example, platelet-rich plasma (PRP) is a blood-derived product, where the blood is centrifuged to allow density separation of its components [12, 13]. There are multiple ways of preparing PRP: manual and commercial. Depending on the protocol used, the final preparation will vary [14]. This autologous product can then be used within the same patient to treat a variety of conditions. Several classification systems have been published to further describe the various PRP products based on the concentrations of the specific contents of the final product and the methods used to obtain them [12]. The most recent classification systems generally include an absolute number for platelet concentration, the presence of leukocytes or white blood cells and neutrophils, red blood cell presence, and if activation of platelets occurs synthetically by exogenous agents or naturally [15]. Some versions have additional detail about the preparation of the product and how it was applied to the patient [16]. Overall, various terminologies describing PRP and methods of classifying the numerous types of products exist, but no nomenclature standardization for PRP has been established.

Bone marrow aspirate concentrate (BMAC) is obtained by the centrifugation of bone marrow aspirate (BMA). This process concentrates the mononucleated cells and increases the ratio of stem and progenitor cells. Oftentimes BMAC is referred as a "stem cell" therapy; however, it is important to mention that only 0.01 to 0.0001% of the heterogenous nucleated cells present in BMAC is actually a stem or progenitor cell [17]. The composition describes the cellular makeup of the aspirate that is obtained from an average individual, which has been reported to be mainly neutrophils and erythroblasts in addition to other cell types, and low percentage of stem and progenitor cells (Tables 2.1 and 2.2) [18, 19].

The International Society for Cellular Therapy proposes minimal criteria to define human mesenchymal stem cells (MSCs). First, MSC must be plastic-adherent when maintained in standard culture conditions. Second, MSC must express CD105, CD73, and CD90, and lack expression of CD45, CD34, CD14 or CD11b, CD79alpha or CD19, and HLA-DR surface molecules [20]. Third, MSC must differentiate to osteoblasts, adipocytes, and chondroblasts in vitro. Recently,

Table 2.1 Summary of growth factors in bone marrow aspirate concentrate (BMAC)

Growth factor/cytokine	Principle action	Signaling pathway
TGF β1, TGF β2, TGF β3	Chondrocyte proliferation + differentiation	SMAD-2 and SMAD-3
BMP-2	Chondrocyte proliferation, matrix synthesis, and hypertrophy	SMAD-1, SMAD-5, SMAD-8, TAK-1
BMP-7	Increase ECM production	Mitogen-activated kinases (JNK, P38, ERK1/2)
IL-8	Inflammatory response; MSC homing to site of injury; increased VEGF production; chondrocyte hypertrophy	Mitogen-activated kinase; P38
VEGF	Promotes angiogenesis to sub-chondral, bone and supports cartilage growth	HIF-1, Runx2
PDGF	Wound healing, collagen synthesis, angiogenesis, suppression of IL-1β, enhanced BMP signaling	ERK 1/2, downregulation of NF-kB signaling
IGF-1	Increased synthetic and metabolic activity-increased collagen and proteoglycan synthesis Chondrogenic differentiation	PI-3 K, ERK 1/2
FGF-2	Chondrogenic differentiation, MSC homing	ERK 1/2, STAT1/P21
FGF-18	Chondrogenic differentiation, enhanced BMP signaling	
L-1/IL-1β	Inflammatory response-cell migration/ recruitment to site of injury	Mitogen-activated kinases (JNK, P38, ERK1/2)

JNK C-Jun N-terminal kinase, *ERK* extracellular signal-related kinases, *TAK-1* TGF-β-activating kinase 1 (TAK-1), *STAT1* signal transducer and activator of transcription-1, *PI-3K* phosphoinositide 3-Kinase, *Runx2* runt-domain transcription factor family-2, *HIF-1* hypoxia inducible factor-1, *NK-kB* nuclear factor kappa beta

as per Arnold Caplan, it is more appropriate to call MSCs as medicinal signaling cells, as these cells in vivo respond to the injury or disease by secreting bioactive factors that have immuno-modulatory effect, providing promising therapeutic options [21].

The DOSES tool for describing cell therapies (Fig. 2.1) must be utilized by clinicians, researchers, regulators, and industry professionals to improve transparency and to allow clinicians and patients to understand the characteristics of current and future cell preparations [22–24]. Adipose-derived therapies have recently receive more attention in the field due to its ease of use and the basic science finding that among all potential sources it has one of the highest concentrations of stem and progenitor cells per nucleated cells [25]. However, not all adipose-derived therapies are equal since some of them only employ mechanical fragmentation of fat, while others employ enzymatic digestion, and culture expansion techniques. Micro-fragmented fat (MF) is a term used to describe the minimally manipulated product of the mechanical breakdown of adipose tissue into tiny particles without requiring additives or ex vivo expansion. This process creates a product that has stem and progenitor cells and claims to maintain the vascular niche and extracellular matrix. Differently, stromal vascular fraction (SVF) refers to the product obtained from the enzymatic digestion of adipose tissue with collagenase and then centrifugation to remove adipocytes and free fat, producing a heterogeneous mixture of cells that includes stem and progenitor cells [26]. Overall, these methods result in products that have been referred as adipose-derived stem cells (ASCs) or adipose-derived mesenchymal stem cells (AMSCs), and they are currently being studied for their efficacy in treatments of various orthopaedic conditions.

The use of amniotic tissue-derived products is an orthobiologic treatment that has also recently gained interest. Amniotic membranes (AM) and amniotic fluid are types of products within this area of biologics that contain many elements potentially useful for orthopaedic regenerative medicine [27]. AM is obtained from the placenta of donors that have undergone a cesarean section and stored for later use by cryopreservation or dehydration. AM contains

Table 2.2 Cellular characterization of bone marrow aspirate (BMA) and bone marrow aspirate concentrate (BMAC)

	n	Median (range)
Pre-spin measures	24	97.8 (75.2–99.4)
Viability, %	25	38.5 (26.0–57.5)
MNCs, %	25	6100 (1950–27,000)
Total MNCs/uL	25	3.2 (0.04–21.0)
HSCs, %	25	0.03 (0.00–0.60)
MSCS, %	25	198 (0–2673)
Total MNCS x	25	13.0 (3.9–62.8)
MSCS, %	25	3.33 (0.17–4.44)
WBCs, 1000/uL	25	32.0 (1.6–38.2)
RBCs, mil/uL	25	95 (7–399)
HCTs, %	25	
Platelet, 1000/uL		
Post-spin measures	22	97.0 (85.4–99.6)
Viability, %	23	56.2 (25.8–87.9)
MNCs, %	23	16,000 (2900–210,000)
Total MNCs/uL	23	4.4 (1.2–14.0)
HSCs, %	23	0.05 (0.0–09)
MSCS, %	23	688 (8.7–28,980)
Total MNCS x	23	31.4 (5.6–97.2)
MSCS, %	23	0.96 (0.63–3.65)
WBCs, 1000/uL	23	8.5 (3.5–1515)
RBCs, mil/uL	23	422 (52–1515)
HCTs, %	23	4,620,000
Platelet, 1000/uL	23	(174,000-130,200,000)
Total HSCs injected	23	34,400 (435–1,449,000)
Total MSCs injected		

HCT hematocrit, *HSC* hematopoietic stem cell, *Mil* million, *MNC* mononuclear cell, *MSC* mesenchymal stem cell, *n* number of patient samples analyzed, *RBC* red blood cell, *WBC* white blood cell

an epithelial layer, a thick basement membrane, and avascular mesenchymal tissue and consists of amniotic epithelial cell and amniotic mononuclear mesenchymal cells. This product is an injectable therapy containing many biologically active compounds, including amniotic fluid and cells, and recent studies have shown that it has the ability to provide effective healing in orthopaedics [28].

2.3 Regulation in Different Part of the World

There are a variety of regulatory principles that exist around the world that govern the implementation of orthobiologic products in treating patients. The agency that regulates the use of these products in the United States is the Food and Drug Administration (FDA) while in Europe is EMA (European Medicines Agency).

Not all orthopaedic providers adhere to the standards outlined by the FDA and EMA and concerns regarding "rogue stem cell" clinics remain, as described by Murray and colleagues in a recent publication [29]. The process between the creation of new products and their ultimate approval for appropriate use takes a considerable amount of time and consists of multiple, complex steps to ensure that it is safe, ethical, and cost-effective. This process is described in detail in the previously referenced paper by Murray and colleagues [29]. In brief, the product is reviewed by the local institution review board (IRB) and the FDA, then preclinical animal studies are performed followed by clinical human trials in multiple phases, and several application submissions are required for the eventual approval and marketing of the product. Specific to the regulation of biologic products, the FDA uses a three-tiered (category 1, 2, and 3) approach based on considerable risk of the product, amount of manipulation on the product, how it is used, and if it is combined with additional substances [30]. In general, the FDA categorizes biologic products as either: human cells, tissues, and cellular and tissue-based products (HCT/Ps), regenerative medicine advanced therapies (RMAT), or as medical devices requiring 510(k), or premarket approval. HCT/Ps comprise most of our available orthopaedic regenerative products and are defined as products containing or consisting of human cells or tissues that are intended for implantation, transplantation, infusion, or transfer into a human recipient. Category 1 products are lowest risk and require no HCT/P oversight. Category 2 products are "lower risk" and are regulated under section 361, and Category 3 products are "higher risk" products and are regulated under section 351. Products regulated under section 361 are considered "minimally manipulated and intended for homologous use and are not subject to formal premarket approval prior to marketing (compared to 351 products). Minimal manipulation means:

1. For structural tissue, processing that does not alter the original relevant characteristics of the

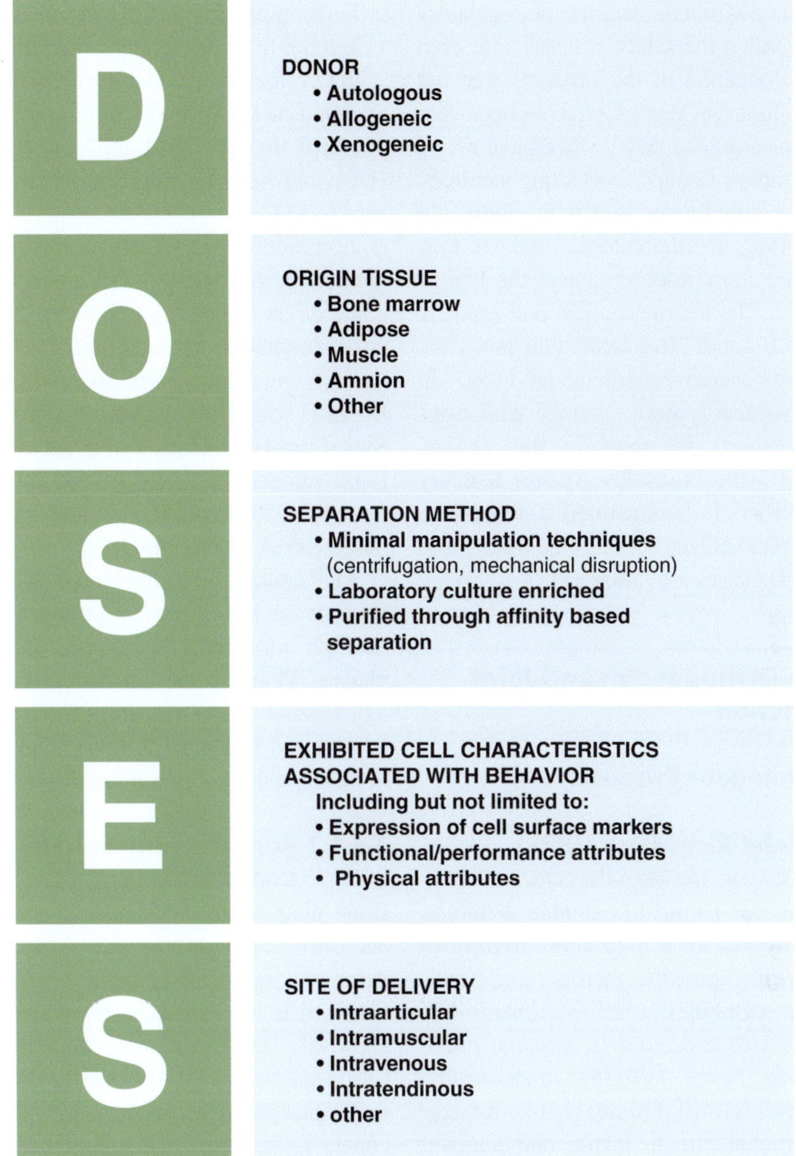

Fig. 2.1 Summary of the "DOSES" cell therapy communication tool [22]

tissue relating to the tissue's utility for reconstruction, repair, or replacement; and

2. For cells or nonstructural tissues, processing that does not alter the relevant biological characteristics of cells or tissues.

In Europe, the European Medicines Agency (EMA) is responsible for regulating orthobiologics and the process of approval is essentially the same as the United States. The main difference in these two systems is the regulation of funding, where more leniency exists in advancing research in the European Union (EU). In Australia, the Therapeutic Goods Administration (TGA) is responsible for orthobiologics regulation and the process is also similar to the United States using a three-tiered system. Finally, in Canada, the process of product approval closely reflects that of the US system and the regulation agency is Health Canada under the Canadian *Food and Drugs Act*

[31]. In Russia and India, there are no regulatory standards regarding the safety, efficacy, and even type of cell contained in the products that are administered. In recent years, China has been very proactive in ensuring the safety and ethical use of stem cell therapies. Groups conducting stem cell research are required to register their studies. In Japan, regulation of orthobiologic products falls under a similar framework as that of the United States and EU. The Pharmaceutical and Medical Devices (PMD) Act of 2013 created the new classification of "regenerative medicine products.". In Mexico, the regulatory body charged with oversight of biological therapies is the Federal Commission for the Protection Against Sanitary Risk (COFEPRIS). In South America, there is an absence of a standardized set of guidelines, but in general, the FDA rules are followed [7].

2.4 Bio-Orthopaedics and Joint Function

2.4.1 Autologous Products

2.4.1.1 Platelet-Rich Plasma (PRP)

PRP in the last years has been the center of attention regarding non-surgical injectable therapies. It is known to contain a high concentration of α-granules with growth factors and anti-inflammatory cytokines such as insulin-like growth factor 1 (IGF-1), IGF-2, vascular endothelial growth factor (VEGF), transforming growth factor-β(TGF-β), fibroblast growth factor (FGF), endothelial growth factor, and platelet-derived growth factors (PDGF). PRP can be obtained from the patient on the same day as the injection is given and is processed through minimal steps, making it both cost-effective and convenient for treatment in patients with OA. To date, randomized controlled trials have demonstrated safety and superior efficacy of PRP than HA in knee osteoarthritis at 12 months. Better outcomes have been reported in younger patients or with mild-to-moderate OA without malalignment, smokers, or obesity. Initial research suggests that leukocyte-poor platelet-rich plasma (LP-PRP) may have stronger efficacy for intra-

articular application. PRP has been shown to provide relief from pain and inflammation associated with OA, making it a viable treatment in the management of *OA* [32–37]. Some new studies suggest that the combined application of PRP with HA could have a synergistic effect on treatment for OA [32].

Some new studies suggest that the combined application of PRP with HA could have a synergistic effect on treatment for OA [38]. Based on our experience, patients with mild knee OA treated with three intra-articular injections of LP-PRP with HA (Cellular Matrix, Regen Lab, Switzerland) 1 month apart, showed significant improvement in pain relief (VAS) and knee function (KOOS Symptom, Pain and Quality Of Life parameter at 12 months.

PRP contains several bioactive agents that can mediate the tissue healing process after an injury through both the inflammatory and remodeling phases. Platelets are involved in homeostasis, aggregation, and clot formation steps, which finally lead to the scaffold formation, necessary to enhance meniscus healing [39].

2.4.1.2 Bone Marrow Aspirate Concentrate (BMAC)

Bone marrow aspirate concentrated (BMAC) is classified through the US Food and Drug Administration (FDA) as a 361 product and, hence, it is not subject to premarket review and approval. The regulatory foundation of the European Union (EU) similar to the US system, finds that processes such as centrifugation, are considered as minimal manipulation. BMAC has progenitor cells and growth factors with reparative, homing, and trophic properties causing them to migrate to areas of damage. Once at the site of injury, they release numerous factors that can help in healing and inflammation modulation. Recently, Cassano et al. found an increased concentration of Interleukin 1 Receptor Antagonist (IL-1RA), which, in combination with the other constituents, may provide anti-inflammatory and immunomodulatory effects [40].

Few studies have demonstrated patient safety and improved clinical outcomes after BMAC treatment for OA; however, there is a paucity of

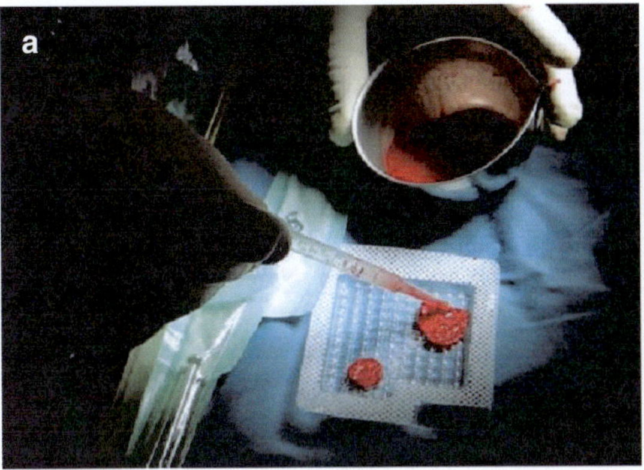
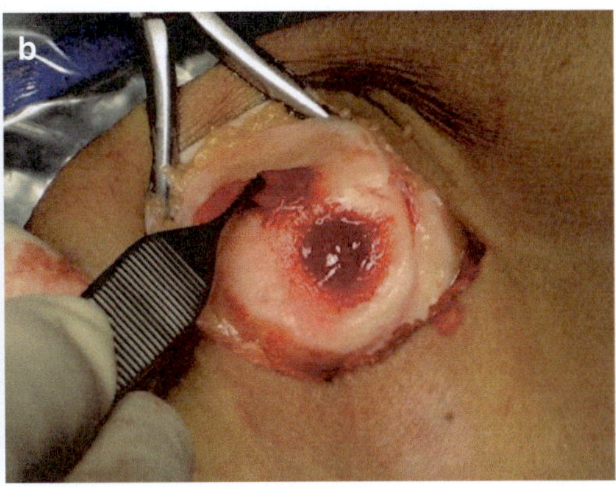

Fig. 2.2 HA-BMAC Technique. (a) The prepared concentrate of bone marrow was placed on the hyaluronic-based scaffold. (b) After a few minutes, the activated BMAC was absorbed by the scaffold, creating a sticky implant that is easy to apply onto a left-sided patellar bi-facetal chondral lesion

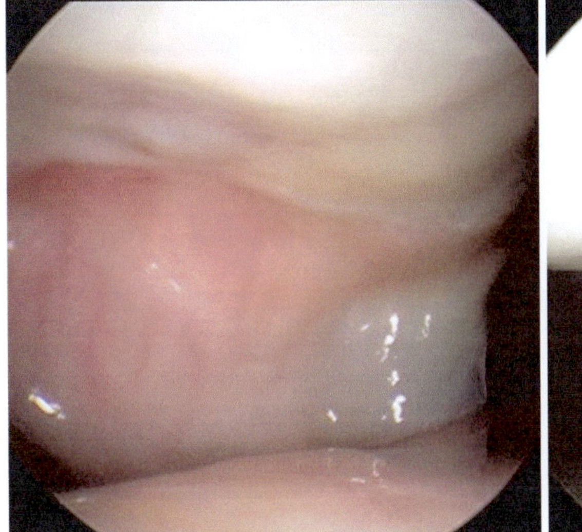
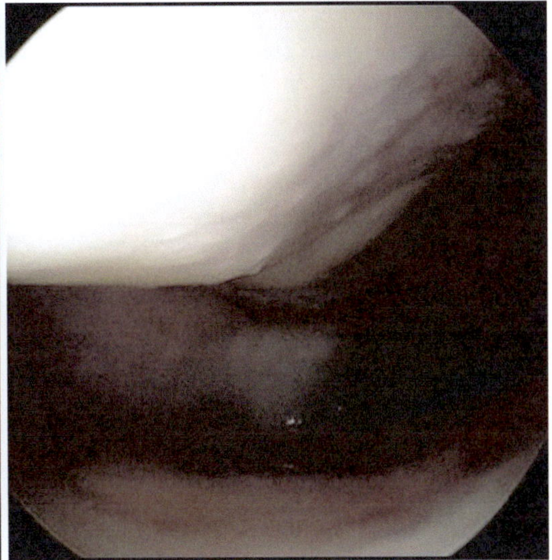

Fig. 2.3 Second-look arthroscopy showing a good quality of a regenerative cartilage tissue at both the patellar facets

high-level studies or randomized trials with joint osteoarthritis.

Gobbi et al. [41] in a prospective study concluded that the repair of full-thickness cartilage injury in the knee with an hyaluronic acid-based scaffold with bone marrow aspirate concentrate (HA-BMAC) scaffold provides good clinical outcomes at long-term follow-up in the treatment of small to large lesions, single or multiple lesions, and lesions in 1 or 2 compartments, as well as in cases of associated lesion treatment (Figs. 2.2 and 2.3) [42]. While good outcomes can be expected among patients more than 45 years of age, outcomes may be comparatively more successful in younger patients [43, 44]. We believe that 1-stage cartilage repair in the knee with a hyaluronic acid-based scaffold embedded with mesenchymal stem cells sourced from bone marrow aspirate concentrate has a prominent role in treating chondral defects because this is a simple technique that could improve the care of patients and be cost-effective in the near future [43, 45]. High-quality randomized controlled trials are necessary to directly compare all cartilage restoration procedures [46].

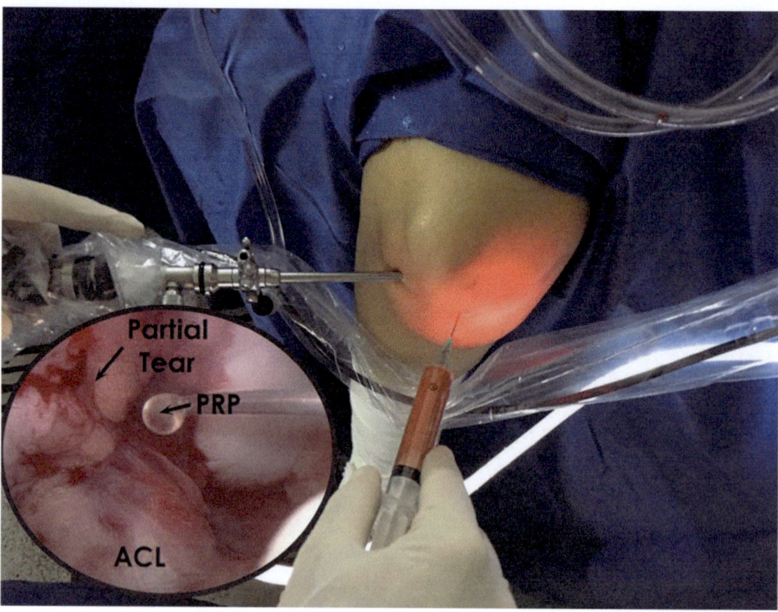

Fig. 2.4 Image showing the arthroscopic and surgeon's view of an infiltration with bone marrow aspirate concentrate (BMAC and platelet-rich plasma (PRP) in a partial lesion of the anterior cruciate ligament (ACL)

In our experience based on the quantification of Colony Forming Units (CFUs) in 25 patients, we did not find any correlation between the clinical outcomes and the number of CFUs [44].

An intriguing explanation for these results may come from the new vision of MSCs recently proposed by Caplan as "Medicinal Signaling Cells." According to this concept, MSCs, rather than participating in tissue formation, work as site-regulated "drugstores" in vivo by releasing trophic and immunomodulatory factors and are activated by local injury [47].

The injection of autologous BMAC and PRP using fluoroscopic or arthroscopic guidance with good clinical outcomes and MRI showing the ACL healing was documented in patients with partial (Figs. 2.4 and 2.5) and complete ACL tears with less than 1-cm retraction [48].

Gobbi et al. [49] concluded that primary ACL repair combining PRP and BMAC to treat select cases of knee instability secondary to incomplete ACL rupture demonstrated good long-term outcomes in athletes, with high rates of restoration of knee stability and returned to preinjury athletic activities. The potential benefits of these biological augmentation approaches for partial ACL tears are improved healing, better proprioception, and a faster return to sport and activities of daily living when compared with standard reconstruction procedures. However, long-term studies with larger cohorts of patients and with technique validation are necessary to assess the real effect of these approaches [50].

2.4.1.3 Adipose Cellular Therapy

Adipose-derived stromal cell (ASC) therapy, also known as an adipose stromal vascular fraction or autologous micro-fragmented adipose tissue (AMAT) has gained recent popularity as a minimally manipulated product. Adipose tissue, which is typically structured with consistent vascularity, has been increasingly recognized as a reliable source of these cells. Compared with BMAC, it has advantages in that it is procured in much larger quantities and is a larger source of MSCs.

A randomized controlled trial (RCT) in patients with mild-to-moderate knee osteoarthritis (OA) demonstrated a significant reduction in Western Ontario and McMaster Universities Osteoarthritis Index (WOMAC), an improvement in Lysholm score, and significant pain reduction (VAS) [51].

Although promising, these studies have been insufficient to conclude the efficacy of ASC therapy to adopt it into standard practices.

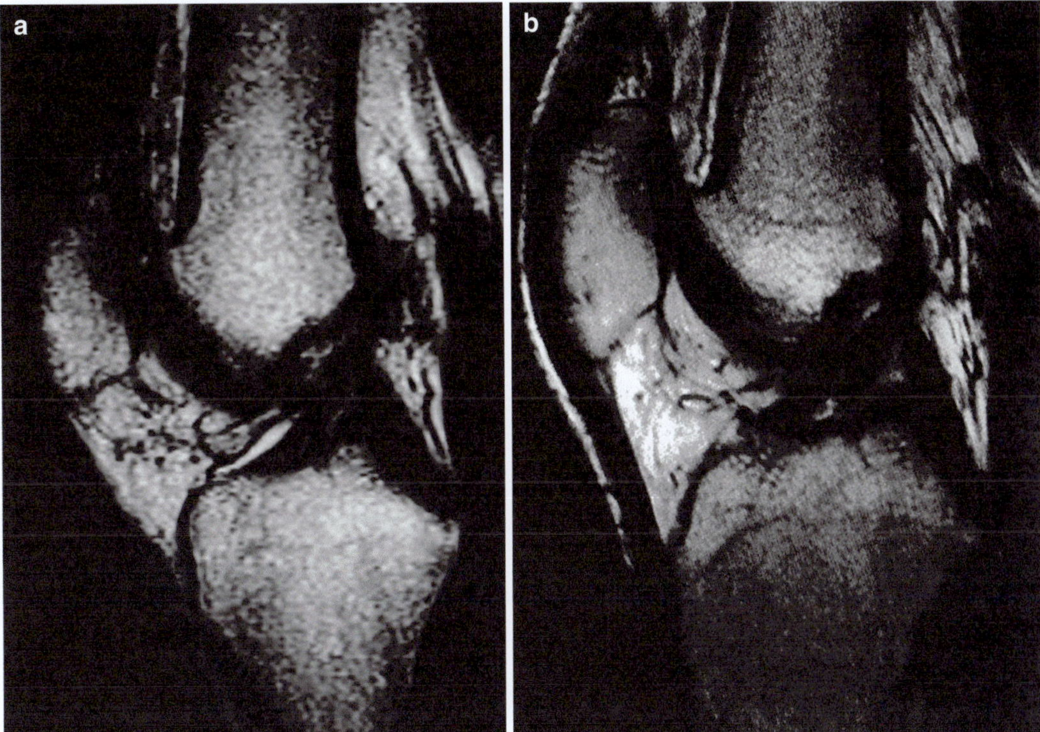

Fig. 2.5 (**a**) Magnetic resonance image (MRI), sagittal view, of the knee with hypertensive and heterogeneous ACL signal, compatible with a partial injury. (**b**) MRI of the same patient at 8 months post injection of BMAC and PRP intraligamentary with hypointense, homogeneous signal, and isointense with the LCP, compatible with ACL healing

A clinical study reported the repair of a grade II meniscal tear following a percutaneous injection of autologous adipose stem cell (ASCs) along with PRP, hyaluronic acid, and CaCl2 [52].

The use of mesenchymal stem cells seems to stimulate the regeneration of meniscal tissue and appears to be a promising approach to restore as much meniscal tissue as possible [53]. However, these regenerative technologies still need to be optimized to support their use.

2.4.2 Allogeneic Products

2.4.2.1 Amniotic Cellular Therapy

Amnion, chorion, amniotic fluid, and the umbilical cord are different placental tissues that have been investigated. An emerging new allogenic orthobiologic option, amniotic tissue, has also been shown to be a source of bioactive components. They are reported to contain growth factors, cytokines, and vasoactive peptides that modulate inflammation. Besides, they contain amniotic epithelial cells and amniotic mononuclear undifferentiated stromal cells, which have chondrogenic and osteogenic differentiation capacity. Amnions (AM) are also rich sources of hyaluronic acid and proteoglycans, which could play a role in the potential therapeutic relief of OA.

[54] Currently, there are several commercially available formulations of AM that differ based on content as well as how they were preserved. In a pilot study, using human amniotic suspension allografts (ASA) for knee OA, Vines, JB et al. concluded that a single intra-articular injection of ASA is feasible in patients with knee OA [55].

Current literature contains evidence that is insufficient to conclude the efficacy of this treatment.

2.5 Conclusions

The application of biological therapies has the potential to facilitate the healing mechanism of tissues with limited healing potential such as tendons, cartilage, meniscus, and ligaments. It is recommended that physicians and institutions offering biologic therapies establish patient registries for surveillance and quality assessments. The DOSES tool for describing cell therapies must be utilized to improve transparency and to better understand the characteristics of biological preparations. Additional studies are needed to identify optimal formulations and in defining the ideal dose and timing of ortho-biologics in the field of bio-orthopaedics.

References

1. Piuzzi NS, et al. Proceedings of the signature series symposium "cellular therapies for orthopaedics and musculoskeletal disease proven and unproven therapies-promise, facts and fantasy," International Society for Cellular Therapies, Montreal, Canada, May 2, 2018. Cytotherapy. 2018;20(11):1381–400.
2. Piuzzi NS, et al. Cellular therapies in orthopedics: where are we? Surg Technol Int. 2017;31:359–64.
3. Ng MK, Mont MA, Piuzzi NS. Analysis of readability, quality, and content of online information available for "stem cell" injections for knee osteoarthritis. J Arthroplasty. 2020;35(3):647–651.e2.
4. Piuzzi NS, et al. The stem-cell market for the treatment of knee osteoarthritis: a patient perspective. J Knee Surg. 2018;31(6):551–6.
5. Ramkumar PN, et al. Cellular therapy injections in today's orthopedic market: a social media analysis. Cytotherapy. 2017;19(12):1392–9.
6. AAOS. Orthobiologics. 2020. Accessed 7 May 2010.
7. Gobbi A, Espregueira-Mendes J, Lane J, Karahan M. Bio-orthopaedics. A new approach, vol. 696. Berlin/Heidelberg: Springer/ISAKOS; 2017. p. XXI. ISBN 3662541815, 9783662541814.
8. Chu CR, et al. Optimizing clinical use of biologics in orthopaedic surgery: consensus recommendations from the 2018 AAOS/NIH U-13 conference. J Am Acad Orthop Surg. 2019;27(2):e50–63.
9. Chahla J, et al. Intra-articular cellular therapy for osteoarthritis and focal cartilage defects of the knee: a systematic review of the literature and study quality analysis. J Bone Joint Surg Am. 2016;98(18):1511–21.
10. Dallo I, Irlandini E. State of the art for injections in orthopaedics. ISAKOS Newsletter. 2020
11. Gobbi A, Espregueira-Mendes J, Lane J, Karahan M. Bio-orthopaedic. A new approach, vol. XIX. Berlin/Heidelberg: Springer; 2017. p. 696.
12. Rossi LA, et al. Classification systems for platelet-rich plasma. Bone Joint J. 2019;101-B(8):891–6.
13. Piuzzi NS, et al. Platelet-rich plasma for the treatment of knee osteoarthritis: a review. J Knee Surg. 2017;30(7):627–33.
14. Franco D, et al. Protocol for obtaining platelet-rich plasma (PRP), platelet-poor plasma (PPP), and thrombin for autologous use. Aesthet Plast Surg. 2012;36(5):1254–9.
15. Mautner K, et al. A call for a standard classification system for future biologic research: the rationale for new PRP nomenclature. PM R. 2015;7(4):S53–9.
16. Lana J, et al. Contributions for classification of platelet rich plasma—proposal of a new classification: MARSPILL. Regen Med. 2017;12(5):565–74.
17. Piuzzi NS, et al. Bone marrow cellular therapies: novel therapy for knee osteoarthritis. J Knee Surg. 2018;31(1):22–6.
18. Piuzzi NS, et al. Bone marrow-derived cellular therapies in orthopaedics: part II: recommendations for reporting the quality of bone marrow-derived cell populations. JBJS Rev. 2018;6(11):e5.
19. Piuzzi NS, et al. Bone marrow-derived cellular therapies in orthopaedics: part I: recommendations for bone marrow aspiration technique and safety. JBJS Rev. 2018;6(11):e4.
20. Dominici M, et al. Minimal criteria for defining multipotent mesenchymal stromal cells. The International Society for Cellular Therapy position statement. Cytotherapy. 2006;8(4):315–7.
21. Caplan AI. Mesenchymal stem cells: time to change the name! Stem Cells Transl Med. 2017;6(6):1445–51.
22. Murray IR, et al. International expert consensus on a cell therapy communication tool: DOSES. J Bone Joint Surg Am. 2019;101(10):904–11.
23. Chahla J, et al. A call for standardization in platelet-rich plasma preparation protocols and composition reporting: a systematic review of the clinical orthopaedic literature. J Bone Joint Surg Am. 2017;99(20):1769–79.
24. Piuzzi NS, et al. Variability in the preparation, reporting, and use of bone marrow aspirate concentrate in musculoskeletal disorders: a systematic review of the clinical orthopaedic literature. J Bone Joint Surg Am. 2018;100(6):517–25.
25. Mantripragada VP, et al. Native-osteoarthritic joint resident stem and progenitor cells for cartilage cell-based therapies: a quantitative comparison with respect to concentration and biological performance. Am J Sports Med. 2019;47(14):3521–30.
26. Kubrova E, et al. Injectable biologics. Am J Phys Med Rehabil. 2020:1.
27. Emara AK, Anis H, Piuzzi NS. Human placental extract: the feasibility of translation from basic science into clinical practice. Ann Transl Med. 2020;8(5):156.
28. Huddleston HP, et al. Amniotic product treatments: clinical and basic science evidence. Curr Rev Musculoskelet Med. 2020;13(2):148–54.

29. Murray IR, et al. Rogue stem cell clinics. Bone Joint J. 2020;102-B(2):148–54.
30. FDA. Code of federal regulations—title 21, volume 7 [Internet]. Code of federal regulations code of federal regulations. 1 Apr 2017. Available from: https://www.accessdata.fda.gov/scripts/cdrh/cfdocs/cfcfr/CFRSearch.cfm?fr=640.34
31. Grieshober JA, Fakunle E, Gambardella RA. Orthobiologics: regulation in different parts of the world. In: Gobbi A, et al., editors. Bio-orthopaedics. Berlin/Heidelberg: Springer; 2017. p. 47–63.
32. Cole BJ, et al. Hyaluronic acid versus platelet-rich plasma: a prospective, double-blind randomized controlled trial comparing clinical outcomes and effects on intra-articular biology for the treatment of knee osteoarthritis. Am J Sports Med. 2017;45(2):339–46.
33. Gobbi A, et al. Platelet-rich plasma treatment in symptomatic patients with knee osteoarthritis: preliminary results in a group of active patients. Sports Health. 2012;4(2):162–72.
34. Gobbi A, Fishman M. Platelet-rich plasma and bone marrow-derived mesenchymal stem cells in sports medicine. Sports Med Arthrosc Rev. 2016;24(2):69–73.
35. Gobbi A, Malchira S, Karnatzikos G. Clinical application of platelet rich plasma—our philosophy. ISAKOS Newsletter. 2011
36. Gobbi A, Somanna MS, Karnatzikos G, Platelet rich plasma (PRP) in osteoarthritis—our early experience. ICRS Newsletter. 2011.
37. Gobbi A, Malchira S, Karnatzikos G. Platelet rich plasma (PRP) in osteoarthritis. Minerva Ortop Traumatol. 2011;
38. Chen WH, et al. Synergistic anabolic actions of hyaluronic acid and platelet-rich plasma on cartilage regeneration in osteoarthritis therapy. Biomaterials. 2014;35(36):9599–607.
39. Haunschild ED, et al. Platelet-rich plasma augmentation in meniscus repair surgery: a systematic review of comparative studies. Arthroscopy. 2020;36(6):1765–74.
40. Cassano JM, et al. Bone marrow concentrate and platelet-rich plasma differ in cell distribution and interleukin 1 receptor antagonist protein concentration. Knee Surg Sports Traumatol Arthrosc. 2018;26(1):333–42.
41. Gobbi A, Whyte GP. Long-term clinical outcomes of one-stage cartilage repair in the knee with hyaluronic acid-based scaffold embedded with mesenchymal stem cells sourced from bone marrow aspirate concentrate. Am J Sports Med. 2019;47(7):1621–8.
42. Gobbi A, et al. One-step cartilage repair with bone marrow aspirate concentrated cells and collagen matrix in full-thickness knee cartilage lesions: results at 2-year follow-up. Cartilage. 2011;2(3):286–99.
43. Gobbi A, Dallo I, Kumar V. Editorial commentary: biological cartilage repair technique-an "effective, accessible, and safe" surgical solution for an old difficult biological problem. Arthroscopy. 2020;36(3):859–61.
44. Gobbi A, Karnatzikos G, Sankineani SR. One-step surgery with multipotent stem cells for the treatment of large full-thickness chondral defects of the knee. Am J Sports Med. 2014;42(3):648–57.
45. Dallo I, Gobbi A, Sherman SL Cells and scaffolds for cartilage regeneration: ICRS Newsletter Summer issue. 2020 29.
46. Gobbi A, Lane JG, Dallo I. Editorial commentary: cartilage restoration-what is currently available? Arthroscopy. 2020;36(6):1625–8.
47. Caplan AI, Correa D. The MSC: an injury drugstore. Cell Stem Cell. 2011;9(1):11–5.
48. Centeno CJ, et al. Anterior cruciate ligament tears treated with percutaneous injection of autologous bone marrow nucleated cells: a case series. J Pain Res. 2015;8:437–47.
49. Gobbi A, Whyte GP. Long-term outcomes of primary repair of the anterior cruciate ligament combined with biologic healing augmentation to treat incomplete tears. Am J Sports Med. 2018;46(14):3368–77.
50. Dallo I, et al. Biologic approaches for the treatment of partial tears of the anterior cruciate ligament: a current concepts review. Orthop J Sports Med. 2017;5(1):2325967116681724.
51. Jones IA, et al. A randomized, controlled study to evaluate the efficacy of intra-articular, autologous adipose tissue injections for the treatment of mild-to-moderate knee osteoarthritis compared to hyaluronic acid: a study protocol. BMC Musculoskelet Disord. 2018;19(1):383.
52. Pak J, et al. Safety reporting on implantation of autologous adipose tissue-derived stem cells with platelet-rich plasma into human articular joints. BMC Musculoskelet Disord. 2013;14:337.
53. Pak J, et al. Current use of autologous adipose tissue-derived stromal vascular fraction cells for orthopedic applications. J Biomed Sci. 2017;24(1):9.
54. Riboh JC, et al. Human amniotic membrane-derived products in sports medicine: basic science, early results, and potential clinical applications. Am J Sports Med. 2016;44(9):2425–34.
55. Vines JB, et al. Cryopreserved amniotic suspension for the treatment of knee osteoarthritis. J Knee Surg. 2016;29(6):443–50.

Cost of Disability

3

Gwenllian F. Tawy and Leela C. Biant

3.1 Introduction

Isolated areas of osteo-chondral damage can be caused by sporting injury, trauma, or development of defects such as osteochondritis dissecans. Damage to articular cartilage and its underlying subchondral bone can progress to establishment of osteoarthritis (OA). As a result, it is commonly mistaken as a burden on the elderly. However, damage to the osteo-chondral (OC) unit can result in a personal, economic, and societal burden of disability at any age.

OC injuries mainly affect the weight-bearing joints such as the hip, knee, and ankle. Focal OC damage resulting from trauma or sporting injury can occur in any joint; however, insult to the knee and ankle are particularly prevalent [1]. Sport-induced OC injury may be complicated by ligamentous or other soft tissue injury [1, 2]. Therefore, injury to the articular cartilage is one of the most common findings at consultation with a sports physician [3].

Irrespective of etiology, damage to the OC unit can cause joint pain and limitation of function with impact on activities of daily living and work. Symptom relief, treatment, and rehabilitation are therefore priorities for individuals living with damage to the OC unit.

OC injuries have direct impact on the lives of patients, caregivers, dependents, health systems, wider society, and the economy.

3.2 Incidence and Prevalence

The years 2000–2010 were declared as the Bone and Joint decade. Endorsed by the World Health Organization and the United Nations, this decade was dedicated to better understanding the worldwide burden of major musculoskeletal conditions [4]. Osteoarthritis was one of the chosen major conditions.

Undiagnosed or untreated isolated OC injuries can progress to established osteoarthritis; however, very few national and international organizations include OC injury in their annual musculoskeletal health reports [1, 4–7]. As a result, statistics on the incidence and prevalence of degenerative, sport, and traumatic injuries to the OC unit are unknown [8].

It is likely that cases of isolated OC damage are included in the data that registries and health-care organizations publish on osteoarthritis, as the symptoms can be similar and they may be viewed as an earlier stage of OA. However, extracting the relevant information from these reports may be

G. F. Tawy
University of Manchester, Manchester, UK
e-mail: gwenllian.tawy@manchester.ac.uk

L. C. Biant (✉)
University of Manchester, Manchester, UK

Manchester University NHS Foundation Trust, Manchester, UK
e-mail: leela.biant@manchester.ac.uk

© ISAKOS 2022
A. Gobbi et al. (eds.), *Joint Function Preservation*, https://doi.org/10.1007/978-3-030-82958-2_3

impossible, given the clinical difficulty in demarcating when an OC injury becomes established osteoarthritis. Even in a scenario where distinctions could be drawn between both patient populations, the resulting figures would underrepresent the true population with OC injury. OC injury prevalence is further underestimated by underreporting of mild symptoms. Fourteen percent of asymptomatic athletes undergoing Magnetic Resonance Imaging (MRI) were shown to have focal OC damage [9]. Furthermore, surgical treatment of an unrelated orthopaedic problem frequently uncovers existing OC damage. For example, full-thickness cartilage lesions are reported in approximately 10% of surgical treatments for femoroacetabular impingement [8].

Despite these limitations, national and international registries can provide vital information on the prevalence and response to treatment of OC damage. Registries for reporting joint restoration by arthroplasty mandated in some countries and commonplace in most developed countries. In 2016, the International Cartilage Regeneration and Joint Preservation Society (ICRS) established a worldwide patient registry for long-term reporting of cartilage damage and treatment outcomes. The inaugural Annual Report was published in February 2019 [10]. Five hundred and thirty-five cases were recorded in the registry; only 9% of which had been carried out prior to 2018. According to the Annual Report, the age of individuals undergoing cartilage treatment ranged between 16 and 89 (mean age: 52 ± 18 years). This highlights the diversity of the population seeking treatment for cartilage damage [10].

Research literature of OC injury predominantly focuses on the knee. It is estimated that 900,000 Americans are affected by cartilage injuries to the knee annually [11]. In 1997, a review of 31,516 knee arthroscopies revealed that grade IV cartilage lesions in patients under 40 years old accounted for 5% of all cases [12]. In a 2002 review of 1000 consecutive knee arthroscopies; 61% were found to have chondral or OC lesions of any type, and 19% had focal chondral or OC lesions [13]. Of those with focal chondral or OC defects, 61% could be related to a traumatic injury to the joint [13]. More recent studies have supported the findings in these reviews, estimating that 5–10% of cartilage injuries in the knee are full-thickness lesions that may require intervention [8]. This is increased to 36% in professional athletes [14].

In the USA, up to 200,000 surgical procedures are carried out on chondral injuries of the knee per year [15, 16]. Excluded from these statistics are all non-surgical procedures on the knee such as injections, and all types of procedures on joints other than the knee. The scale of the problem is likely to be similar worldwide.

The prevalence of conditions which affect the musculoskeletal system increase with age [4]. Worldwide aging populations and increasing life expectancies mean that we will see the prevalence of OC injuries and OA rising [4]. These trends will particularly effect westernized countries, where obesity and low-manual work are more common. However, the burden may be greater felt in developing countries, where life expectancies are increasing but access to treatment may not be universal [4]. The economic and human burden of OC injuries are therefore far reaching and widespread [1].

3.3 Cost to Patients

Not all isolated OC injuries are symptomatic, but those that are can have a similar magnitude of pain and dysfunction as established osteoarthritis [17]. Symptoms can have an all-encompassing effect on an individual's life. Resulting loss of work and mobility financially impacts the lives of family and caregivers of those affected. The cost of disability therefore extends beyond the individual themselves.

3.3.1 Work

The ICRS's 2019 Annual Report stated that patients undergoing surgical repair of cartilage damage had a mean age of 35, with the oldest patient being 57. Those undergoing surgical cartilage repair are therefore of working age [10].

The symptoms associated with OC injury can make it difficult for individuals to work in a manual job or a job that requires prolonged standing. According to the Labour Force Survey, 6.9 million working days were lost in Great Britain in 2018–2019 due to musculoskeletal conditions, with each affected individual talking 14 days off, on average [18]. This accounted for 29% of all illness-related lost working days in 2018–2019 [18].There were no statistical differences between male and female rates, but being over the age of 45 was a risk factor [18].

It is difficult to estimate the average number of days that individuals with OC injury take off each year. However, de Windt stated that patients who undergo diagnostic arthroscopy or microfracture (MF) are expected to take 5 days off work following surgery [19]. This is increased to 15 days following autologous chondrocyte implantation (ACI), high tibial osteotomy or total knee arthroplasty. According to Aae, these could amount to indirect costs of €1075–€3225 ($1200–3500) per patient (based on data from Statistics Norway, 2016) [20]. These costs may be doubled or tripled annually in cases where a secondary procedure follows a failed initial procedure and a diagnostic arthroscopy.

Time spent away from work can have therefore a significant impact on an individual's income, especially if they are self-employed, are paid on an hourly basis, or are not entitled to sick-pay. This can put pressure on the individual, especially if they have dependents.

Previously healthy individuals who take short-term sick leave (fewer than 30 days a year) are at greater risk of reporting long-term sick leave and short-term unemployment within 5 years [21]. However, the study found the individuals to be at no greater risk of long-term unemployment or receiving disability allowances [21]. These trends may not be true for professional athletes where impact of OC injuries can be devastating, culminating in the end of a profession [9, 22].

From a societal perspective, additional loss could be attributed to poor productivity at work due to pain. According to de Windt et al. this could amount to 25 working days in patients undergoing MF and up to 40 days for those under-

going ACI [19]. This could lead to additional losses of €374.25–598.80 ($400–650) per patient.

OC damage caused by trauma or osteochondritis dissecans may impact the education of children and adolescents [23]. Lifelong implications may ensue from an inability to attend school or college due to ongoing symptoms or treatments. In the USA, college scholarships can be based on sports participation. If the student is injured, the bursaries that enable continuation of studies may not be renewed.

3.3.2 Mental Health

Research has shown that up to 85% of individuals living with chronic pain suffer from symptoms of depression [24]. The prevalence of mental health conditions like depression specifically in individuals with pain arising from OC injuries is unknown.

A recent systematic review on the experiences of individuals living with general knee pain (mostly caused by OA) showed the symptoms drastically change the way individuals live their lives [25]. Key problems included social disruption, isolation, and dependency on others. These are all risk factors for the decline of mental health.

For professional athletes, developing an injury during their career, or having to retire early due to injury, has been associated with greater risk of depression, anxiety, stress, and financial instability [22]. Young retired athletes diagnosed with OA, who likely suffered OC injuries to their joint during their career, are at greater risk of developing anxiety and/or depression [22, 26]. As training towards a professional sport starts at an increasingly younger age, anxiety and depression may become more commonly reported in children and young adults injured during sports participation [27].

It is important for orthopaedic clinicians to be cognizant of the implications of the physical problem of OC damage on the mental health of their patients. Screening and monitoring with validated patient reported questionnaires can be helpful.

3.3.3 Comorbidities

Damage to the OC unit commonly involves injury to neighboring joint tissues. This is particularly prevalent in sports injuries [1, 2]. For example, meniscal injury is commonly reported in athletes [9, 14]. Meniscal injuries leave the underlying cartilage less protected and alter joint biomechanics; this is detrimental to force distribution through the OC unit. Ligamentous injuries are another common cause or coexisting condition of OC injury [9, 28]. Instability from ligament disruption can exacerbate a new or existing OC injury and compromise any endogenous or surgical repair. In other traumatic cases, OC injury may be caused by dislocation of the patella or tibiofemoral joint. Such incidences can also tear the joint ligaments, rendering the joint unstable and at greater risk of future dislocation. Repeated patellar dislocation can worsen OC damage to the joint. These concomitant joint injuries complicate surgical repair and may lengthen recovery for patients. Chronic OC damage may precede osteoarthritis [14, 29]. Further detriment may be caused by prolonged favoring of the contralateral limb. This can lead to the development of contralateral musculoskeletal conditions, including OA [30].

Long-term restriction of mobility due to joint pain can predispose patients to cardiovascular diseases [31]. The cost of disability associated with OC injury may therefore impact an individual's quality of life for many years beyond initial insult to the joint. In some cases, an OC injury may have lifelong health (and health cost) implications.

3.4 Cost to Healthcare Providers

OC injuries are an economic burden to healthcare providers, most notably due to the prevalence of the condition and costs associated with diagnosis and some treatments. Late diagnoses and failed treatments also leave patients at greater risk of developing osteoarthritis. Management of end-stage osteoarthritis is costly, so it is beneficial to invest in early effective treatments of painful OC damage that may delay or remove the need for joint replacement.

3.4.1 Diagnosis

Diagnosing OC lesions and defects can be difficult where they are asymptomatic or associated with other soft tissue injury [1]. According to Ondrésik, clinical assessment is combined with radiography and arthroscopy to accurately diagnose an OC injury [1].

One 30-minute consultation at an orthopaedic outpatient clinic has been estimated to cost over $300 in the UK National Health Service (NHS) [32, 33]. In the private sector, this cost may only cover the physician fees, excluding the additional costs of other staff salaries, administrative support, and overheads. In countries like the UK or Netherlands, patients may visit a General Practitioner before being referred to an orthopaedic clinician. This comes at an additional cost (approximately £30 ($40–50) in the UK NHS—one of the cheapest systems) to the healthcare provider per visit [34].

Plain radiographs of the knee (anteroposterior, lateral, and skyline views) are usually requested at a cost of $150–$250 [35]. A recent study found that plain radiographs altered clinical management in 48% of sports medicine patients over the age of 40, but only 3% of those who were younger than 40 [35]. As a result, the true cost of a clinically useful plain radiograph may be between $400 and $7600, depending on whether the individual is under or over 40 years old [35].

In the event where a clinician suspects injury to the OC unit, MRI is the easily accessible imaging technique of choice for most clinicians [36–38]. The cost of an MRI varies and is dependent on the location (joint) and type of scanner used. The average cost of one MRI scan is estimated at $400–$850 [39].

Arthroscopic assessment is used by orthopaedic surgeons to visualize the OC damage directly. Despite recent improvements in MRI technology, this remains the most precise method of classifying OC injury, as MRI can produce false-positive and false-negative findings [39]. According to

Voigt et al., this is problematic as it means that a significant number of patients are required to undergo both procedures to gain an accurate diagnosis of the size of the defect and the quality of the tissue in it. In fact, their study showed that 1,397,304 MRI scans were performed on the knee in the USA in 2010 prior to 694,377 arthroscopies. However, arthroscopy does have the advantage of being therapeutic as well as a diagnostic, unlike MRI.

Considering the roles physician consultations and radiography play in supporting the need for an arthroscopy, the initial diagnosis of OC injury is likely to cost a minimum of $1000 per patient. This estimated cost would be significantly higher in cases where patients are required to make multiple visits or undergo multiple investigations.

An outpatient diagnostic arthroscopy of the knee costs $2000–3000 in operating room fees [33, 40] . Anesthesia and surgeon fees can add a further $4000 to the initial cost [33]. Thus, the total cost of diagnosing an OC injury may be $8000–$10,000 per patient in some countries. Based on previous research, it can be estimated that 10% of patients undergoing an arthroscopy of the knee have a full-thickness (OC) injury [12, 13]. As such, diagnosing OC injuries costs the USA over $560 million each year.

3.4.2 Conservative Treatment

In the first instance, conservative methods may be used to address the symptoms of pain and stiffness caused by injury to the OC unit. These include topical and oral medication, in particular paracetamol (acetaminophen) and non-steroidal anti-inflammatory agents (NSAIDs) such as ibuprofen and diclofenac [1]. According to the National Institute of Clinical Excellence (NICE (UK)), one tube of topical NSAIDs costs £5.40 ($7) [41]. However, topical NSAIDs are significantly more expensive in the USA; one ten-day supply is reported to cost $65, whereas a 10-day course of oral ibuprofen would cost a mere $3 [42, 43]. According to Manoukian and colleagues, this creates a significant barrier to patients who wish to be treated by topical

NSAIDs to avoid the side-effects oral NSAIDs cause, particularly as topical NSAIDs are by prescription only in the USA [42, 43].

Conversely, oral pain relief is likely to be more expensive than topical relief in the UK, particularly as proton-pump inhibitors may be co-prescribed, adding approximately £7 ($10) to the annual cost per patient [41]. Drug-related side-effects may further increase costs for the healthcare provider [43].

Although many oral and topical analgesics may be paid for by the healthcare provider or health insurer, it should also be noted that many patients will buy these products without a prescription, out of pocket.

Opioids may be prescribed for pain relief in cases where over-the-counter drugs have failed. In the USA, it is estimated that at least $500 million dollars are spent each year on opioid treatment for pain relief in osteoarthritis [44]. However, little is known about cost associated with prescription drugs for pain relief in patients with OC injury. There are significant personal, health economic, and societal costs from long-term opioid use.

Other conservative treatments of OC injury include bracing and physiotherapy [1]. A recent study on the cost of physiotherapy in patients with hip and knee osteoarthritis in the Netherlands showed that a 12-week course cost the healthcare provider €241–451 ($300–500) per patient [45]. These excluded secondary costs following the 12-week intervention, such as further physiotherapy and other forms of continued treatment. Costs of physiotherapy can be significantly higher in other countries.

However, noninvasive treatment of symptomatic OC lesions is infrequently successful, especially long term [1]. As such, invasive approaches are inevitable in many cases.

According to the ICRS Patient Registry, injection into the joint was the most common treatment procedure for older patients with cartilage damage in 2018 [10]. Altogether, the Registry reports over 25 different knee injection therapies. Stem cell injections and platelet-rich plasma injections were the most common approaches in the ICRS Registry; however, steroid injections or hyaluronic acid-based preparations are the most commonly employed injections in the community [10]. A study by Rosen et al. states that one injection of hyaluronic acid costs roughly $300 [46]. Added to this would be the cost of the physician visit and administration associated with the procedure—an additional $100–$500. However, the treatment has been found to be cost-effective for osteoarthritis. The average cost per quality-adjusted life-year (QUALY) has been estimated at around $8000. The QUALY is a statistic that is often quoted in health economics and is calculated by multiplying the difference in quality of life before and after the treatment by the individual's life expectancy [47]. The quality of life is usually calculated from a validated patient reported outcome measure. A treatment is usually deemed cost-effective if the result is <$50,000/QUALY [46]. In the case of Rosen's study, injections of HA into the joint were found to be more cost-effective than conservative approaches (pain relief and physiotherapy), which was calculated to be $10,716.67/QUALY.

3.4.3 Surgical Treatment

Arthroscopic assessment and surgical debridement of the OC damage is the most appropriate initial approach to surgical management of the defect [1]. This type of arthroscopic intervention of the knee is the most common orthopaedic surgery in the USA [48]. According to Lubowitz and colleagues, knee arthroscopies are cost-effective, with a cost of $5783/QUALY [47].

Where repair of an OC injury is required, other surgical approaches may be considered. The most common approach to small isolated defects (<2–5 cm^2) is MF [33]. This approach is cheap and performed arthroscopically.

In 2015, Miller and colleagues performed a cost-analysis of MF as a treatment for cartilage lesions of the distal femur. They reported that a primary MF procedure cost $7720 per patient (based on an academic medical/surgical center in the USA) [33]. The cost included anesthesia ($720), operating room fees ($3200), and surgeon fees ($3300) [33]. This was estimated to

increase to $8120 if the procedure was a secondary procedure.

MF failure was reported in 28.6% of all patients ($n = 8/64$). Thus, return visits and secondary treatments increased the cost of MF to $8769 after 1 year and to $10,483 after 10 years. The costs of returning patients to their previous level of sport following MF were significantly greater ($16,953 at 1 year; $38,000 at 3 years; $10,483 at 10 years), suggesting that the cost of treating OC injury is higher in professional athletes [33].

Similar cost-analysis of MF were carried out in 2016 and 2017 [19, 20]. According to de Windt, the total treatment cost for MF in a Dutch setting was €6081 ($6500); half of which was attributed to societal costs. In the second study, the direct cost was found to be similar at €3254 (roughly $3500), despite including twice-weekly physiotherapy for 12 weeks at €30 ($35) per visit (a cost which was excluded in Miller's research) [20]. Similar costs have been reported in the UK [49].

Although these three studies concluded that MF is cost-effective in small lesions, very few long-term studies are available to support its use as a long-term treatment for cartilage repair [49, 50]. A recent study on the long-term outcome of MF reported a failure rate of 66%, with the mean duration until failure being as short as 4 years [50].

The high failure rates of MF procedures, combined with poor functional outcome and pain relief mean that patients can be dissatisfied post-operatively. Dissatisfied patients require prolonged care, and in many cases, surgical re-intervention to alleviate their continuing symptoms. This has a significant effect on the financial cost of treating OC injuries [20, 33]. Microfracture to the base of a defect can also render salvage second surgery of the defect less likely to succeed.

Other methods of cartilage repair, such as mosaicplasty/osteo-chondral autograft transfer (OAT), osteo-chondral allograft, ACI, and other cell therapies may provide better outcomes and longer survival rates than MF (~50% failure rates within 10 years) [50, 51].

Miller and colleagues compared the costs of MF to OAT. Due to the complexity of the surgery in comparison to MF, initial procedure costs were unsurprisingly more expensive by OAT - $10,320-$11,222 for OAT by arthroscopy, and $10,120–$11,020 for OAT by open approach. The most expensive procedures were those that were secondary procedures [33]. Failure was reported in 12.5% of OAT patients. As a result, the cost increases per patient at 1 and 10 years were not as significant as for MF ($10,612 at 1 year and $11,479 at 10 years). OAT was also found to be significantly more cost-effective at returning patients to their sport than MF ($11,427.84 at 1 year; $12,856.42 at 3 years; $32,141.05 at 10 years).

For lesions greater than 2 cm^2, ACI may be a more suitable treatment [19]. Unlike MF and OAT, ACI is a two-stage procedure, making it naturally more expensive. Recent cost analyses of ACI in Norway have shown the direct cost of a primary procedures to be €11,031–24,085 ($12,000–25,000), where surgery and material, cell culture, hospital stay, and physiotherapy following the procedure were the greatest expenditures [19, 20]. However, these costs may underestimate the global average. For example, in the UK, the cost of cells alone may be as high as £16,000 ($20,000), and ACI procedures in the USA have been reported to cost over $66,000 [49]. As with other approaches, the costs continue to rise post-operatively; Aae and colleagues found that the cost of primary ACI of the distal femur increased by 5% at 5 years [20]. Re-intervention is also common (~15%), and the direct cost of ACI has been shown to be slightly greater as a secondary procedure (€11,211 compared to €11,031) [19, 20].

Despite the high direct costs, NICE reported in 2017 that there is efficacy and economic evidence to support the use of ACI in certain patients with OC injuries. The long-term outcomes of ACI have been shown to be better than alternative methods of repairing OC lesions. As such, the method is more cost-effective than MF in the long term. In fact, the cost per QUALY gained in relation to MF is estimated to be $45,000–60,000 [49]. A repeat ACI is also recommended when the primary ACI fails, as this has been found to be more cost-effective than secondary intervention of MF (~$17,000/QUALY vs ~$20,000/QUALY) [49].

Matrix-induced autologous chondrocyte implantation (MACI) is a form of third-generation ACI. Early research is predicting MACI to be more cost-effective than MF and earlier versions of ACI due to its lower failure rates. MACI has been estimated to cost ~$20,000–40,000 [49]. Again, the majority of the cost is attributed to the cell culture. Its cost per QUALY gained compared to MF is estimated at ~$60,000 [49].

In additional to the direct cost of the surgical intervention, it should be borne in mind that postoperative treatments, pain relief, and follow-up clinics further economically burden healthcare providers and patients. For example, the ICRS Annual Report in 2018 stated that 99% of patients who underwent surgical intervention for cartilage injury were prescribed follow-on treatment; of these 94% were prescribed physiotherapy [10]. The duration and frequency of physiotherapy following surgery for OC injury would vary from one clinic to another. However, twice-weekly for 12–24 weeks have been recommended by experienced clinicians, adding around $2400 to each treatment cost [19, 20].

3.5 Conclusions

Estimating the cost of disability following injury to the OC unit is difficult, as the prevalence of the condition in any given population is unknown. However, our understanding of how common OC injuries are will improve over coming years, as national and international patient registries become more commonplace. At present, we know that OC injuries affect individuals of all ages and have serious and potentially lifelong consequences on the individual's physical and mental health both short- and long term. For the majority of patients, their ability to work may be affected, thus also bringing financial burdens. This is especially significant for professional athletes.

Due to the complicated nature of OC injury, an accurate diagnosis typically involves multiple visits to sports or orthopaedic physicians. It is estimated that diagnosing an OC injury could cost up to $10,000 per patient.

Treating OC injuries are currently extremely expensive due to the multimodal approach taken. Treatment often combines different forms of conservative treatment before progressing to surgical treatment, burdening healthcare providers across the globe [52]. However, our understanding of the outcomes of these treatments is improving each year. As such, we are moving towards a future where personalizing treatment plans based on the likelihood of success will be the norm. Getting the treatment right the first time will significantly reduce the cost of disability of OC injury per patient. This would have immediate economic benefits for healthcare providers, as well as long-term societal benefits. Most importantly however, it would significantly improve the patient's quality of life, enabling individuals to return to play and work sooner. Successful and rapid treatment of an OC injury could also reduce the likelihood of the development of established osteoarthritis in the joint, thus preventing the need for longer term and more invasive and costly treatment.

References

1. Ondrésik M, Oliveira JM, Reis RL. Advances for treatment of knee OC defects. In: Oliveira JM, et al., editors. Osteochondral tissue engineering: challenges, current strategies, and technological advances. Cham: Springer; 2018. p. 3–24.
2. Hughes RJ, Houlihan-Burne DG. Clinical and MRI considerations in sports-related knee joint cartilage injury and cartilage repair. Semin Musculoskelet Radiol. 2011;15(1):69–88.
3. Gomoll AH, et al. The subchondral bone in articular cartilage repair: current problems in the surgical management. Knee Surg Sports Traumatol Arthrosc. 2010;18(4):434–47.
4. Woolf AD, Pfleger B. Burden of major musculoskeletal conditions. Bull World Health Organ. 2003;81(9):646–56.
5. European Musculoskeletal Conditions Surveillance and Information Network. Musculoskeletal health in Europe: report v5.0. 2012.
6. World Health Organization. 6.12 Osteoarthritis, in priority medicines for Europe and the world 2013 update. 2013.
7. Bhosale AM, Richardson JB. Articular cartilage: structure, injuries and review of management. Br Med Bull. 2008;87(1):77–95.
8. Løken S, et al. Epidemiology of cartilage injuries 'injuries'. In: Doral MN, Karlsson J, editors. Sports injuries: prevention, diagnosis, treatment and rehabilitation. Berlin/Heidelberg: Springer; 2015. p. 1–12.
9. Steinwachs MR, Engebretsen L, Brophy RH. Scientific evidence base for cartilage injury and repair in the athlete. Cartilage. 2012;3(1 Suppl):11S–7S.
10. Conley, C., McNicholas, M., Biant, L, The ICRS Patient Registry: annual report February 2019. 2019. p. 1–28.
11. Zaslav KR. Management of a 37-year-old man with recurrent knee pain. J Clin Outcomes Manag. 1999;6:46–7.
12. Curl WW, et al. Cartilage injuries: a review of 31,516 knee arthroscopies. Arthroscopy. 1997;13(4):456–60.
13. Hjelle K, et al. Articular cartilage defects in 1,000 knee arthroscopies. Arthroscopy. 2002;18(7):730–4.
14. Flanigan DC, et al. Prevalence of chondral defects in athletes' knees: a systematic review. Med Sci Sports Exerc. 2010;42(10):1795–801.
15. Merkely G, Ackermann J, Lattermann C. Articular cartilage defects: incidence, diagnosis, and natural history. Oper Tech Sports Med. 2018;26(3):156–61.
16. Farr J, et al. Clinical cartilage restoration: evolution and overview. Clin Orthop Relat Res. 2011;469(10):2696–705.
17. Heir S, et al. Focal cartilage defects in the knee impair quality of life as much as severe osteoarthritis: a comparison of knee injury and osteoarthritis outcome score in 4 patient categories scheduled for knee surgery. Am J Sports Med. 2010;38(2):231–7.
18. Health and Safety Executive. Work related musculoskeletal disorder statistics (WRMSDs) in Great Britain, 2019. 2019. p. 11.
19. de Windt TS, et al. Early health economic modelling of single-stage cartilage repair. Guiding implementation of technologies in regenerative medicine. J Tissue Eng Regen Med. 2017;11(10):2950–9.
20. Aae TF, et al. Microfracture is more cost-effective than autologous chondrocyte implantation: a review of level 1 and level 2 studies with 5 year follow-up. Knee Surg Sports Traumatol Arthrosc. 2018;26(4):1044–52.
21. Hultin H, et al. Short-term sick leave and future risk of sickness absence and unemployment—the impact of health status. BMC Public Health. 2012;12(1):861.
22. Mannes ZL, et al. Prevalence and correlates of psychological distress among retired elite athletes: a systematic review. Int Rev Sport Exerc Psychol. 2019;12(1):265–94.
23. Bauer KL. Osteochondral injuries of the knee in pediatric patients. J Knee Surg. 2018;31(5):382–91.
24. Bair MJ, et al. Depression and pain comorbidity: a literature review. Arch Intern Med. 2003;163(20):2433–45.
25. Wride JM, Bannigan K. 'If you can't help me, so help me god I will cut it off myself…' the experience of

living with knee pain: a qualitative meta-synthesis. Physiotherapy. 2018;104(3):299–310.

26. Turner AP, Barlow JH, Heathcote-Elliott C. Long term health impact of playing professional football in the United Kingdom. Br J Sports Med. 2000;34(5):332–6.

27. Palisch AR, Merritt LS. Depressive symptoms in the young athlete after injury: recommendations for research. J Pediatr Health Care. 2018;32(3):245–9.

28. Harris JD, et al. Survival and clinical outcome of isolated high tibial osteotomy and combined biological knee reconstruction. Knee. 2013;20(3):154–61.

29. Takeda H, et al. Prevention and management of knee osteoarthritis and knee cartilage injury in sports. Br J Sports Med. 2011;45(4):304–9.

30. Benedetti MG, et al. Muscle activation pattern and gait biomechanics after total knee replacement. Clin Biomech (Bristol, Avon). 2003;18(9):871–6.

31. Mathieu S, et al. Cardiovascular profile in osteoarthritis: a meta-analysis of cardiovascular events and risk factors. Joint Bone Spine. 2019;86(6):679–84.

32. Meirick T, et al. Determining the prevalence and costs of unnecessary referrals in adolescent idiopathic scoliosis. Iowa Orthop J. 2019;39(1):57–61.

33. Miller DJ, et al. Microfracture and osteochondral autograft transplantation are cost-effective treatments for articular cartilage lesions of the distal femur. Am J Sports Med. 2015;43(9):2175–81.

34. NHS England. Missed GP appointments costing NHS millions. 2019. [cited 2020 May 6]. Available from: https://www.england.nhs.uk/2019/01/missed-gp-appointments-costing-nhs-millions/

35. Alaia MJ, et al. The utility of plain radiographs in the initial evaluation of knee pain amongst sports medicine patients. Knee Surg Sports Traumatol Arthrosc. 2015;23(8):2213–7.

36. Marlovits S, et al. Definition of pertinent parameters for the evaluation of articular cartilage repair tissue with high-resolution magnetic resonance imaging. Eur J Radiol. 2004;52(3):310–9.

37. Marlovits S, et al. Magnetic resonance observation of cartilage repair tissue (MOCART) for the evaluation of autologous chondrocyte transplantation: determination of interobserver variability and correlation to clinical outcome after 2 years. Eur J Radiol. 2006;57(1):16–23.

38. Schreiner MM, et al. The MOCART (Magnetic Resonance Observation of Cartilage Repair Tissue) 2.0 knee score and atlas. Cartilage. 2019:1947603519865308.

39. Voigt JD, Mosier M, Huber B. Diagnostic needle arthroscopy and the economics of improved diagnostic accuracy: a cost analysis. Appl Health Econ Health Policy. 2014;12(5):523–35.

40. Marsh JD, et al. Cost-effectiveness analysis of arthroscopic surgery compared with non-operative management for osteoarthritis of the knee. BMJ Open. 2016;6(1):e009949.

41. National Institute for Health and Care Excellence. Costing tools: osteoarthritis implementing the NICE guideline on osteoarthritis (CG177). 2014. p. 1–29.

42. Rogers NV, Rowland K. An alternative to oral NSAIDs for acute musculoskeletal injuries. J Fam Pract. 2011;60(3):147–8.

43. Manoukian MAC, et al. Topical administration of ibuprofen for injured athletes: considerations, formulations, and comparison to oral delivery. Sports Med Open. 2017;3(1):36.

44. Katz JN, et al. Cost-effectiveness of nonsteroidal anti-inflammatory drugs and opioids in the treatment of knee osteoarthritis in older patients with multiple comorbidities. Osteoarthr Cartil. 2016;24(3):409–18.

45. Kloek CJJ, et al. Cost-effectiveness of a blended physiotherapy intervention compared to usual physiotherapy in patients with hip and/or knee osteoarthritis: a cluster randomized controlled trial. BMC Public Health. 2018;18(1):1082.

46. Rosen J, et al. Cost-effectiveness of different forms of intra-articular injections for the treatment of osteoarthritis of the knee. Adv Ther. 2016;33(6):998–1011.

47. Lubowitz JH, Appleby D. Cost-effectiveness analysis of the most common orthopaedic surgery procedures: knee arthroscopy and knee anterior cruciate ligament reconstruction. Arthroscopy. 2011;27(10):1317–22.

48. Behery OA, et al. What are the prevalence and risk factors for repeat ipsilateral knee arthroscopy? Knee Surg Sports Traumatol Arthrosc. 2019;27(10):3345–53.

49. Mistry H, et al. Autologous chondrocyte implantation in the knee: systematic review and economic evaluation. Health Technol Assess. 2017;21(6):1–294.

50. Solheim E, Hegna J, Inderhaug E. Long-term survival after microfracture and mosaicplasty for knee articular cartilage repair: a comparative study between two treatments cohorts. Cartilage. 2018;11(1):71–6.

51. Nawaz SZ, et al. Autologous chondrocyte implantation in the knee: mid-term to long-term results. J Bone Joint Surg Am. 2014;96(10):824–30.

52. Gobbi A, Bathan L. Biological approaches for cartilage repair. J Knee Surg. 2009;22(01):36–44.

Joint Homeostasis of the Knee: Role of Senescence, Hormones, Cells, and Biological Factors in Maintaining Joint Health

John Mitchell, Haylie Lengel, Verena Oberlohr, Andrew Eck, Kaitlyn E. Whitney, William S. Hambright, and Johnny Huard

4.1 Introduction

The knee is one of the largest diarthrodial joints in the human body that allows flexion, extension and limited rotational movements [1]. In normative knee joints, without trauma or genetic predispositions, controlled biomechanics, fluid biology and metabolic balance largely contribute to the joint's homeostatic regulation [2, 3]. Given the essential mechanical role of the knee, it is considered the most susceptible to injury and the most common weight-bearing joint to progress to osteoarthritis (OA) [4–6]. Despite improvements in surgical interventions to re-establish joint bio-

mechanics and local homeostatic functions, there is a growing body of evidence suggesting that knee injury and reconstructive/reparative surgical interventions compromise macro- and micronetworks to an extent that permanently impairs adaptive remodeling mechanisms [7, 8]. While several risk factors have been identified in maintaining joint homeostasis (i.e., patient demographics, nutrition, and cognitive factors, etc.) [9–11], biomechanical and physiological mechanisms provide insight into homeostatic disruption and the pathogenic development of knee joint diseases, such as OA [12].

Knee OA is a chronic disease that is fundamentally characterized by pain, local inflammation, increased matrix-degrading enzyme expression [13–15] and cartilage degeneration [16]. The incidence of knee OA increases with age due to chronic use, gradual cartilage decline, general wear and tear, and progressive chronic pro-inflammatory state associated with natural aging. Knee OA can also manifest in younger individuals through high-impact activities following trauma, such as ligament and meniscal tears. Post-traumatic osteoarthritis (PTOA) is considered an accelerated version of OA, separated into acute and chronic phases [17]. However, both PTOA and idiopathic age-related OA (AAOA) share similar etiological drivers (Fig. 4.1). This includes the upregulation of

J. Mitchell · W. S. Hambright
Center for Regenerative Sports Medicine, Steadman Philippon Research Institute, Vail, CO, USA

H. Lengel · V. Oberlohr
ProofPoint Biologics, The Steadman Clinic, Vail, CO, USA

A. Eck
Department of Orthopedic Surgery, University of Texas Health Science Center at Houston, Houston, TX, USA

K. E. Whitney (✉) · J. Huard
Center for Regenerative Sports Medicine, Steadman Philippon Research Institute, Vail, CO, USA

ProofPoint Biologics, The Steadman Clinic, Vail, CO, USA
e-mail: kwhitney@thesteadmanclinic.com; jhuard@sprivail.org

© ISAKOS 2022
A. Gobbi et al. (eds.), *Joint Function Preservation*, https://doi.org/10.1007/978-3-030-82958-2_4

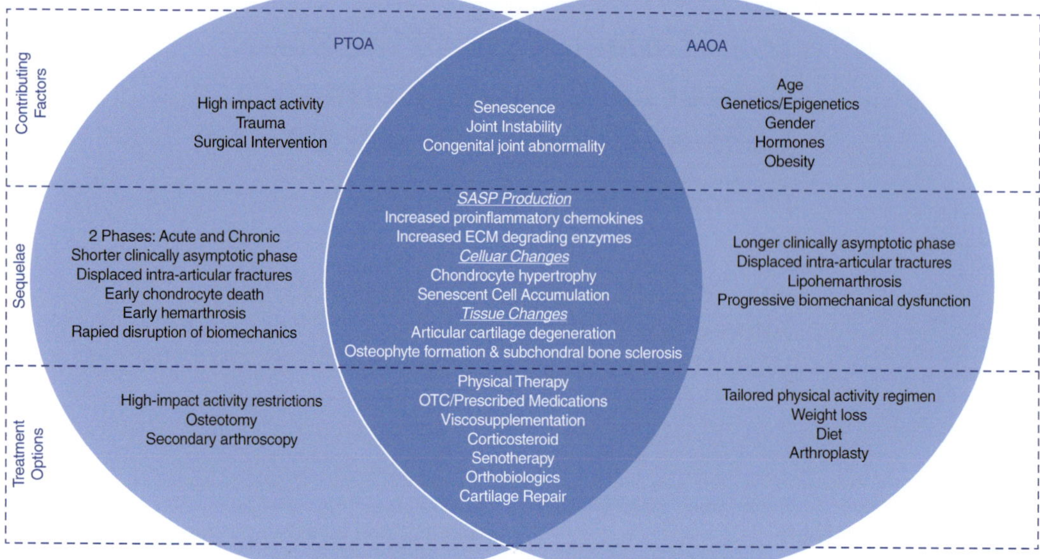

Fig. 4.1 Defining the differences between post-traumatic osteoarthritis (PTOA) and age-associated osteoarthritis (AAOA). This figure highlights the overlapping relationship and independent contributing factors, disease sequelae features, and treatment modalities for each OA condition

cartilage degrading enzymes, inflammation, cartilage fibrillation, and eventual subchondral bone sclerosis [16]. Perhaps one of the most common overlapping mechanisms shared among both age-related and PTOA is the accumulation of senescent cells that seem to play a role in the functional decline of chondrocytes and potentiate the pro-inflammatory and pro-degenerative state of the osteoarthritic joint [18–22]. Thus, targeting senescence and the reduction of the senescence-associated secretory phenotype (SASP) offers an appealing and emerging treatment strategy to restore homeostasis in the knee joint.

This chapter is organized to first provide an overview of the anatomy, physiology, and biomechanics of the knee, followed by an explanation of biomechanical and physiological homeostatic mechanisms. This includes progenitor cells, primary cells, infiltrating cells, senescent cells, vitamins, hormones, and proteins that are involved in normative homeostasis and disrupted (e.g., PTOA, AAOA) conditions. Lastly, the chapter will conclude with current orthobiologic modalities and pharmacologic strategies that are being used to restore homeostasis in knee OA.

4.2 Overview of Knee Anatomy and Physiology

Understanding the anatomy, physiology, and cellular matrices that compose the knee joint is paramount to effectively research, correct, and promote health. The knee joint is the major motor joint of the lower extremity and acts as a gliding hinge to support body weight, absorb shock of force from the feet, and assist with lower extremity movement [23]. It is composed of three bones and two joints; the femur superiorly, the tibia inferiorly, and the patella anteriorly. The tibio-femoral joint allows transmission of force from the femur to the tibia [23]. The patellofemoral joint serves as the extensor mechanism of the knee and acts to dissipate forward momentum of the body during movement [24]. Surrounding the osseous bodies are the components that generally cushion (cartilage and meniscus), lubricate (synovium and articular surfaces), connect (ligaments and tendons), and mobilize the knee (flexors and extensors) [25].

During flexion-extension, articular surfaces of the femur roll over the tibial surface. This union, known as the tibiofemoral joint, can be thought of as a double wheel, as the femur rolls over two curved groves of the tibial surface [26]. The menisci make up for non-congruent areas of this union. These crescent shaped lamellae help correct for incongruencies by increasing contact area, in turn adding stability, motion control [27], and improving shock absorption by transmitting 70–99% of the tibiofemoral compressive load [28]. It is important to note that the menisci are mostly hypovascularized, with the outer third having little blood supply and minimal ability to heal, while the inner portion relies on diffusion from synovial fluid for nutrients [29]. The synovial membrane is adherent to the menisci and the rest of the joint capsule, a fibrous sleeve that connects the distal end of the femur to the proximal border of the tibia [25]. The synovial sac plays a role in cushioning and lubrication and is thought to contain a biological milieu which stimulates cartilage repair in the acute period following traumatic injury.

Moreover, the knee is stabilized by various muscles, ligaments, tendons, in addition to cartilage. Collectively, these stabilize the knee structurally and allow for movement in three rotational planes, consisting of flexion-extension, varus-valgus, and internal-external rotation as well as three translational movements, including anterior-posterior, medial-lateral, and compression-distraction [26].

There are four main ligaments which combat strain and torque placed on the knee to maintain stability. This includes the anterior cruciate ligament (ACL), posterior cruciate ligament (PCL), medial collateral ligament (MCL), and lateral collateral ligament (LCL) [25]. The ACL works to resist anterior translation and prevent hyperextension [30]. The PCL is the strongest ligament in the knee and works to resist posterior translation, and to a lesser extent, tibial rotation. The LCL primarily resists varus force and the MCL primarily resists valgus force [31]. Collectively, these soft tissues protect the joint from compressive forces while maintaining functional stability and are thus critical elements to maintaining joint homeostasis through strain and injury prevention [28]. The amount of mechanical loading itself

has a strong influence on the composition of the meniscus although the exact mechanisms are not fully understood [32]. However, this is likely, in part, due to the intrinsic and extrinsic complexity of mechanobiological factors.

Mechanobiological response involves mechanical stresses that change the biological milieu within the joint. Ligaments can respond using tenocytes, which are fibroblast type cells, through numerous modalities, including intercellular communication pathways, gap junctions, norepinephrine activation of adrenoceptors, and ATP pathways [33]. Through these pathways, ligaments can respond to stress by regulating collagen synthesis and forming a stiffer, stronger tendon [34]. The response of the bone similarly uses extracellular and intracellular signaling pathways that are largely driven by the osteocyte. When this cell is activated by mechanical stress, it produces signaling molecules to activate osteoblasts, osteoclasts, and other effector cells in response to local factors and load [35]. Components of the knee joint must be able to adapt and respond to fluctuating forces exerted on the joint. The structure of the meniscus is controlled through a balance of anabolic and catabolic processes modulated by genetic and biochemical factors including cytokines, chemokines, and growth factors including IL-1, TNF-alpha, and TGF-beta-1 [36]. Thus, biomechanics of the knee joint play a pivotal role in maintaining homeostasis and once disrupted, become a prominent driver in the pathological sequelae of joint dysfunction and tissue degeneration.

4.2.1 Articular Cartilage and Subchondral Bone Interface

Subchondral bone is effectively the foundation on which articular cartilage rests. The articular cartilage, subchondral plate, and trabecular bone function together to form the osteochondral unit. This region of the knee plays a critical role in the distribution of mechanical load across the joint through the gradual transition of stress and strain to curtail injury [37]. Immediately above the subchondral plate lies a calcified layer of cartilage, which was originally thought to be impenetrable. However, extensive evidence suggests that crosstalk between subchondral bone and articular cartilage occurs and regulates homeostatic functions of the joint. Uncalcified cartilage frequently penetrates the calcified cartilage into subchondral bone, allowing for molecular diffusion between the cortical plate and deep calcified zone. The articular cartilage consists of four zones: the superficial, middle, deep, and calcified zone, each with their own distinct cell types. Subchondral bone consists of the cortical plate and cancellous bone. To maintain these cell types, it has been reported that notch signaling is critical for progenitor cell survival and maintenance of an undifferentiated state [38]. Channels and cracks that form during osteoarthritis have also been identified as pathway mechanisms allowing for the transfer of pathologically associated signaling molecules including VEGF, TGF-β, DKK-1, MMPs, and inflammatory cytokines [39, 40]. These molecules play a role in the restructuring of bone and cartilage as OA progresses, but are also crucial for homeostasis. Relative to articular cartilage, subchondral bone is highly vascularized and is responsible for more than half of the nutrient supply and hydration to associated articular cartilage. In fact, the exchange that occurs at this site is critical in providing growth factors, nutrients, and oxygen to the extracellular matrix (ECM) producing articular cartilage chondrocytes. Interestingly, this communicative behavior is stimulated by both biochemical and mechanical properties.

For example, physical loading of cartilage is important for homeostasis and can counter the catabolic effect of inflammatory cytokines. Overloading of the subchondral bone is thought to contribute to OA, where the loss of articular cartilage and degradation of the subchondral bone manifests [39]. It is worth mentioning the new field of mechanogenetics, a field based on the genetic expression of tissues under mechanical forces (mechanobiology). Mechanobiologic signals play critical roles in the regulation of cellular responses under both physiological and pathological conditions. The Hueter-Volkmann "law"

observes that large compression on joints reduces normal growth and lesser than normal forces causes overgrowth. This law lays the foundation of mechanobiology and mechanogenetics [41].

The knee joint is the largest joint in the body and incredibly complex being a compound joint of tibiofemoral and patellofemoral components. Thus, various tissues maintain homeostasis of the joint both structurally and physiologically. Below, we discuss critical tissues regulating normative joint health and function including a brief overview of important cell and tissue-specific signaling factors within the joint capsule.

4.3 Etiological Factors in Homeostatic Disruption

4.3.1 Pathogenesis of Knee Osteoarthritis

The exact pathophysiology of knee OA is poorly understood, but it is widely believed to be caused by multiple factors including genetics, epigenetics, and biomechanics. Biomechanically speaking, articular chondrocytes in the extracellular matrix (ECM) respond to loading via different clustering patterns along force vectors relative to the biomechanics of gait [42]. Pathologic loading patterns can initiate remodeling of the ECM which can change the ratio of collagen to proteoglycan, leading to loss of functional capacity, stiffening of cartilage, and increased inflammation [43–45]. Eventually, the load is transferred to the subchondral bone leading to global pathologic changes of the osteochondral tissues compromising the elastic modulus of the subchondral bone [46]. At the cartilage level, failure of the tissue can occur due to aging or trauma, leading to an imbalance between synthesis and degradation and extra-cartilaginous factors. Type II collagen fibrils (consisting of 90–95% of total collagen) are cleaved by collagenases and degraded by proteolysis while aggrecans are degraded by aggrecanases [47]. Aberrant activation of collagenases and aggrecanases occur during aging and often follow trauma, leaving cartilage unable to properly respond due to cellular senescence and abnormal hypertrophic differentiation. These changes result in an altered extracellular matrix with increased water content decreased elasticity. Chondrocytes, osteocytes, and the synovium produce various pro-inflammatory cytokines, most importantly IL-1ß, which contribute to the degradation of the cartilage. Although it is not clear exactly how certain factors contribute to OA-symptomology, there are several significant pro-inflammatory factors associated with degenerative and inflammatory process, including, but not limited to, matrix metalloproteinases (MMPs) -1, -3, -8, -12, and -13, interleukins (ILs)-1β, -6, and -8, -17, -18, leukemia inhibitory factor (LIF), oncostatin M (OSM), tumor necrosis factor alpha (TNF-α), and prostaglandins [48, 49]. For certain biomarkers, such as MMPs, there is evidence that levels increase in synovial fluid at early stages of OA, which is thought to promote early ECM destruction [50–53]. However, the diagnostic and prognostic promise of OA biomarkers is limited, given the periodic chronicity of factors and lack of longitudinal studies focused on individual biomarkers [54]. Nonetheless, the pro-inflammatory milieu present in the osteoarthritic joint ultimately leads to pain, dysfunction, cartilage damage, synovitis, osteophyte formation, and eventual sclerosis of the subchondral bone, which are all clinical hallmarks of OA.

4.3.2 Knee Homeostasis and Disrupted Conditions: Cell-Specific Functions

4.3.2.1 Meniscal and Ligament Cells

The meniscus of the knee joint is located between the femoral condyle and tibial plateau of the knee. It functions in the force transmission, shock absorption, lubrication, joint stability, and proprioception of the knee joint. The inner avascular region and vascularized outer region make up the two major sections of the tissue. The composition of these sections varies, with the inner resembling articular cartilage, and the outer fibrocartilage representing the cellular makeup and structural ECM components. Characterization of meniscal cells is inconsistent overall, with reported evidence for the presence of fibrocytes, fibroblasts, meniscus cells, fibro-chondrocytes, and chondrocytes [55]. Controversy aside, three general populations of cells can be found in the meniscus. The first population of cells in the outer regions is similar in appearance and behavior to fibroblasts and has an oval, fusiform shape with long cell extensions to facilitate cell to cell communication as well as communication with the extracellular matrix. The extracellular matrix is composed mostly of type I collagen, but also contains small amounts of glycoproteins and collagen types III and V, making the "fibroblast-like" meniscal cell ECM slightly unique from the chondrocyte-derived type II collagen-enriched hyaline articular cartilage [56]. Conversely, the second population of cells in the inner region of the menisci is rounder and surrounded by a matrix of type II collagen and lesser amount of type I collagen. The relative abundance of type II collagen in the inner region is like that of articular cartilage classified by fibro-chondrocytes or chondrocyte-like cells. Finally, the third population found in the superficial zone of the meniscus is flattened, fusiform, and absent of extensions. It is thought that these cells are progenitor cells for the meniscus and are speculated to possess therapeutic potential to differentiate and replace damaged cells [29]. Overall, the meniscal microenvironment is unique in cellular and ECM content, yet responsible for a variety of critical functions in the knee.

Ligaments appear to be relatively complex at the microscopic level, composed of seemingly few cells, fibroblast and fibrocytes, scattered in a longitudinally aligned fibrous matrix. These cells produce the matrix made of dense regular connective tissue while only making up a small portion of the total ligament. Due to their lack in numbers, at first it would seem that fibroblasts and fibrocytes are physically and functionally isolated. However, this is not the case, as fibroblasts form spindle-like cytoplasmic extensions that extend for long distances to connect to other cells. This forms a complex three-dimensional architecture comprised of triple helical collagen molecules which form crosslinked fibers [57, 58]. There is also evidence that these cells incorporate the use of gap junctions to communicate with neighboring cells to coordinate cellular and metabolic responses throughout the tissue. This includes small molecules for chemical communication that are postulated to be involved in homeostasis and the maintenance of the extracellular matrix [59].

4.3.2.2 Local Primary and Progenitor Cells

Cellular infiltration, cytokine production, and inflammatory activation of articular chondrocytes, synoviocytes, and other joint tissues are common in OA [60]. Joint injury causes cellular and molecular alterations of the joint tissue, impacting joint biomechanics and as such, is a well-established risk factor for the development of PTOA [61]. Disruption of the ACL, meniscus, and intra-articular fracture result in hemarthrosis, chondrocyte death, bone bruising, and release of inflammatory mediators [60]. These acute events subsequently trigger a chronic remodeling of the cartilage and other joint tissues [62]. A cascade of metabolic changes is set in motion by the post-traumatic inflammatory phase, lasting hours up to approximately 2 months [63]. Within the first 2 weeks after trauma, three overlapping phases have been observed: the first is characterized by cell death and inflammatory events; the next is a subacute phase marked by persistent inflammation; and a late phase presenting with increased matrix degradation [64].

Trauma resulting in cracks or fissures of the cartilage surface can result in the release of cartilage extracellular matrix molecules [65]. A reduction in cellularity of chondrocytes, responsible for mediating cartilage homeostasis, reduces the reparative and regenerative capabilities of cartilage [63]. The remaining viable chondrocytes are thought to be activated by enhanced cellular metabolism and generation of oxygen radicals, matrix-degrading enzymes, and inflammatory mediators [17]. A series of biomechanical and physiochemical changes to the tissue lead to significant alterations in the chondrocyte's ability to express proteins involved in metabolic pathways and leading to cell death

[63]. Collagen rupture and loss of rapid glycosaminoglycans (GAG) induces cartilage swelling, chondrocyte necrosis, and apoptosis, all of which are irreversible events contributing to PTOA [66]. This is further supported by a study revealing a higher percentage of apoptotic cells in the cartilage of patients with intra-articular fracture compared to patients with osteoarthritis and rheumatoid arthritis [67, 68]. The release or degradation of metalloproteinase (MMP), collagen-type II and other proteins often accompany the continued GAG loss involved in cartilage injury [69, 70]. Elevated MMPs and collagen-type II peptides, proteoglycan degradation, and bone marker release are early signs of articular cartilage degradation and appear in the synovial fluid of patients suffering knee injury [71]. Mesenchymal progenitor cells (MPCs) have high proliferative potential with the ability to differentiate into various mesenchymal lineages including bone, cartilage, fat, tendon, and stromal tissue [72, 73]. These cells are present in trabecular bone, bone marrow, adipose, and synovial tissues [74–76] and have the ability to repair damaged bone and cartilage. It is believed that synovial fluid MPCs can originate from cartilage, synovial lining cells or vascular, multipotential pericytes of the synovium [77, 78]. In a study comparing PCs in synovial fluid to MPCs present in bone marrow, synovial fluid was collected from the knees of 100 patients with arthritis, including rheumatoid arthritis, osteoarthritis, and other arthropathies [79]. Regardless of the intrinsic or extrinsic risk factors that progress PTOA or AAOA etiologies, polyclonal and single cell-derived cultures of synovial fluid fibroblasts revealed the presence of tripotential MPC with similar phenotypes to uncultured bone marrow MPCs, despite inhabiting a diseased joint [79]. The synovial fluid of osteoarthritis patients was characterized by an elevated concentration of MPCs, suggesting that their origin is likely from disrupted joint structures. This may provide insight into the role of MPCs in disrupted homeostasis, and more specifically, on the inflammatory and regenerative phases of tissue healing within a diarthrodial joint [79].

4.3.2.3 Joint Infiltrating Cells

Synovial inflammation is related to acute cellular infiltration and strongly correlated with the extent of joint injury [63]. It has been hypothesized that a complement proteolytic cascade and toll-like receptors are activated and work in conjunction with the cytokine/chemokine first line of immunity [60]. Macrophage content and activation is increased in the synovium of PTOA and ligament injuries [80, 81]. Furthermore, these synovial macrophages along with chondrocytes actively produce complement components and inhibitors [82, 83]. This complement proteolytic cascade is an essential mechanism for clearing pathogens and damaged cells [84]. Neutrophils and leukocytes subsequently destroy macrophages by releasing proteolytic enzymes and thereby propagating the inflammatory cycle [85]. Lymphocytic perivascular infiltrates are also present in patients with meniscal tears and often associated with knee pain [81]. The literature suggests that patients with acute knee injuries also had increased levels of inflammatory mediators, including cytokines and tumor necrosis factor (TNF) [63], though cellular infiltration in humans prior to OA onset needs further investigation [60].

The rupture of blood vessels in the joint is a common consequence of injury resulting in hemorrhage into the immediate area and an organization of red blood cells and cellular debris to form a hematoma [86]. Intra-articular bleeding following ACL rupture and meniscal tear has been shown to damage chondrocytes and as such is an important factor in the genesis of PTOA. In vitro studies have demonstrated that human cartilage exposed to mononuclear cells and red blood cells exhibited an irreversible inhibition of proteoglycan synthesis [87]. Furthermore, intra-articular bleeding and plasma extravasation severely compromise the synovial fluid in its lubricating function. Lubricin, a primary joint lubricant, is degraded by neutrophil-derived enzymes and its synthesis is suppressed by inflammatory mediators within post-traumatic synovium [88]. Activated neutrophils in acute hemarthrosis produce reactive oxygen species (ROS), elastase and lysosomal enzymes that degrade proteoglycans [89]. GAG synthesis is reversibly suppressed by mononuclear cells, however, in the presence of red blood cells and in part, oxygen radicals, this synthesis inhibition becomes irreversible [90]. Blood-induced damage appears to be further mediated by hemoglobin degradation products, such as deoxyhemoglobin, methemoglobin, and hemosiderin [62]. In the presence of hemarthrosis, synovial cells phagocytize erythrocytes and hemoglobin which results in synovial hypertrophy and siderosis [62]. Hemarthrosis is a common complication of hemophilia and as such various studies have been conducted on this patient population to obtain a better understanding of hemarthrosis pathophysiology.

In a recent cross-sectional study that included 4343 males with hemophilia, limited joint range of motion was positively associated with the number of bleeding episodes and history of orthopedic procedures [91]. Repeated episodes of hemarthrosis result in synovial hypertrophy, phagocytosis of cellular debris and iron, and release of hydrolytic enzymes into the joint space. In hypertrophic synovial fluid, inflammation causes chondrolysis and fibrous adhesions, often progressing into a destructive, osteoarthritis-generating condition [87].

4.3.2.4 Senescent Cells and Related Secretory Factors

Cellular dysregulation in different tissues of the joint capsule is thought to play a critical role in the gradual deterioration of articular cartilage leading to mechanical failure and eventual pathology. For example, cell intrinsic effects including genomic instability, telomere shortening, dysregulated nutrient sensing, upregulated pro-inflammatory signaling, mitochondrial dysfunction, and loss of proteases are all characteristic of AAOA and idiopathic OA [18, 20, 21]. Importantly, these features are established hallmarks of cellular senescence. In the aged joint, senescent chondrocytes, synoviocytes, and synovial macrophages have all been found to be present and severely elevated in OA [18, 20, 21]. Senescence is a cell state defined by loss of proliferative capacity, increased metabolic activity, cellular enlargement, and resistance to apoptosis. Senescent cells promote disease and tissue dysfunction through the release of degenerative factors and pro-inflammatory mediators (including MMPs and ADAMTS4-5, IL-6, IL-1β, and TNF-α) otherwise known as the senescence-associated secretory phenotype (SASP) [92]. In the joint, it is thought that senescence can be induced by a variety of age-associated extrinsic (mechanical stress, damage signals, inflammatory factors) and intrinsic signals (reactive oxygen species, DNA damage, mitochondrial dysfunction) that lead to the production of an SASP that importantly initiates senescence in neighboring cells, thereby potentiating loss of homeostatic potential and promoting disease [92, 93]. Indeed, senescent cells have been found in tissues of the arthritic joint in murine systems and in discarded cartilage isolated from arthroscopy procedures [18, 21, 93–95]. Senescent chondrocytes have also been demonstrated to co-localize with osteoarthritic lesions, strongly implicating senescence, and ostensibly SASP, in promoting disease [95]. Further empirical preclinical evidence for the contribution of senescent cells in joint disease is the finding that intra-articular injection of senescent cells into the knee joint can elicit OA-like symptomatology in mice including inflammation, osteophyte formation, and loss of proteoglycan content in cartilage [96]. In addition, using genetically engineered mice, it has also been found that p16^{INK4a} (an established senescence marker) expressing chondrocytes accumulated following ACL injury associated with marked increases in SASP factors like MMP-13, IL-6, and IL-1β [19]. Finally, genetic ablation of p16^{INK4a} expressing chondrocytes in mice was found to reduce the development of OA symptoms and SASP factors [19]. Thus, senescence seems to significantly promote joint deterioration with age and following trauma.

4.3.3 Knee Homeostasis and Disrupted Conditions: Hormones

Hormones play a critical role in the homeostatic regulation of systems throughout the body, both systemically and locally. Joint homeostasis is in part maintained through the actions of specific hormones, including: estrogen (Estradiol (E2)), parathyroid hormone (PTH), growth hormone (GH), ghrelin (GHRL), and α-melanocyte-stimulating hormone (α-MSH). The effect of each hormone and its role in joint homeostasis and function in PTOA and AAOA conditions are described below along with results from the referenced studies, reported in Table 4.1.

4.3.3.1 Estradiol (E2)

The role of E2 in joint homeostasis became evident when a clear trend of osteoarthritis in postmenopausal women emerged [109]. E2 is thought to have anti-inflammatory properties and a chondroprotective effect resulting from the binding and subsequent activation of estrogen receptors (ERs) on target tissues [109–111], predominantly ERβs in cartilage, synovium, and cancellous bone [109]. The binding action of E2 to ERs allows for activation of signaling pathways that, in turn, regulate the expression of signaling molecules, namely calcium (Ca2+) and pro-inflammatory cytokines, IL-6 and TNF-α [109, 112]. Regulation of signaling pathway activation by E2 explains the correlation between heightened levels of pro-inflammatory cytokines, found in E2-deficient women suffering from OA, and likely contributes on a large scale, to the pathogenesis of OA [97, 98, 112]. Adequate levels of E2 are required for homeostatic maintenance of intracellular signaling molecules and thus joint homeostasis.

4.3.3.2 Parathyroid Hormone

Parathyroid hormone (PTH) is a major regulator of calcium-phosphate and Vitamin D (VitD) homeostasis in bones and plays a catabolic and osteoanabolic role. The parathyroid is stimulated to upregulate the release of PTH by low plasma [Ca$_2^+$], circulating PTH then follows two pathways: binding to receptors on the kidneys, triggering the release of VitD or stimulating bone to increase resorption of Ca$_2^+$. Both pathways lead to an increase in plasma [Ca$_2^+$] and restoration of normal secretion of PTH [113]. PTH stimulates bone resorption by activating the NFκB signaling pathway, causing increased osteoclast formation

Table 4.1 Summarized results of key hormones assessed in patients with knee osteoarthritis (OA) and healthy donors

Hormone	Summarized results	Sample type	Patient type	References
Ghrelin (GHRL)	Deficient levels of GRHL in synovial fluid (SF) are negatively correlated with both: radiographic severity of PTOA; and SF concentrations of pro-inflammatory cytokine, TNF-α. Plasma concentrations of TNF-α and SF concentrations of TNF-α were significantly higher in OA patients than in healthy control groups.	Plasma, SF	OA, healthy	[97–99]
Growth hormone (GH)	GH deficiency is thought to have a protective effect against OA; patients with GH deficiency were found to have a significantly lower incidence of OA than patients with acromegaly and patients with normal serum levels of GH analysis of plasma GH levels in a population of OA patients revealed significantly elevated levels.	Serum	OA, acromegaly, healthy	[100, 101] [68].
Parathyroid hormone (PTH)	There is not a significant correlation between serum PTH levels and incidence of radiographic OA, joint space narrowing, or progression of OA. PTH plays an indirect role in the pathogenesis of OA through modulation of VitD metabolism.	Serum	OA, healthy	[102–104]
Estradiol (E2)	Women with E2 deficiency carry a significantly higher risk for the development of knee OA as compared to women with normal levels of E2; serum levels of E2 are found to be significantly lower in women presenting with OA, than in healthy controls. E2 deficiency may contribute to the pathogenesis of OA though regulation of plasma IL-6 levels.	Serum	Postmenopausal women, OA, healthy	[97, 105, 106]
α-Melanocyte-stimulating hormone (α-MSH)	A statistically significant correlation between increasing radiographic severity and decreased levels of α-MSH levels in SF in OA and PTOA patients lead to the hypothesis that α-MSH may serve as an important biomarker in determining disease severity of PTOA and AAOA.	SF	OA, healthy	[107, 108]

and activity [114]. Conversely, PTH acts in an osteoanabolic manner by decreasing the expression of SOST/Sclerostin on osteocytes, thus allowing bone formation [114]. This cycle allows bone catabolism to equal bone anabolism and therefore, homeostasis.

4.3.3.3 Growth Hormone

Growth hormone (GH) plays an indirect role in joint homeostasis though the mechanism is not well defined. GH is produced by the pituitary and then released into the bloodstream, circulating GH then triggers the liver to synthesize and release IGF-1 [100, 115]. This release mechanism is controlled by negative feedback—as [GH] increases, so does [IGF-1], signaling to the

pituitary to slow the production of GH [115]. Normal levels of GH and IGF-1 contribute to joint homeostasis through anabolic effects on chondrocytes and the ability to stimulate growth and repair of adult cartilage [100, 115, 116]. IGF-1 stimulates DNA synthesis, cell replication, and glycosaminoglycan and proteoglycan synthesis [116, 117] and has also been shown to have a role as an anti-inflammatory agent through its ability to inhibit the actions of pro-inflammatory cytokine IL-6 [100]. Chronically heightened levels of GH are linked to high levels of joint inflammation, decreased receptor affinity to IGF-1, and the development of OA [101, 115–117]; further, OA chondrocytes were found to be unresponsive to stimulation by IGF-1 [100, 115].

4.3.3.4 Ghrelin

Ghrelin (GHRL), a peptide hormone secreted from enteroendocrine cells of the GI tract, has been recently discovered to play an important chondroprotective role in joint homeostasis, primarily through the regulation of chondrocyte metabolism [98, 118]. GHRL is expressed in cartilaginous tissue and in chondrocytes, where it is thought to act as a growth factor, playing an important role in chondrocyte differentiation [98, 99, 118]. Through suppression of the NFκB signaling pathway, GHRL inhibits the inflammatory and degenerative effects of pro-inflammatory cytokines, such as TNF-α, IL-6, and IL-1β, thus preventing chondrocyte apoptosis and aiding in the maintenance of healthy articular cartilage [98, 99, 118, 119].

4.3.3.5 α-Melanocyte-Stimulating Hormone

α-Melanocyte-stimulating hormone (α-MSH) is a melanocortin peptide hormone thought to have anti-inflammatory and protective effects on articular cartilage. Though the exact mechanism behind these effects is not clear, α-MSH has been found to block the TNF-α-induced expression of MMP-13, an enzyme that targets cartilage for degradation. This is achieved by decreasing p38 phosphorylation, causing subsequent inhibition of the NFκB signaling pathway [107, 120, 121]. α-MSH is found in both synovial fluid (SF) and plasma and has been proven to reduce the secretion of pro-inflammatory cytokines, TNF-α, IL-6, and IL-8 in mixed synoviocyte cultures, thus exerting an anti-inflammatory effect [122]. While there is a correlation between lowered synovial fluid concentrations of α-MSH and OA disease progression and severity [107, 108], more research is needed to define the role of α-MSH in joint homeostasis.

4.3.4　Knee Homeostasis and Disrupted Conditions: Vitamins

4.3.4.1 Vitamin D (25(OH)D)

Vitamin D deficiency (VDD) occurs in a large percent of the worldwide population. In fact, approximately 50% of the US population suffers from VDD or Vitamin D insufficiency [123].

Vitamin D (VitD) is believed to play an essential role in the pathology of various diseases, including OA. Results on the effect of VDD on OA vary greatly, as the mechanism remains unknown; although, studies have consistently found that a large percent of patients presenting with OA also have VDD [124–126]. However, VDD has not been shown to play a significant role in the progression of radiographic OA though there may be a moderate correlation [127–131]. Studies that manipulated daily intake of VitD found that low dietary intake of VitD is linked to a significantly higher risk of OA progression [131, 132], while supplementation with VitD led to a decreased risk OA progression and lessened joint disability [133, 134]. VDD may be associated with exacerbated pain levels as a result of OA and a poorer quality of life [127, 135–137].

4.3.4.2 Vitamin K (Phylloquinone)

Vitamin K deficiency (VKD) is a common issue thought to play a role in the development of OA. In human chondrocytes, Vitamin K (VitK) is essential for carboxylation of matrix Gla protein (MGP) and Gas-6. VKD leads to under-carboxylation of these proteins, causing dysfunction of said proteins and in turn, affecting chondrocyte differentiation, endochondral ossification, and mineralization of chondrocytes [138–140]. It is suggested that these adverse effects contribute to the development of OA, as demonstrated by the findings that VKD is associated with an increased prevalence of OA manifestations, including osteophyte formation, development of cartilage lesions, articular cartilage damage, and joint space narrowing [138, 140]. VKD is associated with an increased risk for the development of OA [138, 139].

4.4　Interventional and Pharmacologic Strategies to Restore Homeostasis

4.4.1　Interventional Orthobiologics

Orthobiologics are commonly used for musculoskeletal tissue repair and regeneration to accelerate functional recovery, minimize joint failure, and restore joint homeostasis. Novel innovations have developed over the past decade to replicate complex musculoskeletal structures. In adherence to the Food and Drug Administration's (FDA) HCT/P provisions, many forms of autologous and allogenic orthobiologics are minimally manipulated to retain highly concentrated biological properties that are naturally receptive and have been shown to stimulate remodeling pathways in various types of damaged musculoskeletal tissue [141]. Common orthobiologics that require minimal resources for production include platelet-rich plasma (PRP), bone marrow concentrate (BMC), lipoaspirate, and prolotherapy [142]. PRP is more widely used for knee OA, perhaps due to readily attainable growth factors, cytokines, and other bioactive factors at physiologic proportions [143, 144]. Although current practice guidelines are neither for nor against the use of PRP for the treatment of knee OA, several reports have shown that a single or series of PRP injections attenuate symptoms and improve joint function in symptomatic knee OA patients [144–148]. However, patients with early to moderate knee OA have been more therapeutically responsive than those with end-stage OA [145]. It is also unclear whether the mechanism(s) of action in PRP restores homeostatic function to the joint or delays the onset of OA [149]. Further in vitro studies and randomized clinical trials with adequate control groups are necessary to better understand the biological mechanisms of PRP in vivo. Other orthobiologis are manufactured or significantly modified with advanced technology for production, including isolated and expanded cell therapies, biological scaffolds, vehicles for drug/protein delivery (i.e., nanoparticles, microspheres), and exosomes [142]. Specifically, exosomes have garnished attention as potential therapeutic agents. Exosomes are a small subset of, lipid-bilayer enclosed cell-derived particles (extracellular vesicles), with the purpose of facilitating cell to cell communication, and are typically 30–100 nm in diameter [150]. Exosomes are comprised of a protein mosaic lipid-bilayer that stabilizes biological fluids, including peripheral blood and synovial fluids, as well as proteins, mRNA, miRNA, and few small noncoding RNAs. Given their small size, stability, bioactive content, and cell specificity, exosomes can serve as a transport vehicle in stem cell therapies [150]. Many studies focus on the therapeutic potential of stem cell-derived exosomes. One study injected MSC-derived exosomes into a collagenase-induced OA model and observed that this intervention protected mice from joint degradation [151]. Overall, MSC-derived exosome treated defects showed enhanced gross appearance and improved histological scores in another study using a rat model [152]. While newer evidence is promising, further in vivo models are warranted to determine the therapeutic efficacy of exosomes for musculoskeletal repair and regeneration.

4.4.2 Senotherapies

A promising new intervention either alone or in combination with cell-based therapies is the use of senolytic drugs. The understanding of senescent cells and their tissue degrading SASP factors in promoting age-related musculoskeletal pathology is becoming quite clear [153, 154]. Furthermore, targeting senescent cells either genetically (in preclinical models) or pharmacologically has proven very effective in mitigating symptoms of OA at the histological and radiographic level. Thus, translation into clinical studies is promising and underway. There are several clinical trials from our group and others using oral and intra-articular (IA) delivery modalities of senolytic agents for the treatment of Kellgren-Lawrence grade II-IV knee OA (NCT03513016, NCT04129944, NCT04210986). While the benefit of IA injection of senolytic agents is obvious, targeting local senescent cells directly within the joint capsule including chondrocytes, synoviocytes, and even infiltrated macrophages, oral administration of senolytic drugs may also have utility. Systemic delivery of senolytics may dampen the significant inflammatory immune response via SASP inhibition during advanced OA, especially in older patients whose baseline inflammatory state is likely higher in accordance with the inflammaging hypothesis [155]. This is particularly important when considering the administration of senolytic drugs in coordination with other biologic interventions such as BMC, PRP, or isolated MSCs. Several senolytic drugs have been shown to decrease SASP production and robustly improve MSC function through the elimination of senescent cells (including MSCs) [156, 157]. Thus, it stands to reason that senolytic treatment prior to BMC or PRP harvest may result in a superior orthoregenerative product containing reduced levels of pro-inflammatory factors and improved progenitor cell function. Future studies using senolytic drugs should interrogate differences in delivery modalities (intra-articular vs. oral administration), and the potential benefit as part of a combinatorial treatment approach with biologics. Nonetheless, the link between cellular senescence and OA pathogenesis is strong, which highlights senolytic drugs as a very appealing and innovative approach to prevent or treat OA.

Senescence is a cell state defined by loss of proliferative capacity, increased metabolic activity, and importantly, resistance to apoptosis. Several senolytic compounds that selectively target and inhibit anti-apoptotic pathways in senescent cells have been recently identified and shown to kill senescent cells in vitro and in vivo without affecting quiescent or proliferating cells [157]. Thus, senolytic drug use is an innovative and appealing approach for the treatment of OA because they target senescent cells directly, thereby inducing cell death and abrogating systemic SASP factors [157]. Importantly, the safety and efficacy of several senolytic drugs to treat chronic diseases have been demonstrated in several preclinical studies and more recently in Phase I-II clinical trials for OA from our group and others. For example, the senolytic drug Dasatinib is an FDA-approved drug for leukemia with few side effects while other senolytic drugs like quercetin and fisetin are naturally occurring plant flavonoids tolerable at relatively high doses [153, 157, 158]. More importantly, many senolytic compounds target several different anti-apoptotic pathways, allowing for a multi-hit approach [153, 157, 158]. Senolytic drugs are also appealing because they require only intermittent administration as only brief disruption of

anti-apoptotic pathways is sufficient to kill senescent cells [158]. There are also a few reports demonstrating the efficacy of senolytic drugs in reducing disease phenotypes in PTOA [16, 19, 23]. Another benefit to using senolytic drugs is that they can be readily incorporated into established clinical practice via intra-articular delivery and are effective via oral administration as well [19]. OA is a debilitating and costly joint disease that affects millions of individuals each year, for which there are currently no available disease-modifying therapies [16, 93, 159]. Senolytic drugs may offer a promising new approach for the treatment of not only OA symptoms, but a fundamental driver of pathogenesis, senescent cells, and their SASP.

4.4.3 Other Pharmacologic Treatment Modalities

Fibrosis and, more specifically, arthrofibrosis is a result of extracellular matrix factors that forms primarily in the remodeling phase of musculoskeletal healing [160]. The role of losartan, a selective angiotensin II type 1 (AT1) receptor antagonist medication for hypertension, has been extensively evaluated as a fibrotic neutralization blocker for skeletal muscle repair [161, 162], and more recently to improve cartilage repair mechanisms [163, 164] (NCT04212650) and delay the progression of OA [165]. Losartan's primary mechanism of action blocks the binding of angiotensin II to the AT1 receptor to control blood pressure, but also has been shown to downregulate TGF-β1 expression via the TGF-β1/Smad pathway [162, 166, 167], and directly affects the endothelial-to-mesenchymal transition (EndMT) in myocardial tissue [168] and diarthrodial joints (*unpublished data*). In fact, recent findings by Utsunomiya et al. [164] suggests that biological marrow stimulation (BMS) combined with losartan not only inhibits TGF-β1 expression through the Smad2/3 pathway, but also reduce primary cells in the bone marrow that contribute to the development of fibrotic tissue. Thus, inhibition of the TGF-β1/Smad-dependent pathway using targeted agents may be a promising strategy to downregulate EndMT and, ultimately, fibrosis formation in musculoskeletal tissue. Another compelling approach is to prophylactically treat patients with losartan prior to orthobiologic treatment to reduce TGF-β1 levels and potentially primary cells that contribute to fibrosis development.

Chronic unregulated disruption of joint homeostasis principally manifests as pain and joint dysfunction caused by elevated inflammation and structural decline of articular surfaces and subchondral bone. Again, these attributes are classical symptoms of OA, which is by definition, a pathological state of deregulated joint homeostasis. Most medications for controlling symptoms, namely pain, include acetaminophen, topical and oral NSAID's, and steroid injections. In addition, intra-articular injection of visco-supplementation like hyaluronan and related derivatives has been shown to be effective in pain relief, but there is conflicting evidence on whether these produce any functional improvement [169].

Research has been done on popular disease-modifying agents for other diseases including colchicine, hydroxychloroquine, and tumor necrosis factor inhibitors in their efficacy on OA, but these trials have not produced positive results [170]. New medications looking to act on nociceptive nerve fibers are being investigated. Tanezumab is a monoclonal antibody that targets nerve growth factor (NGF) and in preliminary trials has shown superior analgesia when combined with NSAIDs than NSAID monotherapy [170]. Topical capsaicin has been shown to be beneficial for treating OA pain and current trials are focusing on intra-articular capsaicin to achieve better pain control with early preliminary positive results. Given the integral role of IL-1 in OA, IL-1 receptor antagonists have been studied without much success in human trials. Strontium ranelate is also being studied due its decrease in IL-1 and proteinases, with limited results so far [170]. The latest research shows that a major reason that there has not been an effective disease-modifying agent is due to the heterogenous nature of OA and the need to identify and address each patient's specific phenotype [171].

4.5 Conclusion

In this chapter, the etiological differences between PTOA and AAOA were introduced and the homeostatic functions in these prevalent conditions and normative knee joints were reviewed. The role of biomechanics, senescence, hormones, vitamins, and cellular functions provide important roles in maintaining joint homeostasis and in disrupted conditions, such as PTOA/AAOA. Understating the homeostatic functions of the knee undoubtedly will help guide researchers and clinicians to maintain and treat those affected by PTOA and AAOA. Moreover, orthobiologic therapies and senotherapies, or a combination thereof, have the potential to attenuate symptoms and restore homeostatic functions, though further evidence is necessary to elucidate their therapeutic efficacy, both individually and in combination.

References

1. Fuss FK. Anatomy of the cruciate ligaments and their function in extension and flexion of the human knee joint. Am J Anat. 1989;184(2):165–76.
2. Hui AY, McCarty WJ, Masuda K, Firestein GS, Sah RL. A systems biology approach to synovial joint lubrication in health, injury, and disease. Wiley Interdiscip Rev Syst Biol Med. 2012;4(1):15–37.
3. Gatenholm B, Brittberg M. Neuropeptides: important regulators of joint homeostasis. Knee Surg Sports Traumatol Arthrosc. 2019;27(3):942–9.
4. Helmick CG, Felson DT, Lawrence RC, Gabriel S, Hirsch R, Kwoh CK, et al. Estimates of the prevalence of arthritis and other rheumatic conditions in the United States. Part I. Arthritis Rheum. 2008;58(1):15–25.
5. Lawrence RC, Felson DT, Helmick CG, Arnold LM, Choi H, Deyo RA, et al. Estimates of the prevalence of arthritis and other rheumatic conditions in the United States. Part II. Arthritis Rheum. 2008;58(1):26–35.
6. Fleming BC, Hulstyn MJ, Oksendahl HL, Fadale PD. Ligament injury, reconstruction and osteoarthritis. Curr Opin Orthop. 2005;16(5):354–62.
7. Chu CR, Andriacchi TP. Dance between biology, mechanics, and structure: a systems-based approach to developing osteoarthritis prevention strategies. J Orthop Res. 2015;33(7):939–47.
8. Barenius B, Ponzer S, Shalabi A, Bujak R, Norlen L, Eriksson K. Increased risk of osteoarthritis after anterior cruciate ligament reconstruction: a 14-year

follow-up study of a randomized controlled trial. Am J Sports Med. 2014;42(5):1049–57.

9. Bleker LS, de Rooij SR, Roseboom TJ. Malnutrition and depression in pregnancy and associations with child behaviour and cognitive function: a review of recent evidence on unique and joint effects (1). Can J Physiol Pharmacol. 2019;97(3):158–73.

10. Santangelo KS, Radakovich LB, Fouts J, Foster MT. Pathophysiology of obesity on knee joint homeostasis: contributions of the infrapatellar fat pad. Horm Mol Biol Clin Investig. 2016;26(2):97–108.

11. Andriacchi TP, Favre J, Erhart-Hledik JC, Chu CR. A systems view of risk factors for knee osteoarthritis reveals insights into the pathogenesis of the disease. Ann Biomed Eng. 2015;43(2):376–87.

12. Lories RJ. Joint homeostasis, restoration, and remodeling in osteoarthritis. Best Pract Res Clin Rheumatol. 2008;22(2):209–20.

13. Wojdasiewicz P, Poniatowski LA, Szukiewicz D. The role of inflammatory and anti-inflammatory cytokines in the pathogenesis of osteoarthritis. Mediat Inflamm. 2014;2014:561459.

14. Kapoor M, Martel-Pelletier J, Lajeunesse D, Pelletier JP, Fahmi H. Role of proinflammatory cytokines in the pathophysiology of osteoarthritis. Nat Rev Rheumatol. 2011;7(1):33–42.

15. Goldring MB, Marcu KB. Cartilage homeostasis in health and rheumatic diseases. Arthritis Res Ther. 2009;11(3):224.

16. Jeon OH, David N, Campisi J, Elisseeff JH. Senescent cells and osteoarthritis: a painful connection. J Clin Invest. 2018;128(4):1229–37.

17. Kramer WC, Hendricks KJ, Wang J. Pathogenetic mechanisms of posttraumatic osteoarthritis: opportunities for early intervention. Int J Clin Exp Med. 2011;4(4):285–98.

18. Martin JA, Brown TD, Heiner AD, Buckwalter JA. Chondrocyte senescence, joint loading and osteoarthritis. Clin Orthop Relat Res. 2004;427 Suppl:S96–103.

19. Jeon OH, Kim C, Laberge RM, Demaria M, Rathod S, Vasserot AP, et al. Local clearance of senescent cells attenuates the development of post-traumatic osteoarthritis and creates a pro-regenerative environment. Nat Med. 2017;23(6):775–81.

20. Loeser RF. Aging and osteoarthritis: the role of chondrocyte senescence and aging changes in the cartilage matrix. Osteoarthr Cartil. 2009;17(8):971–9.

21. Martin JA, Brown T, Heiner A, Buckwalter JA. Post-traumatic osteoarthritis: the role of accelerated chondrocyte senescence. Biorheology. 2004;41(3–4):479–91.

22. Childs BG, Durik M, Baker DJ, van Deursen JM. Cellular senescence in aging and age-related disease: from mechanisms to therapy. Nat Med. 2015;21(12):1424–35.

23. Zheng W, Feng Z, You S, Zhang H, Tao Z, Wang Q, et al. Fisetin inhibits IL-1beta-induced inflammatory response in human osteoarthritis chondrocytes through activating SIRT1 and attenuates the progression of osteoarthritis in mice. Int Immunopharmacol. 2017;45:135–47.

24. Abid M, Mezghani N, Mitiche A. Knee joint biomechanical gait data classification for knee pathology assessment: a literature review. Appl Bionics Biomech. 2019;2019:7472039.

25. Flandry F, Hommel G. Normal anatomy and biomechanics of the knee. Sports Med Arthrosc Rev. 2011;19(2):82–92.

26. Hirschmann MT, Muller W. Complex function of the knee joint: the current understanding of the knee. Knee Surg Sports Traumatol Arthrosc. 2015;23(10):2780–8.

27. Zhou T. Analysis of the biomechanical characteristics of the knee joint with a meniscus injury. Healthc Technol Lett. 2018;5(6):247–9.

28. Senter C, Hame SL. Biomechanical analysis of tibial torque and knee flexion angle: implications for understanding knee injury. Sports Med. 2006;36(8):635–41.

29. Makris EA, Hadidi P, Athanasiou KA. The knee meniscus: structure-function, pathophysiology, current repair techniques, and prospects for regeneration. Biomaterials. 2011;32(30):7411–31.

30. Bates NA, Myer GD, Shearn JT, Hewett TE. Anterior cruciate ligament biomechanics during robotic and mechanical simulations of physiologic and clinical motion tasks: a systematic review and meta-analysis. Clin Biomech (Bristol, Avon). 2015;30(1):1–13.

31. Varelas AN, Erickson BJ, Cvetanovich GL, Bach BR Jr. Medial collateral ligament reconstruction in patients with medial knee instability: a systematic review. Orthop J Sports Med. 2017;5(5):2325967117703920.

32. McNulty AL, Guilak F. Mechanobiology of the meniscus. J Biomech. 2015;48(8):1469–78.

33. Wall ME, Dyment NA, Bodle J, Volmer J, Loboa E, Cederlund A, et al. Cell Signaling in tenocytes: response to load and ligands in health and disease. Adv Exp Med Biol. 2016;920:79–95.

34. Magnusson SP, Heinemeier KM, Kjaer M. Collagen homeostasis and metabolism. Adv Exp Med Biol. 2016;920:11–25.

35. Klein-Nulend J, Bakker AD, Bacabac RG, Vatsa A, Weinbaum S. Mechanosensation and transduction in osteocytes. Bone. 2013;54(2):182–90.

36. Whitney KE, Bolia I, Chahla J, Utsunomiya H, Evans TA, Provencher M, et al. Physiology and homeostasis of musculoskeletal structures, injury response, healing process, and regenerative medicine approaches. Bio-orthopaedics. 2017:71–85.

37. Pan J, Zhou X, Li W, Novotny JE, Doty SB, Wang L. In situ measurement of transport between subchondral bone and articular cartilage. J Orthop Res. 2009;27(10):1347–52.

38. Milner LA, Bigas A. Notch as a mediator of cell fate determination in hematopoiesis: evidence and speculation. Blood. 1999;93(8):2431–48.

39. Findlay DM, Kuliwaba JS. Bone-cartilage cross-talk: a conversation for understanding osteoarthritis. Bone Res. 2016;4:16028.
40. Sharma AR, Jagga S, Lee SS, Nam JS. Interplay between cartilage and subchondral bone contributing to pathogenesis of osteoarthritis. Int J Mol Sci. 2013;14(10):19805–30.
41. Gao J, Williams JL, Roan E. Multiscale modeling of growth plate cartilage mechanobiology. Biomech Model Mechanobiol. 2017;16(2):667–79.
42. Kumar D, Manal KT, Rudolph KS. Knee joint loading during gait in healthy controls and individuals with knee osteoarthritis. Osteoarthr Cartil. 2013;21(2):298–305.
43. Cox TR, Erler JT. Remodeling and homeostasis of the extracellular matrix: implications for fibrotic diseases and cancer. Dis Model Mech. 2011;4(2):165–78.
44. Maldonado M, Nam J. The role of changes in extracellular matrix of cartilage in the presence of inflammation on the pathology of osteoarthritis. Biomed Res Int. 2013;2013:284873.
45. Musumeci G. The effect of mechanical loading on articular. Cartilage. 2016:154–61.
46. Stewart HL, Kawcak CE. The importance of subchondral bone in the pathophysiology of osteoarthritis. Front Vet Sci. 2018;5:178.
47. Troeberg L, Nagase H. Proteases involved in cartilage matrix degradation in osteoarthritis. Biochim Biophys Acta. 2012;1824(1):133–45.
48. Liao W, Li Z, Li T, Zhang Q, Zhang H, Wang X. Proteomic analysis of synovial fluid in osteoarthritis using SWATH-mass spectrometry. Mol Med Rep. 2018;17(2):2827–36.
49. Blasioli DJ, Kaplan DL. The roles of catabolic factors in the development of osteoarthritis. Tissue Eng Part B Rev. 2014;20(4):355–63.
50. Julovi SM, Yasuda T, Shimizu M, Hiramitsu T, Nakamura T. Inhibition of interleukin-1beta-stimulated production of matrix metalloproteinases by hyaluronan via CD44 in human articular cartilage. Arthritis Rheum. 2004;50(2):516–25.
51. Wang CT, Lin YT, Chiang BL, Lin YH, Hou SM. High molecular weight hyaluronic acid downregulates the gene expression of osteoarthritis-associated cytokines and enzymes in fibroblast-like synoviocytes from patients with early osteoarthritis. Osteoarthr Cartil. 2006;14(12):1237–47.
52. Li H, Wang D, Yuan Y, Min J. New insights on the MMP-13 regulatory network in the pathogenesis of early osteoarthritis. Arthritis Res Ther. 2017;19(1):248.
53. Mueller MB, Tuan RS. Anabolic/catabolic balance in pathogenesis of osteoarthritis: identifying molecular targets. PM R. 2011;3(6 Suppl 1):S3–11.
54. Kraus VB, Hargrove DE, Hunter DJ, Renner JB, Jordan JM. Establishment of reference intervals for osteoarthritis-related soluble biomarkers: the FNIH/OARSI OA biomarkers consortium. Ann Rheum Dis. 2017;76(1):179–85.
55. Nakata K, Shino K, Hamada M, Mae T, Miyama T, Shinjo H, et al. Human meniscus cell: characterization of the primary culture and use for tissue engineering. Clin Orthop Relat Res. 2001;391 Suppl:S208–18.
56. Zellner J, Pattappa G, Koch M, Lang S, Weber J, Pfeifer CG, et al. Autologous mesenchymal stem cells or meniscal cells: what is the best cell source for regenerative meniscus treatment in an early osteoarthritis situation? Stem Cell Res Ther. 2017;8(1):225.
57. Benjamin M, Ralphs JR. The cell and developmental biology of tendons and ligaments. Int Rev Cytol. 2000;196:85–130.
58. Lo IK, Chi S, Ivie T, Frank CB, Rattner JB. The cellular matrix: a feature of tensile bearing dense soft connective tissues. Histol Histopathol. 2002;17(2):523–37.
59. Chi SS, Rattner JB, Sciore P, Boorman R, Lo IK. Gap junctions of the medial collateral ligament: structure, distribution, associations and function. J Anat. 2005;207(2):145–54.
60. Lieberthal J, Sambamurthy N, Scanzello CR. Inflammation in joint injury and post-traumatic osteoarthritis. Osteoarthr Cartil. 2015;23(11):1825–34.
61. Gelber AC, Hochberg MC, Mead LA, Wang NY, Wigley FM, Klag MJ. Joint injury in young adults and risk for subsequent knee and hip osteoarthritis. Ann Intern Med. 2000;133(5):321–8.
62. Lotz MK, Kraus VB. New developments in osteoarthritis. Posttraumatic osteoarthritis: pathogenesis and pharmacological treatment options. Arthritis Res Ther. 2010;12(3):211.
63. Punzi L, Galozzi P, Luisetto R, Favero M, Ramonda R, Oliviero F, et al. Post-traumatic arthritis: overview on pathogenic mechanisms and role of inflammation. RMD Open. 2016;2(2):e000279.
64. Anderson DD, Chubinskaya S, Guilak F, Martin JA, Oegema TR, Olson SA, et al. Post-traumatic osteoarthritis: improved understanding and opportunities for early intervention. J Orthop Res. 2011;29(6):802–9.
65. Otsuki S, Brinson DC, Creighton L, Kinoshita M, Sah RL, D'Lima D, et al. The effect of glycosaminoglycan loss on chondrocyte viability: a study on porcine cartilage explants. Arthritis Rheum. 2008;58(4):1076–85.
66. Olsson O, Isacsson A, Englund M, Frobell RB. Epidemiology of intra- and peri-articular structural injuries in traumatic knee joint hemarthrosis—data from 1145 consecutive knees with subacute MRI. Osteoarthr Cartil. 2016;24(11):1890–7.
67. Kim HT, Lo MY, Pillarisetty R. Chondrocyte apoptosis following intraarticular fracture in humans. Osteoarthr Cartil. 2002;10(9):747–9.
68. Murray MM, Zurakowski D, Vrahas MS. The death of articular chondrocytes after intra-articular fracture in humans. J Trauma. 2004;56(1):128–31.
69. Guo D, Ding L, Homandberg GA. Telopeptides of type II collagen upregulate proteinases and damage

cartilage but are less effective than highly active fibronectin fragments. Inflamm Res. 2009;58(3):161–9.

70. Polur I, Lee PL, Servais JM, Xu L, Li Y. Role of HTRA1, a serine protease, in the progression of articular cartilage degeneration. Histol Histopathol. 2010;25(5):599–608.

71. Sward P, Frobell R, Englund M, Roos H, Struglics A. Cartilage and bone markers and inflammatory cytokines are increased in synovial fluid in the acute phase of knee injury (hemarthrosis)— a cross-sectional analysis. Osteoarthr Cartil. 2012;20(11):1302–8.

72. Friedenstein AJ, Chailakhyan RK, Latsinik NV, Panasyuk AF, Keiliss-Borok IV. Stromal cells responsible for transferring the microenvironment of the hemopoietic tissues: cloning in vitro and retransplantation in vivo. Transplantation. 1974;17(4):331–40.

73. Owen M. Marrow stromal stem cells. J Cell Sci Suppl. 1988;10:63–76.

74. Sottile V, Halleux C, Bassilana F, Keller H, Seuwen K. Stem cell characteristics of human trabecular bone-derived cells. Bone. 2002;30(5):699–704.

75. Noth U, Osyczka AM, Tuli R, Hickok NJ, Danielson KG, Tuan RS. Multilineage mesenchymal differentiation potential of human trabecular bone-derived cells. J Orthop Res. 2002;20(5):1060–9.

76. Zuk PA, Zhu M, Mizuno H, Huang J, Futrell JW, Katz AJ, et al. Multilineage cells from human adipose tissue: implications for cell-based therapies. Tissue Eng. 2001;7(2):211–28.

77. Doherty MJ, Ashton BA, Walsh S, Beresford JN, Grant ME, Canfield AE. Vascular pericytes express osteogenic potential in vitro and in vivo. J Bone Miner Res. 1998;13(5):828–38.

78. Bianco P, Riminucci M, Gronthos S, Robey PG. Bone marrow stromal stem cells: nature, biology, and potential applications. Stem Cells. 2001;19(3):180–92.

79. Jones EA, English A, Henshaw K, Kinsey SE, Markham AF, Emery P, et al. Enumeration and phenotypic characterization of synovial fluid multipotential mesenchymal progenitor cells in inflammatory and degenerative arthritis. Arthritis Rheum. 2004;50(3):817–27.

80. Moradi B, Rosshirt N, Tripel E, Kirsch J, Barie A, Zeifang F, et al. Unicompartmental and bicompartmental knee osteoarthritis show different patterns of mononuclear cell infiltration and cytokine release in the affected joints. Clin Exp Immunol. 2015;180(1):143–54.

81. Pessler F, Dai L, Diaz-Torne C, Gomez-Vaquero C, Paessler ME, Zheng DH, et al. The synovitis of "non-inflammatory" orthopaedic arthropathies: a quantitative histological and immunohistochemical analysis. Ann Rheum Dis. 2008;67(8):1184–7.

82. Rosenthal AK, Gohr CM, Ninomiya J, Wakim BT. Proteomic analysis of articular cartilage vesicles from normal and osteoarthritic cartilage. Arthritis Rheum. 2011;63(2):401–11.

83. Gobezie R, Kho A, Krastins B, Sarracino DA, Thornhill TS, Chase M, et al. High abundance synovial fluid proteome: distinct profiles in health and osteoarthritis. Arthritis Res Ther. 2007;9(2):R36.

84. Ehrnthaller C, Ignatius A, Gebhard F, Huber-Lang M. New insights of an old defense system: structure, function, and clinical relevance of the complement system. Mol Med. 2011;17(3–4):317–29.

85. Bozdech Z. Post-traumatic synovitis. Acta Chir Orthop Traumatol Cechoslov. 1976;43(3):244–7.

86. Knight KL. Effects of hypothermia on inflammation and swelling. J Athl Train. 1976;11:7–10.

87. Lee CA, Kessler CM, Varon D, Martinowitz U, Heim M, Rodriguez-Merchan EC. The destructive capabilities of the synovium in the haemophilic joint. Haemophilia. 1998;4(4):506–10.

88. Elsaid KA, Fleming BC, Oksendahl HL, Machan JT, Fadale PD, Hulstyn MJ, et al. Decreased lubricin concentrations and markers of joint inflammation in the synovial fluid of patients with anterior cruciate ligament injury. Arthritis Rheum. 2008;58(6):1707–15.

89. Borsiczky B, Fodor B, Racz B, Gasz B, Jeges S, Jancso G, et al. Rapid leukocyte activation following intraarticular bleeding. J Orthop Res. 2006;24(4):684–9.

90. Roosendaal G, Vianen ME, Marx JJ, van den Berg HM, Lafeber FP, Bijlsma JW. Blood-induced joint damage: a human in vitro study. Arthritis Rheum. 1999;42(5):1025–32.

91. Soucie JM, Cianfrini C, Janco RL, Kulkarni R, Hambleton J, Evatt B, et al. Joint range-of-motion limitations among young males with hemophilia: prevalence and risk factors. Blood. 2004;103(7):2467–73.

92. Coppe JP, Desprez PY, Krtolica A, Campisi J. The senescence-associated secretory phenotype: the dark side of tumor suppression. Annu Rev Pathol. 2010;5:99–118.

93. McCulloch K, Litherland GJ, Rai TS. Cellular senescence in osteoarthritis pathology. Aging Cell. 2017;16(2):210–8.

94. Martin JA, Buckwalter JA. Human chondrocyte senescence and osteoarthritis. Biorheology. 2002;39(1–2):145–52.

95. Martin JA, Buckwalter JA. The role of chondrocyte senescence in the pathogenesis of osteoarthritis and in limiting cartilage repair. J Bone Joint Surg Am. 2003;85-A(Suppl 2):106–10.

96. Xu M, Bradley EW, Weivoda MM, Hwang SM, Pirtskhalava T, Decklever T, et al. Transplanted senescent cells induce an osteoarthritis-like condition in mice. J Gerontol A Biol Sci Med Sci. 2017;72(6):780–5.

97. Sharma P, Singh N, Singh V, Singh S, Singh HV, Gupta S. Tumor necrosis factor alpha (TNF-α) and estrogen hormone in osteoarthritic female patients. Indian J Clin Biochem. 2006;21(1):205–7.

98. Zou YC, Li HH, Yang GG, Yin HD, Cai DZ, Liu G. Attenuated levels of ghrelin in synovial fluid is

related to the disease severity of ankle post-traumatic osteoarthritis. Biofactors. 2019;45(3):463–70.

99. Qu R, Chen X, Wang W, Qiu C, Ban M, Guo L, et al. Ghrelin protects against osteoarthritis through interplay with Akt and NF-κB signaling pathways. FASEB J. 2018;32(2):1044–58.

100. Ekenstedt KJ, Sonntag WE, Loeser RF, Lindgren BR, Carlson CS. Effects of chronic growth hormone and insulin-like growth factor 1 deficiency on osteoarthritis severity in rat knee joints. Arthritis Rheum. 2006;54(12):3850–8.

101. Bagge E, Edén S, Rosén T, Bengtsson BA. The prevalence of radiographic osteoarthritis is low in elderly patients with growth hormone deficiency. Acta Endocrinol. 1993;129(4):296–300.

102. Lee S. Endogenous parathyroid hormone and knee osteoarthritis: a cross-sectional study. Int J Rheum Dis. 2016;19(3):248–54.

103. Zhang FF, Driban JB, Lo GH, Price LL, Booth S, Eaton CB, et al. Vitamin D deficiency is associated with progression of knee osteoarthritis. J Nutr. 2014;144(12):2002–8.

104. Zhang FF, Driban JB, Lo GH, Price LL, Booth S, Eaton CB, Lu B, Nevitt M, Jackson B, Garganta C, Hochberg MC, Kwoh K, TE MA. Serum levels of vitamin D and parathyroid hormone and progression of knee OA: the osteoarthritis initiative study. Osteoarthr Cartil. 2013;21

105. Afzal S, Khanam A. Serum estrogen and interleukin-6 levels in postmenopausal female osteoarthritis patients. Pak J Pharm Sci. 2011;24(2):217–9.

106. Hussain SM, Cicuttini FM, Bell RJ, Robinson PJ, Davis SR, Giles GG, et al. Incidence of total knee and hip replacement for osteoarthritis in relation to circulating sex steroid hormone concentrations in women. Arthritis Rheumatol. 2014;66(8):2144–51.

107. Guan J, Li Y, Ding LB, Liu GY, Zheng XF, Xue W, et al. Synovial fluid alpha-melanocyte-stimulating hormone may act as a protective biomarker for primary knee osteoarthritis. Discov Med. 2019;27(146):17–26.

108. Liu G, Chen Y, Wang G, Niu J. Decreased synovial fluid α-melanocyte-stimulating-hormone (α-MSH) levels reflect disease severity in patients with posttraumatic ankle osteoarthritis. Clin Lab. 2016;62(8):1491–500.

109. Roman-Blas JA, Castañeda S, Largo R, Herrero-Beaumont G. Osteoarthritis associated with estrogen deficiency. Arthritis Res Ther. 2009;11(5):241.

110. Bay-Jensen AC, Slagboom E, Chen-An P, Alexandersen P, Qvist P, Christiansen C, et al. Role of hormones in cartilage and joint metabolism: understanding an unhealthy metabolic phenotype in osteoarthritis. Menopause. 2013;20(5):578–86.

111. Jung JH, Bang CH, Song GG, Kim C, Kim JH, Choi SJ. Knee osteoarthritis and menopausal hormone therapy in postmenopausal women: a nationwide cross-sectional study. Menopause. 2018;26(6):598–602.

112. Pfeilschifter J, Köditz R, Pfohl M, Schatz H. Changes in proinflammatory cytokine activity after menopause. Endocr Rev. 2002;23(1):90–119.

113. Khundmiri SJ, Murray RD, Lederer E. PTH and vitamin D. Compr Physiol. 2016;6(2):561–601.

114. Silva BC, Costa AG, Cusano NE, Kousteni S, Bilezikian JP. Catabolic and anabolic actions of parathyroid hormone on the skeleton. J Endocrinol Investig. 2011;34(10):801–10.

115. Denko CW, Boja B, Moskowitz RW. Growth promoting peptides in osteoarthritis: insulin, insulin-like growth factor-1, growth hormone. J Rheumatol. 1990;17(9):1217–21.

116. Hauser RA. The regeneration of articular cartilage with prolotherapy. J Prolotherapy. 2009;1(1):39–44.

117. Smith RL, Palathumpat MV, Ku CW, Hintz RL. Growth hormone stimulates insulin-like growth factor I actions on adult articular chondrocytes. J Orthop Res. 1989;7(2):198–207.

118. Wu J, Wang K, Xu J, Ruan G, Zhu Q, Cai J, et al. Associations between serum ghrelin and knee symptoms, joint structures and cartilage or bone biomarkers in patients with knee osteoarthritis. Osteoarthr Cartil. 2017;25(9):1428–35.

119. Liu J, Cao L, Gao X, Chen Z, Guo S, He Z, et al. Ghrelin prevents articular cartilage matrix destruction in human chondrocytes. Biomed Pharmacother. 2018;98:651–5.

120. Capsoni F, Ongari AM, Lonati C, Accetta R, Gatti S, Catania A. α-Melanocyte-stimulating-hormone (α-MSH) modulates human chondrocyte activation induced by proinflammatory cytokines. BMC Musculoskelet Disord. 2015;16:154.

121. Grässel S, Opolka A, Anders S, Straub RH, Grifka J, Luger TA, et al. The melanocortin system in articular chondrocytes: melanocortin receptors, pro-opiomelanocortin, precursor proteases, and a regulatory effect of alpha-melanocyte-stimulating hormone on proinflammatory cytokines and extracellular matrix components. Arthritis Rheum. 2009;60(10):3017–27.

122. Böhm M, Apel M, Lowin T, Lorenz J, Jenei-Lanzl Z, Capellino S, et al. α-MSH modulates cell adhesion and inflammatory responses of synovial fibroblasts from osteoarthritis patients. Biochem Pharmacol. 2016;116:89–99.

123. Nair R, Maseeh A. Vitamin D: the "sunshine" vitamin. J Pharmacol Pharmacother. 2012;3(2):118–26.

124. Goula T, Kouskoukis A, Drosos G, Tselepis AS, Ververidis A, Valkanis C, et al. Vitamin D status in patients with knee or hip osteoarthritis in a Mediterranean country. J Orthop Traumatol. 2015;16(1):35–9.

125. Heidari B, Heidari P, Hajian-Tilaki K. Association between serum vitamin D deficiency and knee osteoarthritis. Int Orthop. 2011;35(11):1627–31.

126. Mermerci Başkan B, Yurdakul FG, Aydın E, Sivas F, Bodur H. Effect of vitamin D levels on radiographic knee osteoarthritis and functional status. Turk J Phys Med Rehabil. 2018;64(1):1–7.

127. Muraki S, Dennison E, Jameson K, Boucher BJ, Akune T, Yoshimura N, et al. Association of vitamin D status with knee pain and radiographic knee osteoarthritis. Osteoarthr Cartil. 2011;19(11):1301–6.

128. Vaishya R, Vijay V, Lama P, Agarwal A. Does vitamin D deficiency influence the incidence and progression of knee osteoarthritis?—a literature review. J Clin Orthop Trauma. 2019;10(1):9–15.

129. Lane NE, Gore LR, Cummings SR, Hochberg MC, Scott JC, Williams EN, et al. Serum vitamin D levels and incident changes of radiographic hip osteoarthritis: a longitudinal study. Study of osteoporotic fractures research group. Arthritis Rheum. 1999;42(5):854–60.

130. Felson DT, Niu J, Clancy M, Aliabadi P, Sack B, Guermazi A, et al. Low levels of vitamin D and worsening of knee osteoarthritis: results of two longitudinal studies. Arthritis Rheum. 2007;56(1):129–36.

131. Bergink AP, Uitterlinden AG, Van Leeuwen JP, Buurman CJ, Hofman A, Verhaar JA, et al. Vitamin D status, bone mineral density, and the development of radiographic osteoarthritis of the knee: the Rotterdam study. J Clin Rheumatol. 2009;15(5):230–7.

132. McAlindon TE, Felson DT, Zhang Y, Hannan MT, Aliabadi P, Weissman B, et al. Relation of dietary intake and serum levels of vitamin D to progression of osteoarthritis of the knee among participants in the Framingham study. Ann Intern Med. 1996;125(5):353–9.

133. Joseph GB, McCulloch CE, Nevitt MC, Neumann J, Lynch JA, Lane NE, et al. Associations between vitamin C and D intake and cartilage composition and knee joint morphology over 4 years: data from the osteoarthritis initiative. Arthritis Care Res (Hoboken). 2019;

134. Hung M, Bounsanga J, Voss MW, Gu Y, Crum AB, Tang P. Dietary and supplemental vitamin C and D on symptom severity and physical function in knee osteoarthritis. J Nutr Gerontol Geriatr. 2017;36(2–3):121–33.

135. Alkan G, Akgol G. Do vitamin D levels affect the clinical prognoses of patients with knee osteoarthritis? J Back Musculoskelet Rehabil. 2017;30(4):897–901.

136. Javadian Y, Adabi M, Heidari B, Babaei M, Firouzjahi A, Ghahhari BY, et al. Quadriceps muscle strength correlates with serum vitamin D and knee pain in knee osteoarthritis. Clin J Pain. 2017;33(1):67–70.

137. Manoy P, Yuktanandana P, Tanavalee A, Anomasiri W, Ngarmukos S, Tanpowpong T, et al. Vitamin D supplementation improves quality of life and physical performance in osteoarthritis patients. Nutrients. 2017;9(8)

138. Neogi T, Booth SL, Zhang YQ, Jacques PF, Terkeltaub R, Aliabadi P, et al. Low vitamin K status is associated with osteoarthritis in the hand and knee. Arthritis Rheum. 2006;54(4):1255–61.

139. Misra D, Booth SL, Tolstykh I, Felson DT, Nevitt MC, Lewis CE, et al. Vitamin K deficiency is associated with incident knee osteoarthritis. Am J Med. 2013;126(3):243–8.

140. Shea MK, Kritchevsky SB, Hsu FC, Nevitt M, Booth SL, Kwoh CK, et al. The association between vitamin K status and knee osteoarthritis features in older adults: the health, aging and body composition study. Osteoarthr Cartil. 2015;23(3):370–8.

141. FDA. Regulatory considerations for human cell, tissues, and cellular and tissue-based products: minimal manipulation and homologous use. 2017.

142. Whitney KE, Liebowitz A, Bolia IK, Chahla J, Ravuri S, Evans TA, et al. Current perspectives on biological approaches for osteoarthritis. Ann N Y Acad Sci. 2017;1410(1):26–43.

143. Sanchez M, Filardo G, Yoshioka T. Platelet rich plasma and Orthopedics: why, when, and how. Biomed Res Int. 2015;2015:949720.

144. Filardo G, Kon E, Roffi A, Di Matteo B, Merli ML, Marcacci M. Platelet-rich plasma: why intra-articular? A systematic review of preclinical studies and clinical evidence on PRP for joint degeneration. Knee Surg Sports Traumatol Arthrosc. 2015;23(9):2459–74.

145. Southworth TM, Naveen NB, Tauro TM, Leong NL, Cole BJ. The use of platelet-rich plasma in symptomatic knee osteoarthritis. J Knee Surg. 2019;32(1):37–45.

146. Gobbi A, Karnatzikos G, Mahajan V, Malchira S. Platelet-rich plasma treatment in symptomatic patients with knee osteoarthritis: preliminary results in a group of active patients. Sports Health. 2012;4(2):162–72.

147. Campbell KA, Saltzman BM, Mascarenhas R, Khair MM, Verma NN, Bach BR Jr, et al. Does intra-articular platelet-rich plasma injection provide clinically superior outcomes compared with other therapies in the treatment of knee osteoarthritis? A systematic review of overlapping meta-analyses. Arthroscopy. 2015;31(11):2213–21.

148. Mascarenhas R, Saltzman BM, Fortier LA, Cole BJ. Role of platelet-rich plasma in articular cartilage injury and disease. J Knee Surg. 2015;28(1):3–10.

149. Sundman EA, Cole BJ, Karas V, Della Valle C, Tetreault MW, Mohammed HO, et al. The anti-inflammatory and matrix restorative mechanisms of platelet-rich plasma in osteoarthritis. Am J Sports Med. 2014;42(1):35–41.

150. Li Z, Wang Y, Xiao K, Xiang S, Li Z, Weng X. Emerging role of exosomes in the joint diseases. Cell Physiol Biochem. 2018;47(5):2008–17.

151. Cosenza S, Ruiz M, Toupet K, Jorgensen C, Noel D. Mesenchymal stem cells derived exosomes and microparticles protect cartilage and bone from degradation in osteoarthritis. Sci Rep. 2017;7(1):16214.

152. Zhang S, Chu WC, Lai RC, Lim SK, Hui JH, Toh WS. Exosomes derived from human embryonic mesenchymal stem cells promote osteochondral regeneration. Osteoarthr Cartil. 2016;24(12):2135–40.

153. Kirkland JL, Tchkonia T. Cellular senescence: a translational perspective. EBioMedicine. 2017;21:21–8.

154. Baar MP, Perdiguero E, Munoz-Canoves P, De Keizer PL. Musculoskeletal senescence: a moving target ready to be eliminated. Curr Opin Pharmacol. 2018;40:147–55.

155. Franceschi C, Bonafe M, Valensin S, Olivieri F, De Luca M, Ottaviani E, et al. Inflamm-aging. An evolutionary perspective on immunosenescence. Ann N Y Acad Sci. 2000;908:244–54.

156. Kawakami Y, Hambright WS, Takayama K, Mu X, Lu A, Cummins JH, et al. Rapamycin rescues age-related changes in muscle-derived stem/progenitor cells from progeroid mice. Mol Ther Methods Clin Dev. 2019;14:64–76.

157. Kirkland JL, Tchkonia T, Zhu Y, Niedernhofer LJ, Robbins PD. The clinical potential of senolytic drugs. J Am Geriatr Soc. 2017;65(10):2297–301.

158. Zhu Y, Tchkonia T, Pirtskhalava T, Gower AC, Ding H, Giorgadze N, et al. The Achilles' heel of senescent cells: from transcriptome to senolytic drugs. Aging Cell. 2015;14(4):644–58.

159. AAOS. Management of osteoarthritis of the hip: evidence-based clinical practice guideline. 2017:106–208.

160. Wynn TA, Vannella KM. Macrophages in tissue repair, regeneration, and fibrosis. Immunity. 2016;44(3):450–62.

161. Garg K, Corona BT, Walters TJ. Therapeutic strategies for preventing skeletal muscle fibrosis after injury. Front Pharmacol. 2015;6:87.

162. Kobayashi M, Ota S, Terada S, Kawakami Y, Otsuka T, Fu FH, et al. The combined use of losartan and muscle-derived stem cells significantly improves the functional recovery of muscle in a young mouse model of contusion injuries. Am J Sports Med. 2016;44(12):3252–61.

163. Huard J, Bolia I, Briggs K, Utsunomiya H, Lowe WR, Philippon MJ. Potential usefulness of losartan as an antifibrotic agent and adjunct to platelet-rich plasma therapy to improve muscle healing and cartilage repair and prevent adhesion formation. Orthopedics. 2018;41(5):e591–e7.

164. Utsunomiya H, Gao X, Deng Z, Cheng H, Nakama G, Scibetta AC, et al. Biologically regulated marrow stimulation by blocking TGF-beta1 with losartan Oral administration results in hyaline-like cartilage repair: a rabbit osteochondral defect model. Am J Sports Med. 2020;48(4):974–84.

165. Thomas M, Fronk Z, Gross A, Willmore D, Arango A, Higham C, et al. Losartan attenuates progression of osteoarthritis in the synovial temporomandibular and knee joints of a chondrodysplasia mouse model through inhibition of TGF-beta1 signaling pathway. Osteoarthr Cartil. 2019;27(4):676–86.

166. Terada S, Ota S, Kobayashi M, Kobayashi T, Mifune Y, Takayama K, et al. Use of an antifibrotic agent improves the effect of platelet-rich plasma on muscle healing after injury. J Bone Joint Surg Am. 2013;95(11):980–8.

167. Wu M, Peng Z, Zu C, Ma J, Lu S, Zhong J, et al. Losartan attenuates myocardial endothelial-to-mesenchymal transition in spontaneous hypertensive rats via inhibiting TGF-beta/Smad Signaling. PLoS One. 2016;11(5):e0155730.

168. Wylie-Sears J, Levine RA, Bischoff J. Losartan inhibits endothelial-to-mesenchymal transformation in mitral valve endothelial cells by blocking transforming growth factor-beta-induced phosphorylation of ERK. Biochem Biophys Res Commun. 2014;446(4):870–5.

169. Taruc-Uy RL, Lynch SA. Diagnosis and treatment of osteoarthritis. Prim Care. 2013;40(4):821–36. vii

170. Ghouri A, Conaghan PG. Update on novel pharmacological therapies for osteoarthritis. Ther Adv Musculoskelet Dis. 2019;11:1759720X19864492.

171. Van Spil WE, Kubassova O, Boesen M, Bay-Jensen AC, Mobasheri A. Osteoarthritis phenotypes and novel therapeutic targets. Biochem Pharmacol. 2019;165:41–8.

"Preparing the Soil": Optimizing Metabolic Management in Regenerative Medicine Procedures

5

Lucas Furtado da Fonseca, José Fábio Lana,
Silvia Beatriz Coutinho Visoni,
Anna Vitoria Santos Lana, Eleonora Irlandini,
and Gabriel Ohana Marques Azzini

5.1 Concepts of Systemic Inflammation

5.1.1 Osteoarthritis and Inflammation

For a long time, osteoarthritis (OA) has been considered a disease affecting the hyaline articular cartilage alone, while at present, it is believed that all articular tissues, including the subchondral bone, the ligaments, the synovium, and the joint capsule, participate collectively, to varying degrees, in the development of such disorder.

Extensive research has shown that metabolic syndrome is tightly linked to osteoarthritis and inflammation, a process which appears to primarily occur in the subchondral bone via the incidence of bone marrow lesions (BMLs). Numerous studies identify obesity, dyslipidemia, insulin resistance, and hypertension as the top metabolic risk factors, the so-called deadly quartet. These factors are responsible for the disruptive physiological processes that culminate in detrimental alterations within the subchondral bone, cartilage damage and, overall, the predominant proinflammatory joint microenvironment.

More recent studies have shown that osteoarthritis (OA) tissue and synovial fluid have abnormally high levels not only of plasma proteins, but also of complement components and cytokines, and that chondrocytes and synovial cells in OA produce or overproduce many of the inflammatory mediators (e.g., IL-1β, TNF, and nitric oxide (NO)) that are characteristic of inflammatory arthritis.

5.1.2 Meta-Inflammation

The inflammatory state that accompanies the metabolic syndrome does not completely fit into the classical definition of acute or chronic inflammation, as it is not accompanied by infection. There is no massive tissue injury and the dimension of the inflammatory activation is

L. F. da Fonseca (✉) · J. F. Lana · S. B. C. Visoni ·
G. O. M. Azzini
IOC–Instituto do Osso e da Cartilagem/The Bone and Cartilage Institute, Indaiatuba, SP, Brazil

A. V. S. Lana
UniMAX Faculdade de Medicina, Indaiatuba, Brazil

E. Irlandini
OASI Bioresearch Foundation, Milan, Italy
e-mail: e.irlandini@oasiortopedia.it

© ISAKOS 2022
A. Gobbi et al. (eds.), *Joint Function Preservation*, https://doi.org/10.1007/978-3-030-82958-2_5

also not large. So it is often called "low-grade" chronic inflammation or "meta-inflammation," meaning metabolically triggered inflammation. In this way, meta-inflammation is characterized by increased serum levels of proinflammatory cytokines like TNF-α, IL-6, and IL-1β. Besides that hyperglycemia and hyperlipidemia might also fuel meta-inflammation.

Different studies demonstrated the connection between long standing type 2 diabetes (T2D) and accelerated OA progression with higher rates of synovial inflammation and joint pain. We know that metabolic syndrome (MS) and T2D contributes to decreased multipotency of MSCs by generating advanced glycation products (AGEs), oxidative stress and inflammation, which can suppress proliferation, induce apoptosis and increase the production of intracellular reactive oxygen species (ROS). Increased apoptosis and ROS accumulation may be partially responsible for the reduced differentiation potential observed in MS cells.

Collectively, the excess of glucose results in alterations of the cell metabolism and morphology, as well as in ECM structure (reduction in collagen synthesis and proteoglycan catabolism). Inside the joint, there is usually an overexpression of GLUT-1, thus potentially leading to glucose toxicity, which may be even more pronounced in the presence of high glucose levels within the joint, as it occurs in type 2 diabetes. Also, high glucose levels have shown to decrease the chondrogenic differentiation of mesenchymal stem cells either derived by the bone marrow, the muscle, or the adipose tissue in vitro. When it comes to the synovium, hyperglycemia has shown to stimulate the release of the vascular endothelial growth factor (VEGF) by human synovial cells. Synovial neo-angiogenesis leads to the local recruitment of proinflammatory cells.

The inflammation in OA is distinct from that in rheumatoid arthritis and other autoimmune diseases: it is chronic, comparatively low-grade, and mediated primarily by the innate immune system. In addition to local inflammation in the joint, systemic inflammation might also have an important role in OA pathogenesis. For instance, obesity is known to predispose individuals to OA—possibly not only by increasing the mechanical load on joints, but also by causing chronic, systemic inflammation through inflammatory mediators (such as adipokines and other proinflammatory cytokines) that are produced by adipose tissue and released into the bloodstream. The involvement of proinflammatory adipokines released by the white adipose tissue in obese adults leads to cartilage degradation, synovial inflammation, and osteophyte formation.

The main adipokines whose role has been evaluated in association with OA are leptin, adiponectin, visfatin, and resistin. For instance, leptin production and its receptor by chondrocytes is increased in OA, and this adipokine has been shown to promote cell apoptosis and the release of metalloproteinases. Adiponectin levels, on its turn, in the synovial fluid harvested from patients with knee OA have been positively associated with aggrecan degradation, while plasma adiponectin concentration was associated with joint pain in female patients and promote the expression of MMP-3, MMP-9, MMP-13, PGE$_2$, nitric oxide synthase 2, IL-6, and monocyte chemoattractant protein-1 (MCP-1) in chondrocytes, thus apparently showing a catabolic and proinflammatory effect.

Adipokines can also cause changes the surrounding environment of the joint cells, thereby changing in the fate of the cells—inducing cell senescence. There are also mesenchymal stem/progenitor cells in articular cartilage that undergo cell senescence. Studying the response of these cells to these new adipokines in the joint environment will be a new direction for the treatment of OA. While the infrapatellar fat pad is a major source of adipokines in knee synovial fluid, adipocytes also accumulate in the bone marrow during aging and obesity. Adipokines can act as SASPs (senescence-associated secretory phenotype factors) that participate in cellular senescence of chondrocytes, but they also regulate energy metabolism impacting bone remodeling. Thus, adipokines are closely related to the metabolic syndrome and degenerative pathological changes in cartilage and bone during OA.

5.2 Sleep Quality and Its Role in Homeostasis

Another point of paramount importance that we must assess in our patient is the quality of sleep. Controlled, experimental studies on the effects of acute sleep loss in humans have shown that mediators of inflammation are altered by sleep loss. Elevations in these mediators have been found to occur in healthy, rigorously screened individuals undergoing experimental vigils of more than 24 h, and have also been seen in response to various durations of sleep restricted to between 25 and 50% of a normal 8 h sleep amount.

The circadian rhythm orchestrates many cellular functions, such as cell division, cell migration, metabolism, and numerous intracellular biological processes. The physiological changes during sleep are believed to promote a suitable microenvironment for stem cells to proliferate, migrate, and differentiate. These effects are mediated either directly by circadian clock genes or indirectly via hormones and cytokines. Hormones, such as melatonin and cortisol, are secreted in response to neural optic signals and act in harmony to regulate many biological functions during sleep.

Recent findings have demonstrated melatonin that enhances osteogenesis and chondrogenesis and inhibits adipogenesis. Melatonin protects against oxidative stress-induced apoptosis in MSCs. Melatonin attenuates intracellular ROS generation to improve cell viability and enhances MSCs differentiation into other lineages. Melatonin also plays important roles in the regulation of the ESCs proliferation and differentiation.

Stimulatory actions of melatonin on bone formation and its inhibitory effects on bone restoration have been reported in a number of studies. It has been suggested that the osteoblast enhancing function of melatonin is mediated by its direct action on the differentiation and proliferation of the bone-forming cells. Moreover, enhancement in the bone alkaline phosphatase levels and mineralization, promotion of the synthesis of collagen type I, increase in the bone mass, and facilitation of new bone growth and osteointegration, are among the positive functions of melato-

nin on bone. These make melatonin an appealing molecule in bone regeneration. Both in vitro and in vivo studies have examined melatonin's potential to influence bone repair.

Interestingly, melatonin improved wound closure and triggered osteogenesis markers such as BMP-2 and -4, osteocalcin and runt-related transcription factor 2 (Runx2) in a dose-dependent fashion. Melatonin counteracted the reduction of cell proliferation by iron overload in BMSCs via reversing the upregulation of p53, ERK, and p38 protein expression in cells.

5.3 Metabolic Syndrome, Synovium, and Subchondral Bone Alterations

Although it has long been thought that osteoarthritis was limited to the cartilage component of the joint, other studies indicate that the disease may originate from the harmful alterations that occur primarily in the subchondral bone, especially via means of vascular pathology. Since metabolic risk factors are manageable to a certain extent, it is therefore possible to decelerate the progression of OA and mitigate its devastating effects on the subchondral bone and subsequent articular cartilage damage.

The early changes that occur beneath the articular cartilage at the osteochondral junction are highly relevant as they become possible mediators of pain and structural progression in OA and may aggravate pathology elsewhere in the joint. These modifications include augmented subchondral bone thickness, diminished flexibility, and trabecular bone density beneath the subchondral plate. This exposes the subchondral bone and its nerves to imbalanced biochemical and biomechanical influence. Continuous biomechanical and biochemical stress applied to articular cartilage contributes to chondropathy. This subsequently promotes additional subchondral bone alterations, such as microfractures which, in turn, may aggravate pain. This interaction would then trigger a positive feedback loop as a result of multiple unsuccessful attempts to repair cartilage and bone tissue, eventually resulting in OA.

It is understood that OA pathogenesis is mainly attributed to both excessive joint loading and the subsequent irregular biomechanical and biochemical patterns, such as hormone and cytokine dysregulation which arise from increased adipose tissue, a rich source of proinflammatory endocrine factors. One of the most prominent features of obesity is the manifestation of low-grade systemic inflammation, affecting many organs and anatomical structures. It is believed that the elevated adipokine expression from adipose tissue elicits direct and downstream effects which lead to the destruction and remodeling of the joint as whole.

Dyslipidemia is strongly involved the pathophysiology of OA by aggravating subchondral bone damage due to BMLs. BMLs are known to be associated with knee pain and structural alterations in the knee of OA patients and subsequently culminate in increased joint space narrowing and cartilage erosion in symptomatic populations. It was observed that greater levels of total cholesterol and triglycerides were associated with the incidence of BMLs in knees free of BMLs at baseline. On the other hand, HDL cholesterol, tended to be inversely related with BMLs, proposing a rather protective role for this specific lipoprotein.

Speaking of inflammation, there is recent evidence connecting the expression of synovial tumor necrosis factor alpha (TNF-α) and insulin resistance to OA pathology. A causal role for TNF-α in OA, especially in obesity- and diabetes-related OA, has been proposed. It was found that elimination of the TNF-α gene significantly reduced high-fat diet-induced development of osteophytes and synovial hyperplasia. Unsurprisingly, a clinical study reported that diabetic rheumatoid arthritis patients benefited from anti-TNF-α therapy, which significantly ameliorated insulin resistance, β-cell function in pancreas and insulin signaling.

On a cellular perspective, enriched insulin receptor expression in murine and human synovia has also been previously identified. Much like other insulin-sensitive tissues, the synovium is susceptible to the insulin resistance syndrome, which normally occurs in parallel with meta-inflammation.

Osteoblasts from sclerotic regions of subchondral bone were capable of switching the profile of chondrocytes toward a catabolic and antianabolic phenotype, as illustrated by a reduction in aggrecan production, but also by an upregulation of MMP production.

The synovium plays an important role in the increased expression of osteoclast differentiation factors, such as RANK-L, are known to occur in the synovial membrane of chronically diseased OA joints, and could contribute to increased bone resorption. The inhibition of RANK-L-dependent osteoclast formation may constitute a potential target to prevent the skeletal changes seen in OA. It is well appreciated that inflammatory mediators are produced by joint tissues in OA, and that these proinflammatory cytokines increase osteoclastogenesis and bone resorption, especially in subchondral bone leading to weakened architecture. An inflamed synovium or bone marrow may stimulate osteoclast-mediated bone resorption through a range of proinflammatory cytokines, such as IL-1, IL-6, TNF-α, and TGF-β. As examples, the biomarker of bone resorption (CTX-I) was shown to be highly correlated to systemic inflammation as measured by high sensitive C-reactive protein.

The most obvious and perhaps cheapest strategy to protect the subchondral bone and bring OA to a halt would be to simply modify lifestyle habits since the majority of MS risk factors are quite manipulable.

5.4 Complementary Exams Before the Regenerative Treatment

A very important aspect in the pre-treatment evaluation is the complementary exams. It will be discussed some of the tests that are considered extremely important for giving a physician a good characterization of patient's metabolic state.

5.4.1 C-Reactive Protein

C-reactive protein (CRP) is an acute inflammatory protein that increases up to 1000-fold at sites of infection or inflammation. CRP is synthesized primarily in liver hepatocytes but also by smooth muscle cells, macrophages, endothelial cells, lymphocytes, and adipocytes. Ultrasensitive C-reactive protein should be used as an important marker of inflammation in the body, and a decrease in its plasma levels reflects a positive result in regenerative treatment.

5.4.2 Blood Count

This is a simple and inexpensive test capable of providing important data such as the level of hemoglobin, leukocytes, and platelets. A low level of hemoglobin directly affects the body's regenerative capacity, as it is essential for transporting oxygen to areas where there is increased inflammatory activity, often suffering from a local state of tissue hypoxia. Leukocytes and platelets are important cellular components for different forms of regenerative treatment, such as platelet-rich plasma and bone marrow aspirate. Therefore, its quantification becomes important for a good choice of therapeutic modality.

5.4.3 Homocysteine

Elevated homocysteine levels cause osteoblast dysfunction via mitochondrial oxidative damage. This leads to a higher occurrence of fractures and a higher risk of osteoporosis. High levels of homocysteine negatively affects wound healing and is also considered an important inflammation marker. Homocysteine inhibits the synthesis of insoluble collagen fibrils by interfering with normal cross-linking. From the perspective of cartilage homeostasis, these changes in matrix organization interfere with chondrocyte-mediated mineralization potentially altering the function and properties of calcified cartilage. The transformation of homocysteine into methionine requires some cofactors such as vitamin B12,

B6, and B9. In that way, replacing these vitamins seems to be quite reasonable.

5.4.4 Serum Protein Electrophoresis

A serum protein electrophoresis is a simple method that allows proteins to be separated from human plasma into fractions. Its interpretation brings useful information to the regenerative doctor. Thus, it is important for the investigation and diagnosis of several diseases by dosage of albumin, alpha-1-globulin, alpha-2-globulin, beta-globulin, and gamma globulin. Albumin is a general health biomarker and its loss is associated with poor healing capacity. Alfa-globulin fractions have increased levels in inflammatory, infectious, and immune processes. The increase in beta-globulin is observed in situations of disturbance of lipid metabolism or iron deficiency anemia. The decrease or absence in the gamma fraction indicates congenital or acquired immunodeficiencies. Its increase suggests a polyclonal increase in immunoglobulins associated with inflammatory, neoplastic (multiple myeloma and lymphoproliferative disorders), or infectious conditions. The knowledge of these patterns helps the physician to assemble the general patient status.

5.4.5 Osteocalcin

As osteocalcin is produced by osteoblasts, it is often used as a marker for the bone formation process. It has been observed that higher serum osteocalcin levels are relatively well correlated with increases in bone mineral density during treatment with anabolic bone formation drugs for osteoporosis, such as Teriparatide. In many studies, osteocalcin is used as a preliminary biomarker on the effectiveness of a given drug on bone formation.

In its carboxylated form (vitamin K2 dependent), it binds calcium directly and thus concentrates in bone, but recent evidence has revealed that it does play an important role beyond bone mineralization. In its uncarboxylated form, osteocalcin acts as a hormone in the body, signaling in the pancreas, fat, muscle, testicles, and brain to improve metabolic state.

5.4.6 Alkaline Phosphatase

Most of the alkaline phosphatase (ALP) iso-enzymes are derived from the bones and liver. High levels of ALP are often encountered during routine blood investigation in elderly patients. Osteoporosis may increase its blood levels up to 3–5 times normal. Bone pathology causes of elevated alkaline phosphatase include Paget's disease, hyperparathyroidism, osteomalacia, metastatic bone disease, and a recent fracture. By these reasons, it is imperative to investigate bone turnover especially when considering to use MSCs for regenerative purposes.

5.4.7 Ferritin

Along with Homocysteine, the iron storage protein ferritin is also a well-known inflammatory marker. It correlates with biomarkers of cell damage, with biomarkers of hydroxyl radical formation (and oxidative stress) and with the presence and/or severity of numerous diseases. It is important to know that 95% of patients with high levels of ferritin in their blood have increased inflammatory activity in the body. Only 5% of these individuals have high levels of ferritin due to large amounts of iron in the body. To differentiate these two groups, the measurement of transferrin saturation (protein responsible for transporting iron in the blood) are used. With transferrin saturation at levels below 45% in a patient with high levels of ferritin, the systemic inflammatory state is confirmed.

5.4.8 Hormone's Screening

The analysis of thyroid function should also be investigated for regenerative therapy. Hypothyroidism can cause the healing process to slow down and may directly affect the outcomes. Therefore, achieving a hormonal balance becomes essential.

Many studies demonstrate the beneficial effects of thyroid hormones on increasing the biochemical content of cells, more specifically, enhancing the collagen production in cultured chondrocytes. Other studies are being conducted to evaluate the potential of thyroid hormones to enhance the functional properties of articular chondrocytes, which remains somewhat understudied.

The anabolic effect of testosterone on bone and cartilage is well known. This effect, however, is not unique result of a single action of testosterone on the tissues. Testosterone does stimulate mRNM expression of osteoprotegerin and thereby inhibits osteoclastogenesis much like DHEA and TGF-beta do. There appears to be a combined effect on bone by testosterone, IGF-1, and estradiol.

Testosterone is also very important for the maintenance or recovery of the muscle mass, something that has a direct influence on the outcome of various therapies. Low testosterone levels, in both men and women, favor an increase in muscle catabolism and increased levels of body fat. Testosterone stimulation increases the proliferation and preserves stemness of mesenchymal stem cells and endothelial progenitor cells suggesting that, besides other factors, the hormone may engineer these cells and increase their therapeutic potential.

Dehydroepiandrosterone (DHEA), on its turn, is a 19-carbon steroid hormone that is classified as an adrenal androgen. DHEA has been shown to antagonize catabolic mediators of cartilage and may exert protective effects in OA, including suppressing matrix metalloproteinases (MMPs) and inducing cartilage restoration. Author's recent research showed that DHEA demonstrated beneficial effects on OA by influencing the balance between the aggrecanases and tissue inhibitors of metalloproteinase-3 (TIMP-3) in cartilage tissues, suggesting that DHEA might protect articular cartilage from degeneration at the molecular level.

Finally, consistent data have been showing that when estrogens are absent, it results in high bone resorption, and hence, an osteoporosis-like phenotype and worsening of microscopic OA features may develop both in men and women. Estradiol is also able to protect chondrocytes from oxidative damage. Systemically increased turnover of cartilage is found more frequently in postmenopausal women than in the premenopausal population. Taken together, these data indicate that sex hormone homeostasis plays a vital role in the regulation of musculoskeletal health.

5.5 The Importance of Diet Under the Regenerative Perspective

Literature has been recently showing the importance of the diet on bone and cartilage health, particularly when it comes to the comparison of the fat and carbohydrates diet. Studies indicate that the bone formation increase in high-fat diet-fed rat due to osteoblast activity. On the other hand, high sugary diet resulted in bone marrow adipose (BMA) expansion and the alteration of the bone marrow microenvironment, along with the proinflammatory environment, which could contribute to a negative effect on bone metabolism.

Moreover, studies showed that fructose-induced MS decreased the ex vivo osteogenic potential of marrow stromal cells (MSC) and increased the ex vivo adipogenic potential of MSC, which was related to a reduction in Runt-related transcription factor (Runx2) and an increase in Peroxisome Proliferator Activator Receptor γ (PPARγ) expression under basal (undifferentiated) conditions. These data suggest fructose-induced MS resulted in the deleterious alterations in bone microarchitecture, and in the re-ossification of bone lesions, and that these changes might be involved in the differentiation of adipogenic/osteogenic commitment of MSC by modulating the ratio of Runx2/PPARγ.

Higher dietary inflammatory are also associated with higher prevalence of radiographic symptomatic knee osteoarthritis, higher serum interleukin (IL)-6, and tumor necrosis factor (TNF)-R2 levels, thus suggesting a close relationship between diet and inflammatory parameters and osteoarthritis progression.

Among the anti-inflammatory diets, Mediterranean diet has been showing an interesting effect on the reduction of symptoms. Participants with a higher adherence to Mediterranean diet had a significantly lower prevalence of knee OA compared to those with lower adherence. Mediterranean-style diet is an established healthy-eating diet pattern that has consistently demonstrated to have beneficial effects on musculoskeletal, cardiovascular, metabolic, and cognitive diseases. This type of diet may influence a reduction in oxidative stress markers and have been reported to influence the onset of OA though providing increasing levels of collagen type II and aggrecan expression while inhibiting apoptosis-related proteins expression, providing a chondroprotective effect.

5.6 The Gut-Joint Axis

The gut microbiome is a key regulator of bone health that affects postnatal skeletal development and skeletal involution. Alterations in microbiota composition and host responses to the microbiota contribute to pathological bone loss, while changes in microbiota composition that prevent, or reverse, bone loss may be achieved by nutritional supplements with prebiotics and probiotics. The notion that the gut microbiome is a bone mineral density (BMD) regulator in health and disease is supported by an established correlation in humans between microbiome diversity and osteoporosis.

Recently, short-chain fatty acids (SCFAs), which are generated by fermentation of complex carbohydrates (and dietary fibers) in the intestine, have emerged as key regulatory metabolites produced by the gut microbiota. SCFAs inhibit recruitment and activation of macrophages and neutrophils through a reduction in proinflammatory cytokine production. It directly induces metabolic reprogramming of osteoclast precursors (downregulating essential osteoclast genes). Indirect effects of SCFAs may account for their Treg-inducing capacity: Tregs were shown to suppress osteoclast differentiation via their secretion of antiosteoclastic cytokines.

However, it is estimated that the current average consumption of fibers among adults in the

USA is half the recommended amount of 30 g per day to be consumed as part of a healthy diet. To counter that, supplementation with pre- and probiotics has emerged as a good alternative. Probiotics are defined as viable microorganisms that confer health benefits when administered in adequate quantities, while prebiotics are nondigestible fermentable food ingredients that promote the growth of beneficial microbes and/or promote beneficial changes in the activity of the microbiome.

Increasing evidence indicates that probiotics positively affect skeletal health in humans. Early trials showed that ingestion of kefir fermented milk for 6 months caused an increase in BMD in men, while treatment with *Lactobacillus casei shirota* improved distal radius fracture healing in elderly men and women. A 1-year-long trial in older women revealed evidence of a favorable change in bone mass in response to probiotic supplementation, and in a study in Japanese women, the probiotic *Bacillus subtilis* C-3102 increased total hip bone BMD by decreasing bone resorption. In humans, prebiotics increase BMD in adolescents and decrease bone turnover in postmenopausal women. The mechanism of action of prebiotics in bone is complex, but emerging evidence has shown that bacterial metabolic pathways, including those that function in the generation of SCFAs, are involved.

Another interesting topic is that gut-derived LPS (lipopolysaccharide, which is a major component of the outer membranes of Gram-negative bacteria) can provoke generalized proinflammatory responses in infected hosts and can augment adipose macrophage accumulation, skewing the polarization of alternatively activated M2 macrophages toward proinflammatory M1 phenotypes. Moreover, studies have shown that physical exercise could modulate gut microbiome composition (Lactobacillus and Bifidobacterium), boosting intestinal mucosal immunity (reduction of LPS effects via suppression of TLR signaling), increasing the Bacteroidetes–Firmicutes ratio, modifying the bile acid profile, and improving the production of SCFAs.

The exploration of approaches for restoring a healthy microbiota, especially increasing the amount of specific commensal microbiota that antagonize proinflammatory microbes and maintain the intestinal mucosa barrier, is an important future direction for OA treatment. Nutritional supplementation with prebiotics and probiotics that increase SCFAs production and exercises may represent an effective, safe, and inexpensive modality in the treatment of metabolic bone disorders.

5.7 Drug Strategies to Target Bone and Cartilage in Osteoarthritis

5.7.1 Antiresorptives Drugs

Bisphosphonate treatment in surgical models of OA resulted in a 50% decrease in disease severity scores, and protection of bone and cartilage from pathological changes. It has also been shown that in OVX rats alendronate significantly attenuated cartilage erosion by inhibiting subchondral bone loss. Strassle et al. demonstrated that when bisphosphonate zolendronate was applied to the monoiodoacetate model of painful arthritis in rats, it protected against subchondral bone loss, cartilage degradation and, importantly, also pain. In fact, a secondary analysis identified those with the highest levels of cartilage degradation as assessed by CTX-II levels to be those in whom OA progressed the most. CTX-II was the same marker that was influenced by bisphosphonate therapy. To date, CTX-II may be the best validated marker for progression of OA.

Calcitonin significantly affected trabecular structure and prevented subchondral bone resorption and trabecular thinning, which was speculated to be a major factor in the degradation of the above cartilage layer. Calcitonin may act different from that of other antiresorptives, as calcitonin has been shown to have direct and indirect actions on articular cartilage on human OA chondrocytes. Oral formulation of calcitonin-inhibited bone and cartilage degradation, evidenced by the biochemical markers CTX-I and CTX-II, respectively.

5.7.2 Bone Anabolic Drugs

PTH is presently the only bone anabolic treatment accepted by the FDA in the USA, and the European Medicines Agency (EMA). PTH stimulates osteoblasts to synthesize bone. Interestingly, chondrocytes and osteoblasts are from the same mesenchymal cell (MSC) lineage, which suggests that PTH might also affect chondrocytes anabolically.

It has also been showed that PTH[1–32]-inhibited expression of type X collagen in MSCs from OA patients in a time-dependent manner. In parallel, PTH[1–32]-stimulated expression of type II collagen, a marker of chondrogenic differentiation and cartilage repair. These results indicate that PTH may be chondroprotective by inhibiting hypertrophy and cartilage calcification. Altogether, different preclinical and clinical studies with various bone drugs indicate that a carefully selected antiresorptive or anabolic treatment could provide clinical benefits in OA for a selected patient population.

5.7.3 Antihypertensive Drugs

Hypertension is often treated with L-type voltage-operated calcium channel blocking drugs, nifedipine being among the most classical ones. Nifedipine had positive effects on the production of collagen type II and proteoglycans in both cell types, implying potentially beneficial anabolic responses in articular cartilage. These results highlight a potential link between antihypertensive drugs and cellular changes that occur in chondrocytes in OA cartilage.

5.8 Hormone Balance and Its Effect on Bone and Cartilage Health

Estrogen is the major hormonal regulator of bone metabolism in both women and men and is important for maintaining bone formation at the cellular level. In young men, the most significant hormonal determinants of the BMD of the hip and of the cortical thickness of the femoral neck are 17β-estradiol and IGF-1 while in aged men (over 60 years) the BMD was not correlated with IGF-1 at any site but only at 17β-estradiol. On the other hand, long-term androgen deficiency can result in a decrease in the calcium content of both tibia and lumbar vertebrae.

A serum workup of growth hormones such as GH and IGF-1 is also necessary for complete analysis of bone homeostasis. GH and IGF-1 are fundamental in achieving a normal longitudinal bone growth and mass during the postnatal period and, in association with sex steroids, play a major role in bone growth and development. For instance, recently it was demonstrated that low serum IGF-1 levels were associated with an increased risk of fractures of about 40%. GH is the major determinant of stimulation of progenitor cells and interacts with progenitor cells in adipose tissue and cartilage and IGF-1 stimulate a subsequent clonal expansion. Osteoblasts and chondrocytes have receptors for GH and the administration of GH at physiological doses exerts a direct action on osteoblasts, stimulating cell proliferation and differentiation. IGF-1 reduces osteoblast apoptosis and promotes osteoblastogenesis.

Thyroid hormones exert widespread and complex actions in almost all tissues during development, throughout childhood and in adults. Both receptor isoforms (TRa1 and TRb1) are present in growth plate chondrocytes, bone marrow stromal cells, and bone-forming osteoblasts. T3 stimulates synthesis and post-translational modification of type I collagen, induces expression of alkaline phosphatase, and regulates synthesis and secretion of the bone matrix proteins osteopontin and osteocalcin. T3 also promotes bone matrix remodeling by stimulating expression of matrix metallo-proteinases-9 and -13. Furthermore, thyroid hormones regulate key pathways involved in osteoblast proliferation and differentiation, inhibiting osteoclastostogenesis by regulating osteoprotogerin, a decoy receptor that ultimately inhibits receptor activator of nuclear factor-kB ligand (RANK-L). From a clinical point of view, in hypothyroidism there is reduced bone turnover which leads to prolonged period of secondary mineralization. Conversely, in thyrotoxicosis, high bone turnover osteoporosis is due to shortening of the remodeling cycle with uncoupling of the activities of osteoclast and osteoblasts that results in a loss of about 10% of mineralized bone per cycle. Consistent with histomorphometry data, population studies have demonstrated that hypothyroidism is associated with a two- to three-fold increased risk of fracture, while thyrotoxicosis is an established cause of osteoporosis and fragility fracture.

5.9 The Role of Hormones in Tendinopathies

One of the most important hormones to pursue optimal levels is insulin. Is also very mandatory to monitor the levels of fasting glucose and HbA1C. It has been shown that at humoral level, higher levels of interleukin-1 beta (IL-1-β) and advanced glycated end-products (AGEs) have been showed in tendons of diabetic patients. Diabetes increases chronic inflammation, TNF-a, IL1-β, and AGEs in the tendons, increasing the risk of rupture and tendinopathies. High glucose concentration upregulates the expression of MMP-9 and MMP-13 in tenocytes.

Sex hormones also play a vital role in tendon healing. Protein analysis has been showing that estrogen and progesterone upregulate gene expression for the proteoglycans aggrecan, biglycan, decorin, and versican in tendons. In addition, estrogen deficiency negatively affects tendon metabolism and healing, reducing proliferation rate, increasing apoptosis and altering tendons composition in terms of collagen I, aggrecan and elastin. Mainly because tendons express the estrogen receptors α and β. Normal and diseased tendons of both male and female patients expressed both estrogen receptors. Conversely, physiological high concentration of estrogen in young athletes is coupled to enhanced joint laxity may enhance the risk of injuries.

Androgens administration reduces MMP expression in tendons, positively affecting tissue remodeling during different training programs. In vitro, progressive increasing concentration of testosterone has direct effects on male human tenocytes, increasing cell number after 48 and 72 h of treatment. But testosterone abuse administration can lead to the alterations of biomechanical properties of tendons, reduction of elastic properties, tendon dysfunction, and fibrosis, with a higher incidence of spontaneous tendon ruptures. Actually, the balancing of both estrogen and testosterone in physiological concentration seems to be important for tendon health and physical function, whereas very low or high concentrations of endogenous or exogenous administrated sex hormones may lead to an enhanced risk of injuries and inadequate adaptation to mechanical loading.

GH/IGF-1 system is closely linked to collagen synthesis and connective tissue maturation. A study on short-term explant culture of the deep flexor tendon in rabbits showed that tenocytes increase their ability to repair and to regenerate ECM if they are cultured with recombinant human insulin-like growth factor (rhIGF-I). In humans, increased GH availability stimulates collagen synthesis in skeletal muscle and tendon, potentially increasing the cross-sectional area of tendons, but without any effect upon myofibrillar protein synthesis, which represents a controversy on literature.

Lastly, the effects of thyroid hormones in tendons are mediated by receptors (TR)-α and -β that seem to be ubiquitous. In particular, T3 and T4 play an antiapoptotic role on tenocytes and cell proliferation, causing an increase in vital tenocytes isolated from tendons in vitro and a reduction of apoptotic ones; they are also able to influence extracellular matrix proteins secretion in vitro from tenocytes, enhancing collagen production. From a clinical point of view, tendinopathy can be the presenting complaint in hypothyroidism, which causes accumulation of glycosaminoglycans (GAGs) in the ECM, and symptomatic relief can be obtained by appropriate management of the primary thyroid deficiency.

Further Reading

Bagger YZ, Tanko LB, Alexandersen P, et al. Oral salmon calcitonin induced suppression of urinary collagen type II degradation in postmenopausal women: a 223 new potential treatment of osteoarthritis. Bone. 2005;37:425–30.

Berardi AC, Oliva F, Berardocco M, la Rovere M, Accorsi P, Maffulli N. Thyroid hormones increase collagen I and cartilage oligomeric matrix protein (COMP) expression in vitro human tenocytes. Muscles Ligaments Tendons J. 2014;4(3):285–91.

Biver E, Berenbaum F, Valdes AM, et al. Gut microbiota and osteoarthritis management: an expert consensus of the European society for clinical and economic aspects of osteoporosis, osteoarthritis and musculoskeletal diseases (ESCEO). Ageing Res Rev. 2019;55:100946. https://doi.org/10.1016/j.arr.2019.100946.

Cerda B, Perez M, Perez-Santiago JD, Tornero-Aguilera JF, Gonzalez-Soltero R, Larrosa M. Gut microbiota modification: another piece in the puzzle of the benefits of physical exercise in health? Front Physiol. 2016;7:51.

Collectively, we found an increased apoptosis and suppressed autophagy in TSH stimulated primary chondrocytes (falar então que o hipot pode causar degeneração articular)

Dam EB, Byrjalsen I, Karsdal MA, et al. Increased urinary excretion of C-telopeptides of type II collagen (CTX-II) predicts cartilage loss over 21 months by MRI. Osteoarthr Cartil. 2009;17:384–9.

de Sire A, de Sire R, Petito V, et al. Gut-joint axis: the role of physical exercise on gut microbiota modulation in older people with osteoarthritis. Nutrients. 2020;12(2):574. Published 22 Feb 2020. https://doi.org/10.3390/nu12020574.

Elkhenany H, AlOkda A, El-Badawy A, El-Badri N. Tissue regeneration: impact of sleep on stem cell regenerative capacity. Life Sci. 2018;214:51–61. https://doi.org/10.1016/j.lfs.2018.10.057.

Felson DT, Anderson JJ, Naimark A, Walker AM, Meenan RF. Obesity and knee osteoarthritis. The Framingham study. Ann Intern Med. 1988;109:18–24.

Felson DT, Chaisson CE, Hill CL, et al. The association of bone marrow lesions with pain in knee osteoarthritis. Ann Intern Med. 2001; https://doi.org/10.7326/0003-4819-134-7-200104030-00007.

Felson DT, McLaughlin S, Goggins J, et al. Bone marrow Edema and its relation to progression of knee osteoarthritis. Ann Intern Med. 2003; https://doi.org/10.7326/0003-4819-139-5_part_1-200309020-00008.

Garnero P, Aronstein WS, Cohen SB, et al. Relationships between biochemical markers of bone and cartilage degradation with radiological progression in patients with knee osteoarthritis receiving risedronate: the knee osteoarthritis structural arthritis randomized clinical trial. Osteoarthr Cartil. 2008;16:660–6.

Hansen M, Kjaer M. Sex hormones and tendon. Adv Exp Med Biol. 2016;920:139–49. https://doi.org/10.1007/978-3-319-33943-6_13.

Henriksen K, Bollerslev J, Everts V, et al. Osteoclast activity and subtypes as afunction of physiology and pathology—implications for future treatments of osteoporosis. Endocr Rev. 2011;32:31–63.

Karsdal MA, Bay-Jensen AC, Lories RJ, et al. The coupling of bone and cartilage turnover in osteoarthritis: opportunities for bone antiresorptives and anabolics as potential treatments? Ann Rheum Dis. 2014a;73(2):336–48.

Karsdal MA, Bay-Jensen AC, Lories RJ, et al. The coupling of bone and cartilage turnover in osteoarthritis: opportunities for bone antiresorptives and anabolics as potential treatments? Ann Rheum Dis. 2014b;73(2):336–48. https://doi.org/10.1136/annrheumdis-2013-204111.

Kornicka K, Houston J, Marycz K. Dysfunction of mesenchymal stem cells isolated from metabolic syndrome and type 2 diabetic patients as result of oxidative stress and autophagy may limit their potential therapeutic use. Stem Cell Rev. Rep. 2018;14(3):337–45. https://doi.org/10.1007/s12015-018-9809-x.

Liu Y, Ding W, Wang HL, et al. Gut microbiota and obesity-associated osteoarthritis. Osteoarthr Cartil. 2019;27(9):1257–65. https://doi.org/10.1016/j.joca.2019.05.009.

Locatelli V, Bianchi VE. Effect of GH/IGF-1 on bone metabolism and Osteoporsosis. Int J Endocrinol. 2014;2014:235060. https://doi.org/10.1155/2014/235060.

Majidinia M, Reiter RJ, Shakouri SK, et al. The multiple functions of melatonin in regenerative medicine. Ageing Res Rev. 2018;45:33–52. https://doi.org/10.1016/j.arr.2018.04.003.

Mullington JM, Simpson NS, Meier-Ewert HK, Haack M. Sleep loss and inflammation. Best Pract Res Clin Endocrinol Metab. 2010;24(5):775–84. https://doi.org/10.1016/j.beem.2010.08.014.

Mwale F, Yao G, Ouellet JA, et al. Effect of parathyroid hormone on type X and type II collagen expression in mesenchymal stem cells from osteoarthritic patients. Tissue Eng Part A. 2010;16:3449–55.

Oestergaard S, Sondergaard BC, Hoegh-Andersen P, et al. Effects of ovariectomy and estrogen therapy on type II collagen degradation and structural integrity of articular cartilage in rats: implications of the time of initiation. Arthritis Rheum. 2006;54:2441–51.

Oliva F, Berardi AC, Misiti S, Falzacappa CV, Iaconeand A, Maffulli N. Thyroid hormones enhance growth and counteract apoptosis in human tenocytes isolated from rotator cuff tendons. Cell Death Dis. 2013;4:e705.

Oliva F, Piccirilli E, Berardi AC, Frizziero A, Tarantino U, Maffulli N. Hormones and tendinopathies: the current evidence. Br Med Bull. 2016a;117(1):39–58. https://doi.org/10.1093/bmb/ldv054.

Oliva F, Piccirilli E, Berardi AC, Tarantino U, Maffulli N. Influence of thyroid hormones on tendon homeostasis. Adv Exp Med Biol. 2016b;920:133–8. https://doi.org/10.1007/978-3-319-33943-6_12.

Quach D, Britton RA. Gut microbiota and bone health. Adv Exp Med Biol. 2017;1033:47–58. https://doi.org/10.1007/978-3-319-66653-2_4.

Robinson WH, Lepus CM, Wang Q, et al. Low-grade inflammation as a key mediator of the pathogenesis of osteoarthritis. Nat Rev Rheumatol. 2016;12(10):580–92. https://doi.org/10.1038/nrrheum.2016.136.

Segovia-Silvestre T, Bonnefond C, Sondergaard BC, et al. Identification of the calcitonin receptor in osteoarthritic chondrocytes. BMC Res Notes. 2011;4:407.

Strassle BW, Mark L, Leventhal L, et al. Inhibition of osteoclasts prevents cartilage loss and pain in a rat model of degenerative joint disease. Osteoarthr Cartil. 2010;18:1319–28.

Tabung FK, Steck SE, Zhang J, et al. Construct validation of the dietary inflammatory index among postmenopausal women. Ann Epidemiol. 2015;25:398–405.

Tian L, Yu X. Fat, sugar, and bone health: a complex relationship. Nutrients. 2017;9(5):506. Published 17 May 2017. doi: https://doi.org/10.3390/nu9050506

Varga F, Rumpler M, Zoehrer R, Turecek C, Spitzer S, Thaler R, Paschalis EP, Klaushofer K. T3 affects expression of collagen I and colla- gen cross-linking in bone cell cultures. Biochem Biophys Res Commun. 2010;402(2–3):180–5.

Veronese N, Stubbs B, Noale M, et al. Adherence to a Mediterranean diet is associated with lower prevalence of osteoarthritis: data from the osteoarthritis initiative. Clin Nutr. 2017;36(6):1609–14. https://doi.org/10.1016/j.clnu.2016.09.035.

Veronese N, Shivappa N, Stubbs B, et al. The relationship between the dietary inflammatory index and prevalence of radiographic symptomatic osteoarthritis: data from the osteoarthritis initiative. Eur J Nutr. 2019;58(1):253–60. https://doi.org/10.1007/s00394-017-1589-6.

Wang J, Wang Y, Gao W, et al. Diversity analysis of gut microbiota in osteoporosis and osteopenia patients. PeerJ. 2017;5:e3450. Published 15 Jun 2017. https://doi.org/10.7717/peerj.3450.

Williams GR. Thyroid hormone actions in cartilage and bone. Eur Thyroid J. 2013;2(1):3–13. https://doi.org/10.1159/000345548.

Xie C, Chen Q. Adipokines: new therapeutic target for osteoarthritis? Curr Rheumatol Rep. 2019;21(12):71.

Xin W, Yu Y, Ma Y, et al. Thyroid-stimulating hormone stimulation downregulates autophagy and promotes apoptosis in chondrocytes. Endocr J. 2017;64(7):749–57. https://doi.org/10.1507/endocrj.EJ16-0534.

Zaiss MM, Jones RM, Schett G, Pacifici R. The gut-bone axis: how bacterial metabolites bridge the distance. J Clin Invest. 2019;129(8):3018–28. Published 15 Jul 2019. https://doi.org/10.1172/JCI128521.

Zieliene I, Bernotiene E, Rakauskiene G, et al. The antihypertensive drug nifedipine modulates the metabolism of chondrocytes and human bone marrow-derived mesenchymal stem cells. Front Endocrinol (Lausanne). 2019;10:756. Published 8 Nov 2019. https://doi.org/10.3389/fendo.2019.00756.

Bone Anatomy and Healing Process of a Fracture

Umile Giuseppe Longo, Giovanna Stelitano, Vincenzo Candela, and Vincenzo Denaro

6.1 Bone Structure and Functions

Bone is the skeletal system element which provides body shape, mechanical supports for muscles and soft tissues, allowing the movement [1]. Its mechanical characteristics of stiffness and strength give it the ability to tolerate heavy load without failure. Bones participation in mineral homeostasis and endocrine metabolism is widely recognized, thanks to their composition of mineral crystals and calcium, essential ions for physiological mechanisms [2]. Bone is a complex structure made by cellular elements, extracellular matrix (ECM), and lipids. It is composed of 20% of water, while the remaining part consists respectively of 35% and 70% of organic and inorganic substances [3]. The cellular population includes osteoblasts, osteocytes, osteoclasts, and osteogenic precursors. Osteoblasts perform the critical role of collagen producers [3]. They represent the essential cellular shape because of their implication into synthesis and regulation of the ECM and in blood-calcium homeostasis, acting as mechanosensors for bones. Osteocytes can

be considered as the mature form of osteoblasts, placed in the lacunae: it is no coincidence that they share the same mesenchymal precursor cell [4]. Osteoclasts, instead, derived from the macrophage-monocyte line. In essence, their primary role is to produce proteolytic enzymes even if they are involved in bone resorption and osseous fracture healing.

Bone ECM is made of an organic and inorganic phase. The organic one is predominantly made of type I collagen fibers, noncollagenous elements like proteoglycans, glycoproteins, osteonectin, fibronectin, osteopontin, osteocalcin, and phospholipids [5, 6]. In reverse, the inorganic substance is characterized by the almost whole presence of crystalline mineral salts and calcium structured in hydroxyapatite [7]. Bones own a complex vascularization system which plays a critical function in the maintenance of osseous integrity. It is possible to distinguish a periosteal and a medullary circulation [8]. The first one is responsible for blood supply of the periosteum and the higher cortical area, while the medullary flow provides the vascular supply to the bone marrow and the lower cortical area [9].

Two different kinds of bone exist: the cancellous and cortical bone. The first type, also known as trabecular bone because of its trabeculae units, is placed the flat and cuboidal bones and in the extremities of long bones [10]. The cortical bone, instead, made of cylindrical structures called osteons or Haversian systems and of many

U. G. Longo (✉) · G. Stelitano · V. Candela
V. Denaro
Department of Orthopaedic and Trauma Surgery,
Campus Bio-Medico University, Rome, Italy
e-mail: g.longo@unicampus.it;
g.stelitano@unicampus.it; v.candela@unicampus.it;
denaro@unicampus.it

© ISAKOS 2022
A. Gobbi et al. (eds.), *Joint Function Preservation*, https://doi.org/10.1007/978-3-030-82958-2_6

lamellae around the Haversian canal, is present in the external surfaces of the long bones. The cortical bone has a mature form defined lamellar bone that is typically not detectable in the cortical osseous zone [11]. In the healing process, a transitory irregular structure made of disorganized collagen fibers and casually dispersed crystals precedes the lamellar bone apposition.

6.2 Bony Fracture

Fracture consists of the effect of single or repeated excessive loads to which the bone is subjected. It causes a deterioration of bone continuity which brings to pathological deformation, failure of bone structure, impairment of bone function, and discomfort. Bone surface separation makes a cavitation, causing critical soft-tissue injury [12]. Even if the fracture of bone mainly represents a mechanical event, it involves complicated biological responses like the vascular and the molecular ones [13]. The rupture of blood vessels both within the bone and the periosteum can be observed [14, 15]. Blood vessels damage owns an essential impact on fracture healing because the blood supply is essential to restore the adequate environment for bone formation and resorption, contributing to the diffusion of freed biochemical factors which support the healing [16]. After fracture, a decrease of nearly 50% of bone cortical circulation has been observed. The cause of this reduction seems to be the vasoconstriction in the periosteal and the medullary vascular structures, consequence of the injury [17, 18]. Throughout the repair process, however, a growing hyperemia occurs in the nearby intraosseous and extraosseous vascularization, touching the acme in 14 days. Consequently, blood provision in the callus area slowly declines encore. Meantime, the disorder of the medullary system causes a reversal of the physiological centripetal circulation. It is important to consider that the vascularization of callus is essential and has a leading function in the outcome of healing. Bone can only develop if sustained by a substantial blood supply: cartilage will not survive in the lack of adequate vascular-

ization. Many factors could influence blood supply. They are listed as follows [19]:

1. The mechanism of injury: the energy of the trauma, the direction, the distribution of strengths in the fracture zone could cause the kind of fracture and the presence of related soft-tissue lesions. This last point plays a crucial role, considering that the blood provision to the callus zone come from the adjacent soft tissues.
2. Initial patient management: it is essential to transport the patient avoiding the mobilization at the fracture site that could worsen the injury.
3. Patient resuscitation: critical conditions as hypovolemia, hypoxia, and coagulopathy could raise bone and soft-tissue injuries.
4. Comorbidities: peripheral vascular disease and diabetes (thanks to its microangiopathy) can compromise blood supply.

6.3 Biology of Fracture Healing

Fracture healing represents a biological mechanism of great importance necessary to wholly replace the lamellar bone to its initial state, reacquiring the primary bone force. Bone represents a specialized tissue: its healing doesn't presuppose fibrous scar development. Two different kinds of fracture healing were identified: primary (direct) and secondary (indirect) mechanisms.

6.3.1 Direct (Primary) Bone Healing

Direct mechanism is a biological process of osseous remodeling quicker than the secondary one [19]. This healing process includes intramembranous bone production and direct cortical remodeling, but not bone callus apposition.

It necessarily occurs in a condition of high stability, if rigid internal fixation diminishes the mobilization of the fracture parts, decreasing interfragmentary strain and, in practice, is the rarest type [17, 20]. Under these conditions, the osteons of the Haversian system can travel along the length of the bone crossing the fracture zone and connect the lacking space putting down cylinders of bone, known as "cutting cones." Thus, the formation of numerous osteons heals the fracture. This mechanism is supported by the generation of new blood vessels, together with endothelial and perivascular mesenchymal cells, that represent the osteoprogenitors for osteoblasts [21]. Osteonal activity grows near the fracture site, and this phenomenon, which probably plays a crucial role in direct fracture healing, is identified as "regional acceleratory phenomenon" (RAP). The mechanism of RAP is still unclear, but it seems to be regulated by the equal molecular pathways observed in [22]. The complete healing process is achieved in a period among some months or some years, and the whole healing is considered happened if the two parts of the cortical bone are fused together, reestablishing connection.

6.3.2 Indirect (Secondary) Bone Healing

Indirect bone healing represents the normal way of bone restore, and it occurs in a condition of relative stability (flexible fixation methods). This mechanism remembers the embryological bone process of development and comprises intramembranous and endochondral bone apposition [23]. It is marked by the formation of bone callus. For this reason, it can also be identified as endochondral ossification, or callus healing. It comprises four stages with different characteristics, even if there is a seamless passage from one phase to another. The four phases of the secondary healing process are as follows:

– Inflammation
– Soft callus formation
– Hard callus formation
– Remodeling

6.3.3 Inflammation

The inflammatory response represents the first mechanism that starts immediately after a frac-

ture. It usually lasts about 1–7 days after fracture when fibrosis, cartilage, or bone formation occurs [21]. The release of powerful cytokines, result of soft-tissue injury and platelets degranulation, triggers the inflammatory cascade. Thus, vasodilatation, hyperemia, migration, and proliferation of inflammatory cells take place. Inflammation is characterized by the hematoma formation, inflammatory exudation from damaged vascular structures and bone necrosis, detected in the final part of fracture segments. The hematoma, made of a network of fibrin, reticulin, and collagen fibril, is consequence of bleeding from the periosteal vessels inside the medullary canal and under the periosteum [24]. Granulation tissue gradually replaced it. Then, platelets aggregation and neo-angiogenesis occur [25]. Vascular damage deprives osteocytes of oxygen at the ends of the fracture sites, and this causes tissue degeneration and/or necrosis. Macrophages phagocytize the degenerated zones and stimulate the reformation phase, thanks to the release of signaling factors and growth factors. During this healing stage, inflammatory molecule levels like interleukin-1 (IL-1), IL-6, IL-11, IL-18, tumor necrosis factor-α (TNF-α) [26], bone morphogenic proteins, bFGF, transforming growth factor-β (TGF-β), platelet-derived growth factor (PDGF), and insulin-like growth factor (IGF) are significantly raised [27]. Each of these mediators has an essential part in the healing mechanism, thanks to their chemotactic effects on other inflammatory cells. They drive migration, recruitment, and proliferation of the cellular elements implicated in healing process [28]. These cellular elements fill the fracture space through the production of granulation tissue. In this stage, a soft callus is generated, and the free mobilization of the osseous fractured parts is diminished. It is important to underline that lymphocytes aren't need to start the wound healing, even if a physiological outcome of tissue repair strictly required an unimpaired cellular immune response [29]. The alteration of lymphocyte immune mechanisms, in fact, could cause a generalized immunosuppression, which raises predisposition to infection and sepsis. Even if the specific onset of post-traumatic immune suppression is still unclear, stress hormones and immunosuppressive mediators (cytokines, prostaglandins, and nitric oxide) stimulate the lymphocytes negatively [30].

On the other hand, macrophages are widely detected in fibrous callus structure and in a part of the fresh created osseous tissue. They are the main responsible for the regulation of the early healing stages, driving the differentiation of progenitors and controlling blood supply.

6.3.4 Soft Callus Formation

The growth of callus marks the soft callus phase, which occurs approximately 2–3 weeks after the fracture. During this period, the bone fragments are no able to move freely and, usually, pain and swelling have decreased [31]. The soft callus formation is marked by the differentiation of the progenitor cellular elements into osteoblasts [32]. Osteoblast are responsible for the intramembranous bone apposition. They create a cuff of woven bone on these areas, far from the fracture site, which fills the intramedullary canal. At the same time, near to the fracture site, mesenchymal progenitors proliferate and move into the callus, becoming fibroblasts, chondrocytes, and generating the extracellular matrix that replaces the hematoma [12, 20]. The neoformation of capillaries inside the callus and the raised vascularization develop. During the last phase of callus production, stability results sufficient to avoid shortening, even if angulation at the fracture elements may yet happen.

6.3.5 Hard Callus Formation

When the soft callus connects the fracture fragments, the hard callus phase begins. This stage lasts 3–4 months until the bone parts result firmly consolidated by new bone [33, 34]. As intramembranous bone formation proceeds, the soft callus which has filled the fracture gap is converted into calcified tissue through to the endochondral ossification process driven by osteoblasts [35]. Hard callus growth commences peripherally, in the region mechanically less stable and in which the

strain is lowest. Slowly, endochondral bone formation progresses in the center of the fracture site. The narrowing fracture space increases the strain until the fracture site is wholly filled by osteoblasts that assume a spiral organization to decrease strain and support woven bone production. The apposition of hard callus finally guarantees the stabilization of the fracture [33]. The endochondral bone ossification in the cartilaginous callus is regulated by several molecular signals (TGF-β2 and -β3, BMPs). Their principal action is both mitogenic and angiogenic, considering that vascularization is essential for bone formation [36].

6.3.6 Remodeling

After the hard callus development, the woven bone strictly links the bone fragments. From this point on, the remodeling stage occurs [37]. During this stage, the woven bone is slowly supplanted by lamellar tissue by the processes of surface erosion and osteonal remodeling. Remodeling can last variably from some months to a number of years, until the original osseous properties, including shape, size, restoration of the medullary canal and biomechanical competency, are restored [38]. The protagonists of this phase are the cellular bone elements. Simultaneously, the osteoclasts remove the woven tissue while osteoblasts substitute the matrix with the lamellar one [39]. Osteoclasts polarized to connect to the mineralized area. They build a ruffled border, where acid and proteinases are put on the resorption area. Bone resorption makes erosive pits on the osseous superficial area called "Howship's lacuna." Once this process is ended, osteoblasts can put down lamellar bone on the eroded surface [40]. Lamellae are distributed parallel to the longitudinal axis of the highest strength with the aim to guarantee the stability and force at the fracture zone. Sufficient strength is claimed to improve osteogenesis and guide the right geometric configuration of osteons. Muscular strengths present during physical activity create mechanical bone forces that stimulate the remodeling [41, 42].

Several proinflammatory signals mediate the remodeling phase. Among these IL-1, IL-6, and IL-11, TNF-α, IL-12, and interferon-γ (IFN-γ) together with the growth hormone and parathyroid hormone have critical part in the healing and strengthening of the fractured callus.

6.3.7 Molecular Signaling

The cascade of molecular pathways in fracture healing is tightly regulated through several growth factors and biological mediators liberated by the damaged bone [30].

The early phases of healing, for example, is marked by the over-expression of genes linked to cell cycle and cell-to-cell signaling. Among these, IL-1 and IL-6, IGF-1 and IGF-2, PDGF, FGF receptor, fibronectin, MMPs, glypican, osteomodulin, osteonectin, tenascin C, cartilage, and collagen represent the most involved molecules.

Genes regulating cell growth and survival are consistently upregulated. They would be responsible for the differentiation of osteogenic precursors and bone matrix apposition [43]. In the healing process, bone morphogenetic proteins (BMPs) play an essential role: they are considered the main inducers of bone development. These factors regulate chondro-osteogenesis, chemotaxis, proliferation, and cellular differentiation and they seem to be also involved in the angiogenesis events. BMPs also control ECM synthesis, and they have an essential function in the recruitment of progenitors [44]. FGF, vascular endothelial growth factor (VEGF) [45], and angiopoietins 1 and 2 are molecular factors implicated in the vascular ingrowth during callus developing. Latest researches have shown the implication of hypoxia-inducible factor-1a (HIF-1a) in bone healing: the hypoxic gradient would regulate mesenchymal stem cell progenitor displacements [46]. Finally, platelets give a contribution to the bone repair [47]. They would induce cell migration, differentiation, neovascularization, and mobilization of inflammatory cellular elements.

6.4 Fracture Healing Complications

Fracture healing can lead to several complications divided, on the basis of temporal criteria, as follows:

- Immediate complications which include systemic hypovolemic shock, major vessels injuries, muscular and tendon tears, articular and local viscera injuries.
- Early complications such as hypovolemic shock, ARDS, fat embolism syndrome [48], deep vein thrombosis, pulmonary syndrome, septicemia, crush and compartment syndrome, infection [49].
- Late complications which comprehend delayed union, nonunion [50], malunion, cross union and avascular necrosis, articular stiffness, Sudeck's dystrophy, osteomyelitis, ischemia, myositis ossificans, and osteoarthritis [51].

Other problems to consider are discomfort, nerves and vessels damages, infection, wound disorders, instability, and hematoma [52]. These conditions occur most frequently in fractures derived from high-impact trauma, while low-energy injuries are rarely associated with severe problems [53]. The soft-tissue integrity, the characteristics of the trauma, fracture comminution and dislocation, contamination, kind of treatment, related damages have different impact on the fracture healing process [54] [55].

References

1. Marolt D, Knezevic M, Novakovic GV. Bone tissue engineering with human stem cells. Stem Cell Res Ther. 2010;1:10.
2. Healy KE, Guldberg RE. Bone tissue engineering. J Musculoskelet Neuronal Interact. 2007;7:328–30.
3. Bigham-Sadegh A, Oryan A. Basic concepts regarding fracture healing and the current options and future directions in managing bone fractures. Int Wound J. 2015;12:238–47.
4. Shapiro F. Bone development and its relation to fracture repair. The role of mesenchymal osteoblasts and surface osteoblasts. Eur Cell Mater. 2008;15:53–76.

5. Wang M, Yang N, Wang X. A review of computational models of bone fracture healing. Med Biol Eng Comput. 2017;55:1895–914.

6. Forriol F, et al. Platelet-rich plasma, rhOP-1 (rhBMP-7) and frozen rib allograft for the reconstruction of bony mandibular defects in sheep. A pilot experimental study. Injury. 2009;40(Suppl 3):S44–9.

7. Giannoudis P, Tzioupis C, Almalki T, Buckley R. Fracture healing in osteoporotic fractures: is it really different? A basic science perspective. Injury. 2007;38(Suppl 1):S90–9.

8. Boskey AL, Coleman R. Aging and bone. J Dent Res. 2010;89:1333–48.

9. Ulstrup AK. Biomechanical concepts of fracture healing in weight-bearing long bones. Acta Orthop Belg. 2008;74:291–302.

10. Feng X, McDonald JM. Disorders of bone remodeling. Annu Rev Pathol. 2011;6:121–45.

11. Salgado AJ, Coutinho OP, Reis RL. Bone tissue engineering: state of the art and future trends. Macromol Biosci. 2004;4:743–65.

12. Doblaré M, García JM. On the modelling bone tissue fracture and healing of the bone tissue. Acta Cient Venez. 2003;54:58–75.

13. Chen G, et al. Simulation of the nutrient supply in fracture healing. J Biomech. 2009;42:2575–83.

14. Kanczler JM, Oreffo RO. Osteogenesis and angiogenesis: the potential for engineering bone. Eur Cell Mater. 2008;15:100–14.

15. van der Wal KG. Chain saw injuries in the mandibulofacial region. Ned Tijdschr Tandheelkd. 1984;91:275–6.

16. Augat P, Simon U, Liedert A, Claes L. Mechanics and mechano-biology of fracture healing in normal and osteoporotic bone. Osteoporos Int. 2005;16(Suppl 2):S36–43.

17. Tsiridis E, Upadhyay N, Giannoudis P. Molecular aspects of fracture healing: which are the important molecules? Injury. 2007;38(Suppl 1):S11–25.

18. Jeon S, et al. Assessment of neovascularization during bone healing using contrast-enhanced ultrasonography in a canine tibial osteotomy model: a preliminary study. J Vet Sci. 2020;21:e10.

19. Hajnovic L, Sefranek V, Schütz L. Influence of blood supply on fracture healing of vertebral bodies. Eur J Orthop Surg Traumatol. 2018;28:373–80.

20. Marsell R, Einhorn TA. The biology of fracture healing. Injury. 2011;42:551–5.

21. Phillips AM. Overview of the fracture healing cascade. Injury. 2005;36(Suppl 3):S5–7.

22. Schindeler A, McDonald MM, Bokko P, Little DG. Bone remodeling during fracture repair: the cellular picture. Semin Cell Dev Biol. 2008;19:459–66.

23. Sathyendra V, Darowish M. Basic science of bone healing. Hand Clin. 2013;29:473–81.

24. LaStayo PC, Winters KM, Hardy M. Fracture healing: bone healing, fracture management, and current concepts related to the hand. J Hand Ther. 2003;16:81–93.

25. Greenbaum MA, Kanat IO. Current concepts in bone healing. Review of the literature. J Am Podiatr Med Assoc. 1993;83:123–9.

26. Glass GE, et al. TNF-alpha promotes fracture repair by augmenting the recruitment and differentiation of muscle-derived stromal cells. Proc Natl Acad Sci U S A. 2011;108:1585–90.

27. Mountziaris PM, Mikos AG. Modulation of the inflammatory response for enhanced bone tissue regeneration. Tissue Eng Part B Rev. 2008;14:179–86.

28. Giannoudis PV, et al. Inflammation, bone healing, and anti-inflammatory drugs: an update. J Orthop Trauma. 2015;29(Suppl 12):S6–9.

29. Kolar P, et al. The early fracture hematoma and its potential role in fracture healing. Tissue Eng Part B Rev. 2010;16:427–34.

30. Lauzon MA, Bergeron E, Marcos B, Faucheux N. Bone repair: new developments in growth factor delivery systems and their mathematical modeling. J Control Release. 2012;162:502–20.

31. Yamagiwa H, Endo N. Bone fracture and the healing mechanisms. Histological aspect of fracture healing. Primary and secondary healing. Clin Calcium. 2009;19:627–33.

32. Liebschner MA. Biomechanical considerations of animal models used in tissue engineering of bone. Biomaterials. 2004;25:1697–714.

33. Cheung WH, Miclau T, Chow SK, Yang FF, Alt V. Fracture healing in osteoporotic bone. Injury. 2016;47(Suppl 2):S21–6.

34. Karladani AH, Granhed H, Kärrholm J, Styf J. The influence of fracture etiology and type on fracture healing: a review of 104 consecutive tibial shaft fractures. Arch Orthop Trauma Surg. 2001;121:325–8.

35. Schlundt C, et al. Macrophages in bone fracture healing: their essential role in endochondral ossification. Bone. 2018;106:78–89.

36. Mathavan N, Raina DB, Tägil M, Isaksson H. Longitudinal in vivo monitoring of callus remodeling in BMP-7- and zoledronate-treated fractures. J Orthop Res. 2020;

37. Oryan A, Moshiri A, Meimandi-parizi AH. Effects of sodium-hyaluronate and glucosamine-chondroitin sulfate on remodeling stage of tenotomized superficial digital flexor tendon in rabbits: a clinical, histopathological, ultrastructural, and biomechanical study. Connect Tissue Res. 2011;52:329–39.

38. Puzas JE, O'Keefe RJ, Schwarz EM, Zhang X. Pharmacologic modulators of fracture healing: the role of cyclooxygenase inhibition. J Musculoskelet Neuronal Interact. 2003;3:308–12. discussion 320–301

39. Marongiu G, Dolci A, Verona M, Capone A. The biology and treatment of acute long-bones diaphyseal fractures: overview of the current options for bone healing enhancement. Bone Rep. 2020;12:100249.

40. Sen C, Prasad J. Exploring conditions that make cortical bone geometry optimal for physiological loading. Biomech Model Mechanobiol. 2019;18:1335–49.

41. Arvidson K, et al. Bone regeneration and stem cells. J Cell Mol Med. 2011;15:718–46.
42. Schroeder JE, Mosheiff R. Tissue engineering approaches for bone repair: concepts and evidence. Injury. 2011;42:609–13.
43. Sandberg MM, Aro HT, Vuorio EI. Gene expression during bone repair. Clin Orthop Relat Res. 1993:292–312.
44. Valiya Kambrath A, Williams JN, Sankar U. An improved methodology to evaluate cell and molecular signals in the reparative callus during fracture healing. J Histochem Cytochem. 2020;68:199–208.
45. Keramaris NC, Calori GM, Nikolaou VS, Schemitsch EH, Giannoudis PV. Fracture vascularity and bone healing: a systematic review of the role of VEGF. Injury. 2008;39(Suppl 2):S45–57.
46. Nakajima F, et al. Spatial and temporal gene expression in chondrogenesis during fracture healing and the effects of basic fibroblast growth factor. J Orthop Res. 2001;19:935–44.
47. Majidinia M, Sadeghpour A, Yousefi B. The roles of signaling pathways in bone repair and regeneration. J Cell Physiol. 2018;233:2937–48.
48. Fukumoto LE, Fukumoto KD. Fat Embolism Syndrome. Nurs Clin North Am. 2018;53:335–47.
49. Altizer L. Compartment syndrome. Orthop Nurs. 2004;23:391–6.
50. Runkel M, Rommens PM. Pseudoarthrosis. Unfallchirurg. 2000;103:51–63. quiz 63
51. Dürr W. Sudeck's disease after radius fracture. Langenbecks Arch Chir Suppl II Verh Dtsch Ges Chir. 1990:693–9.
52. Jha S, Blau JE, Bhattacharyya T. Normal and delayed fracture healing: symphony and cacophony. Horm Metab Res. 2016;48:779–84.
53. Weber B, et al. Systemic and cardiac alterations after long bone fracture. Shock. 2020;54(6):761–73.
54. Szczęsny G. Fracture healing and its disturbances. A Literature Review. Ortop Traumatol Rehabil. 2015;17:437–54.
55. Longo UG, et al. Tissue engineered strategies for pseudoarthrosis. Open Orthop J. 2012;6:564–70.

The Osteochondral Unit

7

Tomoyuki Nakasa and Nobuo Adachi

7.1 Introduction

With the development of various technologies, the concept of "joint preservation" has become increasingly important in the orthopedic field. The strategies for joint preservation aim to achieve two things: one is to prevent or delay development of osteoarthritis (OA), and the other is to treat joints which have already developed OA, to preserve or restore joint function. The articular joint is composed of various structures such as articular cartilage, bone, synovium, ligament, and meniscus, which are responsible for joint function. The main functions of the joint are to achieve smooth movement and to support weight-bearing. Articular cartilage plays a crucial role in these functions. Homeostasis of articular cartilage is maintained by the subchondral bone which underlies the articular cartilage. The interaction between articular cartilage and subchondral bone allows for joint articulation and weight-bearing. Thus, this structure of articular cartilage and subchondral bone, known as "osteochondral unit," should be well understood for joint preservation. Severe damage and degeneration of the osteochondral unit causes limited joint function. Therefore, the most important strategic factor for joint preservation is the restoration of the osteochondral unit. There are many reports of therapeutic strategies, which focused only on the repair of the articular cartilage. However, repair of the osteochondral unit including the subchondral bone is necessary for a good outcome because the subchondral bone plays an important role in articular cartilage function.

7.1.1 Structure of the Osteochondral Unit

The osteochondral unit is composed of hyaline cartilage and subchondral bone. Articular cartilage plays a role in both the absorption of stress and the lubrication of the articular surface. To enable this function, articular cartilage has a unique structure. It is composed of a small number of chondrocytes, has an extracellular matrix (ECM) containing collagens and proteoglycans, and it does not have nerve, blood, and lymphatic vessels. Chondrocytes account for 2% of the total volume of articular cartilage, and mature chondrocytes are responsible for cartilage metabolism including the synthesis and degradation of proteoglycans and collagens. The chondrocyte metabolism is influenced by systemic or local factors such as inflammation and mechanical stress, and cartilage degeneration occurs when the balance between synthesis and degradation is lost.

T. Nakasa (✉) · N. Adachi
Department of Orthopaedic Surgery, Graduate School of Biomedical and Health Sciences, Hiroshima University, Hiroshima, Japan

© ISAKOS 2022

A. Gobbi et al. (eds.), *Joint Function Preservation*, https://doi.org/10.1007/978-3-030-82958-2_7

Articular cartilage has multiple layers within four zones: a superficial/tangential zone, a transitional zone, a deep/radial zone, and a calcified zone. The superficial layer contains "lamina splendens" which is a network of collagen fibers running parallel to the joint surface. There is a glycoprotein called lubricin, which plays an important role in maintaining lubrication. In the deeper part of the superficial zone, known as the tangential zone, chondrocytes are flat, the matrix fibers are parallel to the joint surface, and the content of the proteoglycans is low. In the transitional zone, the matrix fibers are reticulated and become deeply vertical. The morphology of chondrocytes is rounded and large. The radial zone is the major part of the articular cartilage, and chondrocytes here are spherical and numerous. The chondrocytes in the transitional and radial zones have a well-developed rough endoplasmic reticulum and Golgi apparatus, and actively produce cartilage matrix. The radial zone and calcified zone have a boundary between them called a tidemark. The articular cartilage is connected by the calcified cartilage to the subchondral bone, and this calcified cartilage boundary is known as a cement line. This cement line boundary has no crossing collagen fibrils, which makes it the weakest point in the osteochondral unit. Calcified cartilage enables continuous remodeling, which contributes to the natural healing of the base of the cartilage.

Subchondral bone is composed of subchondral spongiosa and a subchondral bone plate which is a metaphyseal trabecular bone. The subchondral bone plate has small holes through which blood vessels penetrate the calcified layer from the subchondral spongiosa. These blood vessels feed the chondrocytes in the calcified layer. One of the most important functions of the subchondral bone is to absorb the load in cooperation with the articular cartilage, and it has a structure that is more effective than the articular cartilage at absorbing loads. This structure of the osteochondral unit with its multiple layers enables to transmission and distribution of the force required for the mechanical adaptation to the joint (Fig. 7.1).

7.1.2 Interaction between Articular Cartilage and Subchondral Bone

In the osteochondral unit, articular cartilage and subchondral bone have crosstalk to maintain homeostasis. The osteochondral unit plays a role in the load-bearing capacity of the joint. Subchondral bone supports the cartilage by distributing the joint force. About 30% of the load is absorbed by normal subchondral bone and 1–3% of that is absorbed by cartilage [1]. Subchondral bone has a greater shock-absorbing ability than cartilage, and it works as the primary shock absorber to support cartilage [2]. However, damage such as microfractures of the subchondral bone changes the elasticity of the bone by abnormal remodeling, which no longer acts as a shock absorber and subsequently leads to cartilage degeneration [3, 4]. The osteochondral unit fully functions as a physiological shock absorber.

Subchondral bone also plays an important role in cartilage nutrition. Cartilage is nourished in two different ways. The superficial zone of cartilage mainly gets nutrition through diffusion via the synovial fluid. On the other hand, vascularity from the subchondral bone nourishes the deep and calcified layer [5–7]. The arteriovenous complex and nerve penetrate the subchondral bone through the canals, which play an important role in the healing of calcified cartilage. These vessels are observed more in the load-bearing areas of the articular cartilage than in other areas [8]. However, nutrients can diffuse to uncalcified cartilage from the subchondral bone [5, 9]. Vessels at the highly load-bearing area provide high blood flow containing nutrients by responding to non-physiological loads, which enable natural healing. However, excessive load on the degenerated cartilage inhibits the supply of nutrients from the subchondral bone to the cartilage [10]. Larger molecules are transported through the canalicular/lacunar network [11, 12]. It is important to keep the subchondral bone in good condition for cartilage nutrition.

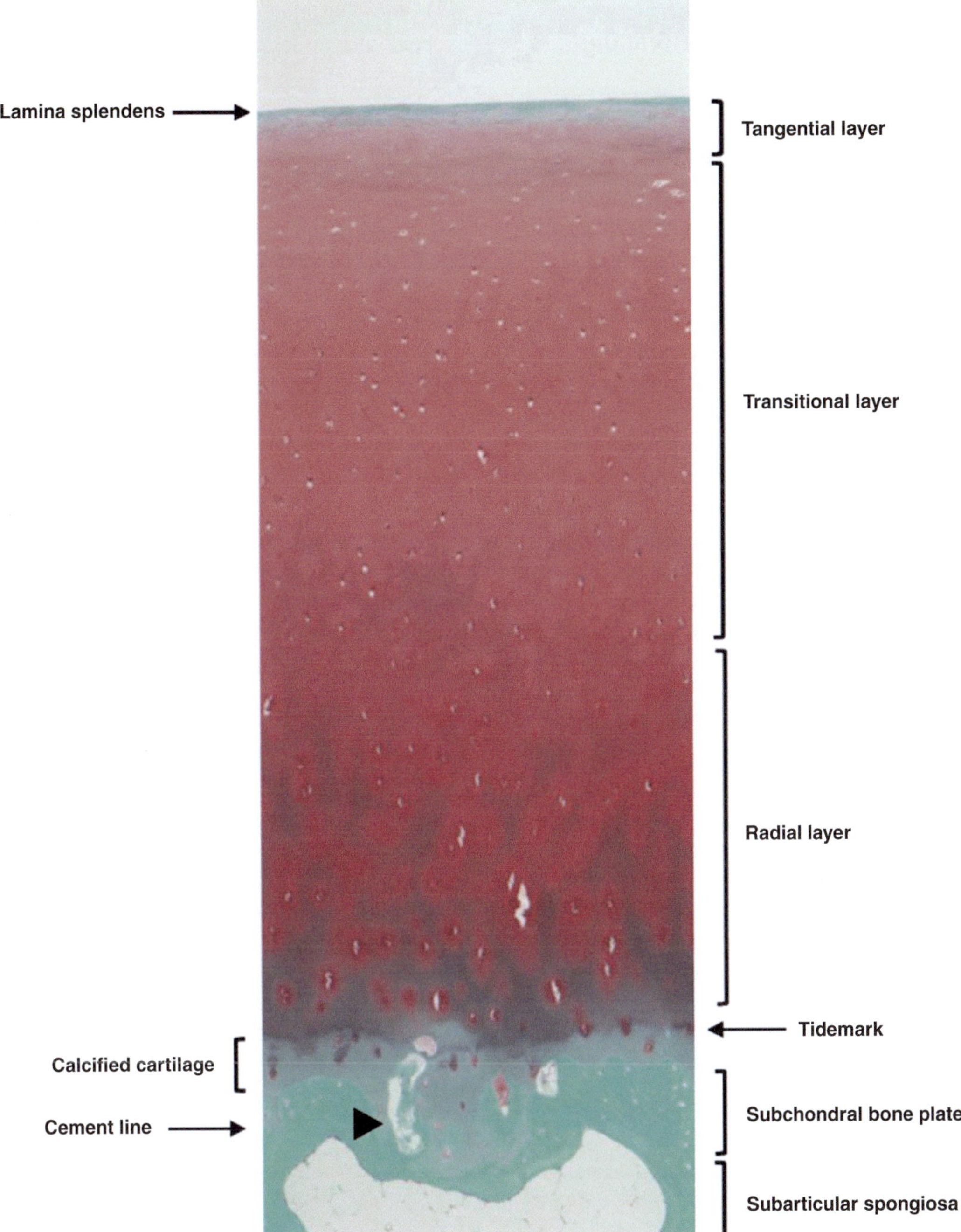

Fig. 7.1 The structure of normal articular cartilage. Arrow head indicates vessel from the subchondral bone into the calcified zone

7.1.3 Pathological Changes of the Osteochondral Unit

Pathological changes of the osteochondral unit include cartilage defect, osteochondral lesion, and osteoarthritis (OA). As for the focal cartilage defect, loading of the focal cartilage defect results in abnormal loading on the surrounding cartilage, which leads to expansion of the cartilage degeneration [13–15]. Degeneration of the surrounding cartilage due to breakdown of the osteochondral unit could be a risk factor for OA progression in the whole joint [16, 17]. In addition, change in the subchondral bone such as sclerosis and cyst formation occur in the focal cartilage defect area. To maintain the function of the osteochondral unit, treatment should achieve the coverage of the subchondral bone with a cartilage layer.

Osteochondral lesion (OCL) including osteochondritis dissecans is a disorder in which the osteochondral fragment is typically detached from its underlying bone. Although the precise mechanism has not been completely elucidated, the mechanism of the progression of OCL has been proposed for the ankle joint. Trauma to the articular cartilage simultaneously damages the subchondral bone by causing injuries such as a micro-fracture and bone bruise. Impaired healing of the subchondral bone may cause the highly pressurized fluid flow through the damaged subchondral bone plate into the subchondral bone to be intermittent. The continuous high fluid pressure can lead to osteonecrosis and bone resorp-tion by activating osteoclasts, subsequently inducing a lytic lesion [18–20]. Decreasing the fluid pressure makes bone resorption stop, which results in bone remodeling around the lytic area. Excessive osteogenesis may result in dense bone and sclerotic change, which inhibits spontaneous bone union between the osteochondral fragment and its underlying bone [21] (Fig. 7.2). Moreover, these subchondral bone changes mean that cartilage homeostasis is no longer maintained, which leads to cartilage degeneration in the osteochondral fragment. According to these mechanisms, the lesion condition including cartilage degeneration can be predicted from the subchondral bone condition on CT images, and appropriate treatment is available [22, 23]. For an unstable lesion, cases with good cartilage condition are fixed [24] if the cartilage degeneration is severe, the osteochondral unit should be replaced such as an osteochondral graft.

Osteochondral unit change in OA is quite complicated. Although various factors are involved in the pathogenesis of OA, alteration of the subchondral bone plays an important role in the progression of OA. In the early phase of OA, while fissure of the articular cartilage extends down to the subchondral bone, the function of the channels through the subchondral bone plate into the non-calcified zone deteriorates due to increased osteoclast activity. Fissures from the articular cavity reach down to the subchondral bone plate, and this enables the continuous flow of joint fluid, cell, and cytokines to the subchon-

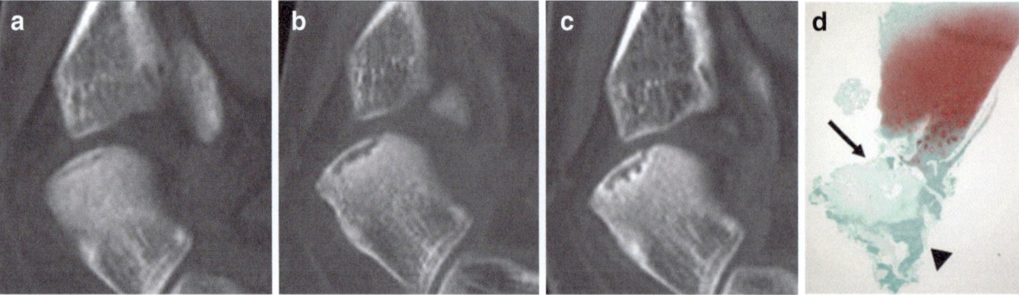

Fig. 7.2 Subchondral bone change during the progression of the osteochondral lesion. (**a**) One month after injury. (**b**) Three months after injury. The fissure between the osteochondral fragment and its underlying bone is enlarged. (**c**) Five months after injury. Sclerotic change of the underlying bone occurs. (**d**) Needle biopsy specimen of the osteochondral fragment and underlying bone with Safranin O staining. Arrow indicates separation site. Arrow head indicates the sclerotic change of the underlying bone

dral bone. These transmissions induce structural changes in the cartilage layers. This disruption of the osteochondral barrier in OA progression occurs in these processes, which causes structural change of the osteochondral junction. Calcification occurs in the non-calcified zone known as the duplication of the tidemark. Beneath the tidemark, ossification and thickening of the subchondral bone plate occur. Angiogenesis at the osteochondral junction occurs accompanying the sensory and sympathetic nerves. Neuropeptides contain sensory nerves which activate osteogenesis, and this leads to subchondral bone sclerosis. Sclerotic change of the subchondral bone alters plasticity, which means that cartilage is no longer able to withstand load-bearing, resulting in cartilage degeneration [25].

In OA progression, prostaglandins, leukotrienes, and growth factors are released by osteoblasts in the subchondral bone, and they reach down into the articular cartilage. Inflammatory and osteoblast stimulation factors released by chondrocytes also induce subchondral bone change including sclerosis [26, 27]. In the early stage of OA, the thickness of the subchondral bone plate decreases and osteoporotic changes progress due to the increase in bone remodeling and vascularity caused by microdamage of the subchondral bone. Microdamage such as microcracks allow catabolic agents to cross the osteochondral junction [28]. In late-stage OA, the thickness of the subchondral bone plate increases due to a high rate of bone turnover induced by microcracks [29, 30]. A high rate of bone turnover also induces calcified cartilage thickness and duplication of the tidemark, which lead to cartilage thinning and degeneration [31]. Subchondral bone change plays a vital role in the pathogenesis of OA (Fig. 7.3).

7.1.4 Treatment of the Osteochondral Unit

7.1.4.1 Non-surgical Treatment

Pain is the first and most frequent phenomenon associated with the abnormalities of the osteochondral unit. Since healthy articular cartilage has no nociceptors, joint pain is mainly recognized by the subchondral bone, joint capsule, ligaments, periosteum, and synovium [32]. Sensory nerves containing neuropeptides are distributed in the joint, and some stimuli irritate the sensory nerve endings, which leads to pain caused by the release of prostaglandins and cytokines [33, 34]. Neuropeptides, such as substance P (SP), calcitonin gene-related peptide (CGRP), and vasoactive intestinal peptide (VIP), are small peptides which play an important role as neurotransmitters of pain. There is an increase of neuropeptides in the subchondral bone, as well as of inflammatory cytokines such as COX2 and TNFα which are part of the OA pathogenesis. Both of these increased factors are a source of OA pain [35]. However, neuropeptides play various roles in maintaining joint homeostasis and tissue repair [36]. Neuropeptides play an important role in bone metabolism and vascularization in the subchondral bone, and these are involved in the pathogenesis of OA [37–40]. In the pathological change of OA, sensory nerve ingrowth occurs in subchondral bone, which induces the pain as well as structural change. It is reported that SP, CGRP, and VIP promote osteogenesis, which may affect the subchondral bone sclerosis during OA progression [41, 42] (Fig. 7.4). These neuropeptides are an obvious target in the treatment of OA progression and pain reliever [41–43]. NGF mAb can alleviate pain in severe OA patients [44]. However, adverse effects such as the rapid progression of OA and osteonecrosis have been reported in clinical trials [45]. Biologic agents against NGF such as fasinumab, tanezumab, and fulranumab have been administered to OA patients to alleviate OA-related pain with good results [46, 47].

During OA progression, osteoclasts induce sensory nerve axonal growth in the subchondral bone [48]. Therefore, osteoclasts could also be targeted in OA treatment. Bisphosphonate targeting of osteoclasts is used to treat osteoporosis by inhibiting bone absorption. Since osteoclasts in the subchondral bone are involved in the progression of OA, bisphosphonates are administered to slow down bone remodeling and to hopefully provide chondroprotection [49]. Several animal

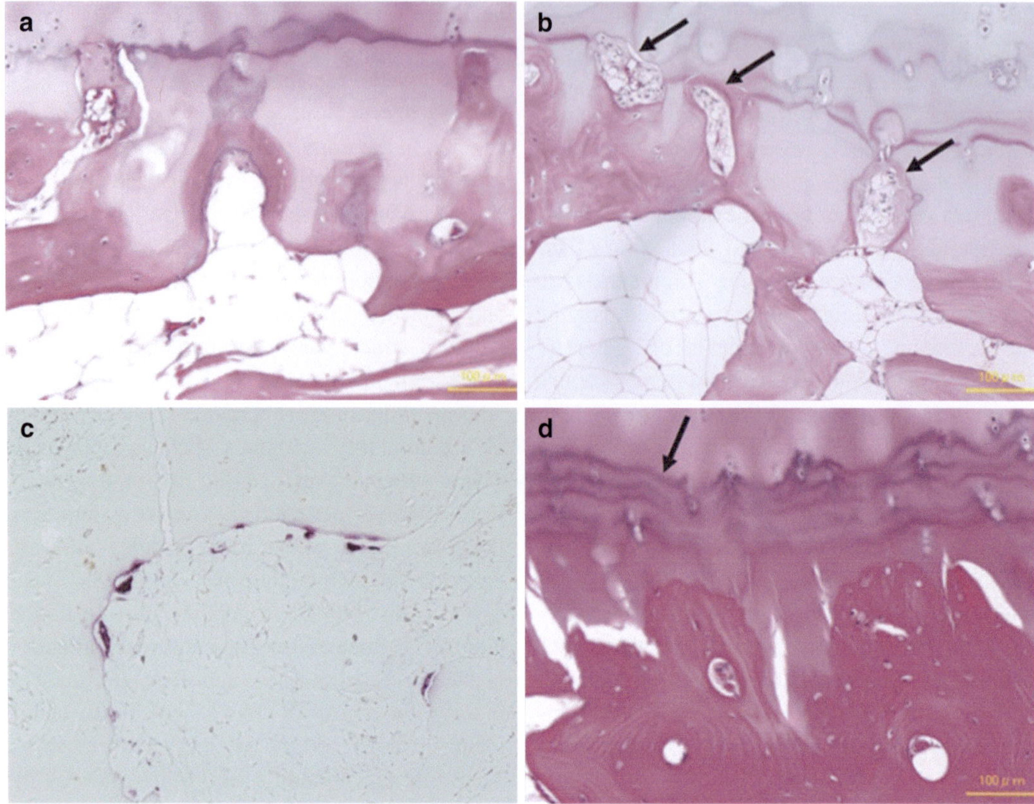

Fig. 7.3 Histological changes of the osteochondral junction in osteoarthritis (OA). (**a**) Thinning of the subchondral bone plate in the early phase of OA. (**b**) Invasion of vessels form the subchondral bone to the cartilage layer. Arrows indicate vessels. (**c**) TRAP staining in the subchondral bone in the early phase of OA. Osteoclasts increase in the subchondral bone. (**d**) Thickened subchondral bone and narrowing of the marrow cavity at end-stage OA. Arrow indicates double tide mark

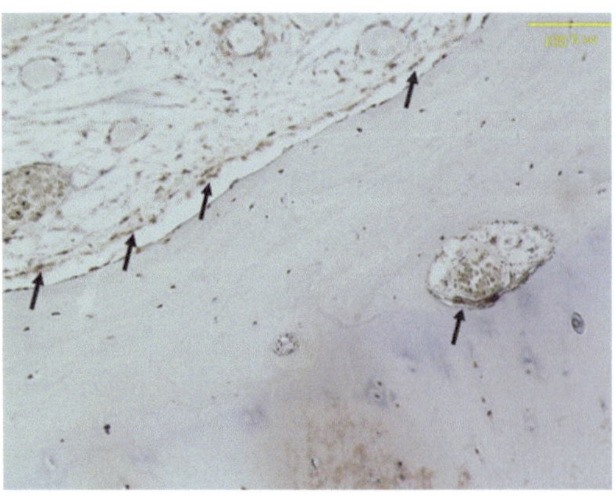

Fig. 7.4 Expression of CGRP in the subchondral bone plate in OA progression. Arrows indicate CGRP

studies have shown that bisphosphonates such as alendronate and zoledronate could attenuate OA progression by the protection of subchondral bone loss and cartilage degeneration [50, 51]. There have been several clinical trials of the administration of bisphosphonates for OA patients. According to these reports, bisphosphonates can reduce various factors: pain, biochemical markers of cartilage degeneration, and bone marrow lesions on MRI [52–56]. However, it is difficult to attenuate the structural deterioration such as joint space narrowing. Hence, the usage of bisphosphonates for the treatment of OA remains controversial.

7.1.5 Surgical Treatment of the Osteochondral Unit

Surgical procedures should be determined based on the specific location in osteochondral unit such as the cartilage layer, subchondral bone, or both. An autologous bone graft is used for a subchondral bone lesion such as a cystic lesion. For a cartilage defect, it is necessary to cover the defect in order to repair the osteochondral unit. Implantation of hyaline-like tissue such as autologous chondrocyte implantation or a similar modified procedure has been performed. A bone marrow stimulation technique such as microfracture can improve the subchondral bone condition by inducing bone remodeling. In addition, this technique induces coverage of the cartilage defect with fibrocartilage. The important thing to keep in mind is the goal of repairing the whole osteochondral unit.

7.1.5.1 Bone Marrow Stimulation

As first-line treatment techniques for the coverage of a full-thickness cartilage defect, a bone marrow stimulation technique such as subchondral bone drilling, abrasion arthroplasty, and microfracture have been commonly performed [57–59]. Although it is a relatively easy technique and good to excellent results in 60–80% of patients have been reported, long-term results were worse due to the durability of the reparative tissue [59–62]. The purpose of bone marrow stimulation technique is to fill the cartilage defect with reparative tissue from the subchondral bone. A clot within the mesenchymal stem cells can enter through a perforation hole in the subchondral bone, depositing itself into the cartilage defect, then gradually differentiating into the fibrocartilaginous tissue [59]. However, it is unclear whether a blood clot infiltrated from the subchondral bone remains in the defect site until it changes into repaired tissue under joint motion or friction stress. Previous animal studies have revealed the early-phase changes in the osteochondral unit histologically after microfracture. In that study, blood clots filled in the microfracture holes but they were not in the cartilage defect. The number of TRAP-positive cells increased until 3 days, and then decreased and localized at the active remodeling site. The microfracture hole diameter became larger until day 14, and then, most holes disappeared by day 28. These histological analyses suggested that cartilage repair by microfracture results in endochondral ossification within the deeper zone of the microfracture holes [63]. It is noteworthy that intralesional osteophytes, which thicken the subchondral bone plate and subchondral bone cyst after microfracture, possibly lead to the deterioration and failure [60, 62]. Excessive bone remodeling of the subchondral bone after microfracture may induce intralesional osteophytes, and the porotic condition of the subchondral bone or the non-appropriate use of a device to create larger diameter holes may induce subchondral bone cysts. It is important to perform a microfracture according to the subchondral bone condition, in order to avoid adverse effects.

Due to concerns about the property of the covered fibrous tissue such as fragility after microfracture, an autologous matrix-induced chondrogenesis (AMIC) has been developed, which combines microfracture with coverage by a collagen membrane [64]. A collagen membrane can be expected to function as a scaffold and to achieve stabilization of the fibrin clot which contains MSC from the bone marrow. Good clinical results of AMIC have been reported [65, 66]. However, there is not enough evidence to prove that AMIC is effective compared to the established procedures [67].

7.1.5.2 Autologous Chondrocyte Implantation

Since the initial report in 1994, autologous chondrocyte implantation (ACI) has been performed to replace damaged cartilage with chondrocytes [68]. This procedure has the advantage of being repaired by hyaline-like cartilage. In the first generation of ACI, healthy articular cartilage is harvested from the non-weight-bearing area, and chondrocytes are isolated by enzymic digestion, then subsequently cultured in a monolayer for 4 weeks. Proliferated chondrocytes are transplanted into the cartilage defect under the periosteum patch. Although this procedure has provided good clinical results, there are several issues which need to be addressed. First, the two-step surgical technique, whereby chondrocytes are harvested and then implanted, is not ideal. Second, the isolation of chondrocytes by enzymic digestion is problematic since it is harmful to cells. Moreover, a periosteum should be harvested to patch the defect site. A periosteum patch has been replaced by the collagen membrane to avoid harvesting a periosteum. Recently, a traditional ACI technique, the so-called matrix-assisted chondrocyte implantation (MACI), was developed by combining the matrix- and tissue-engineering technique. This technique allows a three-dimensional culture of chondrocytes which provides clinical results superior those of traditional ACI [69]. To reconstruct the osteochondral unit, the effect of hyaline-like cartilage tissue on the subchondral bone and vice versa is an important issue. ACI is one of the options after a failed microfracture. However, the failure rate of ACI after microfracture is reported to be 3 to 8 times higher because a microfracture is detrimental to the condition of the subchondral bone for ACI [62, 70, 71].

The two step-surgery of ACI is streamlined into one-step surgery by performing the technique of, minced cartilage implantation. This involves cartilage fragments being loaded onto the scaffold and implanted into the cartilage defect without culture [72, 73]. An animal study using rabbits has shown that better subchondral bone healing occurs after minced cartilage implantation compared to healing after the ACI for the osteochondral defect as well as good cartilage repair. This suggests that minced cartilage secretes factors that induce bone and cartilage formation [74]. Interaction between the subchondral bone and implanted cartilage tissue should be analyzed in the future. Juvenile particulated cartilage allografts have been also available for cartilage defects, with good clinical outcomes in knee and ankle joints at the short-term follow-up have been reported [75, 76]. However, this procedure has some disadvantages including the potential risk of disease transmission, high cost, and the inability to preserve the product once open.

7.1.6 Future Direction

For cartilage repair, coverage of the cartilage defect by a bioabsorbable scaffold and biologics such as bone marrow aspirate concentrate is recommended as one-step surgery. Previous reports demonstrated that chondrocytes or multipotent cells have the potential for good cartilage repair, but these require much time and money, because this two-step surgery means that these cells require cell culture for expansion. Bone Marrow Aspirate Concentrate (BMAC) has attracted attention as a multipotent bone marrow cell source, which can be a one-step cell-based procedure for cartilage repair [77]. Good long-term clinical results from the one-step cartilage repair technique have been reported, using a scaffold and BMAC for a full-thickness cartilage defect. The number of one-step procedure reports using biologics is expected to increase in the future [78]. Considering the interaction (crosstalk) between cartilage and subchondral bone, targeting the subchondral bone for the repair of the osteochondral unit will increase. Subchondroplasty is used to repair subchondral bone lesions such as bone marrow lesions. This procedure fills the subchondral region with calcium phosphate [79, 80]. Biological remodeling of the subchondral bone requires, biologics such as BMAC and PRP to be injected into the subchondral bone region to enhance the repair of the osteochondral unit [81]. Less invasive techniques for biological repair of the osteochondral unit will be developed in the future. Moreover, as tissue-engineering develops, an osteochondral plug could be created using a tissue-engineering technique [82]. The osteochondral unit should be considered as a unit of the treatment for joint preservation.

7.2 Conclusion

The osteochondral unit should be a focus of joint preservation. Cartilage and subchondral bone maintain a communicative crosstalk within the osteochondral unit and support the homeostasis.

In terms of pathological change, interaction between cartilage and subchondral bone occurs after cartilage injury or during cartilage degeneration. Therefore, we should carefully consider the subchondral bone condition in the treatment of cartilage, with the aim of restoring the function of osteochondral unit for the joint preservation.

References

1. Madry H, van Dijk CN, Mueller-Gerbl M. The basic science of the subchondral bone. Knee Surg Sports Traumatol Arthrosc. 2010;18(4):419–33.
2. Malekipour F, Whitton C, Oetomo D, Lee PVS. Shock absorbing ability of articular cartilage and subchondral bone under impact compression. J Mech Behav Biomed Mater. 2013;26:127–35.
3. Ma JX, He WW, Zhao J, et al. Bone microarchitecture and biomechanics of the necrotic femoral head. Sci Rep. 2017;7(1):13345.
4. Lajeunesse D. The role of bone in the treatment of osteoarthritis. Osteoarthr Cartil. 2004;12:S34–8.
5. Pan J, Zhou X, Li W, Novotny JE, Doty SB, Wang L. In situ measurement of transport between subchondral bone and articular cartilage. J Orthop Res. 2009;27(10):1347–52.
6. Arkill KP, Winlove CP. Solute transport in the deep and calcified zone of articular cartilage. Osteoarthritis Cartilage. 2008;16(6):708–14.
7. Berry JL, Thaeler-Oberdoerster DA, Greenwald AS. Subchondral pathways to the superior surface of the human talus. Foot Ankle. 1986;7(1):2–9.
8. Imhof H, Breitenseher M, Kainberger F, Rand T, Trattnig S. Importance of subchondral bone to articular cartilage in health and disease. Top Magn Reason Imaging. 1999;10(3):180–92.
9. Lyons TJ, McClure SF, Stoddart RW, McClure J. The normal human chondro-osseous junctional region: evidence for contact of uncalcified cartilage with subchondral bone and marrow spaces. BMC Musculoskelet Disord. 2006;7:52.
10. Lane LB, Villacin A, Bullough PG. The vascularity and remodeling of subchondral bone and calcified cartilage in adult human femoral and humeral heads. An age- and stress-related phenomenon. J Bone Joint Surg Br. 1977;59(3):272–8.
11. Wang B, Zhou X, Price C, Li W, Pan J, Wang L. Quantifying load-induced solute transport and solute-matrix interaction within the osteocyte lacunar-canalicular system. J Bone Miner Res. 2013;28:1075–86.
12. Calvo E, Castaneda S, Largo R, Fernandez-Valle ME, Rodriguez-Salvanes F, Herrero-Beaumont G. Osteoporosis increases the severity of cartilage damage in the experimental model of osteoarthritis in rabbits. Osteoarthr Cartil. 2007;15(1):69–77.
13. Guettler JH, Demetropoulos CK, Yang KH, Jurist KA. Osteochondral deflects in the human knee – influence of defect size on cartilage rim stress and load redistribution to surrounding cartilage. Am J Sports Med. 2004;32:1451–8.
14. Braman JP, Bruckner JD, Clark JM, Norman AG, Chansky HA. Articular cartilage adjacent to experimental defects is subject to atypical strains. Clin Orthop Relat Res. 2005;430:202–7.
15. Gratz KR, Wong BL, Bae WC, Sah RL. The effects of focal articular defects on cartilage contact mechanics. J Orthop Res. 2009;27:584–92.
16. Segal NA, Anderson DD, Iyer KS, Baker J, Torner JC, Lynch JA, Felson DT, Lewis CE, Brown TD. Baseline articular contact stress levels predict incident symptomatic knee osteoarthritis development in the MOST cohort. J Orthop Res. 2009;27:1562–8.
17. Houck DA, Matthew BA, Kraeutler J, Belk JW, Frank RM, McCarty EC, Bravman JT. Do focal chondral defects of the knee increase the risk for progression to osteoarthritis? A review of the literature. Orthop J Sports Med. 2018;6(10):1–8.
18. Johansson L, Edlund U, Fahlgren A, Aspenberg P. Bone resorption induced by fluid flow. J Biomech Eng. 2009;131(9):094505.
19. Radin EL, Rose RM. Role of subchondral bone in the initiation and progression of cartilage damage. Clin Orthop Relat Res. 1986;213:34–40.
20. Yonetani Y, Nakamura N, Natsuume T, Shiozaki Y, Tanaka Y, Horibe S. Histological evaluation of juvenile osteochondral dissecans of the knee: a case series. Knee Surg Sports Traumatol Arthrosc. 2010;18:723–30.
21. van Dijk CN, Reilingh ML, Zengerink M, van Bergen CJA. Osteochondral defects in the ankle: why painful? Knee Surg Sports Traumatol Arthrosc. 2010;18:570–80.
22. Nakasa T, Adachi N, Kato T, Ochi M. Appearance of subchondral bone in computed tomography is related to cartilage damage in osteochondral lesions of the talar dome. Foot Ankle Int. 2014;35(6):600–6.
23. Nakasa T, Ikuta Y, Yoshikawa M, Sawa M, Tsuyuguchi Y, Adachi N. Added value of preoperative computed tomography for determining cartilage degeneration in patients with osteochondral lesions of the talar dome. Am J Sports Med. 2018;46(1):208–16.
24. Nakasa T, Ikuta Y, Ota Y, Kanemitsu M, Adachi N. Clinical results of bioabsorbable pin fixation relative to the bone condition for osteochondral lesion of the talus. Foot Ankle Int. 2019;40(12):1388–96.
25. Castaneda S, Roman-Blas JA, Largo R, Herrero-Beaumont G. Subchondral bone as a key target for osteoarthritis treatment. Biochem Pharmacol. 2012;83:315–23.
26. Lajeunesse D, Reboul P. Subchondral bone in osteoarthritis: a biologic link with articular cartilage leading to abnormal remodeling. Curr Opin Rheumatol. 2003;15:628–33.
27. Bellido M, Lugo L, Roman-Blas JA, Castaneda S, Caeiro JR, Dapia S, Calvo E, Large R, Herrero-

Beaumont G. Subchondral bone microstructural damage by increased remodeling aggravates experimental osteoarthritis preceded by osteoporosis. Arthritis Res Ther. 2010;12:R152.

28. Burr DB, Gallant MA. Bone remodeling in osteoarthritis. Nat Rev Rheumatol. 2012;8:665–73.

29. Li G, Yin J, Gao J, Cheng TS, Palvos NJ, Zhang C, Zheng MH. Subchondral bone in osteoarthritis: insight into risk factors and microstructural changes. Arthritis Res Ther. 2013;15:223.

30. Kawcak CE, Mcllwraith CW, Norrdin RW, Park RD, James SP. The role of subchondral bone in joint disease: a review. Equine Vet J. 2001;33:120–6.

31. Goldring SR. Role of bone in osteoarthritis pathogenesis. Med Clin North Am. 2009;93:25–35.

32. Grässel SG. The role of peripheral nerve fibers and their neurotransmitters in cartilage and bone physiology and pathophysiology. Arthritis Res Ther. 2014;16:485.

33. Dirmeier M, Capellino S, Schubert T, Angele P, Anders S, Straub RH. Lower density of synovial nerve fibres positive for calcitonin gene-related peptide relative to substance P in rheumatoid arthritis but not in osteoarthritis. Rheumatology. 2008;47:36–40.

34. Wong HY. Neutral mechanisms of joint pain. Ann Acad Med Singap. 1993;22:646–50.

35. Ogino S, Sasho T, Nakagawa K, et al. Detection of pain-related molecules in the subchondral bone of the osteoarthritis knees. Clin Rheumatol. 2009;28:1395–402.

36. Gatenholm B, Brittberg M. Neuropeptides: important regulators of joint homeostasis. Knee Surg Sports Traumatol Arthrosc. 2019;27:942–9.

37. Jones KB, Mollano AV, Morcuende JA, et al. Bone and brain: a review of neural, hormonal, and musculoskeletal connections. Iowa Orthop J. 2004;24:123–32.

38. Qiao Y, Wang Y, Zhou Y, et al. The role of nervous system in adaptive response of bone to mechanical loading. J Cell Physiol. 2019;234:7771–80.

39. Hukkanen M, Kreicbergs A, Brodin E, et al. Substance P- and CGRP-immunoreactive nerves in bone. Peptides. 1988;9:165–71.

40. Fukuda T, Takada S, Xu R, et al. Sema3A regulates bone-mass accrual through sensory innervations. Nature. 2013;497:490–3.

41. Nakasa T, Ishikawa M, Takada T, Miyaki S, Ochi M. Attenuation of cartilage degeneration by calcitonin gene-related peptide receptor antagonist via inhibition of subchondral bone sclerosis in osteoarthritis mice. J Orthop Res. 2016;34(7):1177–84.

42. Kanemitsu M, Nakasa T, Shirakawa Y, Ishikawa M, MIyaki S, Adachi N. Role of vasoactive intestinal peptide in the progression of osteoarthritis through bone sclerosis and angiogenesis in subchondral bone. J Orthop Sci. 2020; In press

43. Benschop RJ, Collins EC, Darling RJ, et al. Development of a novel antibody to calcitonin gene-related peptide for the treatment of osteoarthritis-related pain. Osteoarthr Cartil. 2014;22:578–85.

44. Lane NE, Schnitzer TJ, Birbara CA, Mokhtarani M, Shelton DL, Smith MD, Brown MT. Tanezmab for the treatment of pain from osteoarthritis of the knee. N Engl J Med. 2010;363(16):1521–31.

45. Mullard A. Drug developers reboot anti-NGF pain programmes. Nat Rev Drug Discov. 2015;14(5):297–8.

46. Schnitzer TJ, Ekman EF, Spierings EL, Greenberg HS, Smith MD, Brown MT, West CR, Verburg KM. Efficacy and safety of tanezumab monotherapy or combined with non-steroidal anti-inflammatory drugs in the treatment of knee or hip osteoarthritis pain. Ann Rheum Dis. 2015;74:1202–11.

47. Hochberg MC, Tive LA, Abramson SB, Vignon E, Verburg KM, West CR, Smith MD, Hungerford DS. When is osteonecrosis not osteonecrosis?: Adjudication of reported serious adverse joint events in the tanezmub clinical development program. Arthritis Rheumatol. 2016;68:382–291.

48. Zhu S, Zhu J, Zhen G, Hu Y, An S, Li Y, Zheng Q, Chen Z, Yang Y, et al. Subchondral bone osteoclasts induce sensory innervation and osteoarthritis pain. J Clin Invest. 2019;129(3):1076–93.

49. Davis AJ, Smith TO, Hing CB, Sofat N. Are bisphosphonates effective in the treatment of osteoarthritis pain? A meta-analysis and systemic review. PLoS One. 2013;8:e72714.

50. Zhu S, Chen K, Lan Y, Zhang N, Jiang R, Hu J. Alendronate protects against articular cartilage erosion by inhibiting subchondral bone loss in ovariectomized rats. Bone. 2013;53:340–9.

51. Strassle BW, Mark L, Leventhal L, Piesla MJ, Jian Li X, Kennedy JD, Glasson SS, Whiteside GT. Inhibition of osteoclasts prevents cartilage loss and pain in rat model of degenerative joint disease. Osteoarthr Cartil. 2010;18:1319–28.

52. Bingham CO III, Buckland-Wright JC, Garnero P, Cohen SB, Dougados M, Adami S, Clauw DJ, Spector TD, Pelletier JP, Raynauld JP, et al. Risedronate decreases biochemical markers of cartilage degradation but does not decrease symptoms or slow radiographic progression in patients with medial compartment osteoarthritis of the knee: results of the two-year multinational knee osteoarthritis structural arthritis study. Arthritis Rheum. 2006;54:3494–507.

53. Laslett LL, Dore DA, Quinn SL, Boon P, Ryan E, Winzenberg TM, Jones G. Zoledronic acid reduces knee pain and bone marrow lesions over 1 year: a randomized controlled trial. Ann Rheum Dis. 2012;71:1322–8.

54. Xing RL, Zhao LR, Wang PM. Bisphosphonates therapy for osteoarthritis: a meta-analysis of randomized controlled trials. Springerplus. 2016;5:1704.

55. Valenti MT, Mottes M, Biotti A, Perduca M, Pisani A, Bovi M, Deiana M, Cheri S, Dalle Carbonare L. Clodronate as a therapeutic strategy against osteoarthritis. Int J Mol Sci. 2017;18:2696.

56. Rossini M, Adami S, Fracassi E, Viapiana O, Orsolini G, Povino MR, Idolazzi L, Gatti D. Effects of intra-articular clodronate in the treatment of

knee osteoarthritis: results of a double-blind, randomized placebo-controlled trial. Rheumatol Int. 2015;35:255–63.

57. Pridie E. A method of resurfacing knee joints. J Bone Joint Surg Br. 1959;41:618–9.

58. Johnson LL. Arthroscopic abrasion arthroplasty: a review. Clin Orthop Relat Res. 2001;391:S306–17.

59. Steadman JR, Rodkey WG, Rodrigo JJ. Microfracture: surgical technique and rehabilitation to treat chondral defects. Clin Orthop Relat Res. 2001;391:S361–9.

60. Mithoefer K, Williams RJ III, Warren RF, Potter HG, Spock CR, Jones EC, Wickiewicz TL, Marx RG. The microfracture technique for the treatment of articular cartilage lesions in the knee. A prospective cohort study. J Bone Joint Surg Am. 2005;87:1911–20.

61. Knutsen G, Engebretsen L, Ludvigsen TC, Drogset JO, Grontvedt T, Solheim E, Strand T, Roberts S, Isaksen V, Johansen O. Autologous chondrocyte implantation compared with microfracture in the knee. A randomized trial. J Bone Joint Surg Am. 2004;86A:455–64.

62. Kreuz PC, Steinwachs MR, Erggelet C, Krause SJ, Konrad G, Uhl M, Sudkamp N. Results after microfracture of full-thickness chondral defects in different compartments in the knee. Osteoarthr Cartil. 2006;14:1119–25.

63. Hayashi S, Nakasa T, Ishikawa M, Nakamae A, Miyaki S, Adachi N. Histological evaluation of early-phase changes in the osteochondral unit after microfracture in a full-thickness cartilage defect rat model. Am J Sports Med. 2018;46(12):3032–9.

64. Benthien JP, Behrens P. The treatment of chondral and osteochondral defects of the knee with autologous matrix-induced chondrogenesis (AMIC): method description and recent developments. Knee Surg Sports Traumatol Arthrosc. 2011;19(8):1316–9.

65. Gottschalk O, Altenberger S, Baumbach S, Kricgclstcin S, Drcyer F, Mehlhorn A, et al. Functional medium-term results after autologous matrix-induced chondrogenesis for osteochondral lesions of the talus: a 5-year prospective cohort study. J Foot Ankle Surg. 2017;56:930–6.

66. Schiavone Panni A, Del Regno C, Mazzitelli G, D'Apolito R, Corona K, Vasso M. Good clinical results with autologous matrix induced chondrogenesis (Amic) technique in large knee chondral defects. Knee Surg Sports Traumatol Arthrosc. 2018;26(4):1130–6.

67. Gao L, Orth P, Cucchiarini M, Madry H. Autologous matrix-induced chondrogenesis: a systemic review of the clinical evidence. Am J Sports Med. 2019;47(1):222–31.

68. Brittberg M, Lindahl A, Nilsson A, Ohssson C, Isaksson O, Peterson L. Treatment of deep cartilage defects in the knee with autologous chondrocyte transplantation. N Engl J Med. 1994;331:889–95.

69. Kon E, Delcogliano M, Filardo G, Montaperto C, Marcacci M. Second generation issues in cartilage repair. Sports Med Arthrosc Rev. 2008;16:221–9.

70. Minas T, Gomoll AH, Rosenberger R, Royce RO, Bryant T. Increased failure rate of autologous chondrocyte implantation after previous treatment with marrow stimulation techniques. Am J Sports Med. 2009;37(5):902–8.

71. Pestka JM, Bode G, Salzmann G, Sudkamp NP, Niemeyer P. Clinical outcome of autologous chondrocyte implantation for failed microfracture treatment of full-thickness cartilage defects of the knee joint. Am J Sports Med. 2012;40(2):235–331.

72. Christensen BB, Foldager CB, Jensen J, Lind M. Autologous dual-tissue transplantation for osteochondral repair: early clinical and radiological results. Cartilage. 2015;6(3):166–73.

73. Cole BJ, Farr J, Winalski CS, et al. Outcomes after a single-stage procedure for cell-based cartilage repair: a prospective clinical safety trial with 2-year follow-up. Am J Sports Med. 2011;39:1170–9.

74. Matsushita R, Nakasa T, Ishikawa M, Tsuyuguchi Y, Matsubara N, MIyaki S, Adachi N. Repair of an osteochondral defect with minced cartilage embedded in atelocollagen gel. A rabbit model. Am J Sports Med. 2019;47(9):2216–24.

75. Farr J, Tabet SK, Margerrison E, Cole BJ. Clinical, radiographic, and histological outcomes after cartilage repair with particulated juvenile articular cartilage: a 2-year prospective study. Am J Sports Med. 2014;42:1417–25.

76. Buckwalter JA, Bowman GN, Albright JP, Wolf BR, Bollier M. Clinical outcomes of patellar chondral lesions treated with juvenile particulated cartilage allografts. Iowa Orthop J. 2014;34:44–9.

77. Gobbi A, Whyte GP. One-stage cartilage repair using a hyaluronic acid-based scaffold with activated bone marrow-derived mesenchymal stem cells compared with microfracture: five-year follow-up. Am J Sports Med. 2016;44(11):2846–54.

78. Gobbi A, Whyte GP. Long-term clinical outcomes of one-stage cartilage repair in the knee with hyaluronic acid-based scaffold embedded with mesenchymal stem cells sourced from bone marrow aspirate concentrate. Am J Sports Med. 2019;47(7):1621–8.

79. Sharkey PF, Cohen SB, Leinberry CF, et al. Subchondral bone marrow lesions associated with knee osteoarthritis. Am J Orthop (Belle Mead NJ). 2012;41:413–7.

80. Brimmo OA, Bozynski CC, Cook CR, et al. Subchondroplasty for the treatment of post-traumatic bone marrow lesions of the medial femoral condyle in a pre-clinical canine model. J Orthop Res. 2018;36:2709–17.

81. Oliver HA, Bozynski CC, Cook CR, Kuroki K, Sherman SL, Stoker AM, Cook JL. Enhanced subchondroplasty treatment for post-traumatic cartilage and subchondral bone marrow lesions in a canine model. J Orthop Res. 2019; In press

82. Hu X, Xu J, Li W, Li L, Parungao R, Wang Y, Zheng S, Nie Y, Liu T, Song K. Therapeutic "tool" reconstruction and regeneration of tissue engineering for osteochondral repair. Appl Biochem Biotechnol. 2019; In press

Diagnosis of Cartilage and Osteochondral Defect

Felipe Galvão Abreu, Renato Andrade,
André Tunes Peretti, Raphael F. Canadas,
Rui L. Reis, J. Miguel Oliveira,
and João Espregueira-Mendes

8.1 Introduction

Articular cartilage is a specialized connective tissue composed of hydrated proteoglycans within a matrix of collagen fibrils forming the load-bearing surfaces of all synovial joints. Its highly organized structure provides the biomechanical properties necessary for the tissue to withstand multiple forces created during movement [1–3]. The main functions of articular cartilage in synovial joints are to provide a low-friction surface for motion and to resist tensile, shear, and compressive forces [1, 4, 5]. Cartilage varies in specific composition within the same joint and between different joints, but it consists of the same basic components and structure throughout them all [1]. Composed primarily of water (65–80% of the wet weight), cells, and macromolecules, articular cartilage possesses the unique ability to absorb shock impacts, support heavy and repetitive loads, and withstand wear and tear over the course of a lifetime [6].

F. G. Abreu
Clínica Sportsmed, São José do Rio Preto, Brazil

Hospital Israelita Albert Einstein, São Paulo, Brazil

R. Andrade
Clínica Espregueira - FIFA Medical Centre of Excellence, Porto, Portugal

Dom Henrique Research Centre, Porto, Portugal

Porto Biomechanics Laboratory (LABIOMEP), Faculty of Sports, University of Porto, Porto, Portugal

A. T. Peretti
Clínica Sportsmed, São José do Rio Preto, Brazil

R. F. Canadas
ICVS/3B's–PT Government Associate Laboratory, Braga/Guimarães, Portugal

3B's Research Group—Biomaterials, Biodegradables and Biomimetics, University of Minho, European Institute of Excellence on Tissue Engineering and Regenerative Medicine, Guimarães, Portugal

R. L. Reis · J. M. Oliveira
ICVS/3B's–PT Government Associate Laboratory, Braga/Guimarães, Portugal

3B's Research Group—Biomaterials, Biodegradables and Biomimetics, University of Minho, European Institute of Excellence on Tissue Engineering and Regenerative Medicine, Guimarães, Portugal

The Discoveries Centre for Regenerative and Precision Medicine, University of Minho, Guimarães, Portugal

J. Espregueira-Mendes (✉)
Clínica Espregueira - FIFA Medical Centre of Excellence, Porto, Portugal

Dom Henrique Research Centre, Porto, Portugal

ICVS/3B's–PT Government Associate Laboratory, Braga/Guimarães, Portugal

3B's Research Group—Biomaterials, Biodegradables and Biomimetics, University of Minho, European Institute of Excellence on Tissue Engineering and Regenerative Medicine, Guimarães, Portugal

School of Medicine, University of Minho, Braga, Portugal
e-mail: espregueira@dhresearchcentre.com

© ISAKOS 2022
A. Gobbi et al. (eds.), *Joint Function Preservation*, https://doi.org/10.1007/978-3-030-82958-2_8

Macroscopically, articular cartilage appears as a white, smooth, homogeneous tissue, with the thickness varying, approximately from 2 to 5 mm. When probed, healthy cartilage is firm and resists deformation. Diseased cartilage, in the other hand, is soft, deforms when probed, and may contain visible surface disruptions and erosion. Following injury, articular cartilage has limited healing potential because the cells have minimal mitotic activity and the matrix lacks a vascular supply [7–9].

Herein, in this chapter, we will present the basic concepts required to diagnose chondral and osteochondral lesions, including the physiological characteristics of cartilage, classifications systems, and pinpoint the most common diagnostic procedures.

8.2 Cartilage Zones and Physiological Characteristics

The articular cartilage is a multi-layered composite structure, which is composed of four structural layers (zones): superficial, middle, deep, and calcified (Fig. 8.1). In each of these, the collagen fibrils are differently oriented. This precise arrangement of the tissue components provides specific mechanical properties for each zone [4, 10, 11]. Each zone has unique chondrocyte morphology, arrangement of type II collagen fibers, and levels of proteoglycans and water. This structure creates different mechanical properties for each specific zone [4, 5, 12, 13]. An understanding of zonal structure is important for the development of artificial cartilage constructs or for induction of a cartilage reparative response for future treatment of chondral injuries.

The superficial zone, is the thinnest zone, representing 10–20% of the matrix, and consists of two layers. The top layer is a clear film called the lamina splendens, which contains no cells, little polysaccharide, and few collagen fibrils. The main layer consists of flattened ellipsoid, densely packed, and horizontally arranged chondrocytes that synthesize a matrix with high collagen content [4, 13, 14]. The abundance and parallel organization of collagen to the joint surface permits the superficial zone to provide strength to resist

a

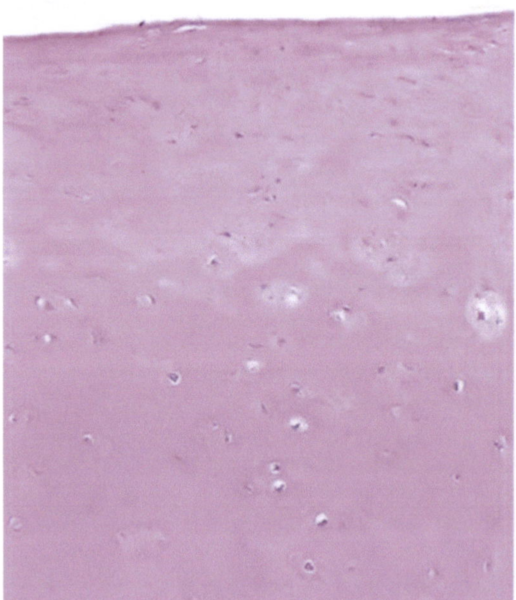

b

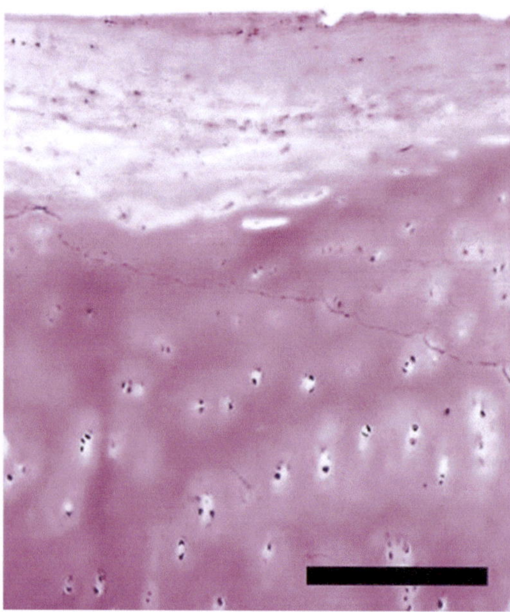

Fig. 8.1 Healthy and osteoarthritic human cartilage. Hematoxylin and eosin stain (H&E) histological staining of explanted cartilage from knee arthroscopy. (**a**) Low-bearing tissue preserved region is demonstrated for health condition, (**b**) while injured inflamed region is represented for osteoarthritic condition. Scale bar = 500 μm

tensile and shear forces. The high concentration of water also provides lubrication and resistance to compression [2, 15]. Removal of this layer, as observed in early cartilage degeneration, results in increased permeability and decreased resistance to tensile forces. This can lead to softening of the cartilage and increased loading of the remainder of the matrix [4].

The transitional zone is the largest zone (40–60% of the matrix) and functions to resist shear and compressive forces. The chondrocytes are spheroidal and synthesize matrix with larger diameter collagen fibrils oriented obliquely to the surface into rotational arches. This arrangement allows the fibers to resist shear forces. The higher proteoglycans and the lower water content of the matrix compared with the superficial zone permit increased compressibility and thus shock absorption and load distribution [1, 2, 13, 15].

The deep zone is of intermediate thickness (30% of the matrix) and functions to resist compressive forces. The chondrocytes are spheroidal, are arranged in vertical columns perpendicular to the surface, and synthesize matrix with the greatest number of proteoglycans. The collagen fibrils are the largest in diameter and are arranged vertically to resist compression, provide stiffness, and anchor the cartilage to the subchondral bone. Removal of the deep vertical fibrils increases the tensile strain in the superficial fibrils and the junction with subchondral bone [1, 2, 13].

The thin calcified cartilage zone between the deep zone and subchondral bone anchors the cartilage to the bone via type X collagen. The tidemark is located in this zone and is the boundary between calcified and uncalcified cartilage [14].

Cartilage is avascular, and chondrocytes in the superficial zones are believed to derive nutrition from synovial fluid. Deeper zones probably obtain nutrition from subchondral bone [5]. The lack of vascular supply, together with the absence of chondrogenic progenitor cells, and decreased mitotic activity, are largely responsible for the difficulty in the healing process of cartilage [14]. Articular cartilage is not innervated, and thus damage does not cause pain (unless it reaches the subchondral bone) which results in many of injuries get undetected [16].

8.3 Chondral and Osteochondral Defects

Articular cartilage lesions are a common pathology of the knee joint. Already since the times of Hippocrates, it has been observed that cartilage once damaged should never heal. So far, the natural history of cartilage lesions remains unpredictable and not well understood [17]. Chondral and osteochondral lesions are frequently observed by arthroscopy and can be diagnosed in 60–66% of all patients submitted to an arthroscopic procedure [6, 18, 19]. Currently, diagnostic arthroscopy is the gold standard for making the diagnosis and, subsequently, choosing the best treatment. Acute and chronic defects occur mainly in people who are exposed to great physical effort, often during sports activity [16, 20–22]. Traumatic lesions occur 7.5 times more frequently than the non-traumatic defects, and the mechanical trauma is the cause of the cartilage damages even in 80% [23, 24].

The term osteochondral lesion is used to describe a spectrum of disease from traumatic osteochondral injury to chronic osteochondritis dissecans (OCD). Lesions may arise from forces applied to the chondral surface in a single traumatic event or over time as the result of repeated minor injury. Underneath the cartilage is the subchondral bone. Together, the articular cartilage and the subchondral bone form the osteochondral unit, which is a functional unit uniquely adapted to assure the transfer of loads across the diarthrodial joint [6]. Damage to the articular cartilage and joint surface may result indirectly from pathologic changes in subchondral bone. A retrospective analysis of 25,124 knee arthroscopies found chondral lesion in 60% of the patients [17]. Of these chondral lesions, 67% were classified as localized focal osteochondral or chondral lesions, 29% as osteoarthritis (OA), and, in 2% of the cases, an OCD was diagnosed. Most commonly, osteochondral lesions are encountered in the talus, femoral condyles, and elbow [5]. In the knee, the medial femoral condyle is the most common location, but patellofemoral lesions can also be frequent [18, 20, 25].

8.4 Classification

Numerous classifications have been proposed to grade cartilage lesions based largely on arthroscopic findings, and less so on magnetic resonance imaging (MRI) findings. Among those using arthroscopy, the most commonly used are the International Cartilage Repair Society (ICRS), Outerbridge and Noyes (Table 8.1). These classification systems describe articular cartilage damage ranging from swelling and signal heterogeneity to fissuring, ulceration, partial-thickness defects, and full-thickness defects with exposure of the subchondral bone.

Most common method of classification is using arthroscopy. Several systems have been developed to grade chondral and osteochondral lesions [26]. Before grading the lesion, the surgeon should debride the defect and defined as a superficial, partial-thickness (chondral), or full-thickness (osteochondral) defect. The ICRS classification system focuses on the lesion depth (graded from 0 to 4) and the area of damage (graded from normal to severely abnormal). Grade 0 is graded as normal cartilage (without damage), Grade I as nearly normal (superficial lesions) and subclassified into soft indentation (IA) and/or superficial fissures and cracks (IB), Grade II as abnormal which describes lesions extending down to <50% of cartilage depth. Grade III is classified as severely abnormal and subclassified according cartilage defects extending down >50% of cartilage depth (IIIA) as well as down to calcified layer (IIIB) and down to but not through the subchondral bone (IIIC), and if blisters are included (IIID). Grade IV is also classified as severely abnormal that extend to the subchondral bone. The arthroscopic ICRS classification system has good inter-observer and intra-observer reliability and a high correlation with histological assessment of depth [27, 28]. The Outerbridge scale classifies cartilage abnormalities based on arthroscopic findings. Grade I includes softening or swelling of the articular cartilage, Grade II describes cartilage fragmentation and fissuring less than 1.5 cm in diameter, Grade III describes cartilage fragmentation and fissuring greater than 1.5 cm in diameter, and Grade IV involves cartilage erosion to bone [29]. The Outerbridge scale also shows reasonable intra- and inter-observer reliability figures [30]. In the Noyes system, Grade 1 depicts an intact cartilage surface, Grade 2A reflects cartilage damage with less than 50% cartilage thickness involved, Grade 2B cartilage defects involve greater than half of the cartilage thickness, and Grade 3 represents full-thickness cartilage defects with exposed subchondral bone [31].

The use of MRI to grade the chondral and osteochondral is not so commonly used. The Whole-Organ Magnetic Resonance Score (WORMS) and Boston-Leeds Osteoarthritis Knee Score (BLOKS) are common classifications of cartilage damage, but are directed for OA and not cartilage defects [32, 33]. The Area Measurement And Depth and Underlying

Table 8.1 Chondral and osteochondral injury classifications

ICRS (MRI)	Outerbridge	Noyes
Grade 0: normal cartilage	Grade 0: normal cartilage	
Grade 1: increased T2 signal in the cartilage	Grade I: softening and swelling of cartilage	Grade 1: intact cartilage surface
Grade 2: partial-thickness defect <50% of normal cartilage thickness	Grade II: cartilage fragmentation and fissuring <1.5 cm diameter	Grade 2A: cartilage surface damaged with <50% thickness involved
		Grade 2B: cartilage defects involve >50% cartilage thickness
Grade 3: partial-thickness defect >50% of normal cartilage thickness	Grade III: fragmentation and fissuring >1.5 cm diameter	Grade 3: bone exposed (3A cortical surface intact, 3B cortical surface cavitation)
Grade 4: full-thickness defect	Grade IV: cartilage erosion to bone	

ICRS International Cartilage Repair Society, *MRI* magnetic resonance imaging

Structures (AMADEUS) score was developed to assess focal chondral or osteochondral defects [34]. It emphasizes the following: (1) size of the cartilage defect area ("area measurement"), (2) cartilage defect morphology/depth ("depth"), and (3) underlying structures with the presence of adjacent osseous defects/subchondral cysts and bone marrow edema-like lesions ("underlying structures") [35]. The ICRS has adopted the classification system for using in MRI assessment [36]. Grade 0 represents normal cartilage, Grade 1 describes increased T2 signal within the cartilage, Grade 2 refers to a partial-thickness defect less than 50% of normal cartilage thickness, Grade 3 represents a partial-thickness defect greater than 50% of normal cartilage thickness, and Grade 4 describes a full-thickness defect.

The ICRS developed an osteochondritis dissecans (OCD) evaluation score system: grade 0, the lesion is stable and the overlying cartilage is normal and intact; grade I includes stable lesions with some softening of the cartilage surface; grade II refers to lesions with partial discontinuity of the cartilage surface; grade III respects to the defect when is unstable due to a complete discontinuity of the osteochondral defect; and a grade IV lesion is an empty defect or a defect with loose fragments [37]. According to this grading system, grade III and IV lesions are unstable and, therefore, may have indication for surgical orthopedic treatment [6, 38].

8.5 Diagnosis

Chondral and osteochondral lesions are more common at the knee joint; thus, we will explore the diagnosis focusing the knee joint. However, beyond the physical exam, the other diagnostic procedures are similar for other human joints.

Patients with symptomatic chondral or osteochondral defects typically present with activity-related joint pain and swelling. Larger lesions often cause catching or locking of the knee. Tibiofemoral defects located at the femoral condyles result in pain at or near the joint line cause impairments in activities such as running or descending stairs. The patellofemoral defects lead to anterior knee pain during movements that include ascending stairs, squatting, prolonged sitting in a flexed position, or getting up from the sited position [39]. These symptoms have poor sensitivity because can also be found in other knee injuries, such as meniscal tears and patellofemoral pain syndrome.

Careful collection and interpretation of the clinical history is crucial before determining the treatment choice. Prior injuries and previous surgeries influence the diagnosis and should be explored [39]. The body mass index, mechanical alignment, occupation, sports participation, associated medical conditions and inflammatory diseases, steroid intake, smoking habits, and responsiveness to rehabilitation are important factors to be taken into account during the diagnosis [40]. There are no definitive signs, but the location, onset and type of pain, aggravating and relieving factors, or any other symptoms such as locking or catching are useful for the diagnosis [41]. Outcome clinical and functional scores such as the ICRS subjective score and the Knee and Osteoarthritis Outcome Score (KOOS) can be helpful to establish a functional baseline.

The degree of the cartilage injury plays a major role in diagnosis because injuries with a grade III and IV are those that are usually indicated for surgical treatment. Three other key factors determine the choice of treatment and should be carefully analyzed during diagnosis, which include defect size, patient age, and sports activity. Defect size is important because will determine the best surgical technique to be used for treating the defect, including those that are small (<2 cm²), large (2–4 cm²), and very large (>4 cm²). Larger lesions are usually treated with more complex surgical procedures. Outcomes related to patient age are still insufficient and inconclusive. It is believed that younger patients (under 30–40 years) benefit more from cartilage repair surgery, but there are conflicting results showing no benefit at all [42–47]. Level of activity also play a fundamental role in the outcomes of cartilage surgery and should be taken into consideration. Patients that are more active often show better outcomes [48]. The type and level of sports competition is also important because the surgeon must plan the injury management with the aim to return safely the athlete to sports activity.

8.5.1 Physical Examination

Cartilage defects can be asymptomatic, that is why it is essential to confirm physical examination findings that are consistent with a cartilage injury before performing any surgical treatment. Physical findings can easily be confounded with meniscal injury or patellofemoral disorders because no physical test can clearly differentiate these conditions. The examination should also focus on detecting potential contraindications to cartilage restoration and determining if concomitant procedures will be required to optimize results [49].

Standing and gait analysis should be performed to evaluate for malalignment or specific gait abnormalities. Rotational deformity in the lower extremity and muscular imbalance have also been shown to alter the forces and biomechanics of the knee joint [50]. Therefore, a thorough examination should include assessment of hip rotation, hamstring and quadriceps strength and flexibility, and foot alignment. Muscle weakness, which can contribute to knee pain, can be assessed by manual muscle strength testing or via instrumented strength testing.

Because ligament insufficiency has been shown to adversely affect cartilage restoration, it is essential to thoroughly examine the anterior cruciate ligament (ACL), posterior cruciate ligament (PCL), collateral ligaments, and the posterolateral and posteromedial corners [51]. To rule out any rotational instability, a pivot shift examination should be performed both in the office and before any surgical procedure, especially when a prior ACL reconstruction is placed in a slightly vertical position.

The clinician should search for effusion signs which is indicative of inflammatory activity. Joint line tenderness may be present in varying degrees of flexion, depending on the location of the cartilage defect, with pain typically worse directly over the cartilage defect. Tenderness at the femoral condyle can help to exclude other causes of injury (e.g., meniscal injury). Pain after pressure over the patella can be indicative of trochlear or patellar defects. Pain during these tests can be described in many ways and may have a mechanical rhythm that worsens gradually after the start of a new activity, especially when a previous trauma is involved. Pain can be exacerbated for numerous reasons, from walking to more intense sport activities [52]. Associated symptoms, such as locking, pseudo-locking, and giving-way can be reported when loose bodies or associated lesions as meniscal tears or ligamentous injuries are present [6].

The Wilson test can be performed for diagnosis of an osteochondral defect. This is a provocative test of the pathology, performed with the patient sitting with the limb hanging on the table. With the knee flexed 90°, the tibia is internally rotated while the patient actively extends the limb. The test is positive when there is a complaint of pain from 30° to the maximum extension and pain relief by performing the same test in external rotation [53].

8.5.2 Imaging

8.5.2.1 Radiography

Radiographic studies are not very helpful in diagnosing cartilage lesions in an early stage, but they can be valuable for detecting osteochondral lesions, osteoarthritis, OCD, loose bodies, and limb malalignment [6].

Plain X-ray films should include standard anteroposterior, flexion weight-bearing anteroposterior (tunnel view), lateral, and Merchant views. Flexion weight-bearing anteroposterior in addition to standard anteroposterior allows better visualization of lesions along the posterolateral aspect of the medial femoral condyle [54].

8.5.2.2 CT Arthrogram

While, for most joints, high-resolution MRI has replaced computed tomography arthrogram (CTA), for the elbow and the ankle joints, the

CTA remains clinically relevant [55–57]. In such cases, cartilage thickness measurements can be more accurate on CTA than on conventional MRI [58]. On MRI, it is very difficult to detect fissural defects that allow communication between intra-articular synovial fluid and cyst. The evaluation of the presence of subchondral cysts is also an important indication for the performance of CTA [59].

8.5.2.3 MRI

Usually conventional MRI is considered as the modality of choice [60]. Using MRI the specificity, sensitivity, positive likelihood ratio (+LR), negative likelihood ratio (−LR), and diagnostic odds ratio values are reasonable for both 1.5-T (0.664, 0.826, 4.222, 0.414, and 9.383, respectively) and 3.0-T MRI (0.702, 0.851, 4.988, 0.304, and 17.765, respectively) [61]. However, in comparison to the knee joint, articular cartilage at the ankle joint is very thin (0.4–2.1 mm), making the assessment of morphological cartilage defects a challenging task [55, 59, 62]. Besides, low sensitivities for detection of osteochondral lesions at the ankle on MRI were reported, which varied between 50% at 1.5 T and 75% at 3.0 T [55, 59].

MRI is a reliable and noninvasive technique to determine volume, thickness, and alterations in the cartilage structure. Quantitative T2-mapping is a proven technique to quantify the water content and collagen fiber orientation of cartilage. Elevated T2-relaxation times are closely associated with a loss of collagen fiber integrity and an altered water content [63]. This has shown to be a sensitive parameter for the evaluation of cartilage degradation [64–66]. However, quantitative MRI does not correlate very well with the arthroscopic ICRS classification [67].

Normal articular cartilage has a homogeneous or laminar appearance with a smooth surface contour. Articular cartilage has intermediate signal on both T1- and T2-weighted images. The sectional imaging not only enables an evaluation of the cartilage, but also of the subchondral bone, which arthroscopically mostly cannot be assessed [68, 69]. Chondral abnormalities are diagnosed on MRI by recognizing a contour defect within the cartilage, focal thinning compared with the thickness of the adjacent cartilage, and/or signal alteration within the cartilage. A secondary sign of cartilage defect includes underlying bone marrow edema, as manifested by increased signal in the subchondral bone on fat-suppressed proton density and T2-weighted images. Subchondral bone marrow edema is a nonspecific finding that may be seen with acute injury (bone contusion or bruise, fracture), mechanical disturbance such as stress response or overlying meniscal tear, and many other conditions, including metabolic and neoplastic lesions. However, a flame-shaped or rounded focus of marrow edema in the subchondral bone should initiate a search for overlying hyaline cartilage abnormality [5]. The ICRS recommended MRI acquisition protocols for articular cartilage are elsewhere described [70].

8.5.2.4 Arthroscopic Evaluation

Arthroscopic assessment is still the gold standard method to assess chondral and osteochondral lesions (Figs. 8.2 and 8.3) the location, grade, size, depth, morphology, and characteristics (monofocal, bifocal, or multifocal) of chondral and osteochondral defects, as well as their reparability. The smooth surfaces can be probed to evaluate the integrity of cartilage (Fig. 8.4). Any fissures and tissue surrounding the cartilage should be carefully probed. The stability of the defect should also be assessed using the probe.

8.5.3 Osteochondritis Dissecans and Osteonecrosis

Both OCD and osteonecrosis may lead to destruction of the articular surface. In the knee, OCD tends to occur on the intercondylar aspect of the medial femoral condyle in young people. These lesions may separate from the surface and form a loose body. The base of these lesions, if debrided, will reveal vascular subchondral bone. As repetitive trauma is thought to be the most frequent cause, the term osteochondral lesion seemed to be more appropriate than the term "osteochondritis" [6, 38, 71].

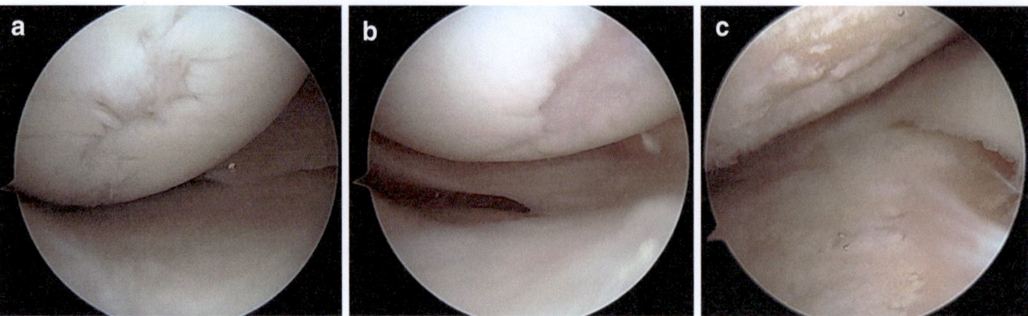

Fig. 8.2 Arthroscopic visualization of femoral chondral and osteochondral lesions. (**a**) Grade III chondral damage at the medical femoral condyle, (**b**) Grade IV osteochondral damage at the medical femoral condyle, (**c**) Grade IV osteochondral damage at the medical femoral condyle and medial tibial plateau

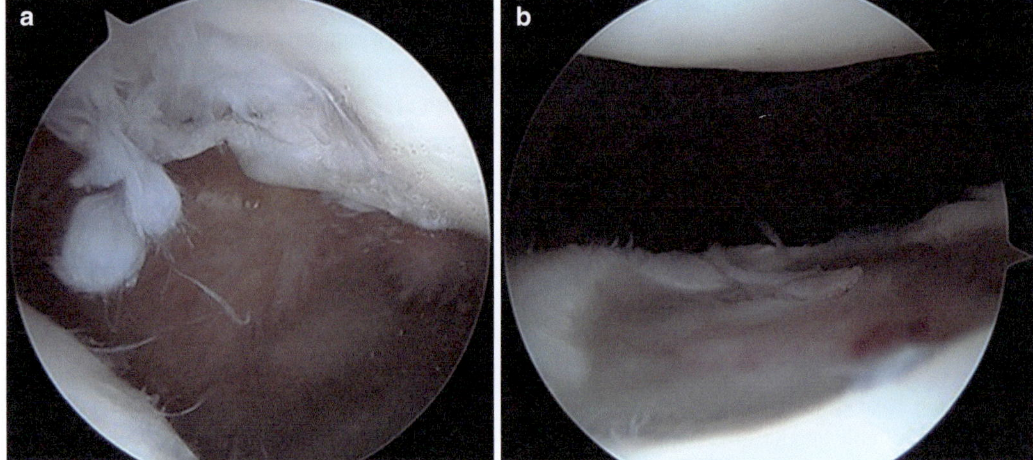

Fig. 8.3 Arthroscopic visualization of patellofemoral chondral and osteochondral lesions. (**a**) Grade III-IV chondral/osteochondral damage at the patella, (**b**) Grade IV osteochondral damage at the femoral trochlea

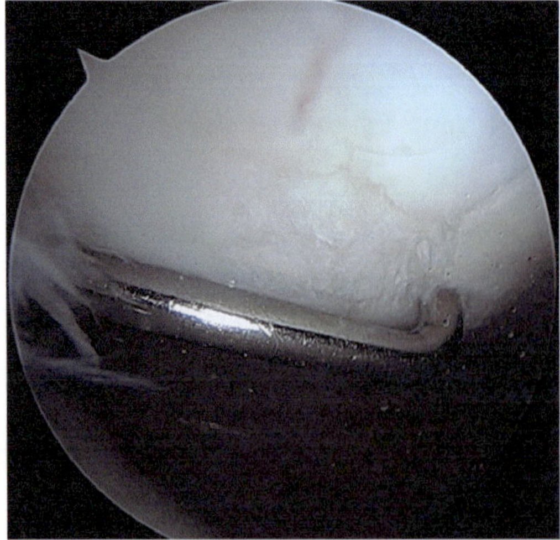

Fig. 8.4 Probing of the smooth surfaces c to evaluate the integrity of cartilage

Performing a complete and accurate physical examination is of fundamental importance, considering that its high sensitivity (90.9%) for the diagnosis of OCD, and reasonable specificity 69%, which is greater than the MRI [72]. Radiographic images of patients with adult OCD show a lesion that typically appears as an area of osteosclerotic bone, with a high-intensity line between defect and epiphysis. Classic radiographic findings include a lucent osseous defect that may have a fragmented or corticated osseous density within the lucency. MRI should be performed to accurately characterize OCD, to evaluate size and location, and to determine the stability of the lesion. An increased signal of the defect on T2-weighted images represents joint fluid surrounding the lesion; irregularity of the articular surface may also be noted. Measurement is generally performed using T1-weighted images. Surrounding bone marrow edema is variable and may represent healing response or irritation from lesion instability, so this finding is nonspecific; however, it is often the case that the more bone marrow edema is present, the more painful the lesion is. An unstable lesion is identified by one or more of the following findings on T2-weighted fat-suppressed images or STIR images: (1) linear high signal intensity surrounding the osteochondral fragment, (2) cystic change interposed between the osteochondral fragment and normal bone, or (3) overlying cartilage defect or fissuring. Intra-articular gadolinium may dissect beneath the osteochondral fragment, also indicating lesion instability [5]. A healed osteochondral lesion will not demonstrate fluid bright signal between the osteochondral fragment and the host bone. Normal bone marrow fat signal will return to the osteochondral fragment once it heals. The overlying articular cartilage may be intact, without contour irregularities, or may exhibit degeneration, thinning, or fraying.

Osteonecrosis results in a similar osteochondral fragment but tends to occur in elderly patients on the weight-bearing aspect of the medial femoral condyle. In distinction to the lesions in OCD, fragments in osteonecrosis separate from a bed of avascular bone. Again, radiographs may reveal a lucent defect at the involved site, but MRI is more reliable for evaluation of these defects. The osteonecrotic fragment has low signal intensity on T1- and T2-weighted images. A curvilinear area of low signal with variable bone edema is characteristic. Although the articular cartilage is initially normal, both processes may lead to detachment of osteochondral loose bodies, fragmentation, and collapse of the articular surface with resultant degenerative changes.

8.6 Conclusions

Understanding the cartilage macro- and microphysiological characteristics is important to recognize the specific mechanical properties of each

of the cartilage tissue zones. There are several classifications of chondral and osteochondral injuries using either MRI or arthroscopy visualization, which are an important step of diagnosis because will help to determine the most suitable treatment. The diagnosis should include a combination of a comprehensive physical examination, imaging procedures and, if needed, arthroscopic evaluation. During the physical examination, the orthopedic surgeon must pay attention to symptoms and patient/defect characteristics. Imaging exams have an important role in identifying any associated injuries, as well as identifying the location and extent of the defects to guide the surgical indications. The diagnostic arthroscopy can further evaluate the location, depth, and extent of damage to refine the surgical strategy. Both OCD and osteonecrosis are also important entities that can lead to destruction of the articular surface. While physical examination may be sufficient to diagnose OCD, the MRI assessment is better suited to diagnose osteonecrosis.

References

1. Carter DR, Beaupre GS, Wong M, Smith RL, Andriacchi TP, Schurman DJ. The mechanobiology of articular cartilage development and degeneration. Clin Orthop Relat Res. 2004;(427 Suppl):S69–77. https://doi.org/10.1097/01.blo.0000144970.05107.7e.
2. Lu XL, Mow VC. Biomechanics of articular cartilage and determination of material properties. Med Sci Sports Exerc. 2008;40(2):193–9. https://doi.org/10.1249/mss.0b013e31815cb1fc.
3. Thomas GC, Asanbaeva A, Vena P, Sah RL, Klisch SM. A nonlinear constituent based viscoelastic model for articular cartilage and analysis of tissue remodeling due to altered glycosaminoglycan-collagen interactions. J Biomech Eng. 2009;131(10):101002. https://doi.org/10.1115/1.3192139.
4. Wong M, Carter DR. Articular cartilage functional histomorphology and mechanobiology: a research perspective. Bone. 2003;33(1):1–13. https://doi.org/10.1016/s8756-3282(03)00083-8.
5. Scott N, Insall JN, Pedersen HB, Math KR, Vigorita VJ, Cushner FD. Insall & Scott surgery of the knee. New York, NY: Elsevier; 2015. p. 146–51.
6. Vilela CA, da Silva MA, Pina S, Oliveira JM, Correlo VM, Reis RL, et al. Clinical trials and Management of Osteochondral Lesions. Adv Exp Med Biol. 2018;1058:391–413. https://doi.org/10.1007/978-3-319-76711-6_18.

7. Furman BD, Olson SA, Guilak F. The development of posttraumatic arthritis after articular fracture. J Orthop Trauma. 2006;20(10):719–25. https://doi.org/10.1097/01.bot.0000211160.05864.14.

8. Sui Y, Lee JH, DiMicco MA, Vanderploeg EJ, Blake SM, Hung HH, et al. Mechanical injury potentiates proteoglycan catabolism induced by interleukin-6 with soluble interleukin-6 receptor and tumor necrosis factor alpha in immature bovine and adult human articular cartilage. Arthritis Rheum. 2009;60(10):2985–96. https://doi.org/10.1002/art.24857.

9. Szczodry M, Coyle CH, Kramer SJ, Smolinski P, Chu CR. Progressive chondrocyte death after impact injury indicates a need for chondroprotective therapy. Am J Sports Med. 2009;37(12):2318–22. https://doi.org/10.1177/0363546509348840.

10. Guilak F, Alexopoulos LG, Upton ML, Youn I, Choi JB, Cao L, et al. The pericellular matrix as a transducer of biomechanical and biochemical signals in articular cartilage. Ann N Y Acad Sci. 2006;1068:498–512. https://doi.org/10.1196/annals.1346.011.

11. Montes GS. Structural biology of the fibres of the collagenous and elastic systems. Cell Biol Int. 1996;20(1):15–27. https://doi.org/10.1006/cbir.1996.0004.

12. Sharma B, Williams CG, Kim TK, Sun D, Malik A, Khan M, et al. Designing zonal organization into tissue-engineered cartilage. Tissue Eng. 2007;13(2):405–14. https://doi.org/10.1089/ten.2006.0068.

13. Youn I, Choi JB, Cao L, Setton LA, Guilak F. Zonal variations in the three-dimensional morphology of the chondron measured in situ using confocal microscopy. Osteoarthr Cartil. 2006;14(9):889–97. https://doi.org/10.1016/j.joca.2006.02.017.

14. Buckwalter JA, Mankin HJ, Grodzinsky AJ. Articular cartilage and osteoarthritis. Instr Course Lect. 2005;54:465–80.

15. Nugent GE, Aneloski NM, Schmidt TA, Schumacher BL, Voegtline MS, Sah RL. Dynamic shear stimulation of bovine cartilage biosynthesis of proteoglycan 4. Arthritis Rheum. 2006;54(6):1888–96. https://doi.org/10.1002/art.21831.

16. Vannini F, Spalding T, Andriolo L, Berruto M, Denti M, Espregueira-Mendes J, et al. Sport and early osteoarthritis: the role of sport in aetiology, progression and treatment of knee osteoarthritis. Knee Surg Sports Traumatol Arthrosc. 2016;24(6):1786–96. https://doi.org/10.1007/s00167-016-4090-5.

17. Widuchowski W, Widuchowski J, Trzaska T. Articular cartilage defects: study of 25,124 knee arthroscopies. Knee. 2007;14(3):177–82. https://doi.org/10.1016/j.knee.2007.02.001.

18. Hjelle K, Solheim E, Strand T, Muri R, Brittberg M. Articular cartilage defects in 1,000 knee arthroscopies. Arthroscopy. 2002;18(7):730–4. https://doi.org/10.1053/jars.2002.32839.

19. Frank RM, Cotter EJ, Hannon CP, Harrast JJ, Cole BJ. Cartilage restoration surgery: incidence rates, complications, and trends as reported by the American Board of Orthopaedic Surgery Part II candidates. Arthroscopy. 2019;35(1):171–8. https://doi.org/10.1016/j.arthro.2018.08.028.

20. Andrade R, Vasta S, Papalia R, Pereira H, Oliveira JM, Reis RL, et al. Prevalence of articular cartilage lesions and surgical clinical outcomes in football (soccer) Players' knees: a systematic review. Arthroscopy. 2016;32(7):1466–77. https://doi.org/10.1016/j.arthro.2016.01.055.

21. Flanigan DC, Harris JD, Trinh TQ, Siston RA, Brophy RH. Prevalence of chondral defects in athletes' knees: a systematic review. Med Sci Sports Exerc. 2010;42(10):1795–801. https://doi.org/10.1249/MSS.0b013e3181d9eea0.

22. Lansdown DA, Cvetanovich GL, Zhang AL, Feeley BT, Wolf BR, Hettrich CM, et al. Risk factors for intra-articular bone and cartilage lesions in patients undergoing surgical treatment for posterior instability. Am J Sports Med. 2020;48(5):1207–12. https://doi.org/10.1177/0363546520907916.

23. Wilder FV, Hall BJ, Barrett JP Jr, Lemrow NB. History of acute knee injury and osteoarthritis of the knee: a prospective epidemiological assessment. The Clearwater Osteoarthritis Study. Osteoarthritis Cartilage. 2002;10(8):611–6. https://doi.org/10.1053/joca.2002.0795.

24. Gersoff WK. Considerations prior to surgical repair of articular cartilage injuries of the knee. Oper Tech Sports Med. 2000;8:86–90.

25. Andrade R, Nunes J, Hinckel BB, Gruskay J, Vasta S, Bastos R, et al. Cartilage restoration of patellofemoral lesions: a systematic review. Cartilage. 2019; https://doi.org/10.1177/1947603519893076.

26. Spahn G, Klinger HM, Hofmann GO. How valid is the arthroscopic diagnosis of cartilage lesions? Results of an opinion survey among highly experienced arthroscopic surgeons. Arch Orthop Trauma Surg. 2009;129(8):1117–21.

27. Brittberg M, Winalski CS. Evaluation of cartilage injuries and repair. J Bone Joint Surg Am. 2003;85-A(Suppl 2):58–69. https://doi.org/10.2106/00004623-200300002-00008.

28. Dwyer T, Martin CR, Kendra R, Sermer C, Chahal J, Ogilvie-Harris D, et al. Reliability and validity of the arthroscopic international cartilage repair society classification system: correlation with histological assessment of depth. Arthroscopy. 2017;33(6):1219–24. https://doi.org/10.1016/j.arthro.2016.12.012.

29. Outerbridge RE. The etiology of chondromalacia patellae. J Bone Joint Surg Br. 1961;43-b:752–7.

30. Brismar B, Wredmark T, Movin T, Leandersson J, Svensson O. Observer reliability in the arthroscopic classification of osteoarthritis of the knee. J Bone Joint Surg Br. 2002;84(1):42–7. https://doi.org/10.1302/0301-620x.84b1.11660.

31. Noyes FR, Stabler CL. A system for grading articular cartilage lesions at arthroscopy. Am J Sports Med. 1989;17(4):505–13. https://doi.org/10.1177/036354658901700410.

32. Peterfy CG, Guermazi A, Zaim S, Tirman PF, Miaux Y, White D, et al. Whole-organ magnetic resonance imaging score (WORMS) of the knee in osteoarthritis. Osteoarthr Cartil. 2004;12(3):177–90. https://doi.org/10.1016/j.joca.2003.11.003.

33. Hunter DJ, Lo GH, Gale D, Grainger AJ, Guermazi A, Conaghan PG. The reliability of a new scoring system for knee osteoarthritis MRI and the validity of bone marrow lesion assessment: BLOKS (Boston Leeds osteoarthritis knee score). Ann Rheum Dis. 2008;67(2):206–11. https://doi.org/10.1136/ard.2006.066183.

34. Jungmann PM, Welsch GH, Brittberg M, Trattnig S, Braun S, Imhoff AB, et al. Magnetic resonance imaging score and classification system (AMADEUS) for assessment of preoperative cartilage defect severity. Cartilage. 2017;8(3):272–82. https://doi.org/10.1177/1947603516665444.

35. Massen FK, Inauen CR, Harder LP, Runer A, Preiss S, Salzmann GM. One-step autologous minced cartilage procedure for the treatment of knee joint chondral and osteochondral lesions: a series of 27 patients with 2-year follow-up. Orthop J Sports Med. 2019;7(6):2325967119853773. https://doi.org/10.1177/2325967119853773.

36. Yulish BS, Montanez J, Goodfellow DB, Bryan PJ, Mulopulos GP, Modic MT. Chondromalacia patellae: assessment with MR imaging. Radiology. 1987;164(3):763–6. https://doi.org/10.1148/radiology.164.3.3615877.

37. Ellermann JM, Donald B, Rohr S, Takahashi T, Tompkins M, Nelson B, et al. Magnetic resonance imaging of osteochondritis dissecans: validation study for the ICRS classification system. Acad Radiol. 2016;23(6):724–9. https://doi.org/10.1016/j.acra.2016.01.015.

38. Chen CH, Liu YS, Chou PH, Hsieh CC, Wang CK. MR grading system of osteochondritis dissecans lesions: comparison with arthroscopy. Eur J Radiol. 2013;82(3):518–25. https://doi.org/10.1016/j.ejrad.2012.09.026.

39. Gomoll AH, Lattermann C, Farr J. "A unifying theory" treatment algorithm for cartilage defects. In: Jack F, Gomoll Andreas H, editors. Cartilage restoration: practical clinical applications. 2nd ed. Cham: Springer International; 2018. p. 39–49.

40. Cole BJ, Pascual-Garrido C, Grumet RC. Surgical management of articular cartilage defects in the knee. Instr Course Lect. 2010;59:181–204.

41. Amaravathi R, Andrade R, Bastos R, Espregueira-Mendes J. The mosaicplasty/OAT procedure: technique, pearls and pitfalls. Asian J Arthrosc. 2019;4(1):15–22.

42. Asik M, Ciftci F, Sen C, Erdil M, Atalar A. The microfracture technique for the treatment of full-thickness articular cartilage lesions of the knee: midterm results. Arthroscopy. 2008;24(11):1214–20. https://doi.org/10.1016/j.arthro.2008.06.015.

43. Vanlauwe J, Saris DB, Victor J, Almqvist KF, Bellemans J, Luyten FP. Five-year outcome of characterized chondrocyte implantation versus microfracture for symptomatic cartilage defects of the knee: early treatment matters. Am J Sports Med. 2011;39(12):2566–74. https://doi.org/10.1177/0363546511422220.

44. Gille J, Schuseil E, Wimmer J, Gellissen J, Schulz AP, Behrens P. Mid-term results of autologous matrix-induced chondrogenesis for treatment of focal cartilage defects in the knee. Knee Surg Sports Traumatol Arthrosc. 2010;18(11):1456–64. https://doi.org/10.1007/s00167-010-1042-3.

45. de Windt TS, Concaro S, Lindahl A, Saris DB, Brittberg M. Strategies for patient profiling in articular cartilage repair of the knee: a prospective cohort of patients treated by one experienced cartilage surgeon. Knee Surg Sports Traumatol Arthrosc. 2012;20(11):2225–32. https://doi.org/10.1007/s00167-011-1855-8.

46. Kreuz PC, Erggelet C, Steinwachs MR, Krause SJ, Lahm A, Niemeyer P, et al. Is microfracture of chondral defects in the knee associated with different results in patients aged 40 years or younger? Arthroscopy. 2006;22(11):1180–6. https://doi.org/10.1016/j.arthro.2006.06.020.

47. Knutsen G, Drogset JO, Engebretsen L, Grøntvedt T, Isaksen V, Ludvigsen TC, et al. A randomized trial comparing autologous chondrocyte implantation with microfracture. Findings at five years. J Bone Joint Surg Am. 2007;89(10):2105–12. https://doi.org/10.2106/jbjs.G.00003.

48. Kon E, Gobbi A, Filardo G, Delcogliano M, Zaffagnini S, Marcacci M. Arthroscopic second-generation autologous chondrocyte implantation compared with microfracture for chondral lesions of the knee: prospective nonrandomized study at 5 years. Am J Sports Med. 2009;37(1):33–41. https://doi.org/10.1177/0363546508323256.

49. Gomoll AH, Farr J, Gillogly SD, Kercher JS, Minas T. Surgical management of articular cartilage defects of the knee. Instr Course Lect. 2011;60:461–83.

50. Collado H, Fredericson M. Patellofemoral pain syndrome. Clin Sports Med. 2010;29(3):379–98. https://doi.org/10.1016/j.csm.2010.03.012.

51. Efe T, Füglein A, Getgood A, Heyse TJ, Fuchs-Winkelmann S, Patzer T, et al. Anterior cruciate ligament deficiency leads to early instability of scaffold for cartilage regeneration: a controlled laboratory ex-vivo study. Int Orthop. 2012;36(6):1315–20. https://doi.org/10.1007/s00264-011-1437-x.

52. Arøen A, Løken S, Heir S, Alvik E, Ekeland A, Granlund OG, et al. Articular cartilage lesions in 993 consecutive knee arthroscopies. Am J Sports Med. 2004;32(1):211–5. https://doi.org/10.1177/0363546503259345.

53. Conrad JM, Stanitski CL. Osteochondritis dissecans: Wilson's sign revisited. Am J Sports Med. 2003;31(5):777–8. https://doi.org/10.1177/03635465030310052301.

54. Harding WG 3rd. Diagnosis of ostechondritis dissecans of the femoral condyles: the value of the lateral x-ray view. Clin Orthop Relat Res. 1977;123:25–6.

55. Schmid MR, Pfirrmann CW, Hodler J, Vienne P, Zanetti M. Cartilage lesions in the ankle joint: comparison of MR arthrography and CT arthrography. Skelet Radiol. 2003;32(5):259–65. https://doi.org/10.1007/s00256-003-0628-y.

56. Cerezal L, de Dios B-MJ, Canga A, Llopis E, Rolon A, Martín-Oliva X, et al. MR and CT arthrography of the wrist. Semin Musculoskelet Radiol. 2012;16(1):27–41. https://doi.org/10.1055/s-0032-1304299.

57. Delport AG, Zoga AC. MR and CT arthrography of the elbow. Semin Musculoskelet Radiol. 2012;16(1):15–26. https://doi.org/10.1055/s-0032-1304298.

58. El-Khoury GY, Alliman KJ, Lundberg HJ, Rudert MJ, Brown TD, Saltzman CL. Cartilage thickness in cadaveric ankles: measurement with double-contrast multi-detector row CT arthrography versus MR imaging. Radiology. 2004;233(3):768–73. https://doi.org/10.1148/radiol.2333031921.

59. Kirschke JS, Braun S, Baum T, Holwein C, Schaeffeler C, Imhoff AB, et al. Diagnostic value of CT arthrography for evaluation of osteochondral lesions at the ankle. Biomed Res Int. 2016;2016:3594253. https://doi.org/10.1155/2016/3594253.

60. Naran KN, Zoga AC. Osteochondral lesions about the ankle. Radiol Clin N Am. 2008;46(6):995–1002. https://doi.org/10.1016/j.rcl.2008.10.001.

61. Cheng Q, Zhao FC. Comparison of 1.5- and 3.0-T magnetic resonance imaging for evaluating lesions of the knee: A systematic review and meta-analysis (PRISMA-compliant article). Medicine (Baltimore). 2018;97(38):e12401. https://doi.org/10.1097/md.0000000000012401.

62. Ba-Ssalamah A, Schibany N, Puig S, Herneth AM, Noebauer-Huhmann IM, Trattnig S. Imaging articular cartilage defects in the ankle joint with 3D fat-suppressed echo planar imaging: comparison with conventional 3D fat-suppressed gradient echo imaging. J Magn Reson Imaging. 2002;16(2):209–16. https://doi.org/10.1002/jmri.10153.

63. Waldenmeier L, Evers C, Uder M, Janka R, Hennig FF, Pachowsky ML, et al. Using cartilage MRI T2-mapping to analyze early cartilage degeneration in the knee joint of young professional soccer players. Cartilage. 2019;10(3):288–98. https://doi.org/10.1177/1947603518756986.

64. Domayer SE, Welsch GH, Dorotka R, Mamisch TC, Marlovits S, Szomolanyi P, et al. MRI monitoring of cartilage repair in the knee: a review. Semin Musculoskelet Radiol. 2008;12(4):302–17. https://doi.org/10.1055/s-0028-1100638.

65. Mosher TJ, Dardzinski BJ. Cartilage MRI T2 relaxation time mapping: overview and applications. Semin Musculoskelet Radiol. 2004;8(4):355–68. https://doi.org/10.1055/s-2004-861764.

66. Apprich S, Welsch GH, Mamisch TC, Szomolanyi P, Mayerhoefer M, Pinker K, et al. Detection of degenerative cartilage disease: comparison of high-resolution morphological MR and quantitative T2 mapping at 3.0 tesla. Osteoarthr Cartil. 2010;18(9):1211–7. https://doi.org/10.1016/j.joca.2010.06.002.

67. Casula V, Hirvasniemi J, Lehenkari P, Ojala R, Haapea M, Saarakkala S, et al. Association between quantitative MRI and ICRS arthroscopic grading of articular cartilage. Knee Surg Sports Traumatol Arthrosc. 2016;24(6):2046–54. https://doi.org/10.1007/s00167-014-3286-9.

68. Grambart ST. Arthroscopic Management of Osteochondral Lesions of the talus. Clin Podiatr Med Surg. 2016;33(4):521–30. https://doi.org/10.1016/j.cpm.2016.06.008.

69. O'Loughlin PF, Heyworth BE, Kennedy JG. Current concepts in the diagnosis and treatment of osteochondral lesions of the ankle. Am J Sports Med. 2010;38(2):392–404. https://doi.org/10.1177/0363546509336336.

70. Brittberg M, Winalski CS. Evaluation of cartilage injuries and repair. J Bone Joint Surg Am. 2003;85(Suppl 2):58–69.

71. Durur-Subasi I, Durur-Karakaya A, Yildirim OS. Osteochondral lesions of major joints. Eurasian J Med. 2015;47(2):138–44. https://doi.org/10.5152/eurasianjmed.2015.50.

72. Kocher MS, DiCanzio J, Zurakowski D, Micheli LJ. Diagnostic performance of clinical examination and selective magnetic resonance imaging in the evaluation of intraarticular knee disorders in children and adolescents. Am J Sports Med. 2001;29(3):292–6. https://doi.org/10.1177/03635465010290030601.

Bone Marrow Edema

9

Massimo Berruto, Daniele Tradati, and Eva Usellini

9.1 Treatment

As previously mentioned, BMLs represent an altered MRI signal underlying different pathology; therefore, the treatment should not be related on images only but it should highly evaluate based on patient symptoms, underlying aetiology, and chance of disease progression.

A post-traumatic BML in a patient sustaining an ACL injury will hardly require treatment, while a more focused BML related to a subchondral insufficiency fracture of the knee (SIFK) could progress to avascular necrosis (AVN) amendable of surgical treatment.

9.1.1 Basic Principles

The treatment of the BMLs should be primarily focused on pain-relief and avoiding lesion progression. Pain control could be achieved by balancing both nonsteroidal anti-inflammatory drugs (NSAID), Cox-1/Cox-2 Inhibitor, and opioid drugs. NSAID act as inhibitors of the cytokine and leukotrienes pathway aiming to stop the biological mechanism sustaining BMLs, while opioids could be more effective on pain control avoiding central pain hyper-sensibilization.

Weight bear should be limited as soon as possible. Repeated microtraumas could sustain or worsening BMLs progression, leading to microfractures occurring in the bone that already underwent to an altered remodelling process. Walking in partial weight bear (15–20 kg) using crutches should be recommended until further MRI assessments demonstrate BMLs regression or pain substantially reduce.

Isometric quadriceps strengthening, range of movement maintenance, and stretching of knee flexors should be highly encouraged in order to avoid further delays in the recovery phase, promoting a faster recovery.

9.1.2 Physical Therapies

9.1.2.1 Hyperbaric Oxygen Therapy

Hyperbaric oxygen therapy (HBOT) has been proposed in the treatment of BMLs due to its role in modulating inflammation and oxidative stress [1].

The interaction between inflammatory factors and OPG/RANK/RANKL homeostasis [2, 3] represents a potential target in order to decrease bone remodelling underlying BMLs.

M. Berruto (✉) · E. Usellini
UOS Chirurgia Articolare del Ginocchio, I Clinica Ortopedica, ASST Gaetano Pini-CTO, Milan, Italy
e-mail: massimo.berruto@fastwebnet.it;
eva.usellini@asst-pini-cto.it

D. Tradati
Orthopaedic Department, IRCCS Ospedale San Raffaele, Milan, Italy

© ISAKOS 2022
A. Gobbi et al. (eds.), *Joint Function Preservation*, https://doi.org/10.1007/978-3-030-82958-2_9

Bosco et al. reported increased ROS levels associated with a decrease in TNF-α and IL-6 plasma values in 23 patients with hip AVN (15 patients staged III according to Ficat classification) treated with HBOT. In a prospective randomized study comparing NSAID to NSAID plus HBOT in 41 patients, Capone et al. reported significant better clinical and radiological outcomes in the HBOT group at 3 months follow-up (WOMAC score 56.4 vs 70.8; BML area reduction 28% vs 55%). Similarly, in a double-blind randomized controlled prospective study including 20 patients with unilateral hip AVN (Ficat stage II) treated with HBOT or compressed air, better outcomes were reported in the HBOT group with stable results at 7 years follow-up and no total hip replacement (THR) conversions [4].

Side effects should be considered too in the cost-benefits analysis. Light-headedness, claustrophobia, and fatigue are frequently reported by patients undergoing HBOT therapy; nevertheless, also major complications such as lung damage, ear and eye disorders, and sinus damage should be considered in specific patients.

9.1.2.2 Extracorporeal Shockwave Therapy

Shockwave therapy (ESWT) represents a feasible alternative in the treatment of BMLs, characterized by a low-profile risk and limited costs. A promoting effect on osteoblast migration, adhesion, and bone formation was previously documented [5–7]; moreover, an increase in pro-angiogenetic (NO, VEGF, FGF) and a decrease in pro-inflammatory factors (il-1, TG-B) was reported by different authors [8].

Kang et al. [9] comparing ESWT and alendronate therapy in 126 patients affected by knee osteoarthrosis (OA) and concomitant BMLs reported a shorter natural course of the disease in favor of the ESWT group in both clinical and radiological scores. Similarly, Vitali et al. [10] comparing ESWT and conservative treatment in 56 patients with no radiological sign of OA reported higher clinical and functional scores at 4 months in the ESWT group.

These results seem to confirm the previous systematic review by Zhang et al. [11] reporting a decrease of BML progression, a reduced surgery demand, and a significant decrease in MRI detected BML area. Despite this evidence, lower benefits were evidenced in advanced stage ON (ARCO grade 3–4) and no synergy with other conservative therapies was observed.

In consideration of safeness of the procedure and the cost-benefits balance, ESWT represents a reliable alternative in patients with comorbidities of when pharmacological is not available [12, 13].

9.1.2.3 Pulsed Electromagnetic Fields

Pulsed electromagnetic fields (PEMF) represent a noninvasive treatment aiming to reduce pro-inflammatory cytokines and oxidative damage [14] while promoting new bone formation thought osteoclast apoptosis and osteoblast migration [15–18]. The huge advantage of this therapy is the possibility to be formed at home by the patient. Few studies focused on the role of PEMF in the treatment of BMLs. A study considering 66 patients with femoral AVN, at mean FU of 28 months, reported about 53% of patients pain-free at 2 months, 94% hip preservation in stage 1 and 2 AVN, and a 20% conversion rate to THR only in patients staged 3 and 4 [19]. Similar results were proposed by Martinelli et al. in a limited cohort of patients with talus BMLs reporting a substantial clinical benefit at 3 months after the treatment and a resolution of the MRI edema in 90% of patients at 3 months FU [20]. Limited adverse effects (nausea, lethargy, headaches, fatigue, and muscle aches) were reported, but caution should be used in pregnant patients and subjects with pacemaker, defibrillator, or implanted cochlear hearing device.

9.2 Pharmacological Therapies

9.2.1 Prostacyclin Derivatives

Prostaglandins have an important role in inflammatory responses and cell differentiation, promoting microcirculation and bone regeneration [21].

Both intraosseous pressure normalization and decrease in local leukotrienes and cytokines concentration have been advocated as possible therapeutical targets of PCD. At the current time, Iloprost represents the most used PCD used in BML treatment.

In a prospective double-blind randomized controlled study comparing Iloprost and tramadol, no significant difference was reported; nevertheless, a greater improvement in BML was reported in the Iloprost group as well as an increased bone formation [22]. In a similar study by the same author, a mean decrease in SSF of 42% and 100% at 3 and 12 months, respectively, was reported in patients treated with Iloprost, while in patients treated with tramadol the decrease was −2.2% and 65.7%, respectively [22].

As reported by Classen et al. [23], a significative improvement in both pain and BML size was reported in 74.8% of patients with hip and knee AVN who underwent PCD treatment; nevertheless, better results should be expected in patients with early stages ON. A conversion rate to joint arthroplasty of 4% and 20% was reported in early stages AVN (ARCO I-II) in comparison to 71% and 100% in stage III and IV, respectively.

The PCD side effects usually occur during the first 30′ after the infusion begin and decrease after the flowing infusions. The most common reported side effects are represented by: moderate or severe headache (39%), nausea (21%), temporary pain increase (21%), and flush or local erythema (21%) [24, 25]. Caution should be used in patients with unstable angina or previous acute myocardial infarction during the last 6 months.

9.2.2 Bisphosphonates

Bisphosphonates (BPs) represent the most studied molecules in the treatment of BMLs, in consideration of their ability to affect the BLMs pathway at multiple levels: the anti-inflammatory activity related to cytokines inhibition (IL-1, IL-12, and TNF-a), the effect on micro circle permeability related to NGF suppression and the reduction of cell apoptosis.

In the last decades, good-to-excellent results were reported by many authors using different molecules such as Alendronate, Clodronate, Pamidronate, Ibandronate, and Zoledronate (TAB) [26–31]. Nevertheless, none of them is formerly labelled for the treatment of BMLs. The only BP labelled to be used in BMLs is represented by the neridronate.

In a randomized double-blind placebo-controlled study by Varenna et al. [32], Neridronate was firstly demonstrated effective in reducing pain and improving life quality in patients affected by Complex Regional Pain Syndrome (CRPS). Furtherly, in a similar study by the same author [33], Neridronate was evaluated in the treatment of BMLs in osteoarthritic knees. A statistical significant decrease in VAS score was observed 10 days after the treatment (from 59.4 + 14.7 to 30.4 + 15.6) and only 13% of patients required additional analgesics therapy, in comparison to 72% in the placebo group. A similar trend was observed in MRI knee WORMS score values, decreasing from 6.3 + 3.0 to 3.7 + 4.2 at 2 months after the treatment.

In a minority of patients minor side effects such as flu-like symptoms, headache, joints pain, and fever could present during the first IV subministration. These symptoms usually persist for 2 or 3 days after the treatment and do not occur at the next subministrations. A prophylactic dose of acetaminophen should be recommended during the first infusion to reduce the incidence and intensity of these symptoms. Uveitis and osteonecrosis of the jaw have been reported as a rare side effect associated with BP therapies; nevertheless, the limited duration of the treatment related to BMLs markedly reduces these risks. Care should be given to patients with poor kidney function or low calcium levels.

9.2.3 TNF-Inhibitors and RANK-L Antibodies

Few studies are available related to the use of TNF-inhibitors usually focused on specific classes of patients.

Positive results were reported in patients affected by rheumatoid arthritis, spondylitis, and psoriatic arthritis [34, 35]; nevertheless, the limited population number doesn't allow to extrapolate clinically relevant conclusions about these molecules. Similarly promising results were reported using RANK-L antibodies, a reduction of BMLs was observed in 93% of patients after a single subministration as long as a significant decrease in VAS score and no adverse effect [36]. In consideration of the high cost of these treatments, a careful evaluation of cost and benefits should be performed.

9.3 Surgical Therapies

The surgical treatment should be focused on:

- removing the underlying mechanism sustaining the edema
- avoid SIFK progression providing structural support to the subchondral bone and promoting bone formation

Compartment overloading due to malalignment, meniscal extrusion, or root lesions could lead to BMLs formation. In patients not responding to conservative treatment and presenting painful BMLs, the surgical treatment should be evaluated. Malalignment should be treated with corrective osteotomies, while meniscal lesion should undergo suture/reinsertion inpatients with low-grade OA (Kelgrenn-Lawrence grade 1–2) and meniscectomy in high-grade OA (Kelgrenn-Lawrence 3–4).

In the occurrence of BMLs and concomitant SIFK, the surgery defined subchondroplasty could represent a reliable solution burned with limited invasivity and side effect. This procedure should be considered in patients with delayed diagnosis of SIFK, therefore at high risk of AVN, or in high demanding patients with early diagnosis of SIFK.

Using a percutaneous approach, a synthetic calcium phosphate bone void filler is injected at the level of the BMLs under fluoroscopy, according to the pre-operative MRI planning. Calcium phosphate endothermically sets in 10 min at 37 °C, without thermal necrosis, and can easily be injected in the trabecular bone.

The filler provides early structural support to the subchondral bone allowing for an early weight bear recover and promoting bone remodelling due to its osteoconductive characteristics [37]. The procedure is usually performed in association to knee arthroscopy in order to treat concomitant meniscal injuries and to exclude the possibility of intra-articular leakage.

This should be considered one of the most relevant risks of the procedure that could lead to pain persistence and critical cartilage deterioration. Other reported side effects are represented by subcutaneous calcium phosphate leakage and pain increase in the early postoperative period. A single case of osteomyelitis related to this technique is reported in literature [38].

In a recent literature review, Astur et al. [39] analyzed the outcomes of 164 patients who underwent subchondral injection of calcium phosphate, reporting significant improvement in knee clinical outcomes, full return to activity after 3 months, and a TKR conversion rate reduced by 70%.

Further studies could be required to define the proper timing and selection criteria in order to maximize the procedure results.

9.4   Algorithm Approach

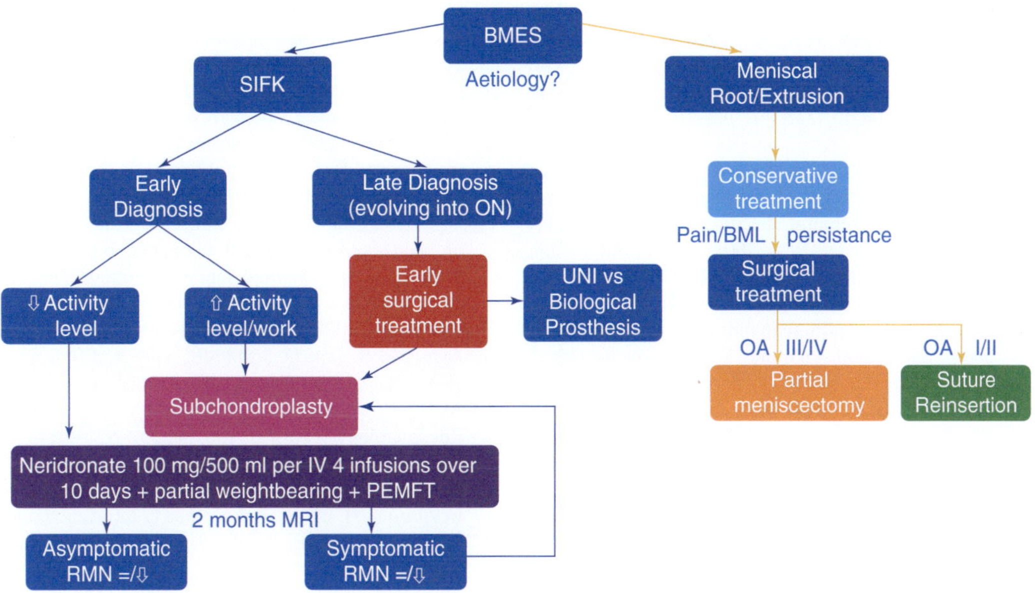

References

1. Bosco G, Vezzani G, Mrakic Sposta S, Rizzato A, Enten G, Abou-Samra A, Malacrida S, Quartesan S, Vezzoli A, Camporesi E. Hyperbaric oxygen therapy ameliorates osteonecrosis in patients by modulating inflammation and oxidative stress. J Enzyme Inhib Med Chem. 2018;33:1501–5. https://doi.org/10.1080/14756366.2018.1485149.

2. Mohamad JA, Kwan MK, Merican AM, Abbas AA, Kamari ZH, Hisa MK, Ismail Z, Idrus RM. Early results of metal on metal articulation total hip arthroplasty in young patients. Med J Malaysia. 2004;59 Suppl F:3–7.

3. Kurokouchi K, Kambe F, Yasukawa K, Izumi R, Ishiguro N, Iwata H, Seo H. TNF-alpha increases expression of IL-6 and ICAM-1 genes through activation of NF-kappaB in osteoblast-like ROS17/2.8 cells. J Bone Miner Res. 1998;13:1290–9. https://doi.org/10.1359/jbmr.1998.13.8.1290.

4. Camporesi EM, Vezzani G, Bosco G, Mangar D, Bernasek TL. Hyperbaric oxygen therapy in femoral head necrosis. J Arthroplast. 2010;25:118–23. https://doi.org/10.1016/j.arth.2010.05.005.

5. d'Agostino C, Romeo P, Lavanga V, Pisani S, Sansone V. Effectiveness of extracorporeal shock wave therapy in bone marrow edema syndrome of the hip. Rheumatol Int. 2014;34:1513–8. https://doi.org/10.1007/s00296-014-2991-5.

6. Wang CJ, Wang FS, Yang KD, Huang CC, Lee MS, Chan YS, Wang JW, Ko JY. Treatment of osteonecrosis of the hip: comparison of extracorporeal shockwave with shockwave and alendronate. Arch Orthop Trauma Surg. 2008;128:901–8. https://doi.org/10.1007/s00402-007-0530-5.

7. Xu JK, Chen HJ, Li XD, Huang ZL, Xu H, Yang HL, Hu J. Optimal intensity shock wave promotes the adhesion and migration of rat osteoblasts via integrin beta1-mediated expression of phosphorylated focal adhesion kinase. J Biol Chem. 2012;287:26200–12. https://doi.org/10.1074/jbc.M112.349811.

8. Wang CJ, Yang YJ, Huang CC. The effects of shockwave on systemic concentrations of nitric oxide level, angiogenesis and osteogenesis factors in hip necrosis. Rheumatol Int. 2011;31:871–7. https://doi.org/10.1007/s00296-010-1384-7.

9. Kang S, Gao F, Han J, Mao T, Sun W, Wang B, Guo W, Cheng L, Li Z. Extracorporeal shock wave treatment can normalize painful bone marrow edema in knee osteoarthritis: a comparative historical cohort study. Medicine (Baltimore). 2018;97:e9796. https://doi.org/10.1097/MD.0000000000009796.

10. Vitali M, Naim Rodriguez N, Pedretti A, Drossinos A, Pironti P, Di Carlo G, Fraschini G. Bone marrow Edema syndrome of the medial femoral condyle treated with extracorporeal shock wave therapy: a clinical and MRI retrospective comparative study. Arch Phys Med Rehabil. 2018;99:873–9. https://doi.org/10.1016/j.apmr.2017.10.025.

11. Zhang Q, Liu L, Sun W, Gao F, Cheng L, Li Z. Extracorporeal shockwave therapy in osteonecrosis of femoral head: a systematic review of now available clinical evidences. Medicine

(Baltimore). 2017;96:e5897. https://doi.org/10.1097/MD.0000000000005897.

12. Wang CJ, Wang FS, Huang CC, Yang KD, Weng LH, Huang HY. Treatment for osteonecrosis of the femoral head: comparison of extracorporeal shock waves with core decompression and bone-grafting. J Bone Joint Surg Am. 2005;87:2380–7. https://doi.org/10.2106/JBJS.E.00174.

13. Ludwig J, Lauber S, Lauber HJ, Dreisilker U, Raedel R, Hotzinger H. High-energy shock wave treatment of femoral head necrosis in adults. Clin Orthop Relat Res. 2001:119–26. https://doi.org/10.1097/00003086-200106000-00016.

14. Pagani S, Veronesi F, Aldini NN, Fini M. Complex regional pain syndrome type I, a debilitating and poorly understood syndrome. Possible role for pulsed electromagnetic fields: a narrative review. Pain Physician. 2017;20:E807–22.

15. He Z, Selvamurugan N, Warshaw J, Partridge NC. Pulsed electromagnetic fields inhibit human osteoclast formation and gene expression via osteoblasts. Bone. 2018;106:194–203. https://doi.org/10.1016/j.bone.2017.09.020.

16. Kubat NJ, Moffett J, Fray LM. Effect of pulsed electromagnetic field treatment on programmed resolution of inflammation pathway markers in human cells in culture. J Inflamm Res. 2015;8:59–69. https://doi.org/10.2147/JIR.S78631.

17. Lin CC, Chang YT, Lin RW, Chang CW, Wang GJ, Lai KA. Single pulsed electromagnetic field restores bone mass and microarchitecture in denervation/disuse osteopenic mice. Med Eng Phys. 2020; https://doi.org/10.1016/j.medengphy.2019.10.004.

18. Martini F, Pellati A, Mazzoni E, Salati S, Caruso G, Contartese D, De Mattei M. Bone morphogenetic protein-2 signaling in the osteogenic differentiation of human bone marrow mesenchymal stem cells induced by pulsed electromagnetic fields. Int J Mol Sci. 2020;21 https://doi.org/10.3390/ijms21062104.

19. Massari L, Fini M, Cadossi R, Setti S, Traina GC. Biophysical stimulation with pulsed electromagnetic fields in osteonecrosis of the femoral head. J Bone Joint Surg Am. 2006;88 Suppl 3:56–60. https://doi.org/10.2106/JBJS.F.00536.

20. Martinelli N, Bianchi A, Sartorelli E, Dondi A, Bonifacini C, Malerba F. Treatment of bone marrow edema of the talus with pulsed electromagnetic fields: outcomes in six patients. J Am Podiatr Med Assoc. 2015;105:27–32. https://doi.org/10.7547/8750-7315-105.1.27.

21. Eriksen EF. Treatment of bone marrow lesions (bone marrow edema). Bonekey Rep. 2015;4:755. https://doi.org/10.1038/bonekey.2015.124.

22. Mayerhoefer ME, Kramer J, Breitenseher MJ, Norden C, Vakil-Adli A, Hofmann S, Meizer R, Siedentop H, Landsiedl F, Aigner N. MRI-demonstrated outcome of subchondral stress fractures of the knee after treatment with iloprost or tramadol: observations in 14 patients. Clin J Sport Med. 2008;18:358–62. https://doi.org/10.1097/JSM.0b013e31817f3e1c.

23. Classen T, Becker A, Landgraeber S, Haversath M, Li X, Zilkens C, Krauspe R, Jager M. Long-term clinical results after Iloprost treatment for bone marrow Edema and avascular necrosis. Orthop Rev (Pavia). 2016;8:6150. https://doi.org/10.4081/or.2016.6150.

24. Baier C, Schaumburger J, Gotz J, Heers G, Schmidt T, Grifka J, Beckmann J. Bisphosphonates or prostacyclin in the treatment of bone-marrow oedema syndrome of the knee and foot. Rheumatol Int. 2013;33:1397–402. https://doi.org/10.1007/s00296-012-2584-0.

25. Disch AC, Matziolis G, Perka C. The management of necrosis-associated and idiopathic bone-marrow oedema of the proximal femur by intravenous iloprost. J Bone Joint Surg Br. 2005;87:560–4. https://doi.org/10.1302/0301-620X.87B4.15658.

26. Agarwala S, Jain D, Joshi VR, Sule A. Efficacy of alendronate, a bisphosphonate, in the treatment of AVN of the hip. A prospective open-label study. Rheumatology (Oxford). 2005;44:352–9. https://doi.org/10.1093/rheumatology/keh481.

27. Varenna M, Zucchi F, Ghiringhelli D, Binelli L, Bevilacqua M, Bettica P, Sinigaglia L. Intravenous clodronate in the treatment of reflex sympathetic dystrophy syndrome. A randomized, double blind, placebo controlled study. J Rheumatol. 2000;27:1477–83.

28. Varenna M, Zucchi F, Binelli L, Failoni S, Gallazzi M, Sinigaglia L. Intravenous pamidronate in the treatment of transient osteoporosis of the hip. Bone. 2002;31:96–101. https://doi.org/10.1016/s8756-3282(02)00812-8.

29. Bartl C, Imhoff A, Bartl R. Treatment of bone marrow edema syndrome with intravenous ibandronate. Arch Orthop Trauma Surg. 2012;132:1781–8. https://doi.org/10.1007/s00402-012-1617-1.

30. Meier C, Kraenzlin C, Friederich NF, Wischer T, Grize L, Meier CR, Kraenzlin ME. Effect of ibandronate on spontaneous osteonecrosis of the knee: a randomized, double-blind, placebo-controlled trial. Osteoporos Int. 2014;25:359–66. https://doi.org/10.1007/s00198-013-2581-5.

31. Muller F, Appelt KA, Meier C, Suhm N. Zoledronic acid is more efficient than ibandronic acid in the treatment of symptomatic bone marrow lesions of the knee. Knee Surg Sports Traumatol Arthrosc. 2020;28:408–17. https://doi.org/10.1007/s00167-019-05598-w.

32. Varenna M, Adami S, Rossini M, Gatti D, Idolazzi L, Zucchi F, Malavolta N, Sinigaglia L. Treatment of complex regional pain syndrome type I with neridronate: a randomized, double-blind, placebo-controlled study. Rheumatology (Oxford). 2013;52:534–42. https://doi.org/10.1093/rheumatology/kes312.

33. Varenna M, Zucchi F, Failoni S, Becciolini A, Berruto M. Intravenous neridronate in the treatment of acute painful knee osteoarthritis: a randomized controlled study. Rheumatology (Oxford). 2015;54:1826–32. https://doi.org/10.1093/rheumatology/kev123.

34. Kuroda T, Wada Y, Kobayashi D, Murakami S, Sakai T, Hirose S, Tanabe N, Saeki T, Nakano M, Narita I. Effective anti-TNF-alpha therapy can induce rapid resolution and sustained decrease of gastroduodenal

mucosal amyloid deposits in reactive amyloidosis associated with rheumatoid arthritis. J Rheumatol. 2009;36:2409–15. https://doi.org/10.3899/jrheum.090101.

35. Marzo-Ortega H, McGonagle D, Rhodes LA, Tan AL, Conaghan PG, O'Connor P, Tanner SF, Fraser A, Veale D, Emery P. Efficacy of infliximab on MRI-determined bone oedema in psoriatic arthritis. Ann Rheum Dis. 2007;66:778–81. https://doi.org/10.1136/ard.2006.063818.

36. Rolvien T, Schmidt T, Butscheidt S, Amling M, Barvencik F. Denosumab is effective in the treatment of bone marrow oedema syndrome. Injury. 2017;48:874–9. https://doi.org/10.1016/j.injury.2017.02.020.

37. Tofighi A, Rosenberg A, Sutaria M, Balata S, Chang J. New generation of synthetic, bioresorbable and injectable calcium phosphate bone substitute materials: alpha-bsm®, Beta-bsm™ and gamma-bsm™. J Biomim Biomater Tissue Eng. 2009;2:39–55.

38. Dold A, Perretta D, Youm T. Osteomyelitis after calcium phosphate subchondroplasty a case report. Bull Hosp Jt Dis (2013). 2017;75:282–5.

39. Astur DC, de Freitas EV, Cabral PB, Morais CC, Pavei BS, Kaleka CC, Debieux P, Cohen M. Evaluation and Management of Subchondral Calcium Phosphate Injection Technique to treat bone marrow lesion. Cartilage. 2019;10:395–401. https://doi.org/10.1177/1947603518770249.

Functional Anatomy of Cartilage and Subchondral Bone in the Joint

10

Iain R. Murray, Taylor E. Ray, Geoff D. Abrams, and Seth L. Sherman

10.1 Introduction

The osteochondral unit plays an integral role in the overall function of a synovial joint. Its unique and complex architecture permits smooth lubricated motion to facilitate human ambulation. Articular hyaline cartilage that lines the bone ends in these joints can withstand high loads while maintaining a near-frictionless articulating interface. Load dispersal is shared with the subchondral bone, the morphology of which reflects adaptations of loading in adults [1].

The articular joint surface is composed of hyaline cartilage connected through a zone of calcified cartilage to the subchondral cortical bone known as the subchondral plate, which gives way to metaphyseal trabecular bone. Articular cartilage, the calcified cartilage, and the underlying subchondral bone form a tight functional association, and are highly interdependent. Alterations in the structure, biomechanics, and physiology of any individual components of this unit results in disruption of joint integrity and loss of function [2]. In this chapter, we describe the anatomy of articular cartilage and the subchondral bone, outlining how this structure contributes to its function.

10.2 Anatomy of Cartilage and Subchondral Bone

10.2.1 Articular Cartilage

Hyaline cartilage of 2 to 4 mm in thickness coats the articular surfaces of synovial joints. It is composed of individual chondrocytes sparsely distributed within a dense extracellular matrix and is avascular, aneural, and alymphatic. The extracellular matrix is principally composed of water, collagen, and proteoglycans with other non-collagenous proteins and glycoproteins present in lesser amounts [3]. The structure of articular cartilage can be divided into histologically identifiable zones based on the general orientation of the collagen fibrils, the morphology and arrangement of the chondrocytes, and the staining properties of the matrix (Fig. 10.1) [2, 4]. The superficial zone faces the synovial fluid with the middle zone, the deep zone, and the calcified zone ultimately transitioning into subchondral bone.

10.2.1.1 Superficial Zone

The superficial zone makes up approximately 10–20% of articular cartilage thickness and protects deeper layers from sheer stresses [5]. This zone consists of a lamina splendins, a clear film

I. R. Murray
The University of Edinburgh, Edinburgh, UK
e-mail: imurray@stanford.edu

T. E. Ray · G. D. Abrams · S. L. Sherman (✉)
Stanford University, Stanford, CA, USA
e-mail: raytay@stanford.edu; geoffa@stanford.edu

© ISAKOS 2022
A. Gobbi et al. (eds.), *Joint Function Preservation*, https://doi.org/10.1007/978-3-030-82958-2_10

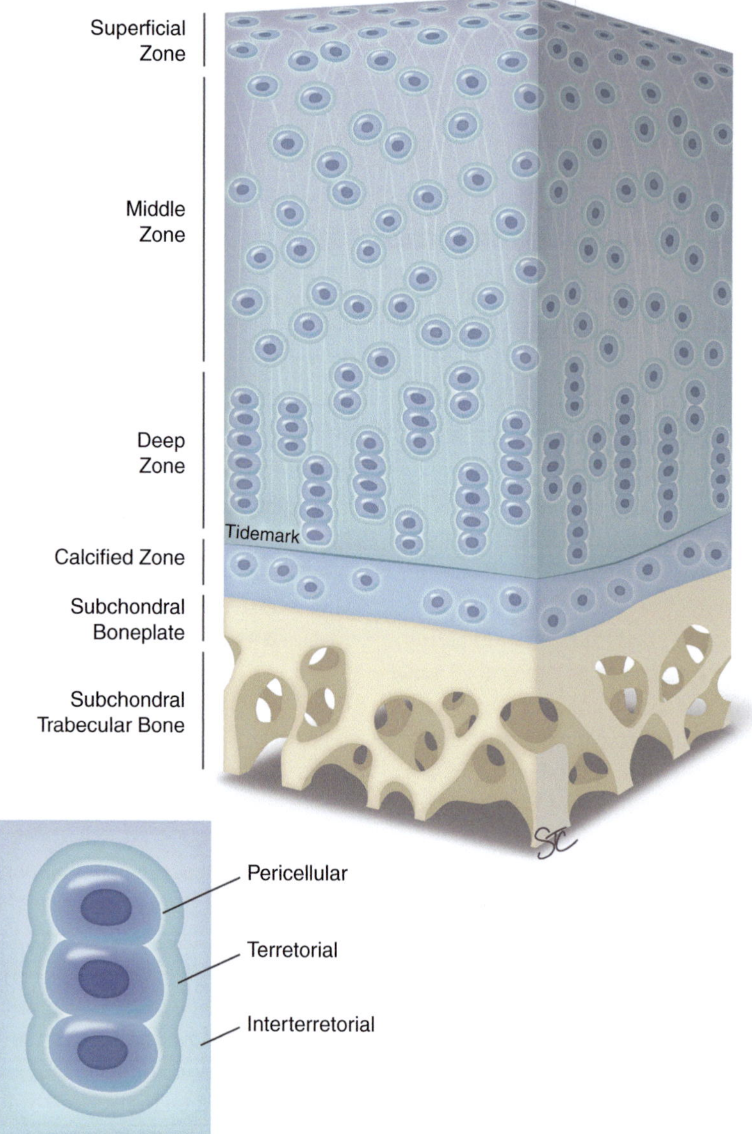

Fig. 10.1 Cellular composition and matrix morphology of articular cartilage and subchondral bone

of small collagen fibrils and cellular layer of flattened chondrocytes one to three cells thick. The collagen fibers of the superficial zone (primarily type II and IX) lie parallel to the articular surface and are tightly packed in an arrangement that resists sheer stresses [5]. It has the highest water and collagen content and the lowest level of proteoglycan synthesis of all the zones. In addition, the superficial zone is also thought to function as a barrier to the passage of large molecules from the synovial fluid. The integrity of this layer is imperative in the protection and maintenance of deeper layers.

10.2.1.2 Middle Zone

Immediately below the superficial zone lies the middle (or transitional) zone. This zone makes up between 40% and 60% of total cartilage volume and contains thicker collagen fibrils and more proteoglycans. Here, the fibers are arranged obliquely and the chondrocytes are more spherical and at a lower density. This zone forms the transition between the shearing forces of the surface layers and the compression forces in the deeper layers.

10.2.1.3 Deep Zone

The deep zone represents approximately 30% of articular cartilage volume and is responsible for providing the greatest resistance to compressive forces. The deep layer is characterized by rounded chondrocytes arranged in columns, a high proteoglycan content, and a radial collagen network. The chondrocytes in zones II and III produce all the components of the extracellular matrix. Although there are no intercellular junctions between chondrocytes, communities of two or more cells come together to form chondrons. Chondrocytes within chondrons share the same pericellular matrix, which differs in its composition and has a higher rate of turnover compared with the inter-territorial extracellular matrix between the chondrons.

10.2.1.4 Calcified Zone

The tidemark is a discrete band of mineralized cartilage between the deep zone and the calcified zone made visible by histological staining. The tidemark represents a calcification front, at which non-mineralized cartilage matrix comes to contain hydroxyapatite [6]. It is cell free and tends to migrate towards the surface with age. Immediately below the tidemark, the calcified cartilage is a 20–250 µm thick transitional zone that forms an interface between cartilage and the stiffer bone. It also functions to anchor the cartilage to subchondral bone and forms a barrier to diffusion from blood vessels supplying the subchondral bone. The small, rounded chondrocytes in the calcified zone are distributed in an extracellular matrix composed of types II and X collagen, glycosaminoglycans, and alkaline phosphatases that contribute to hydroxyapatite mineral deposition in the matrix [7]. The type II collagen fibrils of the non-calcified layers cross the tidemark to be anchored within the calcified cartilage [8]. No collagen fibers cross the boundary between the calcified cartilage and subchondral bone, with the two tissues only held together by three-dimensional interdigitation [9]. These undulations in the interfaces between the articular cartilage, calcified cartilage, and underlying cortical plate help to transform shear stresses into compressive and tensile stresses during joint loading and motion.

10.2.2 Subchondral Bone

Located directly below the calcified cartilage, the subchondral bone is organized into two anatomically distinct regions with unique architectural, mechanical, and biological properties, namely the subchondral bone plate and the subchondral trabecular bone [10, 11]. The subchondral bone plate separates the calcified cartilage from the marrow cavity and is similar to cortical bone at other skeletal sites. Like compact bone, it is composed of osteons consisting of concentric lamellae surrounding the central Haversian canal. The thickness and mineral density of the subchondral bone plate varies by age, weight, location, and stresses applied. In general, the central more heavily loaded contact areas are thicker and more mineralized [7]. The bone in the cortical bone plate merges into a network of trabecular bone that is more porous and metabolically active than cortical bone. The trabeculae are oriented in different directions depending on location, and they provide a unique structural network that is also adapted to the local mechanical influences through the continuous remodeling activity of osteoclasts and osteoblasts [12]. In general, the mean bone strength reduces rapidly with increasing distance from the surface and is higher in men than in women [13].

Subchondral bone receives sensory and sympathetic innervation which modulate bone regeneration, remodeling, and cartilage homeostasis [14]. In addition, the subchondral bone is highly vascular providing metabolic support to the overlying cartilage. Narrow canals and wider ampullae provide connections between the marrow cavity and the calcified cartilage, penetrating across the subchondral plate [15]. These penetrating blood vessels enable signaling molecules and nutrients to reach the deeper layers of cartilage accounting for approximately 50% of the water, oxygen, and glucose requirements of cartilage. Channels are narrower and form a tree-like network in regions where the subchondral plate is thicker, while they tend to be wider and resemble ampullae where the plate is thinner. In the absence of these branching vascular conduits, the cartilage must receive all nutrients through diffusion of synovial fluid [4]. In addition to osteocytes and osteoblasts, the subchondral bone contains mesenchymal cells with multipotent potential in vitro that are thought to migrate from the subchondral bone in the setting of osteochondral injury to contribute to the formation of fibrous repair tissue [16].

10.2.3 The Molecular Organization of Normal Articular Cartilage

The cartilage matrix surrounding chondrocytes in healthy articular cartilage is arranged into regions defined by their architecture and distance from the cell and composition, namely the pericellular, territorial, and inter-territorial regions (Fig. 10.1). The pericellular matrix lies immediately around the cell and is the region where molecules that interact with cell surface receptors are located, for example, hyaluron binds to the receptor CD44 [17]. It contains mainly proteoglycans, as well as glycoproteins and other non-collagenous proteins. The territorial matrix surrounds the pericellular matrix, slightly further from the cell. It is composed mostly of fine collagen fibrils, forming a basket-like network around the cells that is thought to protect them from mechanical stresses [18]. At largest distance from the cell is the inter-territorial matrix which contributes most to the biomechanical properties of articular cartilage [19]. This region is characterized by an abundance of proteoglycans and bundles of large collagen fibrils that are arranged parallel to the surface of the superficial zone, obliquely in the middle zone, and perpendicular to the joint surface in the deep zone.

10.3 Composition of Articular Cartilage

Articular cartilage in adults is composed of an extensive extracellular matrix with sparse distribution of highly specialized cells called chondrocytes that are responsible for the production of this matrix.

10.3.1 Cells

Chondrocytes are the single resident cell type in articular cartilage accounting for only 1–2% of the total cartilage volume [20]. Chondrocytes are highly metabolically active, responsible for the development, maintenance, and repair of the extracellular matrix. During fetal development, chondroblasts derived from mesenchymal progenitors proliferate with the majority of cartilage transforming into bone through endochondral ossification. At skeletal maturity, hyaline cartilage persists at the articular surface with the matrix and synovial fluid environment playing extremely important roles in the maintenance of the phenotype of articular chondrocytes.

The shape, number, and size of chondrocytes vary depending on the anatomical regions of the articular cartilage. Chondrocytes in the superficial zone are flatter and smaller and generally have a greater density than chondrocytes situated in the deeper zones. Individual chondrocytes establish a specialized microenvironment in its immediate vicinity and are responsible for the turnover of extracellular matrix within this area. They must be able to respond to changes in matrix composition by synthesizing appropriate types and amounts of macromolecules. Although these cells have high metabolic activity, they are relatively few in number and so the total activity within cartilage is low [21]. The chondrocyte is trapped within the matrix it produces preventing migration to adjacent areas of cartilage. It is unusual for chondrocytes to form direct connections for communication, but they do respond to stimuli such as mechanical loads, hydrostatic pressures, piezoelectric forces, and growth factors [5]. Chondrocytes are highly dependent on optimal chemical and chemical environments to maintain their health. As they have limited capacity for replication, the intrinsic healing capacity of cartilage is low.

It has not been fully elucidated how chondrocytes obtain nutrition to fuel their metabolism; however, contact between articular cartilage and its vascularized subchondral bone appears to be crucial. In addition, synovial fluid provides chondrocytes with nutrients via diffusion. A double-diffusion barrier requires passage through the synovium first, followed by passage through the extracellular matrix to the chondrocyte. As such, metabolism in articular cartilage is primarily anaerobic in an environment with very low oxygen concentration.

10.3.2 Extracellular Matrix

Water makes up 65–80% of the total weight of healthy articular cartilage. Collagens and proteoglycans are the principal load-bearing macromolecules in articular cartilage. Other classes of molecules found in smaller amounts make up the remaining ECM including lipids, phospholipids, proteins, and glycoproteins (Table 10.1).

10.3.2.1   Water
Water is the most plentiful component of articular cartilage. The relative water concentration decreases from approximately 80% at the superficial zone to 65% in the deep zone [3]. The majority of this water is contained within the pore space of the extracellular matrix, with approximately 30% within the intra-fibrillar space of collagen and a small proportion contained within the intracellular space [22, 23]. This water contains inorganic ions including sodium, calcium chloride, and potassium [24]. The flow of water through the cartilage aids in the transport and distribution nutrients to chondrocytes. Much of the interfibrillar water appears to exist as a gel, and most of it may be moved through the matrix by applying a pressure gradi-

ent across the tissue or by compressing the solid matrix [25]. The small pore size of the ECM causes high frictional resistance against this flow. It is the combination of the frictional resistance to water flow and the pressurization of water within the ECM that is responsible for the compressive strength and the ability of articular cartilage to withstand high joint loads.

10.3.2.2   Collagen
Collagen is the most abundant structural macromolecule in extracellular matrix, making up approximately 60% of the dry weight of cartilage. Type II collagen makes up approximately 90% of the collagen in ECM and forms fibrils and fibers intertwined with proteoglycan aggregates that form an extensive network throughout the territorial and interterritorial matrix. These fibrils vary in diameter, from approximately 20 nm in the superficial zone to 70 to 120 nm in the deep zone. The collagens form a cross-linked network that adds to the three-dimensional stability and tensile properties of articular cartilage. Collagen types IV, V, VI, IX, and XI contribute a minor proportion that help to form and stabilize the type II collagen fibril network. All collagen types contain a region consisting of 3 polypeptide α-chains wound into a triple helix. The amino acid composition of polypeptide chains is primarily glycine and proline, with hydroxyproline providing stability via hydrogen bonds along the length of the molecule. The triple helix structure of the polypeptide chain imparts important shear and tensile properties, which help to stabilize the matrix [5].

Type X collagen is present in the calcified cartilage layer and also forms a meshwork. It is associated with cartilage calcification and is produced by hypertrophied chondrocytes during endochondral ossification. Type 1 collagen is not found in normal articular cartilage but is present following injury in the subsequently formed fibrocartilage.

10.3.2.3   Proteoglycans
Proteoglycans are heavily glycosylated protein monomers accounting for 10–15% of the wet

Table 10.1 Components of the articular cartilage extracellular matrix and their function

Matrix component	Comments	Function
Collagen		
Type II	Principal component of macrofibril	Tensile strength
Type VI	Forms microfibrils in pericellular sites	Unknown
Type IX	Cross-linked to surface of macrofibril	Tensile properties and/or fibril-interfibrillar connections
Type X	Associated with macrofibril and present in pericellular latticework. Only synthesized by hypertrophic chondrocytes. Only usually present in calcified layer	Unclear but may add structural support
Type XI	Present within and on macrofibrils	Nucleates fibril formation
Type XII and XIV	Each is homotrimeric	Probably part of macrofibril
Proteoglycans		
Aggrecan	Majority proteoglycan by mass. Binds hyaluronan by a G1 domain. Most concentrated in deep zone.	Compressive stiffness
Non-aggregating proteoglycans		
Decorin	Has one chondroitin or dermatan sulfate chain near amino-terminus. Equimolar to aggrecan. Concentrated at articular surface and in pericellular sites.	Regulates formation of macrofibrils
Biglycan	Has two chondroitin or dermatan sulfate chains near amino-terminus.	Interacts with collagen VI
Fibromodulin	May contain as many as 4 keratan sulfate chains	Regulates formation of macrofibrils
Lumican	Contains keratan sulfate in immature	Regulates formation of macrofibrils
Perlecan	Located at cell surface. Contains heparan sulfate	Cell-matrix adhesion
Other molecules		
Lubricin	Synthesized by cells of superficial zone	Joint lubrication
Cartilage oligomeric Protein	Five armed molecule of thrombospondin family	Binds type II collagen and may be involved in macrofibril assembly
Link protein	Structure homologous to G1 domains of aggrecan and versican	Stabilizes binding of aggrecan and versican G1 domains to hyaluronan
Maytrix-γ-Carboxyglutamic acid Protein	Pericellular location. Also known as matrix Gla protein	Inhibits calcification
Fibrillin-1	Forms microfibrillar network	Unknown
Hyaluronic acid (hyaluronan)	Forms macromolecular aggregates with aggrecan and/or versican: Binds G1 domain of these molecules and link protein. Interacts with collagen fibril.	Retention of aggrecan and versican in matrix
Chondroadherin (CHAD)	Leucine-rich protein	Cell-matrix binding
CD44	The cell surface receptor for hyaluronic acid	Cell-matrix binding
Chondronectin		Cell-matrix binding
Fibronectin		Cell-matrix binding

weight of cartilage and are responsible for providing compressive strength. There are two major classes of proteoglycans in articular cartilage: large aggregating proteoglycan monomers or aggrecans, and smaller non-aggregating proteoglycans including decorin, biglycan, fibromodulin, lumican, and perlecan. Proteoglycans have a "bristle-brush" structure consisting of a protein core with one or more linear covalently attached glycosaminoglycan chains (chondroitin sulfate, keratan sulfate) that remain separated from each other through charge repulsion (Fig. 10.2).

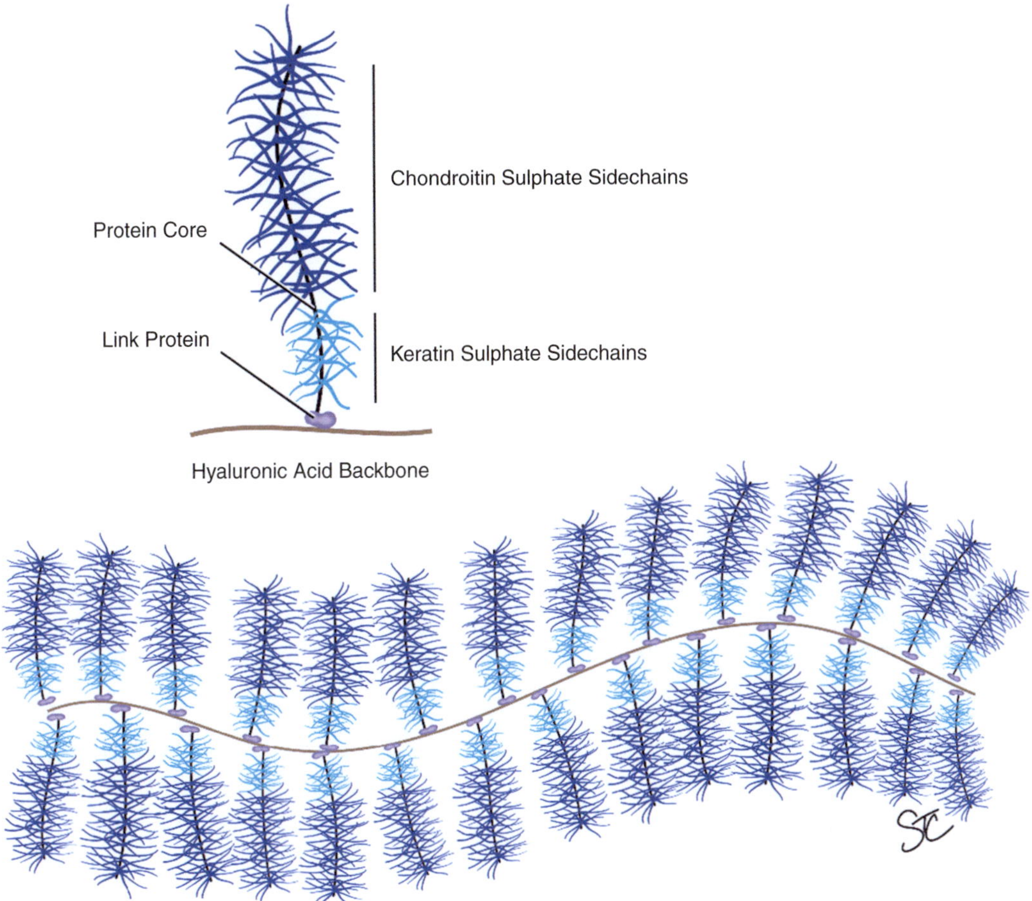

Fig. 10.2 Proteoglycan structure

Aggrecan is the largest and most abundant proteoglycan with over 100 chondroitin sulfate and keratin sulfate chains attached to its core protein. Aggrecans interact with hyaluronic acid to form large proteoglycan aggregates stabilized via link proteins [26]. Aggrecans occupy the interfibrillar space generating high negative charges that attract water generating high osmotic pressures that enable cartilage to resist high compressive loads[7].

Non-aggregating proteoglycans including decorin, biglycan, and fibromodulin are smaller and are characterized by their ability to interact with collagen. Although their protein structure is similar to aggrecan, their glycosaminoglycan composition and function are different. Decorin and biglycan possess one and two dermatan sulfate chains, respectively, whereas fibromodulin possesses several keratin sulfate chains. Decorin and fibromodulin play a role in fibrillogenesis and interfibril interactions by interacting with the type II collagen fibrils in the matrix. Biglycan is principally found close to chondrocytes, where they may interact with collagen VI.

10.3.2.4   Other Non-collagenous Proteins and Glycoproteins

Non-collagenous proteins and glycoproteins are sparsely distributed through the extracellular matrix. Although generally poorly understood, a number are thought to be involved in maintaining structure by acting as adhesives, binding matrix components and chondrocytes. These include, fibronectin that binds to integrin transmembrane receptors, and chondronectin and anchorin that mediate attachments of chondrocytes with collagen fibrils.

Lubricin (previous known as "superficial zone protein") is mucinous glycoprotein synthesized by synoviocytes and articular chondrocytes that is abundant in the synovial fluid, synovial membrane, and superficial zone of articular cartilage [27]. It functions to protect the cartilage surface from protein deposition and cell adhesion, inhibit synovial cell overgrowth, and in preventing cartilage-cartilage adhesion [28].

10.3.2.5 Homeostasis of the Extracellular Matrix

The maintenance and turnover of the extracellular matrix is regulated by chondrocytes and degradative enzymes. In healthy cartilage, a balance is achieved between the degradation of the different macromolecules with their replacement by newly synthesized products. The metabolic activity of the chondrocytes can be altered by a variety of factors within their surrounding chemical and mechanical environment.

Proteinases including metalloproteinases and cathepsins degrade collagen and proteoglycan aggregates as part of the normal turnover of the matrix constituents. The matrix metalloproteinases are classified into collagenases, gelatinases, stromelysins, and membrane-associated metalloproteinases. All are secreted as latent proenzymes that require activation extracellularly. Collagenases cause collagen fibril degradation. Gelatinases degrade denatured type II and type IV collagen and also have significant activity against fibronectin, elastin, and collagen types V, VII, X, and XI. Stromelysin functions to degrade the protein core of aggrecan. Cathepsins are active in the degradation of aggrecan. Balancing this action are the protein inhibitors. Tissue inhibitors of matrix metalloproteinases (TIMPS) are acidic polypeptides that prevent degradation by metalloproteinases.

Matrix components including proteoglycans are produced by chondrocytes and secreted into the ECM. A number of growth factors contribute to the regulation of proteoglycan metabolism although the molecular mechanism by which these proteins exert their effects is not fully understood. Insulin-like growth factors, transforming growth factor-β, interleukin-1, and tumor necrosis factor-α have all been demonstrated to influence proteoglycan metabolism. TGFβ stimulates proteoglycan synthesis and decreases the catabolic activity of interleukin-1 and MMPs. Basic fibroblasts growth factor (bFGF) stimulates DNA synthesis in articular chondrocytes. Insulin growth factor-1 (IGF-1) stimulates extracellular matrix synthesis.

Chondrocyte activity is also influenced by mechanical forces. Joint motion and load are important to maintain normal articular cartilage structure and function. Inactivity of the joint has also been shown to lead to the degradation of cartilage. Regular joint movement and dynamic load is important for the maintenance of healthy articular cartilage metabolism while excess force can damage cells. A further function of the extracellular matrix is therefore to protect chondrocytes from the potentially damaging biomechanical forces transmitted through joints.

10.4 Composition and Metabolism of Subchondral Bone

10.4.1 Cells

Subchondral bone contains at least five cell types critical to the production and turnover of subchondral bone including osteoblasts, osteoclasts, osteocytes, bone lining cells, and osteoprogenitors. Osteoblasts and osteoclasts form bone remodeling units that maintain the integrity of the bone and balance between deposited and resorbed bone. Osteocytes, form an interconnected network in the bone matrix and with cells on the bone surface. As osteocytes are widely distributed throughout cortical and trabecular bone, they are well positioned to respond to local biomechanical influences and soluble mediators, and accordingly regulate bone remodeling and adaptation via cell–cell interactions with osteoclasts and osteoblasts. Bone marrow of the trabecular bone maintains a heterogeneous population of multipotent mesenchymal cells that may function as progenitors for osteochondral cell lineages as well as supporting a trophic environment for hematopoiesis [29, 30].

Table 10.2 Components of the subchondral bone matrix and their function

Matrix component	Comment	Function
Organic matrix		
Collagen	Primarily type 1 collagen	Provides tensile strength
Proteoglycans	Glycosaminoglycan–protein complexes	Contribute to compressive strength
Non-collagenous matrix Proteins	Osteocalcin (bone gamma-carboxyglutamic acid containing protein) Osteonectin	Promote mineralization and bone formation
Growth factors and cytokines	Include TGFβ, IGF, IL-1, IL-6, BMPs	Aid in bone cell differentiation, activation, growth, and turnover
Inorganic matrix		
Calcium hydroxyapatite	$Ca_{10}(PO_4)_6(OH)_2$	Provides compressive strength
Osteocalcium phosphate	$Ca_8H_2(PO_4)_65H_2O$	Provides compressive strength

TGFβ transforming growth factor beta, *IGF* insulin-like growth factor, *IL-1* interleukin-1, *IL-6* interleukin-6, *BMPs* Bone morphogenetic proteins.

10.4.2 Matrix

Bone matrix is composed of organic and inorganic components (Table 10.2). Collagen (primarily type 1) comprises 90% of the organic matrix and provides the tensile strength of bone. The type 1 collagen forms an organized template for mineralization by inorganic components of calcium hydroxyapatite and calcium phosphate that make up 60% of the dry weight of bone. Proteoglycans and other non-collagenous proteins including osteocalcin and osteonectin are also present.

10.4.3 Bone Remodeling and Homeostasis

The subchondral bone plate and subchondral trabecular bone is continuously remodeled throughout life. The adaptive capabilities of subchondral bone reflect Wolff's law which states that the magnitude and direction of applied load determine the internal architecture and external conformation of bone [31]. This response is facilitated through the formative and resorptive activities of osteoblasts and osteoclasts. The rich vascularization and innervation of subchondral bone facilitates a comprehensive and extensive local response to both physiologic and pathologic alterations within the bone.

10.5 Functions of Articular Cartilage and Subchondral Bone

The principal functions of articular cartilage and subchondral bone are to provide a smooth lubricated surface for low friction articulation and to facilitate the transmission of loads to the underlying bone.

10.5.1 Shock Absorption and Load Transmission

Articular cartilage is able to tolerate high cyclic loads, with little or no evidence of degeneration or damage [3]. The subchondral bone plays a key role in mechanically and metabolically supporting the articular cartilage, maintaining joint shape, and absorbing shock [4].

As a shock absorber, articular cartilage is capable of absorbing considerable compressive, tensile, and sheer stresses. The compressive properties of articular cartilage are a direct result of its biphasic nature. Water is the principal component of the fluid phase that also contains the inorganic ions sodium, calcium, chloride, and potassium. The solid phase consists of the extracellular matrix which is permeable and porous. When a load is applied to articular cartilage, the trapped synovial liquid flows through the porous extracellular matrix causing the generation of significant frictional resistance. Even for very small flow speeds, very large drag forces are exerted on the solid matrix, dissipating the stress. When the material starts to deform, the porosity decreases and the permeability is reduced, decreasing the flow rate and increasing the drag forces. The low permeability of articular cartilage prevents fluid from being squeezed out of the matrix [5]. Therefore, cartilage responds to load by increasing the hydraulic pressure and becoming stiffer.

Articular cartilage is viscoelastic and exhibits time-dependent behaviors including creep and stress relaxation. The viscoelastic nature of cartilage is related to the low permeability of the tissue, and the pressurization of the fluid phase that increases as the loading rate is increased. When a constant load is applied to cartilage, the tissue initially deforms rapidly by liberating the confined liquid before the process slows down until equilibrium is reached. Similarly, when cartilage is deformed and held at a constant strain, high stresses are produced due to the hydraulic pressure, which are reduced after liquid flow. This effect is known as stress relaxation and strongly influences the compressive behavior of the material.

The compressive properties of articular cartilage vary along the depth of the tissue and is primarily related to differences in the fluid flow in each zone. Thus, the highly permeable superficial zone is exposed to compressive strains of up to 50%. Fluid flow greatly decreases in the middle and deep zones, resulting in compressive strains of less than 5%. The deformation of cartilage is further attenuated by the subchondral bone that remains impermeable and stabilizes the tissue.

Cartilage is also anisotropic, having different mechanical properties depending on the direction in which it is loaded. Compressive forces generate significant tensile (hoop) stresses within cartilage. Typically, collagenous fibrous tissues show nonlinear tensile load deformation behavior. The orientation of the collagen network is the primary determinant of the tensile behavior cartilage within each zone. Initially, in the "toe region" of the stress–strain curve, a small load causes a large deformation. In cartilage, it has been shown that the initial portion of the "toe region" is caused by the drag force required to slide the collagen meshwork through the proteoglycans; collagen fibers themselves are not initially particularly stretched. As the collagen fibers eventually become taut and assume a uniform structure, the slope of the load deformation curve becomes constant. The organization of the collagen fibers varies in the different zones of articular cartilage therefore influencing the mechanical properties.

Articular cartilage is also able to tolerate the shear stresses that occur with translational and rotational movements of bones. Shear stress occurs when forces are applied parallel to the surfaces of a material. Although the predominant load on articular cartilage is compressive, significant shear stresses are developed particularly in the deep zone near the tidemark. The overall stiffness of articular cartilage in shear is directly proportional to the amount of collagen present in the tissue with proteoglycans not contributing to shear resistance.

10.5.2 Low-Friction Gliding Surface

The second principal function of articular cartilage is to provide a smooth lubricated surface for low friction articulation. Articular cartilage has a very low coefficient of friction, 30 times smoother than most modern joint replacements, and less than 1/5 of that of ice on ice. This coefficient of friction can be lowered further by fluid-film formation, elastic deformation of articular cartilage, synovial fluid and efflux of fluid from cartilage.

The primary mechanisms of lubrication in synovial joints are boundary and fluid-film lubrication. Each type of lubrication comes into play at a different point in the movement of the joint. In boundary lubrication a monolayer of lubricant (likely the glycoprotein lubricin) separates each surface boundary of the joint. This prevents direct articular contact and is most important at rest or under load. In fluid-film lubrication, a thin layer of fluid increases the separation of the two surfaces. Hydrodynamic, squeeze film and elastohydrodynamic forms of fluid-film lubrication have all been described. In hydrodynamic lubrication, the two surfaces are at an angle to each other and the viscosity in the resulting wedge of fluid separates the two surfaces. In squeeze film lubrication, the two surfaces are parallel and move perpendicularly to each other. The viscosity of the incompressible fluid maintains the lubrication with high loads carried for short lengths of time. As the layer of fluid lubricant is forced out, it becomes thinner and the joint surfaces come into contact, but they are still protected by lubricin.

Elastohydrodynamic lubrication occurs as speed increases and is similar to squeeze film, but the yielding articular surfaces create a larger surface area when compressed by the fluid. There is less dissipation of the fluid-film, and therefore the load is sustained for a longer period. This is the predominant lubrication mechanism in synovial joints during dynamic joint function.

Synovial joint, being non-rigid structures, exhibit modified forms of boundary and fluid-film lubrication. When movement begins, boundary lubrication is exhibited at points of close contact of the two surfaces and fluid-film elsewhere.

Two other forms of lubrication are also thought to occur between the static state with boundary lubrication and the elastohydrodynamic lubrication seen at speed: weeping and boosted lubrication. Articular cartilage is variably permeable to fluid, depending on whether it is loaded. As the articular cartilage of the joint slides under compression, fluid is exuded under and in front of the leading edge of the load, enhancing lubrication. As the load decreases after maximum compression, water is once again imbibed and the articular cartilage reforms its shape. In boosted lubrication, the solvent part of the lubricant enters the articular cartilage, which leaves behind the concentrated hyaluronic acid complexes as a lubricant in "trapped pools" of concentrated synovial fluid.

10.6 Conclusions

Articular cartilage and the subchondral bone have uniquely evolved to effectively dissipate forces through joints and provide articulations with extremely low friction that are able to tolerate high cyclic loads. These functions are reliant on their complex anatomical structures made up of cells and extracellular matrix that constantly adapt to the local biomechanical and biological conditions. Alterations in the architecture, biomechanics or physiology of any individual components of this unit results in disruption of joint integrity and loss of function.

References

1. Castañeda S, Roman-Blas JA, Largo R, Herrero-Beaumont G. Subchondral bone as a key target for osteoarthritis treatment. Biochem Pharmacol. 2012;83(3):315–23.
2. Gomoll AH, Madry H, Knutsen G, van Dijk N, Seil R, Brittberg M, et al. The subchondral bone in articular cartilage repair: current problems in the surgical management. Knee Surg Sports Traumatol Arthrosc. 2010;18(4):434–47.
3. Buckwalter JA, Mankin HJ. Articular cartilage: tissue design and chondrocyte-matrix interactions. Instr Course Lect. 1998;47:477–86.
4. Imhof H, Sulzbacher I, Grampp S, Czerny C, Youssefzadeh S, Kainberger F. Subchondral bone and cartilage disease: a rediscovered functional unit. Investig Radiol. 2000;35(10):581–8.
5. Sophia Fox AJ, Bedi A, Rodeo SA. The basic science of articular cartilage: structure, composition, and function. Sports Health. 2009;1(6):461–8.
6. Lyons TJ, Stoddart RW, McClure SF, McClure J. The tidemark of the chondro-osseous junction of the normal human knee joint. J Mol Histol. 2005;36(3):207–15.
7. Imhof H, Breitenseher M, Kainberger F, Rand T, Trattnig S. Importance of subchondral bone to articular cartilage in health and disease. Top Magn Reson Imaging. 1999;10(3):180–92.
8. Menetrey J, Unno-Veith F, Madry H, Van Breuseghem I. Epidemiology and imaging of the subchondral bone in articular cartilage repair. Knee Surg Sports Traumatol Arthrosc. 2010;18(4):463–71.
9. Oegema TR Jr, Carpenter RJ, Hofmeister F, Thompson RC Jr. The interaction of the zone of calcified cartilage and subchondral bone in osteoarthritis. Microsc Res Tech. 1997;37(4):324–32.
10. Goldring SR, Goldring MB. Changes in the osteochondral unit during osteoarthritis: structure, function

and cartilage-bone crosstalk. Nat Rev Rheumatol. 2016;12(11):632–44.

11. Goldring MB, Goldring SR. Articular cartilage and subchondral bone in the pathogenesis of osteoarthritis. Ann N Y Acad Sci. 2010;1192:230–7.

12. Burr DB. Anatomy and physiology of the mineralized tissues: role in the pathogenesis of osteoarthrosis. Osteoarthr Cartil. 2004;12 Suppl A:S20–30.

13. Madry H. The subchondral bone: a new frontier in articular cartilage repair. Knee Surg Sports Traumatol Arthrosc. 2010;18(4):417–8.

14. Schaible HG. Mechanisms of chronic pain in osteoarthritis. Curr Rheumatol Rep. 2012;14(6):549–56.

15. Madry H, van Dijk CN, Mueller-Gerbl M. The basic science of the subchondral bone. Knee Surg Sports Traumatol Arthrosc. 2010;18(4):419–33.

16. Madry H, Gao L, Eichler H, Orth P, Cucchiarini M. Bone marrow aspirate concentrate-enhanced marrow stimulation of chondral defects. Stem Cells Int. 2017;2017:1609685.

17. Heinegard D, Saxne T. The role of the cartilage matrix in osteoarthritis. Nat Rev Rheumatol. 2011;7(1):50–6.

18. Guilak F, Mow VC. The mechanical environment of the chondrocyte: a biphasic finite element model of cell-matrix interactions in articular cartilage. J Biomech. 2000;33(12):1663–73.

19. Mow VC, Guo XE. Mechano-electrochemical properties of articular cartilage: their inhomogeneities and anisotropies. Annu Rev Biomed Eng. 2002;4:175–209.

20. Hunziker EB, Lippuner K, Shintani N. How best to preserve and reveal the structural intricacies of cartilaginous tissue. Matrix Biol. 2014;39:33–43.

21. Bhosale AM, Richardson JB. Articular cartilage: structure, injuries and review of management. Br Med Bull. 2008;87:77–95.

22. Maroudas A, Wachtel E, Grushko G, Katz EP, Weinberg P. The effect of osmotic and mechanical pressures on water partitioning in articular cartilage. Biochim Biophys Acta. 1991;1073(2):285–94.

23. Torzilli PA. Influence of cartilage conformation on its equilibrium water partition. J Orthop Res. 1985;3(4):473–83.

24. Lai WM, Hou JS, Mow VC. A triphasic theory for the swelling and deformation behaviors of articular cartilage. J Biomech Eng. 1991;113(3):245–58.

25. Mow VC, Ratcliffe A, Poole AR. Cartilage and diarthrodial joints as paradigms for hierarchical materials and structures. Biomaterials. 1992;13(2):67–97.

26. Knudson CB, Knudson W. Cartilage proteoglycans. Semin Cell Dev Biol. 2001;12(2):69–78.

27. Gleghorn JP, Bonassar LJ. Lubrication mode analysis of articular cartilage using Stribeck surfaces. J Biomech. 2008;41(9):1910–8.

28. Zappone B, Ruths M, Greene GW, Jay GD, Israelachvili JN. Adsorption, lubrication, and wear of lubricin on model surfaces: polymer brush-like behavior of a glycoprotein. Biophys J. 2007;92(5):1693–708.

29. Murray IR, Peault B. Q&a: mesenchymal stem cells - where do they come from and is it important? BMC Biol. 2015;13(1):99.

30. Murray IR, West CC, Hardy WR, James AW, Park TS, Nguyen A, et al. Natural history of mesenchymal stem cells, from vessel walls to culture vessels. Cell Mol Life Sci. 2013;

31. Frost HM. From Wolff's law to the Utah paradigm: insights about bone physiology and its clinical applications. Anat Rec. 2001;262(4):398–419.

Mapping of the Osteochondral Defect

11

Shanmugasundaram Saseendar,
Saseendar Samundeeswari,
Anandapadmanabhan Jayajothi,
and Asode Ananthram Shetty

It was eerie. I saw myself in that machine.
I never thought my work would come to this.
Isidor Isaac Rabi
(American physicist, won Nobel Prize in Physics (1944) for discovery of nuclear magnetic resonance)

11.1 Introduction

Articular cartilage has limited capacity for spontaneous repair and hence demand an early and accurate diagnosis and intervention. This chapter intends to summarize the various magnetic resonance imaging (MRI) techniques to identify and quantify articular cartilage injury in an orthopedic surgeon's perspective.

S. Shanmugasundaram (✉)
Specialist Knee and Shoulder Surgery and Cartilage
Reconstruction, Department of Orthopaedics,
Apollo Hospital, Muscat, Sultanate of Oman

S. Saseendar
Orthopaedic Surgeon, CARE Sports Injury Clinic,
Puducherry, India

A. Jayajothi
Consultant Radiologist, Department of
Radiodiagnosis, Swansea Bay University Health
Board NHS, Port Talbot, UK

A. A. Shetty
Consultant Orthopaedic Surgeon, Faculty of Health
and Social Sciences, Chatham Maritime, Canterbury
Christ Church University, Kent, UK

11.2 Basics of Magnetic Resonance Imaging

Magnetic resonance (MR) imaging is the most important imaging modality for the evaluation of traumatic or degenerative cartilaginous lesions in the knee [1]. Currently, standardized cartilage-sensitive pulse sequences are available for all joints.

Volumetric (quantitative) MR imaging is likely to become more available to standardized work stations, permitting the longitudinal assessment of cartilage volume over time.

One of the major advantages of MRI is that it allows the manipulation of contrast to highlight different tissue types and also provides multiplanar capability with spatial resolution that approaches that of computed tomography (CT), without the potentially harmful ionizing radiations of radiographs and CT [2].

It is essential to understand the basic principles of the functioning of MRI to understand how cartilage mapping functions.

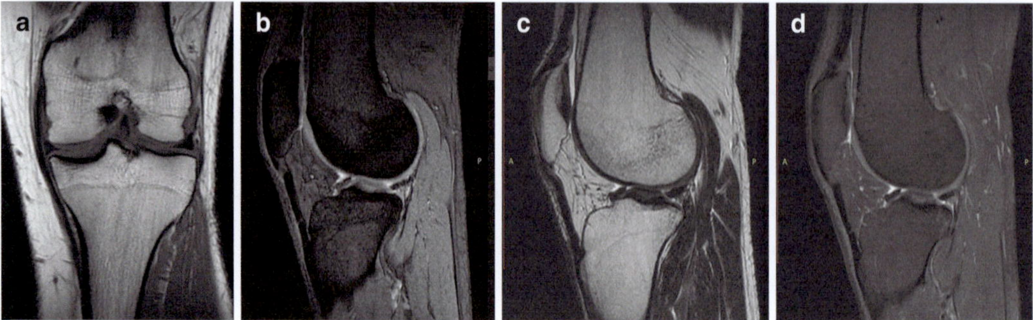

Fig. 11.1 (**a**) Coronal T1, (**b**) Sagittal GRE, (**c**) Sagittal T2, and (**d**) Sagittal PD fat-saturated images of knee joint

MRI scanners can be grouped roughly based on field strength into ultralow-field scanners (<0.1 Tesla (T)), low-field scanners (0.3 to 0.7 T), and high-field scanners (>1.0 T).

Low-field systems have shown poorer diagnostic performance in comparison to high-field systems, especially when assessing partial-thickness cartilage damage [3].

High-field scanners generate higher signal-to-noise images and allow shorter scanning times, thinner scan slices, and smaller fields of view, the most commonly used scanner being the 1.5 T MRI.

(a) **Surface coils** are devices that act as antennae placed close to the joint or limb and markedly improve signal and resolution and help achieve good image quality and visualization of cartilage. Surface coils for wrist, shoulder, knee, and ankle are currently standard.

(b) **How an MRI scan is performed:** The patient is placed in a strong magnetic field many times stronger than the earth's magnetic field. The magnetic force affects the nuclei of elements with odd numbers of protons or neutrons within the field, the most abundant being hydrogen, which is plentiful in water and fat. These hydrogen nuclei, which are essentially protons, align themselves with respect to the strong magnetic field. In this steady state, a radiofrequency (RF) pulse excites the magnetized protons and perturbs the steady state. A receiver coil listens for an emitted RF signal that is generated as these excited protons relax or return to equilibrium. This emitted signal is used to create the MR image.

Musculoskeletal MRI examinations primarily use spin-echo (SE) technique (Fig. 11.1), which produces T1-weighted, proton density (PD), and T2-weighted images. T1 and T2 are tissue-specific characteristics. These values reflect measurements of the rate of relaxation to the steady state. By varying the timing of the application of RF pulses (TR, or repetition time) and the timing of acquisition of the returning signal (TE, or echo time), an imaging sequence can accentuate T1 or T2 tissue characteristics. In most cases, fat has a high signal (bright) on T1-weighted images and fluid has a high signal on T2-weighted images. Structures with little water or fat, such as cortical bone, tendons, and ligaments, are hypointense (dark) in all types of sequences.

Improvements in MR techniques led to the development of a relatively new techniques called the fast spin-echo (FSE) that allows faster imaging, thereby improving patient tolerance and decreasing motion artifacts. Fat signal in FSE images remains fairly intense, requiring fat-suppression techniques, e.g., chemical-shift fat-suppression and short tau inversion recovery (STIR) sequence. These fat-suppression techniques help in the detection of edema in both bone marrow and soft tissue and hence are named "fluid-sensitive" sequences (Fig. 11.2). Another fast imaging method, gradient-echo technique, can be used selectively for cartilage imaging (such as for the glenoid labrum).

Fig. 11.2 Coronal section of knee: STIR image showing marrow edema (arrow) in the medial tibial condyle

The general consensus is IM-weighted sequences have echo times (TE) in the range of 30–60 ms and T2-weighted sequences have TE of 70–80 ms and PD-weighted sequences have TE of 10–30 ms [3]. In general, fat-suppressed, fluid-sensitive IM-weighted FSE sequences have been the most useful standard imaging for cartilage. With IM- and T2-weighted FSE sequences, normal hyaline cartilage is intermediate in signal and fluid is bright, allowing good contrast to identify surface abnormalities as well as pathologies of the cartilage matrix. However, they cannot characterize the severity of cartilage degeneration as validated by histology [4].

Diagnostic performance for cartilage lesions increases when different imaging planes are used in comparison to a single imaging plane alone. Isometric/volume acquisition also reduces time duration and the ability to multiplanar reconstruction of images, thereby reducing the time taken as well as reducing the risk of motion artifacts (Fig. 11.3).

(c) **MRI sequences in cartilage mapping:** MR imaging techniques can be divided into two broad categories based on their usefulness for a) morphologic and b) compositional evaluation. **Morphologic assessment techniques** provide accurate information on the structure of the cartilage and identify fissuring, focal or diffuse, partial- or full-thickness cartilage loss and hence are used for semiquantitative or quantitative assessment of the cartilage. They include conventional SE, GRE, FSE, and more advanced isotropic three-dimensional (3D) SE and GRE sequences. Objective evaluation scores have been proposed to describe focal cartilage defects in the knee, the most commonly used being the Outerbridge score. The score was primarily developed for arthroscopic assessment of the cartilage, but has been modified and extended for use with MRI [5].

Modified Outerbridge grading of cartilage:
- **Grade I:** Focal areas of hyperintensity with normal contour.
- **Grade II:** Swelling/ fraying of articular cartilage extending to surface.
- **Grade III:** Partial-thickness cartilage loss with focal ulceration.
- **Grade IV:** Full-thickness cartilage loss with underlying bone reactive changes.

On the other hand, **compositional assessment techniques** identify changes in the composition of the cartilage with special address to the water content and proteoglycan and collagen content. These techniques include T2 mapping, delayed gadolinium-enhanced MR imaging of cartilage (dGEMRIC), T1rho imaging, sodium imaging, and diffusion-weighted imaging.

11.3 Morphologic Assessment of Cartilage

(a) **Two-dimensional SE and Fast SE Imaging:** 2D or multisection T1-weighted, PD-weighted, and T2-weighted imaging sequences with or without fat-suppression are the most commonly used imaging sequences for the assessment of joint cartilage (Fig. 11.1). **T1-weighted images** show intrasubstance anatomic detail of hyaline cartilage but do not provide good contrast

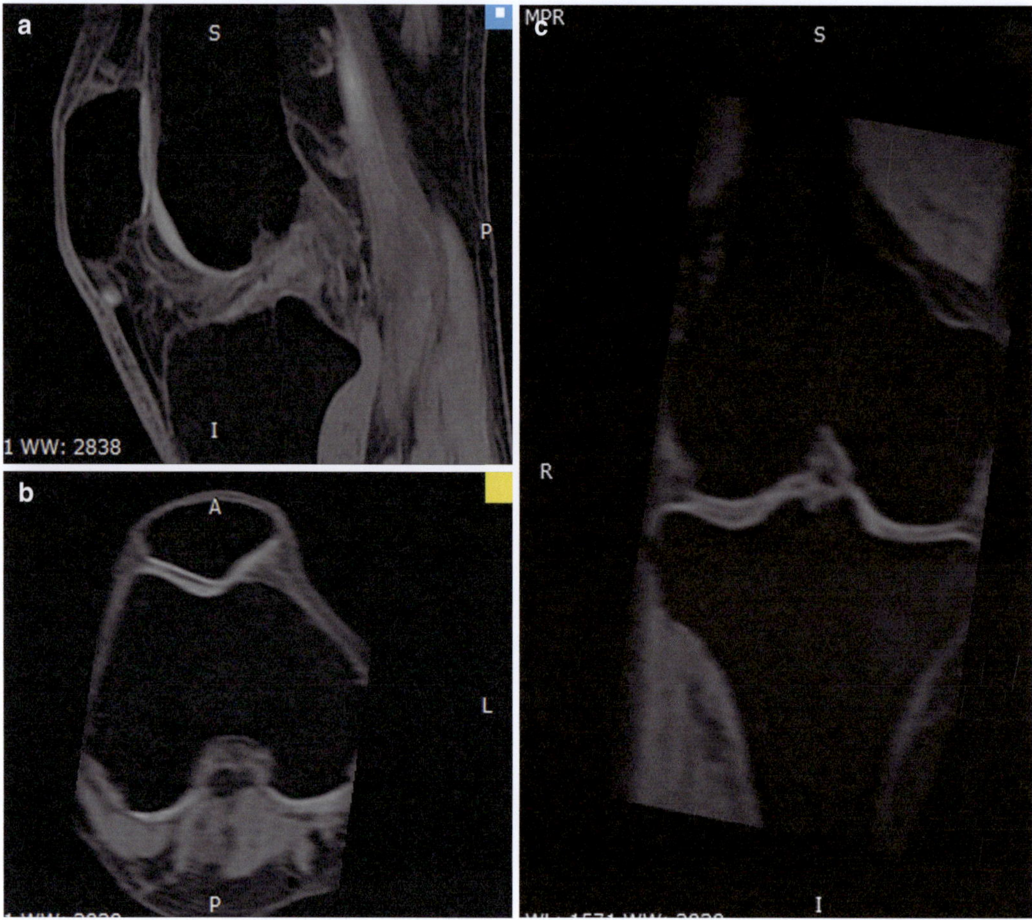

Fig. 11.3 Multiplanar image reconstruction using single isometric volume acquisition: (**a**) sagittal, (**b**) axial, (**c**) coronal

between joint fluid and the cartilage and also carry poor capability for depicting ligament injuries and may lead to overestimation of meniscal abnormalities. **T2-weighted imaging** provides good contrast between the cartilage surface and joint effusion, which is useful for detecting focal areas of delamination or other defects, whereas internal cartilage signals are weakened. **Proton density-weighted imaging** is mostly the main workhorse in MSK imaging, carrying the benefit of both depicting surface cartilaginous defects as well as abnormalities of internal cartilage composition. **Intermediate-weighted sequences** are being used more commonly in recent times. They provide the combination of the contrast advantage of

proton density weighting and also a higher signal intensity in cartilage than standard T2-weighted sequences, allowing better differentiation between cartilage and subchondral bone.

(b) **Two-dimensional fast or turbo SE imaging** sequences are techniques where multiple echoes are acquired with each sequence repetition. Hence, acquisition time is shorter than that with standard SE sequences and signal-to-noise (SNR) and contrast-to-noise (CNR) are higher. It is the technique most often used in clinical practice for the assessment of knee joint abnormalities, including cartilaginous lesions.

(c) **Proton density-weighted** and **T2-weighted FSE imaging techniques** are well suited for morphologic assessments of articular cartilage as well as menisci and ligamentous structures, providing information of a quality comparable to that obtained in surgery [6]. Although fast SE sequences provide excellent SNR and contrast between tissues of interest, 2D fast SE imaging may suffer from anisotropic voxels, section gaps, and partial volume effects. Furthermore, this technique requires the acquisition of image data in multiple planes.

(d) **MR Arthrography**: Direct MR arthrography with use of T1-weighted pulse sequences (Fig. 11.4)[7] following intra-articular injection of gadolinium chelates has been shown to represent a reliable imaging technique for the detection of surface lesions of articular cartilage with high sensitivity and specificity [8]. The injected fluid produces high contrast between joint space, cartilage, and subchondral bone, and at the same time distends the joint and thus, improves the separation of corresponding joint surfaces, such as the chondral surfaces of the femur and the acetabulum at the hip joint. However, this technique is of limited use for osteoarthritis imaging due to its invasive nature.

(e) **Three-dimensional MR Imaging**: These sequences generate isotropic voxels and allow high-quality reformations in any plane. Thus, it may be possible to only obtain one high spatial-resolution image dataset and get the additional planes as reformations. This would potentially save acquisition time and shorten patient examinations substantially. These techniques are considered the standard technique for morphologic evaluations of knee cartilage because they offer higher sensitivity than 2D techniques and provide excellent depiction of cartilaginous defects, comparable to that achieved with arthroscopy [9]. The commonly used 3D imaging sequences include 3D FSE, 3D GRE, 3D SPGR. The terminologies of these sequences could vary depending on the manufacturer, though the technique and imaging parameters remain the same.

(f) **Limitations:**
1. Small focal lesions and fissures are obscured because of the lack of reliable contrast between cartilage and fluid.
2. The gradient-echo sequences are not suited to visualize bone marrow pathology and are very limited in assessing menisci, ligaments, and tendons and are best suited only for quantitative measurement of volume and thickness of cartilage [10],
3. They overestimate cartilage, ligament, and meniscal tear.
4. Long acquisition times may lead to motion artifacts and less accurate measurements although these problems may be less severe with current MR imaging systems.
5. Fourth, the technique is highly vulnerable to susceptibility artifacts. In a recent study, high-resolution images of knee joint cartilage were obtained with an increased SNR, better cartilage-to-fluid contrast, and shorter acquisition time

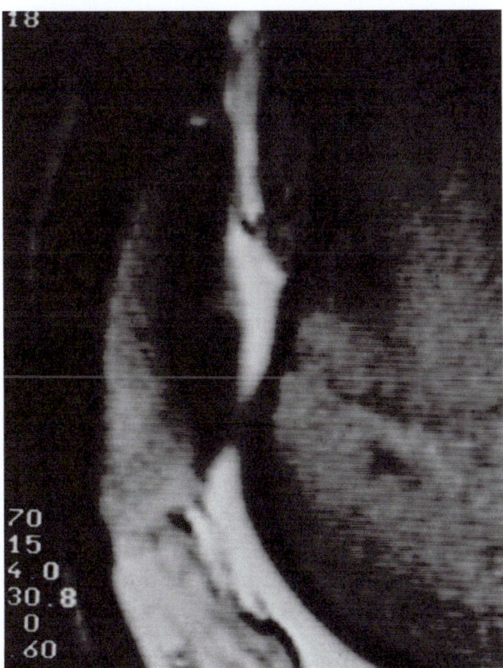

Fig. 11.4 Direct MR arthrography of the knee joint, T1-weighted SE showing hyperintense synovial fluid. There is a cartilage defect within the patella representative of first-stage chondromalacia (Reproduced with permission from Imhof et al. [7])

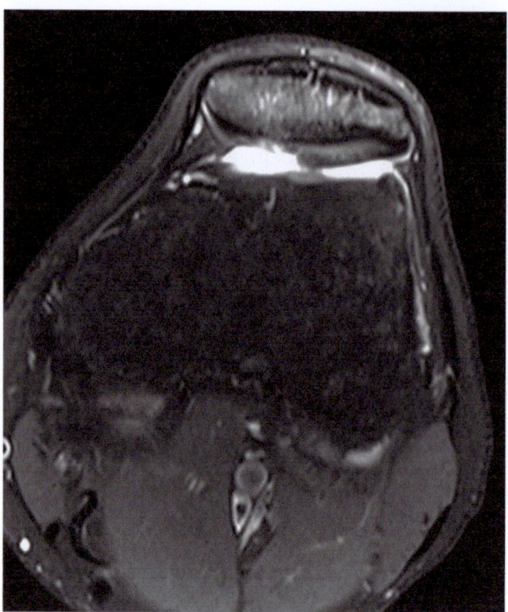

Fig. 11.5 Axial fat-suppressed 3D DESS sequence depicting a partial-thickness cartilage defect in the inter-facetal patellar cartilage (Courtesy: Dr. Ananthram Shetty, Spire Alexandra Hospital, UK, Dr. Stelzeneder, Medical University of Vienna, Austria)

with combined IDEAL and SPGR sequences than with a standard fat-saturated SPGR sequence alone [11].

Other 3D sequences have been described as modifications and improvements over these sequencing techniques for better visualization of the cartilage and improving the SNR and CNR, e.g., fast low-angle shot (FLASH) imaging, 3D-driven equilibrium Fourier transform (DEFT), balanced steady-state free precision (bSSFP), and 3D dual-echo steady-state (DESS) imaging (Fig. 11.5).

11.4 Compositional Assessment of the Cartilage Matrix

In addition to assessing cartilage pathology as well as thickness and volume, recent studies have shown the potential of MRI parameters to reflect changes in biochemical composition of cartilage with early OA.

These techniques include T2 quantification, T1rho quantification, and delayed Gadolinium-enhanced MRI of cartilage (dGEMRIC) [12]. These techniques allow characterization of the cartilage matrix and, potentially, quality before morphological damage occurs.

(a) **T2 Quantification:** This technique is based on the finding that increasing T2 relaxation time is proportional to the distribution of cartilage water and is sensitive to small water content changes [13] and is inversely proportional to the distribution of proteoglycans. Thus, measurement of the spatial distribution of the T2 reflecting areas of increased and decreased water content may be used to quantify cartilage degeneration before morphologic changes are appreciated (Figs. 11.6 and 11.7). Aging is associated with an asymptomatic diffuse increase in T2 of the transitional zone of articular cartilage in the senescent cartilage which is different from the focal increased T2 observed in damaged articular cartilage [14].

(b) **T1rho Quantification:** A different parameter that has been proposed to measure cartilage composition is 3D T1rho relaxation mapping (Fig. 11.8) [13, 15]. Loss of glycosaminoglycans (GAG) is reflected in measurements of T1rho due to less-restricted motion of water protons.

Both T2- and T1rho-measurements carry the benefit of identifying biochemical changes before the actual development of cartilage degeneration in asymptomatic subjects [16] and also being noninvasive and not requiring contrast injection.

(c) **Delayed Gadolinium-enhanced MRI of Cartilage (dGEMRIC):** Cartilage consists of approximately 70% water and the remainder predominantly of type II collagen fibers and GAG. These GAG macromolecules contain negative charges that attract sodium ions (Na+). One of the most commonly used MRI

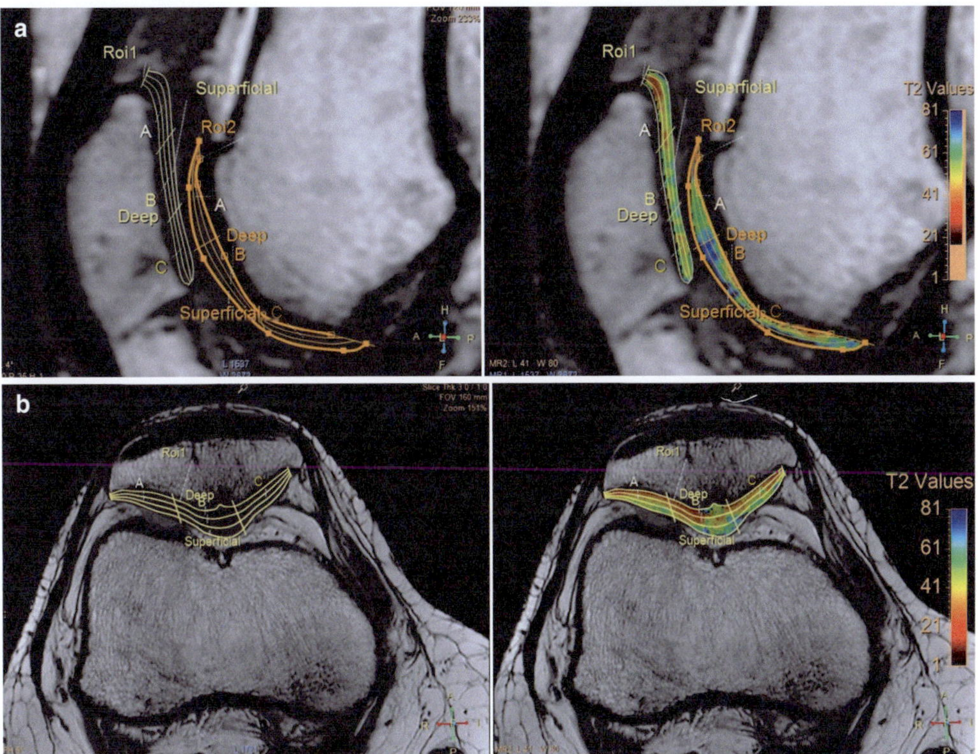

Fig. 11.6 Standard T2 and corresponding T2 mapping shows a normal appearance of the patellar and femoral trochlear articular cartilage. The deep layer of the cartilage appears blue and the superficial layer appears green on T2 mapping images. (Courtesy: Dr. Alvaro Zamorano, Dr. Jorge Diaz, University of Chile Clinical Hospital, Chile)

Fig. 11.7 Sagittal (**a**) and axial (**b**) MRI image of the knee showing the patellar and femoral articular cartilage T2 mapping showing red to orange marking in deeper layer of cartilage (lower T2 relaxation) and the green marking in superficial layer of cartilage (higher T2 relaxation). Zone B of both patellar and femoral articular cartilage show blue regions (higher abnormal T2 relaxation) indicating early cartilage damage (Courtesy: Dr. Raju Vaishya, Dr. Nitin Ghonge, Indraprastha Apollo Hospitals, India)

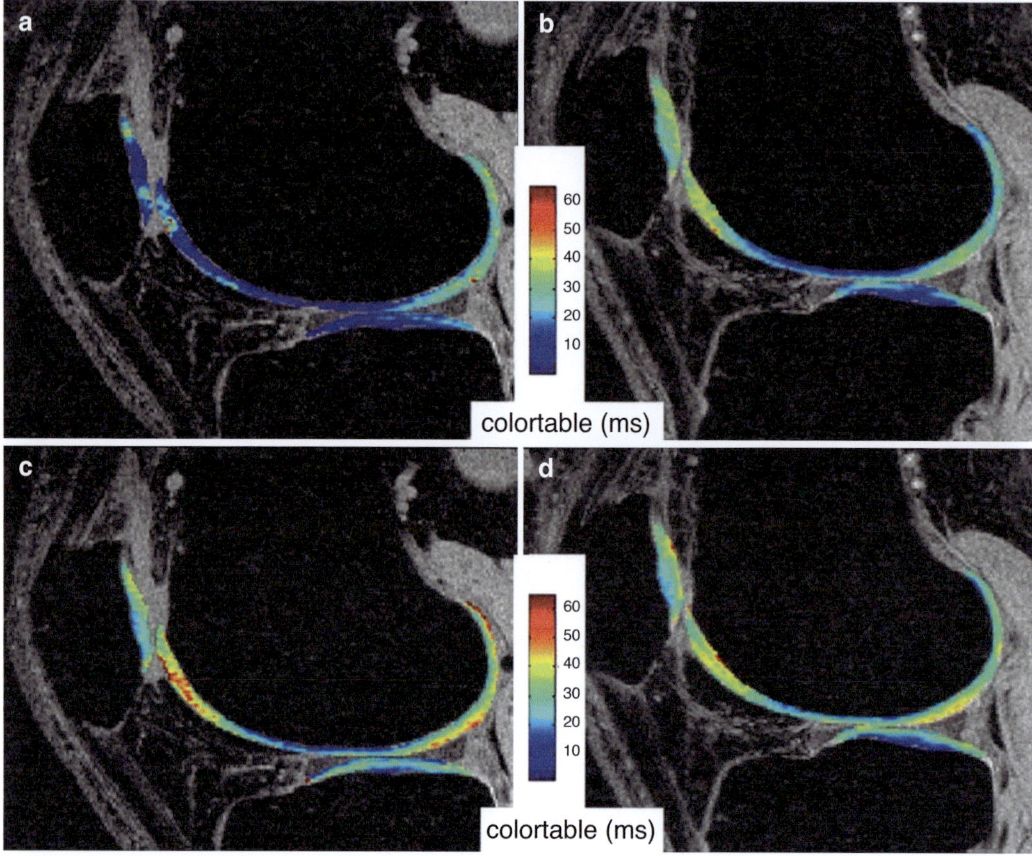

Fig. 11.8 Color-coded T2 (**a**, **b**) and T1rho (**c**, **d**) maps overlaid on a sagittal SPGR image in a 35-year-old male before (**a**, **c**) and after (**b**, **d**) a marathon. After the marathon (**b**), T2 times were significantly increased, mainly in the patella and the trochlea, indicating cartilage edema with increased water content secondary to the physical stress. The patella, femur, and lateral tibia plateau showed only a small increase in T1rho times after the marathon (**d**), indicating only a subtle change in cartilage macromolecular matrix (Reproduced without changes from Link et al. [15] (Licensed under CC by 4.0)

contrast agents Gd-DTPA2 has a negative charge and will therefore not penetrate cartilage in areas of high GAG concentrations. It gets distributed in higher concentrations in areas with lower GAG concentration and thus pathologic cartilage composition. Concentrations of Gd-DTPA2 in cartilage can be measured, reflecting the composition of cartilage (Fig. 11.9) [17, 18]. dGEMRIC measurements of GAG have correlated well with concentrations measured with biochemistry and histology [12, 19].

11.5 Clinical Cartilage Imaging

(a) **Cartilage Imaging in Traumatic Lesions**: Articular cartilage lesions are common after injury especially in the knee. MRI serves as a noninvasive option for the evaluation of the cartilage and other structures of the joint. Early identification of such lesions and cartilage repair when indicated may offer the possibility for patients to avoid the development of osteoarthritis or delay its progression. Newly developed cartilage repair techniques,

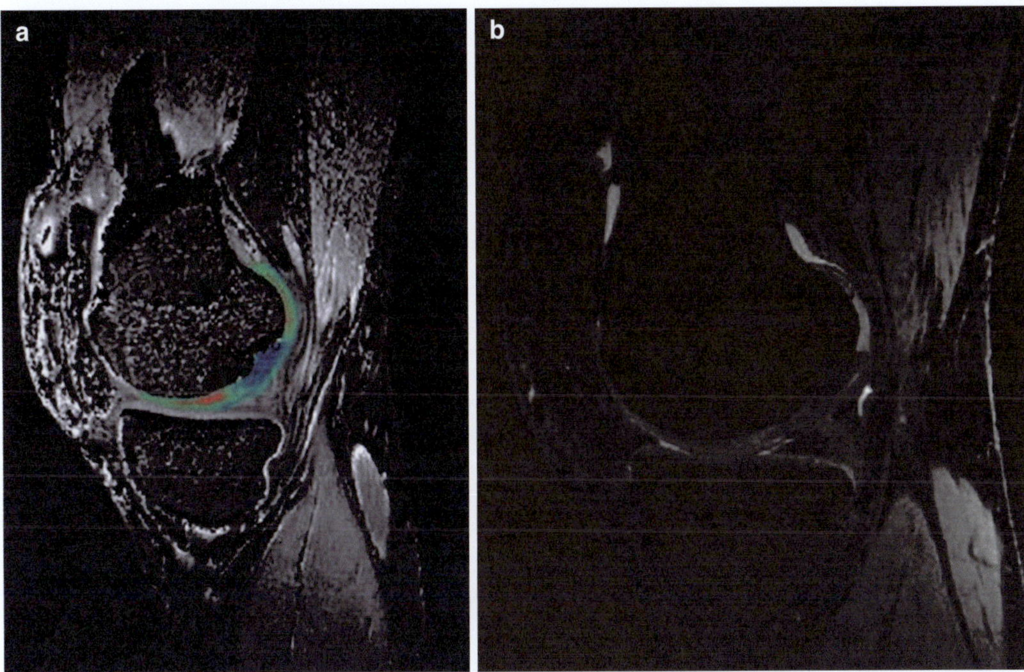

Fig. 11.9 A dGEMRIC image of a matrix-associated ACT 2 years after surgery. (**a**) The cartilage layer of the graft shows different T1 values, representing proteoglycan concentration, compared with hyaline cartilage. (**b**) a 3D-GRE image of the same patient, which shows morphology of cartilage implant with hypointense signal alteration of the cartilage implant in comparison with normal hyaline cartilage (Reproduced without changes from Trattnig et al. [17] (Licensed under CC by 4.0)

including marrow-stimulation techniques, osteochondral grafting, autologous chondrocyte implantation and require high-quality follow-up to assess healing [20].

Adequate preoperative imaging is required to study the lesions carefully. It is important to differentiate between an isolated cartilage injury from an osteochondral fracture as the treatment options and protocol for rehabilitation would vary tremendously. Osteochondral injuries can be recognized by the presence of hyperintense fatty marrow attached to the cartilage fragment or by the absence of the thin, low-signal-intensity subchondral plate between the cartilage and the bone (Fig. 11.10).

(b) **Cartilage Imaging in Osteoarthritis**: Lot of recent research has gone into MR imaging of osteoarthritis. Various noninvasive and invasive regenerative options have been proposed for the treatment of osteoarthritis. Establishing their efficacy would need objective morphological and compositional assessment of the cartilage, in order to assess both the extent of structural cartilage healing and the quality of the regenerate.

MRI, especially T2 mapping, can identify both early osteoarthritis changes characterized by cartilage softening and later by cartilage thinning, and also more severe changes such as subchondral sclerosis, cyst, and osteophyte formation. Both quantitative and qualitative assessment of such lesions is possible. Figure 11.11 depicts T2 mapping of an adult patient with significant medial knee osteoarthritis.

(c) **Cartilage Imaging in Repair**: Hyaline cartilage is an avascular and aneural structure that carries little to no inherent capacity for spontaneous repair [21]. The field of cartilage repair has been rapidly expanding in an attempt to bring about healing of the defect

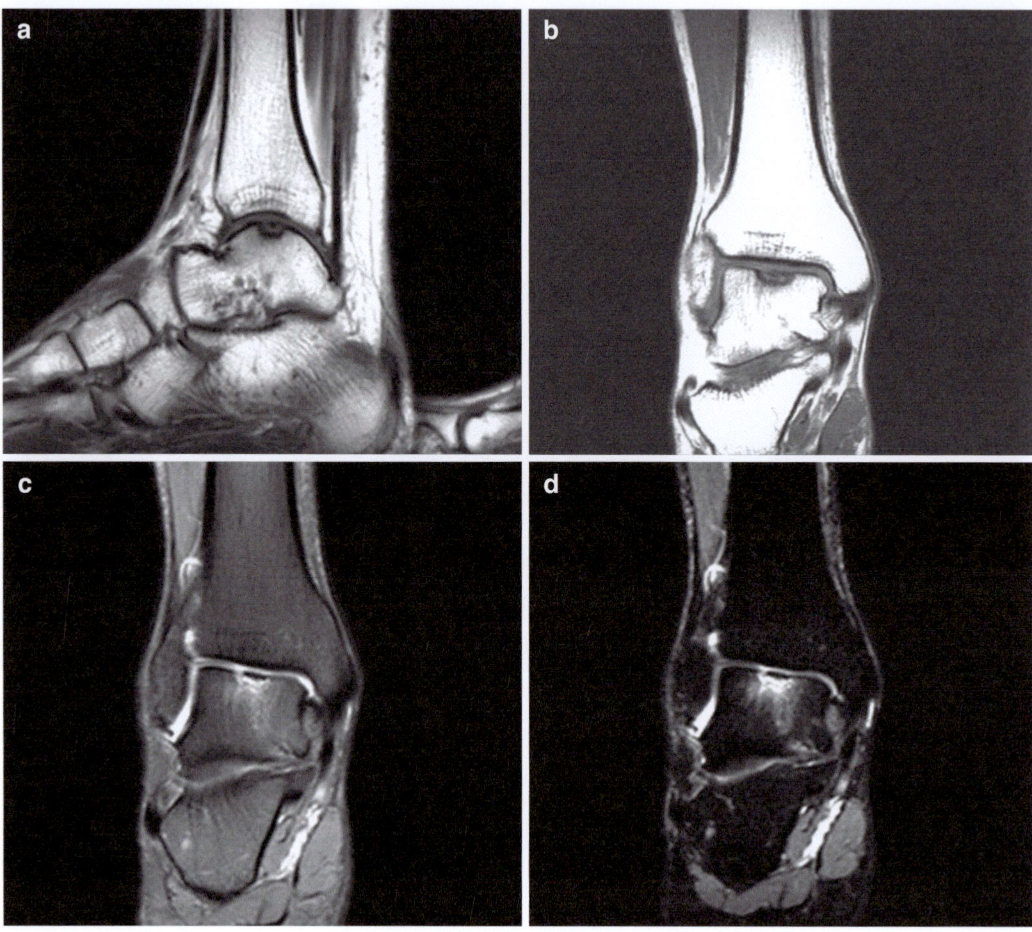

Fig. 11.10 (a) Sagittal T1, (b) Coronal T1, (c) PD fat-saturated, (d) STIR images of the ankle joint demonstrating osteochondral lesion of the talus with an undisplaced fragment and high signal rim around the osteochondral defect typical for a grade III lesion

created by chondral and osteochondral damage. The techniques include simple debridement, abrasion chondroplasty and microfracture, autologous osteochondral transplantation, allograft transplantation, and autologous chondrocyte implantation. The basic biological principles of these methods vary tremendously. Confirming improvement in the structure and composition of the new cartilage tissue would need dedicated MR imaging techniques.

Various scoring systems exist to objectively evaluate the repair tissue. The MOCART classification is the most frequently scoring system based on MRI for postoperative cartilage repair tissue evaluation [22, 23]. It is a 9-part and 29-item scoring system, resulting in a repair tissue score between 0 and 100 points where 100 points indicates the best imaginable score and 0 points indicates the worst imaginable score.

1. **Microfracture**: Microfracture is one of the most popular resurfacing techniques. It consists of debriding calcified cartilage and drilling small holes into the subchondral bone. The principle behind this technique is to allow release of multipotential stem cells from the marrow that would encourage healing of the defect with reparative fibrocartilage. The out-

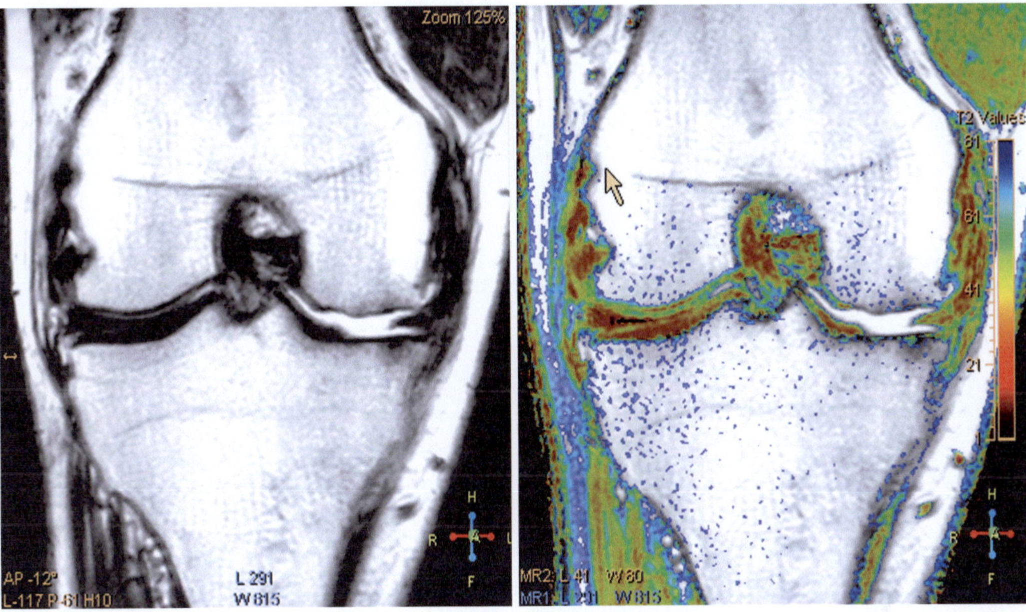

Fig. 11.11 T2 mapping of the knee joint demonstrating degenerative changes with cartilage thinning, subchondral sclerosis, and near-complete loss of medial tibial and femoral articular cartilage with rarefaction (Courtesy: Dr. Manuel Mosquera, Clinica la Carolina, Dr. Ruben Guzman, Clinica el Rosario, Colombia)

comes of this technique have been shown to be dependent on good MRI fill grade in addition to low body mass index (BMI) and short duration of preoperative symptoms [24].

The response to microfracture is characterized by initial hyperintensity due to increased mobility of water in the newly formed matrix [24, 25] in addition to underlying bone marrow edema that decreases progressively. Overgrowth of subchondral bone has also been reported following microfracture along with corresponding thinning of the overlying repair tissue [24, 25].

The significance of subchondral bone overgrowth is not yet certain, but could result from excessive removal of subchondral bone, leading to overstimulation for endochondral ossification [26]. Preservation of the subchondral bone has been emphasized recently in the conceptualization of nanofracture technique. MRI can also help in assessing peripheral integration of the repair tissue. T2 mapping can help in assessing the quality of cartilage repair. T2 mapping following microfracture has usually produced prolonged T2 relaxation

times in comparison to the adjacent and opposite hyaline cartilage (Fig. 11.12) [24].

2. **Osteochondral Autografts and Allografts**: Osteochondral autograft or allograft transplantation consists of harvesting one or more plugs from a less important part of a joint, most commonly the intercondylar notch region, and transferring into the defect in a weight-bearing portion of the joint. Bone-cartilage plug allografts are usually reserved for large defects while autografts are the choice for smaller defects.

In addition to patient reported outcome measures, objective evaluation of repair provides insight into the healing capacity of the technique and possibly the long-term outcomes of the treatment. MRI assessment has largely replaced histologic evaluation of biopsy specimens as the method for objective assessment (Fig. 11.13) [17].

Brown et al. [25] proposed parameters to be assessed in an MR imaging after cartilage transplantation or microfracture: signal intensity of the repair cartilage, presence of delamination, interface with the native cartilage,

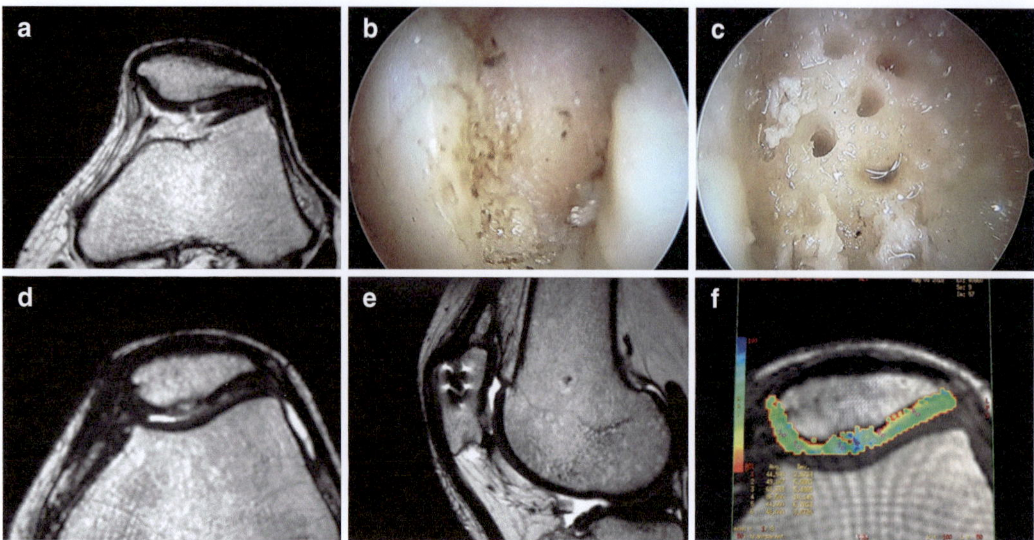

Fig. 11.12 (**a**) Preoperative MRI axial section T2-weighted image of the patellofemoral joint following patella dislocation, showing a linear fissure and delamination in the ridge and lateral facet of patella; arthroscopic image of cartilage defect in the patella before (**b**) and after (**c**) microfracture; postoperative sagittal T2-weighted (**d**), axial (**e**), and T2 mapping (**f**) images demonstrating repair tissue with slightly higher relaxation values (Courtesy: Dr. David Figueroa, Clinica Alemana, Chile)

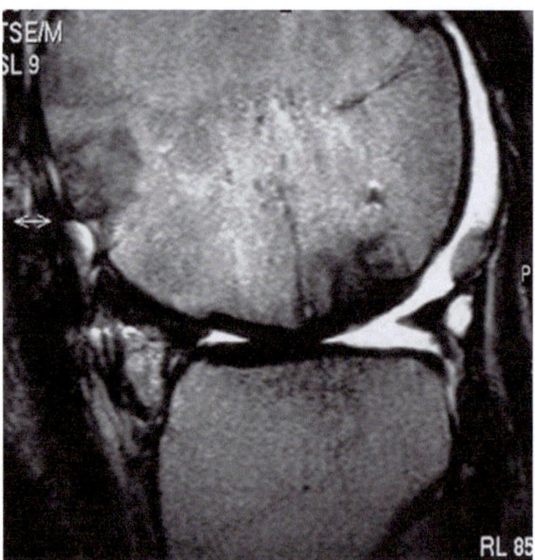

Fig. 11.13 Normal cartilage integration of osteochondral autografts in the weight bearing region of the femoral condyle in a patient 2 years after osteochondral autografts (Reproduced without changes from Trattnig et al. [17] (Licensed under CC by 4.0)

percentage fill of the lesion in coronal and sagittal images, integrity of the articular cartilage in the surrounding environment, including cartilage in the adjacent and opposite surfaces.

Cartilage-sensitive MR imaging and T2 mapping in a canine model showed trabecular osseous integration in 89% of specimens at 6 months. However, on histology, the cartilage showed incomplete or no integration between the host and graft surfaces in both autografts and allografts, asserting that articular cartilage does not regenerate completely across gaps [27]. MRI can also assess the degree of offset of the subchondral plate in relation to the host tissue. Thus, MRI can provide more detailed information than the invasive second-look arthroscopy. T2 relaxation times observed after autologous osteochondral transplantation have been found to be closer to that of the host tissue. MRI can also help in assessing the surface alignment of the graft plug in relation to the rest of the joint surface. Proud plugs are associated with increased contact pressures and formation of subchondral cavitations suggesting excessive motion between the graft and recipient site.

While cartilage-sensitive sequences assess the integrity of the cartilage and its surface, fat-suppressed images help assess the relation of the graft and the host at the subchondral bone. Low signal intensity on all pulse sequences strongly suggests loss of bone viability, which may lead to eventual implant failure. However, care must be taken to avoid mistaking the low-signal-intensity of trabecular compression in a "press fit" fixation for failure of the graft to incorporate. If instrumentation had been used, modification of pulse sequence would be necessary to reduce susceptibility artifacts in the presence of metallic fixation.

3. **Autologous Chondrocyte Implantation**: Autologous chondrocyte implantation is an example of tissue engineering technique for cartilage reconstruction. It consists of three key elements—a matrix scaffold, cells and signaling molecules, including growth factors or genes [21]. An MRI done after ACI would need to assess the following parameters: fill, maturation of tissue, integration with the subchondral bone and integration with adjacent hyaline cartilage. The fill after ACI is consistently better that after microfracture (Fig. 11.14). However, graft hypertrophy is a common complication after ACI and can lead to morbidity. Hypertrophy usually occurs within the first 6 months postoperatively [25].

In the initial few months, the repair cartilage is hyperintense due to the immature matrix and increased mobility of water. This is topped by the low-intensity periosteum [25]. The repair cartilage stays hyperintense until 8 weeks following which there is a transitional phase with lower, more inhomogeneous signal intensity for 3 to 6 months. In the final remodeling phase, the signal approaches that of the host hyaline cartilage [28].

A good integration of the repair tissue with the underlying subchondral bone should lack fluid signals at the deep interface. The presence of persistent fluid signal intensity at this interface suggests impending delamination [29]. Peripheral integration, one of the most important factors in evaluating cartilage

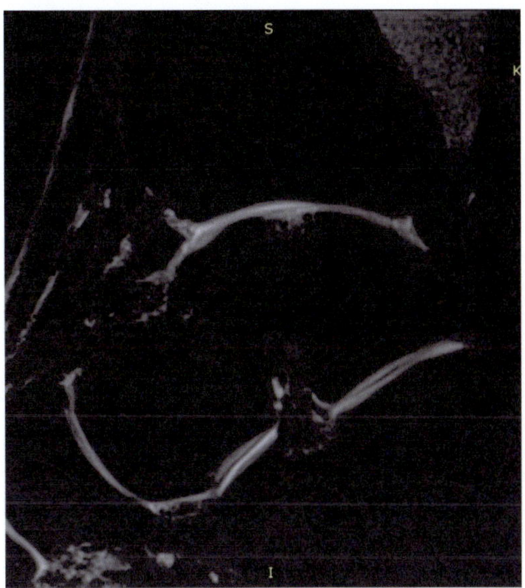

Fig. 11.14 Postoperative sagittal STIR MRI sequence of Fig. 11.10 following autologous chondrocyte implantation, showing good repair tissue

repair, can be evaluated with the help of high-resolution fluid-sensitive pulse sequences [20]. Edge integration is known to take up to 2 years and is seen as a lack of fluid intensity between the native and repair cartilage [28]. Finally, dGEMRIC techniques can be used to evaluate the quality of the repair tissue. The glycosaminoglycan levels have been found to reach levels comparable to the adjacent host hyaline cartilage after 12 months.

References

1. Crema MD, Roemer FW, Marra MD, Burstein D, Gold GE, Eckstein F, Baum T, Mosher TJ, Carrino JA, Guermazi A. Articular cartilage in the knee: current MR imaging techniques and applications in clinical practice and research. Radiographics. 2011;31(1):37–61.
2. Canale ST, Beaty JH. Campbell's operative orthopaedics e-book. Philadelphia, PA: Elsevier Health Sciences; 2012.
3. Link TM. MRI of cartilage: standard techniques. In: Link TM, editor. Cartilage imaging: significance, techniques, and new developments. New York: Springer Science & Business Media; 2011.

4. Saadat E, Jobke B, Chu B, Lu Y, Cheng J, Li X, Ries MD, Majumdar S, Link TM. Diagnostic performance of in vivo 3-T MRI for articular cartilage abnormalities in human osteoarthritic knees using histology as standard of reference. Eur Radiol. 2008;18(10):2292–302.

5. Cole BJ, Malek MM. Articular cartilage lesions: a practical guide to assessment and treatment. New York: Springer Science & Business Media; 2013.

6. Kijowski R, Davis KW, Woods MA, Lindstrom MJ, De Smet AA, Gold GE, Busse RF. Knee joint: comprehensive assessment with 3D isotropic resolution fast spin-echo MR imaging—diagnostic performance compared with that of conventional MR imaging at 3.0 T. Radiology. 2009;252(2):486–95.

7. Imhof H, Nöbauer-Huhmann I, Krestan C, et al. MRI of the cartilage. Eur Radiol. 2002;12:2781–93.

8. Gagliardi JA, Chung EM, Chandnani VP, Kesling KL, Christensen KP, Null RN, Radvany MG, Hansen MF. Detection and staging of chondromalacia patellae: relative efficacies of conventional MR imaging, MR arthrography, and CT arthrography. AJR. Am J Roentgenol. 1994;163(3):629–36.

9. Disler DG, McCauley TR, Kelman CG, Fuchs MD, Ratner LM, Wirth CR, Hospodar PP. Fat-suppressed three-dimensional spoiled gradient-echo MR imaging of hyaline cartilage defects in the knee: comparison with standard MR imaging and arthroscopy. AJR. Am J Roentgenol. 1996;167(1):127–32.

10. Eckstein F, Charles HC, Buck RJ, Kraus VB, Remmers AE, Hudelmaier M, Wirth W, Evelhoch JL. Accuracy and precision of quantitative assessment of cartilage morphology by magnetic resonance imaging at 3.0 T. Arthritis Rheum. 2005;52(10):3132–6.

11. Siepmann DB, McGovern J, Brittain JH, Reeder SB. High-resolution 3D cartilage imaging with IDEAL–SPGR at 3 T. Am J Roentgenol. 2007;189(6):1510–5.

12. Bashir A, Gray ML, Hartke J, Burstein D. Nondestructive imaging of human cartilage glycosaminoglycan concentration by MRI. Magn Reson Med. 1999;41(5):857–65.

13. Liess C, Lüsse S, Karger N, Heller M, Glüer CC. Detection of changes in cartilage water content using MRI T2-mapping in vivo. Osteoarthr Cartil. 2002;10(12):907–13.

14. Mosher TJ, Dardzinski BJ, Smith MB. Human articular cartilage: influence of aging and early symptomatic degeneration on the spatial variation of T2—preliminary findings at 3 T. Radiology. 2000;214(1):259–66.

15. Link TM, Stahl R, Woertler K. Cartilage imaging: motivation, techniques, current and future significance. Eur Radiol. 2007;17:1135–46.

16. Stahl R, Luke A, Li X, Carballido-Gamio J, Ma CB, Majumdar S, Link TM. T1rho, T 2 and focal knee cartilage abnormalities in physically active and sedentary healthy subjects versus early OA patients—a 3.0-tesla MRI study. Eur Radiol. 2009;19(1):132–43.

17. Trattnig S, Millington SA, Szomolanyi P, Marlovits S. MR imaging of osteochondral grafts and autologous chondrocyte implantation. Eur Radiol. 2007;17(1):103–18.

18. Watanabe A, Obata T, Ikehira H, Ueda T, Moriya H, Wada Y. Degeneration of patellar cartilage in patients with recurrent patellar dislocation following conservative treatment: evaluation with delayed gadolinium-enhanced magnetic resonance imaging of cartilage. Osteoarthr Cartil. 2009;17(12):1546–53.

19. Trattnig S, Mlyńarik V, Breitenseher M, Huber M, Zembsch A, Rand T, Imhof H. MRI visualization of proteoglycan depletion in articular cartilage via intravenous administration of Gd-DTPA. Magn Reson Imaging. 1999;17(4):577–83.

20. Potter HG, Foo LF. Articular cartilage. In: Stoller DW, editor. Magnetic resonance imaging in orthopaedics and sports medicine. Philadelphia: Lippincott Williams & Wilkins; 2007.

21. Hunziker EB. Articular cartilage repair: basic science and clinical progress. A review of the current status and prospects. Osteoarthr Cartil. 2002;10(6):432–63.

22. Marlovits S, Singer P, Zeller P, Mandl I, Haller J, Trattnig S. Magnetic resonance observation of cartilage repair tissue (MOCART) for the evaluation of autologous chondrocyte transplantation: determination of interobserver variability and correlation to clinical outcome after 2 years. Eur J Radiol. 2006;57(1):16–23.

23. Schreiner MM, Raudner M, Marlovits S, Bohndorf K, Weber M, Zalaudek M, Röhrich S, Szomolanyi P, Filardo G, Windhager R, Trattnig S. The MOCART (magnetic resonance observation of cartilage repair tissue) 2.0 knee score and atlas. Cartilage. 2019:1947603519865308.

24. Mithoefer K, Williams RJ III, Warren RF, Potter HG, Spock CR, Jones EC, Wickiewicz TL, Marx RG. The microfracture technique for the treatment of articular cartilage lesions in the knee: a prospective cohort study. JBJS. 2005;87(9):1911–20.

25. Brown WE, Potter HG, Marx RG, Wickiewicz TL, Warren RF. Magnetic resonance imaging appearance of cartilage repair in the knee. Clin Orthop Relat Res. 2004;422:214–23.

26. Rubak JM. Reconstruction of articular cartilage defects with free periosteal grafts: an experimental study. Acta Orthop Scand. 1982;53(2):175–80.

27. Glenn RE Jr, McCarthy EC, Potter HG, et al. Comparison of fresh osteochondral autografts and allografts: a canine model. Am J Sports Med. 2006;34(7):1084–93. https://doi.org/10.1177/0363546505284846.

28. Verstraete KL, Almqvist F, Verdonk P, Vanderschueren GE, Huysse W, Verdonk R, Verbrugge G. Magnetic resonance imaging of cartilage and cartilage repair. Clin Radiol. 2004;59(8):674–89.

29. Alparslan L, Minas T, Winalski CS. Magnetic resonance imaging of autologous chondrocyte implantation. Semin Ultrasound CT MR. 2001;22(4):341–51.

Subchondral Bone and Healthy Cartilage

12

Deepak Goyal and Anjali Goyal

12.1 Introduction

The articular cartilage and the osteochondral unit, both have a limited potential to self-heal, and hence it is a big challenge to treat the chondral as well as the osteochondral lesions. Osteochondral (OC) defects account for nearly 5% of all the articular cartilage lesions [1]. Previous treatment strategies have had a strong focus on the exclusive structural repair of the articular cartilage; without the due considerations to the deeper subchondral (SC) pathology [2, 3]. It has been proved beyond doubt that a failure to treat the SC bone is the main reason for the failure of a cartilage repair surgery in OC defects [3, 4]. The articular cartilage and the SC bone are two different tissues with different anatomy, histology, and mechanical properties; but still are one functional unit and are inter-dependent [3]. The SC bone plays an important role in the natural healing of the cartilage through its various properties like the nutritional properties, the

load-bearing properties, and as a warehouse of the mesenchymal cells and the growth factors [3, 5]. The fundamentals of the tightly controlled homeostasis inside the osteochondral unit must be understood properly before determining the treatment strategies for the chondral, the osteochondral, and the subchondral bone lesions [2]. The purpose of this review chapter is to establish the role of SC bone in the maintenance of the health of the cartilage [3, 6].

12.2 The Subchondral Bone and the Cartilage Health

Goyal et al. [3] described the SC bone as the healthy soil for the healthy cartilage and compared the soil-plant equilibrium with the SC bone-cartilage equilibrium. The soil nourishes the plants, hosts its roots, and protects it from the eroding forces of the nature. In a similar way, SC bone also acts as a fertile soil to the overlying cartilage and shares the loads put on the cartilage. If the SC bone is unhealthy, the overlying cartilage will soon get desiccated and will get separated from the bone; the best example is seen in osteochondritis dissecans and osteonecrosis. The SC bone plays three important roles to act as a healthy soil; the nutritional role, the load sharing role and the supplier of the important cells, and the growth factors.

D. Goyal (✉)
Saumya Arthroscopy and Sports Knee Clinic, Ahmedabad, India

Graduate School of Biomedical and Health Sciences, Hiroshima University, Hiroshima, Japan
e-mail: deepak@knee.in

A. Goyal
Department of Pathology, Smt N H L Municipal Medical College, Ahmedabad, India
e-mail: anjali@knee.in

© ISAKOS 2022

141

A. Gobbi et al. (eds.), *Joint Function Preservation*, https://doi.org/10.1007/978-3-030-82958-2_12

12.2.1 Nutritional Role

Traditionally, it is believed that the cartilage gets its nutrition through a diffusion from the synovial fluid and it is the only source of nutrition to the cartilage [3]. However, many historic studies hinted about a few evidences that could suggest the presence of a vascular supply to the cartilage from the SC bone. These suggestions were in the form of the presence of porosities in the SC bone plate [7], the presence of deficiencies in the SC bone plate reaching till the basal cartilage, [8] or the presence of canals penetrating from the SC bone plate to the calcified cartilage, etc. [9] All these studies pointed towards the presence of arterioles going to the basal cartilage and returning as venules. In some other studies, the vascular channels were not only demonstrated between the calcified cartilage and the SC bone but also between the uncalcified cartilage and the SC bone. Lyons et al. [10] showed various villous - like projections penetrating the calcified cartilage and reaching the uncalcified cartilage and vice versa thus confirming another network of channels between the SC bone and the uncalcified cartilage. Apart from the channels between the SC bone and the cartilage, the other modes of transport like diffusion were also identified. A study by Pan et al. [11] demonstrated the diffusion of sodium fluorescence through the uncalcified cartilage using the FLIP (fluorescence loss induced by photobleaching) technique and suggested that the smaller molecules like glucose, prostaglandin E2, nitric oxide, etc. can easily permeate through the uncalcified cartilage into the SC bone and vice versa. All these studies confirm that it is a myth to believe about the existence of an absolute barrier of nutritional supply to the cartilage from the SC bone [3, 9]. There is a reasonable evidence to support that the smaller particles can perfuse through the cartilage in the OC unit while the transportation of the larger molecules occur through the canalicular/lacunar network [12]. It is thus obvious that all these inherent mechanisms play a crucial role in supplying nutrition to the overlying cartilage and in its health (Fig. 12.1).

It has also been studied that the number of these vessels increase in response to the loads on a particular area of the OC unit [3, 9]. Lane et al.

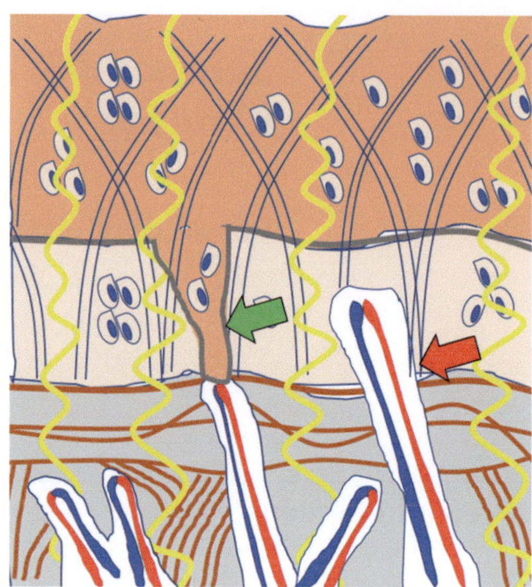

Fig. 12.1 Nutritional supply to the cartilage. The osteochondral complex has various mechanisms to support the nutritional supply to the cartilage in the form of canalicular formations between the subchondral bone and the calcified cartilage [5, 9] (red arrow), in the form of villous like projections from the uncalcified cartilage that dip into the calcified cartilage and reach subchondral bone [10] (green arrow) and in the form of permeation that allows transport of small molecules across various layers of osteochondral complex (yellow wavy lines) [11]

had demonstrated that the number of vascular perforations respond to the physiological stresses by increasing their number and by enhancing the healing potential; and thus support the cartilage to cope up with the extra loads [13]. However, the abnormally high or chronic non-physiological loads may impede the flow of the nutrients from the SC bone and thus don't play a role in the healing [3]. In osteoarthritis, angiogenesis of the SC bone has a negative effect on the entire OC unit. A study by Walsh et al. [14] found that a new microvascular growth and nerve growth takes place in the SC bone space and the uncalcified cartilage, and it had a direct connection with the presence of vascular endothelial growth factor (VEGF) and the nerve growth factor (NGF) in the region. SC bone angiogenesis not only increased the innervation of the overlying cartilage but also facilitated increased crosstalk leading to the cartilage degradation. The SC angiogenesis also stimulated nociceptors growth which can cause an increase in the pain [15]. In

Fig. 12.2 Role of the subchondral bone to support the cartilage. The subchondral bone to the cartilage is like a hard wooden bed to the soft mattress. A child can jump on the mattress without pain or damage because the mattress is supported by the hard bed. The soft, bouncy, and elastic mattress will lose its functional properties without a firm support. Similarly, subchondral bone works as a load-bearing structure to the cartilage allowing painless elastic movements of the joint

summary, the nutritional role of the SC bone to support the health of the cartilage is quite efficient during the physiological limits, and it also responds to the non-physiological loads by angiogenesis. However, a chronic non-physiological loading disrupts the balance and causes a compromise in the nutritional supply to the cartilage.

12.2.2 Load Bearing

12.2.2.1 Physiological Loading

The hyaline cartilage has a unique ability to withstand high loads, while maintaining a near-frictionless articulating interface between the bones. Traditionally, the cartilage is considered as a load-bearing structure of the joint; however, the cartilage with its limited regenerating potential and a weak architecture cannot be the principal load-bearing structure without a firm support of the SC bone. Take an example of a bed and a mattress. A child can jump on the mattress endlessly without causing any damage to the mattress or himself because the mattress is soft, bouncy, and spongy in nature. (Fig. 12.2) But if the hard wooden bed-frame is removed from underneath the mattress, the child won't be able to jump in the same way. The mattress is allowing all the load bearing because there is a firm support underneath, in the form of a wooden bed-frame. Same way, the cartilage is a load-bearing structure because there is a firm SC bone underneath. The presence of a bone marrow edema and the architectural breaks in the SC bone plate and the spongiosa on MRI in post-trauma cases are enough evidences to prove that the SC bone is an equal load-bearing structure [3].

The SC bone is a mineralized, tough, and stiff structure while the cartilage is a non-mineralized, soft, and a viscoelastic structure. The calcified cartilage is an interface that is made up of chondrocytes, vertically oriental collagen II fibers in continuity with the collagen type I fibers of the SC bone and a mineralized front called the tidemark [16]. These structural properties of the calcified cartilage allow two different structures of the OC unit to bear the various tensile, compressive, and the shear forces as a single unit. While the cartilage and the bone have different healing potentials, both must heal as a single unit to counter the physiological forces. This can only be possible when the load distribution is balanced between all the layers of the OC unit. The different properties of the different layers of the OC unit minimize the angular, shear, and the vertical forces applied to the OC unit. In fact, load bearing is actually a load sharing where all the components of the osteochondral unit have a role to play. It should also be noted that the loads are different in the different joints and also different in the different areas of a particular joint. Lower limb joints bear more loads than the upper limb joints, ankle bears more loads than the knee joint because of its proximity to the ground and a smaller size, the medial tibial plateau bears more loads than the lateral tibial plateau due to the physiological varus [3]. Hence, different components of the OC unit may express different load-bearing properties in a particular joint.

Subchondral Bone Spongiosa and Plate

The cancellous bone under the SC bone plate is made up of the compression trabeculae and the traction trabeculae, both respectively architectured to bear the compressive and the tensile stresses on the joint coming from an angular force. Any non-physiological angular load put on these trabeculae can cause a break in the continuity of these trabeculae and also can lead to a fracture. The spongiosa bone plays a crucial role in absorbing the angular forces put on the joint.

The SC bone plate is thicker and denser at the areas where the loads are high. Also, the forces get concentrated on a particular area of the SC bone plate and may not be equally distributed all along the SC bone plate; for example, the deepest portion of the concave bone will have the maximum concentration of the forces, and hence it should be thicker and denser than the adjoining SC bone plate. There are various studies that support this phenomenon. A study by Milz and Putz [17] measured the thickness of the proximal tibial articular SC bone and found it to be the thickest at the most concave part of the medial and the lateral tibial plateau. The thickest part was 900 μm thick while the thinnest part was only 100 μm thick. While the thickness of the bone can be easily measured, density measurement requires a technology called the computed tomography osteoabsorptiometry (CT-OAM) that measures the mineralization of the bone.

Muller-Gerbl et al. [18] analyzed the SC bone plate of the proximal tibial plateau and found that the bone areas that were thicker were also denser. In another study, Muller-Gerbl et al. [5] found more density towards the medial periphery of the medial tibial plateau in patients with tibia vara, which reverted to the center of the medial tibial plateau post valgus-producing osteotomy. It is important to determine if the thickness of the SC bone corelates with the density of the bone or not, which actually correlates the strength of the bone with the mineralization of the bone. It is widely known that lateral patellar surface has a greater contact area than the medial patellar facet and hence has higher loads on the lateral patellar facet [19]. Hoechel et al. [20] studied 20 human patella and mapped their strength distribution. They reported that the mechanical strength had a non-homogenous distribution on the patella and this variable strength had a regular reproducible mirroring picture with the density distribution of the SC bone plate as measured with the CT-OAM. This correlation consistently favored the high strength and density of the SC bone plate in the lateral patellar facet of 19/20 patella. A similar study on the glenoid cavity by Kraljević et al. [21] elicited a similar correlation.

Calcified and Uncalcified Cartilage

The calcified cartilage aids tremendously in the remodeling and thus the healing process of the OC unit. This remodeling process is triggered by the physiological loading and the unloading of the joint. This is possible because the vessels invading the calcified cartilage from the SC bone bring nutrients and osteoblasts to the calcified cartilage, thereby helping in a constant equilibrium at the base of the calcified cartilage [3].

The orientation of the collagen fibrils in different directions in the different layers of the uncalcified cartilage is designed to absorb the various types of forces like a vertical pressure, traction, and the shear forces [9]. This orientation of the fibers must be supported by the rich matrix of water and proteoglycans [3]. The osmotic pressure generated by these proteoglycans draw water into the cartilage in response to the mechanical pressures put on it. The phenomenon of osmosis along with the permeable diffusion form the SC bone through the villous like projections, further helps in bringing the nutrients to the loaded area; thereby helping in maintaining the health of the uncalcified cartilage. An intrinsic healing mechanism exists in the uncalcified cartilage that stimulates the synthesis of the proteoglycans and the collagen fibrils in response to the loading. The hydrophilic proteoglycans can withstand the physiological loads till it reaches the critical capacity to tolerate the loads, beyond which the proteoglycans break causing a structural damage [3, 9–11]. A study by Milz et al. [22] proved that a joint may have a different osseous and chondral congruency. The thickness of the cartilage varies from joint to joint and also in different regions of a single joint. It is known that the smaller joints like ankle and subtalar joints bear more loads than the knee joint, but still have much thinner cartilage than the cartilage of the knee. It is also known that the subtalar and the ankle joints are highly congruent joint as compared to the knee joint. The cartilage is thicker when the joint is less congruent and its varied thickness helps to compensate the difference in the osseous congruency. This adjustment helps in an equal distribution of the loading forces across

the OC unit, with more deformation being allowed by the more malleable uncalcified cartilage [9]. All these load distribution mechanisms help the uncalcified cartilage to repair itself within the physiological limits, but the disintegration starts if the loads are beyond the physiological limits.

Tidemark and Cementing Line

The tidemark is a unique micro-trilaminated structure like an egg basket which separates two important dissimilar subzones mainly the calcified and the uncalcified cartilage. There are tiny linear collage type II fibrils that cross the tidemark vertically, adding to its unique structural strength against the shear forces. The cementing line is so far the weakest structure in the OC unit and hence responsible for many isolated chondral avulsions.

12.2.2.2 Non-Physiological Loading

The SC bone and SC spongiosa must adapt to the mechanical needs of the OC unit and support the overlying cartilage. The morphology of the subchondral bone is a direct expression and adaptation of the past loading history in adults [2]. However, non-physiological loading conditions can damage the SC bone and dysregulate the OC unit. It is crucial to study the response of the SC bone after an acute trauma and in osteoarthritis; as it will give further insight in understanding the integrated relationship between the SC bone and the cartilage.

Post-traumatic bone bruises or a bone marrow edema are often seen on MRI, for example, in the lateral femoral condyle after an acute anterior cruciate ligament injury. The small to moderate insults to the SC bone heal spontaneously and no active treatment is needed for the same. However, moderate to severe SC bone marrow insults may lead to an insufficiency fracture, an impaction fracture or an osteochondral fracture of the SC bone. The extent of the SC bone damage depends on several factors including the abnormal loads put on the bone during the injury, a load distribution between the cartilage and the bone, and the strength of the bony architecture. The SC bone remodeling may not be congruent or may lead to a collapse of the SC bone causing an unsupported overlying cartilage. Damage to the subchondral bone can alter the elastic modulus of the cartilage and thus its force distribution properties, which can lead to the cartilage degeneration even by the physiological loads [2, 23]. Gradually, the injury leads to a local traumatic arthritis and eventually a generalized arthritis of the joint [24, 25]. Thus, a non-physiological traumatic loading of the OC unit can damage the cartilage due to a direct injury, or during the remodeling of the SC bone or due to the subsequent failure of the SC bone.

While an acute trauma is a one-time event, osteoarthritis is a repeated overloading of the osteochondral unit over a longer period of time. The cascade of repair and remodeling attempts on a chronically overloaded joint affects the long-term health of the OC unit and the overall joint [2, 26]. Signals from the subchondral bone have been shown to alter the differentiation potential of the bone marrow stem cells and thereby induce phenotypic degenerative changes towards osteoarthritis [27]. The healthy hyaline cartilage doesn't have nociceptors (pain receptors), and hence it must be intriguing to determine the source of the pain in the presence of cartilage lesions. Clearly, either the pain is coming from the underlying SC bone or is originating from the synovium due to the lesion-induced synovitis [2] [28]. It is a known fact that the osteoarthritis pain reduces in response to unloading or in response to an anti-inflammatory drug; both indicating the source of pain as either the SC bone or the synovial inflammation. A cartilage loss leads to the SC bone getting exposed and at some stage the pain starts. If the SC bone is the source of the pain then, it is unclear at what point the SC damage becomes significant enough to induce pain. As there are no known time indicators when the SC bone's potential is breached to show the pain, the timeline of the treatment strategy cannot be planned in time [2]. Moisio et al. [29] studied MRI cross sections of the osteoarthritis patients and tried to find an association between the SC bone exposure and the knee pain. They concluded that a moderate to severe knee pain correlated with the percentage of the denuded bone.

However, it needs to be studied if there are the presence of more regulators other than the SC bone that decides the extent of pain [30].

12.2.3　Ware House

The SC bone marrow has an abundant supply of the growth factors like IGFs, BMPs, FGFs, and TGF-B that permeates into the overlying cartilage and plays a crucial role in the remodeling and the natural cartilage healing process in association with the chondrocytes. The cells originating from the undifferentiated mesenchymal bone marrow stem cells advances through the calcified cartilage towards the uncalcified cartilage and in the process gets converted into the chondroblasts and then the chondrocytes, respectively. However once in the uncalcified cartilage, the chondrocytes become trapped in their lacuna, sort of becoming prisoners in their own home, and hence are unable to proliferate further. Hence unless the SC bone continues to provide the unlimited supply of the mesenchymal stem cells, the physiological cartilage repair will not take place.

12.2.4　Role of SC Bone in Postoperative Health of the Cartilage

The SC bone and the cartilage, both are vital to each other's existence and function, while maintaining the biomechanical and the physical equilibrium. The SC bone not only supplies cells, growth factors, and nutrition to the cartilage; but also provides a firm and strong support to withstand the shear and compressive forces put on the cartilage. In lieu, the cartilage provides a necessary cover to the SC bone, acts a co-shock absorber and provides an almost frictionless movement between the two bones in the joint. The chondral or the osteochondral surgical restoration surgeries become an extremely complex process because of this unique osteochondral relationship with its differential mechanical strengths and the biological properties [2]. A deep understanding about the relationship between the two components of the OC unit is necessary to understand the impact of the SC on various cartilage repair surgeries.

After a microfracture technique, the role of SC bone is not just to supply mesenchymal stem cells, but much more beyond that. Beck et al. [31] showed that post-MF SC responded in the form of an increased bone volume, an increased trabecular thickness, a decreased trabecular separation, and formation of the cysts that communicated with the MF holes and had a high density of osteoclasts in the peripheral cyst area. Qiu et al. [32] noted that after drilling of the chondral defect, the regenerating SC bone continued to migrate upward towards the cartilage surface and at 32 weeks, the upward migration crossed beyond the limits of the surrounding SC bone; leading to thinning of the already regenerated cartilage. Gradually, the drilling hole that stimulated the cartilage repair process, lead to the destruction of the repaired cartilage as endochondral ossification remained unabetted. Another study by Orth et al. [33] documented a deterioration of the SC bone and the SC spongiosa post-drilling in the form of SC bone cysts, intralesional osteophytes, along with a decreased bone volume and a decreased bone density. A meta-analysis by Goyal et al. [34] concluded that MF gives good results for up to 5 years with the patients having small lesions and low postoperative demands; however, MF is likely to fail beyond 5 years. A study by Mithoefer et al. [35] observed an SC overgrowth in 62% of the patients at 22 months post-op after the MF technique. Patients who showed the SC overgrowth had a much higher failure rate (25%) than the patients without an SC overgrowth (3.1%). All the above studies [31–35] point towards a similar phenomenon. Drilling or microfracture stimulates the SC bone to regenerate the cartilage by supplying the mesenchymal cells, nutrients and by supporting the super-clot; however, the SC bone also successively starts regenerating osseous tissues. If this subchondral ossification process doesn't have a check point; then it will continue to overgrow causing a damage to the overlying recently repaired cartilage. While Mithoefer et al. [35] suggested an over-scrapping or overstimulation of the SC bone plate as the possible reason for the

SC overgrowth, Qiu et al. [32] hypothesized that a properly reconstructed SC bone can act as an efficient check for the further advancement of the regenerating bone.

The advantage of osteochondral cylinder transplantation techniques has a distinct advantage over others, as the technique transfers both the healthy hyaline cartilage and the bone tissues. This helps in repairing the SC bone as well as the cartilage in one step; but unfortunately, has a disadvantage in the form of the graft size limitation. It should also be noted that the thickness of the cartilage is different in different regions of the knee joint, that means an osteochondral cylinder harvested from a non-weight bearing lateral femoral trochlear border may provide a chondral congruency at the weight bearing cartilage defect of the medial femoral condyle but may not provide an osseous congruency. A discrepancy between the donor-host osseous congruency has an impact on the long-term results of OCT procedure or not; should be a subject of further research.

Autologous chondrocytes implantation (ACI) procedure doesn't directly interfere with the SC bone, but still SC bone reactions have been observed post-ACI. The SC bone plate advancement, intralesional osteophytes, and the SC cysts have been reported on the mid- to long-term follow-up of ACI cases [36, 37]. ACI done after a failed MF have also been reported to give poorer results as compared to the ACI done in a fresh case [36]. The disturbed SC bone architecture and the biological environment could be the reason for poor ACI results in failed MF cases, but it is difficult to attribute the reasons behind the disturbed SC bone response after fresh ACI cases. Either the SC bone fails to develop a biological equilibrium with the implanted cells or the defect preparation of the lesion hyper-stimulates the SC bone; leading to an aggravated SC bone response. Goyal et al. [38] reported cases with extra-large osteochondral defects where such large lesions were treated with osseous reconstruction using the iliac crest bone grafting and then a further cartilage repair was done using ACI, collectively named as *the overlay ACI technique*. The long-term results showed good osseous repair and chondral repair with a congruency maintained at both the structures. The bi-mimetic scaffolds have also been used to repair the osteochondral unit. However, these scaffolds failed to show a proper regeneration of the SC bone on CT scan and on MRI, [39] indicating a complex healing mechanism of the OC unit much beyond the concept of just the cells and the scaffolds.

12.3 Conclusion

Osteochondral complex is one unit with the subchondral bone and the cartilage being important and equal partners to each other for the mechanical support and the biological equilibrium. The soil–plant relationship between the SC bone and the cartilage is very crucial to each other. The involvement of one component of the osteochondral unit by a pathology not only reduces function of that component, but it is only a matter of time when the function of the other component will also compromised. The SC bone has an immense role to play in the form of nutrition supply, load bearing, and a source of enormous cells and the growth factors to the cartilage. While it is tempting to do a cartilage repair surgery, a through thought must be given for a possible etiology in the SC bone. A treatment strategy that focuses solely on repairing the cartilage while ignoring the pathology of the SC bone, is bound to fail from day one. A continued research into role of SC bone in the cartilage health and in the cartilage repair surgeries is required that can lead to long-term curative treatment strategies for the joints.

References

1. Menetrey J, Unno-Veith F, Madry H, Van Breuseghem I. Epidemiology and imaging of the subchondral bone in articular cartilage repair. Knee Surg Sports Traumatol Arthrosc. 2010;18(4):463–71.
2. Lepage SIM, Robson N, Gilmore H, Davis O, Hooper A, St John S, et al. Beyond cartilage repair: the role of the osteochondral unit in joint health and disease. Tissue Eng Part B Rev. 2019;25(2):114–25.
3. Goyal D, Goyal A, Adachi N. Subchondral bone: healthy soil for the healthy cartilage. In: Gobbi A, Espregueira-Mendes J, Lane JG, Karahan M, editors. Bio-orthopaedics [Internet]. Berlin/Heidelberg: Springer; 2017. p. 479–86. [cited 14 Apr 2020]. Available from: http://link.springer.com/10.1007/978-3-662-54181-4_38.
4. Fu FH, Soni A. ACI versus microfracture: the debate continues: commentary on an article by Gunnar Knutsen, MD, PhD, et al.: "A randomized multicenter trial comparing autologous chondrocyte implantation with microfracture: long-term follow-up at 14 to 15 years.". J Bone Joint Surg Am. 2016;98(16):e69.
5. Madry H, Dijk CN, Mueller-Gerbl M. The basic science of the subchondral bone. Knee Surg Sports Traumatol Arthrosc. 2010;18(4):419–33.
6. Gomoll AH, Madry H, Knutsen G, van Dijk N, Seil R, Brittberg M, et al. The subchondral bone in articular cartilage repair: current problems in the surgical management. Knee Surg Sports Traumatol Arthrosc. 2010;18(4):434–47.
7. Fisher AGT. Chronic (non-tuberculous) arthritis. London: HK Lewis; 1929.
8. Berry JL, Thaeler-Oberdoerster DA, Greenwald AS. Subchondral pathways to the superior surface of the human talus. Foot Ankle. 1986;7(1):2–9.
9. Imhof H, Breitenseher M, Kainberger F, Rand T, Trattnig S. Importance of subchondral bone to articular cartilage in health and disease. Top Magn Reson Imaging. 1999;10(3):180–92.
10. Lyons TJ, McClure SF, Stoddart RW, McClure J. The normal human chondro-osseous junctional region: evidence for contact of uncalcified cartilage with subchondral bone and marrow spaces. BMC Musculoskelet Disord. 2006;7:52.
11. Pan J, Zhou X, Li W, Novotny JE, Doty SB, Wang L. In situ measurement of transport between subchondral bone and articular cartilage. J Orthop Res. 2009;27(10):1347–52.
12. Wang B, Zhou X, Price C, Li W, Pan J, Wang L. Quantifying load-induced solute transport and solute-matrix interaction within the osteocyte lacunar-canalicular system. J Bone Miner Res. 2013;28(5):1075–86.
13. Lane LB, Villacin A, Bullough PG. The vascularity and remodelling of subchondral bone and calcified cartilage in adult human femoral and humeral heads. An age- and stress-related phenomenon. J Bone Joint Surg Br. 1977;59(3):272–8.
14. Walsh DA, McWilliams DF, Turley MJ, Dixon MR, Fransès RE, Mapp PI, et al. Angiogenesis and nerve growth factor at the osteochondral junction in rheumatoid arthritis and osteoarthritis. Rheumatology (Oxford). 2010;49(10):1852–61.
15. Suri S, Gill SE, Massena de Camin S, Wilson D, McWilliams DF, Walsh DA. Neurovascular invasion at the osteochondral junction and in osteophytes in osteoarthritis. Ann Rheum Dis. 2007;66(11):1423–8.
16. Hoemann CD, Lafantaisie-Favreau C-H, Lascau-Coman V, Chen G, Guzmán-Morales J. The cartilage-bone interface. J Knee Surg. 2012;25(2):85–97.

17. Milz S, Putz R. Quantitative morphology of the subchondral plate of the tibial plateau. J Anat. 1994;185(Pt 1):103–10.

18. Müller-Gerbl M, Dalstra M, Ding M, Linsenmeier U, Putz R, Hvid I. Distribution of strength and mineralization in the subchondral bone plate of human tibial heads. J Biomech. 1998;31:123.

19. Salsich GB, Ward SR, Terk MR, Powers CM. In vivo assessment of patellofemoral joint contact area in individuals who are pain free. Clin Orthop Relat Res. 2003;417:277–84.

20. Hoechel S, Wirz D, Müller-Gerbl M. Density and strength distribution in the human subchondral bone plate of the patella. Int Orthop. 2012;36(9):1827–34.

21. Kraljević M, Zumstein V, Wirz D, Hügli R, Müller-Gerbl M. Mineralisation and mechanical strength of the glenoid cavity subchondral bone plate. Int Orthop. 2011;35(12):1813–9.

22. Milz S, Eckstein F, Putz R. The thickness of the subchondral plate and its correlation with the thickness of the uncalcified articular cartilage in the human patella. Anat Embryol. 1995;192(5):437–44.

23. Ma J-X, He W-W, Zhao J, Kuang M-J, Bai H-H, Sun L, et al. Bone microarchitecture and biomechanics of the necrotic femoral head. Sci Rep. 2017;7(1):13345.

24. Bonadio MB, Filho AGO, Helito CP, Stump XM, Demange MK. Bone marrow lesion: image, clinical presentation, and treatment. Magn Reson Insights. 2017;10:1178623X17703382.

25. Kazakia GJ, Kuo D, Schooler J, Siddiqui S, Shanbhag S, Bernstein G, et al. Bone and cartilage demonstrate changes localized to bone marrow edema-like lesions within osteoarthritic knees. Osteoarthr Cartil. 2013;21(1):94–101.

26. Stiebel M, Miller LE, Block JE. Post-traumatic knee osteoarthritis in the young patient: therapeutic dilemmas and emerging technologies. Open Access J Sports Med. 2014;5:73–9.

27. Leyh M, Seitz A, Dürselen L, Schaumburger J, Ignatius A, Grifka J, et al. Subchondral bone influences chondrogenic differentiation and collagen production of human bone marrow-derived mesenchymal stem cells and articular chondrocytes. Arthritis Res Ther. 2014;16(5):453.

28. Wenham CYJ, Conaghan PG. The role of synovitis in osteoarthritis. Ther Adv Musculoskelet Dis. 2010;2(6):349–59.

29. Moisio K, Eckstein F, Chmiel JS, Guermazi A, Prasad P, Almagor O, et al. Denuded subchondral bone and knee pain in persons with knee osteoarthritis. Arthritis Rheum. 2009;60(12):3703–10.

30. Hunter DJ, Guermazi A, Roemer F, Zhang Y, Neogi T. Structural correlates of pain in joints with osteoarthritis. Osteoarthr Cartil. 2013;21(9):1170–8.

31. Beck A, Murphy DJ, Carey-Smith R, Wood DJ, Zheng MH. Treatment of articular cartilage defects with microfracture and autologous matrix-induced chondrogenesis leads to extensive subchondral bone cyst formation in a sheep model. Am J Sports Med. 2016;44(10):2629–43.

32. Qiu Y-S, Shahgaldi BF, Revell WJ, Heatley FW. Observations of subchondral plate advancement during osteochondral repair: a histomorphometric and mechanical study in the rabbit femoral condyle. Osteoarthr Cartil. 2003;11(11):810–20.

33. Orth P, Goebel L, Wolfram U, Ong MF, Gräber S, Kohn D, et al. Effect of subchondral drilling on the microarchitecture of subchondral bone: analysis in a large animal model at 6 months. Am J Sports Med. 2012;40(4):828–36.

34. Goyal D, Keyhani S, Lee EH, Hui JHP. Evidence-based status of microfracture technique: a systematic review of level I and II studies. Arthroscopy. 2013;29(9):1579–88.

35. Mithoefer K, Venugopal V, Manaqibwala M. Incidence, degree, and clinical effect of subchondral bone overgrowth after microfracture in the knee. Am J Sports Med. 2016;44(8):2057–63.

36. Minas T, Gomoll AH, Rosenberger R, Royce RO, Bryant T. Increased failure rate of autologous chondrocyte implantation after previous treatment with marrow stimulation techniques. Am J Sports Med. 2009;37(5):902–8.

37. Vasiliadis HS, Wasiak J, Salanti G. Autologous chondrocyte implantation for the treatment of cartilage lesions of the knee: a systematic review of randomized studies. Knee Surg Sports Traumatol Arthrosc. 2010;18(12):1645–55.

38. Goyal D. The illustrative overlay autologous chondrocytes implantation (overlay ACI) technique for repair of the extra-large osteochondral defects. In: Goyal D, editor. The illustrative book of cartilage repair [Internet]. Springer International Publishing; 2020. Available from: https://www.springer.com/in/book/9783030471538#aboutBook.

39. Christensen BB, Foldager CB, Jensen J, Jensen NC, Lind M. Poor osteochondral repair by a biomimetic collagen scaffold: 1- to 3-year clinical and radiological follow-up. Knee Surg Sports Traumatol Arthrosc. 2016;24(7):2380–7.

Failure or Delay of Fracture Healing

Macarena Morales, John G. Lane, Fabio Sciarretta, Ignacio Dallo, and Alberto Gobbi

13.1 Introduction

A delay in fracture healing is a recognized and well-described phenomenon in long bone fractures, but when it comes to subchondral bone, the literature fails to identify and describe it as a problem, remaining as an occult, progressive, non-solved pathology. Patients have persistent pain, and prolonged impaired function of the joint. The main risk of the lack of early recognition is that injury continues to the osteochondral unit with exposure to the noxious stimulus, with consequential joint damage and collapse.

13.2 Definition

Classically, a fracture nonunion is defined by the U.S Food and Drug Administration as one that is at least 9 months old and has not shown any signs of healing for three consecutive months [1], or the one proposed by Müller: failure of a fracture to unite after 8 months of nonoperative treatment [2]. When referring to subchondral bone, it is not reasonable to wait that long, as the further fracture progression occurs, the joint is at risk of evolving to irreversible damage. In this situation, the best definition would be a fracture that has no possibility of healing without further intervention.

Anterior cruciate ligament rupture is the most frequently associated cause of articular damage secondary to traumatic bone bruises, with a reported incidence between 56% and 80%.

Graf et al. investigated knee MRIs of 98 patients with ACL injuries. They reported an incidence of 71% with bone bruises, with total and spontaneous resolution after 6 of the injuries [3]. Similar findings were observed by Miller. In a prospective MRI study of 65 patients with MCL injuries. Bone bruises in 45% of their patient population experienced complete resolution 6–12 weeks after the initial injury. These findings suggest a timing of about 6–12 weeks to the normal healing of a bone bruise. When longer healing times are present, one should be concerned and secondary causes of joint injury should be ruled out.

M. Morales
Orthopaedic Arthroscopy Department, Hospital de Mataró, Barcelona, Spain

J. G. Lane (✉)
University of California, San Diego, San Diego, USA

Musculoskeletal and Joint Research Foundation, La Jolla, CA, USA

F. Sciarretta
Clinica Nostra Signora Della Mercede, Rome, Italy

I. Dallo · A. Gobbi
Orthopaedic Arthroscopic Surgery International (O.A.S.I.) Bioresearch Foundation, Gobbi N.P.O., Milan, Italy

© ISAKOS 2022
A. Gobbi et al. (eds.), *Joint Function Preservation*, https://doi.org/10.1007/978-3-030-82958-2_13

In fact, this is the importance of diagnosis of bone marrow traumatic lesions; contusions are reversible, and subchondral fractures can be reversible only if there is no presence of fracture deformity or avulsion of bone fragments.

Reversible lesions are transient osteoporosis of the hip (TOH), regional migratory osteoporosis (RMO), complex regional pain syndrome (CRPS), and insufficiency fractures. All these lesions result from an atraumatic origin. Among irreversible lesions where diagnoses include avascular necrosis (AVN) and spontaneous osteoporosis of the knee (SONK), we include spontaneous insufficiency fractures of the knee (SIFK). SIFK is generally the consequence of overload in malaligned knees, chondral lesions, and meniscal lesions, especially root lesions and meniscal extrusions.

13.3 Risk Factors

Delay or non-healing of subchondral fractures appear to be the consequence of fracture severity, patient comorbidity, and medication use [4]. The correct identification and treatment of risk factors could accelerate recovery in early stage lesions and decrease joint failure progression in advanced stages. Finally by correctly understanding risk factors, it will also identify possible pathways of secondary osteoarthritis and will help consider future treatments regarding articular surface maintenance.

13.3.1 Host Factors

13.3.1.1 Age

Subchondral fractures in general appear to increase with age. This was reported by Pape et al. which found a 3.4% of SIFK in patients between 50 and 65 years, while in older patients the rate increased to 9.4%. No cases were described in patients with younger ages [5].

Elderly patients present with a decreased inflammatory response with less cellular activity and fewer inflammatory cells within the callus, and this causes impaired vascularization and angiogenesis, as usually they suffer from increased proinflammatory status [6].

Other important factors to consider are that older patients tend to take more time from pain onset to the medical evaluation; and tend to have more difficulties in complying with non-weight bearing restrictions impeding proper fracture remodeling [7].

13.3.1.2 Gender

Animal model researchers have failed to establish a clear correlation between gender and nonunion or delayed union fractures [8]. Yamamoto et al. compared the outcomes of males versus females with hip SIF. Of the male patients, 86% healed by non-surgical treatments, while in the female group, (mean 66 years), only 48% healed spontaneously [9]. But this finding could be explained due to hormonal changes correlated to age; estrogen deficit, and osteoporosis seen in this cohort of patients which may produce more difference than gender itself. Overall, gender per se is not considered as a risk factor for delayed fracture healing [10].

13.3.1.3 Metabolic Syndrome

Metabolic syndrome is a systemic proinflammatory and atherogenic state in which the subchondral unit is also affected, as systemically the patient is in a catabolic and antianabolic condition. This syndrome is characterized by hypertension, decreased HDL, hypertriglyceridemia, elevated fasting glucose, and an increased waistline. Patients with this syndrome have a significant impairment of bone healing compared with control groups [11, 12]. It is not a cause of osteoporosis that could cause further insufficiency fractures as previously suggested [12]. Perhaps even though there is scarce literature regarding this topic, metabolic syndrome is a modifiable factor that should be addressed, as it not only affects fracture healing, but it also has been correlated with secondary osteoarthritis [13].

13.3.1.4 Diabetes Mellitus (DM)

DM can potentially cause a twofold increase in delayed union or nonunion compared with control groups [14]. It causes decreased expression of growth factors such as platelet-derived growth factor (PDGF), insulin-like growth factor-1 (IGF-1), vascular endothelial growth factor (VEGF), and transforming growth factor-beta (TEGF-β) among others [15]. This decreases the capability of MSCs to undergo osteoblastic differentiation causing a delay of cell proliferation and abnormal bone remodeling, with significantly decreased density and deteriorated structure [16].

Poor glycemic control with HbA1c levels >7%, as well as peripheral neuropathy, are predictors of impairment of bone healing and osteoarthritis [17]. On the other hand, appropriate maintenance of glucose levels, and direct delivery of insulin improves fracture healing potential to a normal state [18]. Moreover, the direct administration of insulin has provided good results in diabetic wound healing, suggesting that insulin could be a good target for future therapies [19].

13.3.1.5 Nutrition

Nutrition is an important factor that affects bone healing. As previously mentioned, fats, with their proinflammatory state, and sugars with impaired bone formation directly affect bone healing. Proteins also play an important role, as they are crucial for collagen fiber formation, with most fracture healing pathways being regulated by protein mediators such as BMPs [20].

Malnourished protein subjects have experienced a decrease in systemic IGF-I, associated with lower subchondral trabecular mass, and mineral density [21], developing a poor quality callus with reduced strength in comparison to control groups.

One known pathologic pathway is citrulline-arginine-nitric oxide (NO). The lack of arginine supplementation correlates with a down regulation of NO synthesis and impaired healing with lower bone formation rate and mineral apposition rate [22]. Arginine, as well as citrulline supplementation correlates with NO expression, with new bone formation, and muscle mass augmentation [23]. This suggests NO-citrulline-arginine pathway as a treatment target for delayed union.

13.3.1.6 Vitamin D

Vitamin D deficiency correlates directly with bone mineral loss being considered a cause of subchondral fracture, as well as for impaired healing. There is a high percentage of vitamin D deficiency in unexplained fractures with impaired healing; and thus it should be ruled out when no other causes are identified [24]. On the other hand, its supplementation stimulates osteogenesis, through an increase in the production of osteocalcin and osteoclast-mediated bone resorption [25]. Moreover, it does not only aid in fracture healing but also prevents it from forming [26]. These factors make vitamin D supplementation a cost-effective treatment method for fracture prevention [27].

Less frequent but not less important are metabolic diseases resulting in vitamin D deficiency such as thyroid or parathyroid hormone disorders. Affected patients often present with impaired healing. In these cases, medical treatment by itself can aid in fracture resolution [28]. It should be ruled out especially when no other causes of delayed healing are present.

13.3.1.7 Osteoporosis

Histological findings of SIFK present with loss of bone trabeculae and osteoporosis. But actually, there are few studies that support this effect [29]. Low BMI is present only in 16% to 20% of SIFK patients, suggesting local osteoporotic changes are secondary to fracture healing and regional damage rather than the true etiology of the fracture [30]. Moreover, bone turnover markers have been shown to be elevated in aspirates from the local defect, without any significant change in serum concentration [31] supporting the theory of local changes. Thus, in this setting osteoporosis does not appear to be a risk factor for fracture delay healing.

13.3.2 Pharmacological Treatments and Drugs

13.3.2.1 Steroids

Prolonged corticosteroid therapy produces a decrease in mineral density by means of inhibiting the osteogenic differentiation of MSCs and generating osteoblast and osteocyte apoptosis, with a reduction in organic matrix synthesis [31, 32]. This was demonstrated in in vivo rat models with induced fractures [33]. It has also been recorded in asthmatic patient populations with chronic corticoid use [34]. Moreover, steroid local injections have been linked directly as causal factor for insufficiency fractures in the hip and knee [35], as well as its chondrotoxic effects intra-articularly at high doses [36, 37]. In patients with chronic use of steroids, treatment of delayed union with teriparatide has shown to give good results [38, 39].

13.3.2.2 Nonsteroidal Anti-Inflammatory Drugs (NSAID)

NSAIDs anti-inflammatory action is frequently associated with impaired bone healing [40], but meta-analysis of available clinical trials fail to convincingly support this hypothesis [41]. Moreover, recent studies support the safe use of nonselective NSAIDs regarding nonunion [24, 42], in comparison to the use of selective COX-2 inhibitory medications which have shown a strong association with the development of nonunion of fractures [43, 44]. Little is known regarding the effect they could have in healing of subchondral bone. In animal models with induced osteoarthritis, the use of selective cyclooxygenase-2 inhibitor is correlated with reduced bone volume with lower trabecular thickness and increased trabecular fractures in the subchondral bone. These findings suggest that caution must be taken regarding the use of selective COX 2 inhibitors [45].

13.3.2.3 Alcohol Abuse

High doses of Ethanol (>1000 cc/day) inhibit the ossification of newly formed bone, producing a poorly mineralized structure with reduced mechanical stability [46]. This is thought to be through inhibition of Wnt signaling required for normal fracture repair [47]. It has been correlated with AVN in chronic alcoholics, but studies have not shown a significant relationship with delayed union, as it is linked to the high intensity fractures more than to alcohol by itself [4].

13.3.2.4 Smoking

Nicotine is a well-known cause of delayed fracture healing [48–52]. Its vasoconstrictor properties produce a decreased perfusion rate, resulting osteochondral hypoxia and ischemia. It alters macrophage, fibroblast, and osteoblast activity as well as inhibits tissue differentiation and the normal angiogenic response in the early stages of fracture healing [51]. The mechanism of delay is thought to be through the cholinergic anti-inflammatory pathway, as the secretion of TNF-a is inhibited by nicotine [53, 54].

13.3.3 Mechanical

13.3.3.1 Meniscus

Medial meniscus root tears have been identified in approximately 70% of the patients with SIFK lesions. Disruption of the posteromedial meniscus root results in twice the peak pressures with weight bearing when compared with a knee with an intact meniscus [55]. This produces a subsequent alteration of normal knee biomechanics, similar to total meniscectomy. Also, the occurrence of SONK lesions have been described after meniscectomy, supporting it as an etiological factor [56].

13.3.3.2 Varus Alignment

Varus deformity of femorotibial angle of 180° or more has been associated with a poor prognosis for fracture progression [57], while on the other hand unloading of the medial compartment, with high tibial osteotomy, produces a significant reduction in symptoms [58, 59], and correlates with a decrease in the lesion size [60].

These two factors, meniscus deficiency and varus alignment, show a strong correlation with subchondral fracture development with joint overload, making it an imperative target for prevention of further subchondral fracture progression. Thus, this hypothesis is the main concept indicating early non-weight bearing and correct patient compliance.

13.4 Prognostic Factors

The main prognostic factors will be the type of fracture, which is diagnosed primarily by MRI after proper clinical assessment; as well as the time of diagnostic evaluation and the initiation of treatment. Therefore, the main goals will be the prompt identification of the problem and early treatment, even if it is with non-surgical treatment, with non-weight bearing, before irreversible markers of osteoarthritis appear.

13.4.1 Clinical Presentation

The presence of persistent pain is one of the first and most important aspects of the diagnosis, so its presence should raise high clinical suspicion of fracture non-healing. It can also manifest with direct tenderness at the site of the lesion, or as weight bearing instability. In cases of non-traumatic lesions, the intensity of symptoms such as the extent of swelling, or flexion contractures have also been related to worse outcomes [61]. When referring to high impact lesions (i.e., traumatic multiligament knee injuries), they have been correlated with prolonged bone marrow edema with persistence up to 2 years after the lesion with knee OA, making this patient population at high risk in developing delayed healing, or fracture collapse [62].

It is important to consider, especially in older or neuropathic patients, that sensory nerve fibers course mainly through the periosteum and trabecular fractures may not be perceived as painful [63]. In this instance, global patient assessment should be performed in order to identify early markers or risk factors for non-healing fracture progression.

13.4.2 Time of Diagnosis

The time interval to diagnosis is crucial as late diagnosis, or delay of treatment have been shown to be predictive factors for osteoarthritis progression in subchondral fractures [7]. Aglietti compared the time between the onset of symptoms

and osteoarthritis. Of the patients that initiated treatment after less than 6 months, 64% presented with signs of arthritis; 6–12 months 85%; and more than 12 months 95.8% [61]. Regardless of this, the timing between the onset of symptoms and diagnosis with treatment can be as long as 15 months, making early diagnosis and prompt treatment a strong factor to improve in order to avoid further failure progressions.

13.4.3 Bone Marrow Edema (BME)

Quantity or extension of bone marrow edema is not associated with prognosis nor delayed healing, particularly when talking about purely trabecular edema. These type of lesions usually heal without any further intervention. Sometimes the extent of edema can be associated with factors such as lesion severity, and it can persist up to 2 years [64]. In these cases, it correlates with a poor prognosis, and secondary lesions are causal factors of early osteoarthritis (i.e., instability, meniscal tear, chondral lesion) [62, 65]. Particularly edema that extends towards the articular surface, articular disruption, or geographic lesions, suggest osteochondral injury and thus tend to have a worse outcome [64, 66]. A careful and wise interpretation has to be made in order to make correct treatment choices.

13.4.4 Lesion Characteristics

Subchondral and osteochondral lesions have certain characteristics that can be interpreted as progression markers, and thus will help prompt treatment be initiated, even if the symptoms do not correlate with the lesion. Several studies have already correlated the size of the lesion with outcomes in subchondral fractures [61, 67, 68]. Lecouvet described as poor prognosis, a fracture line that is longer than 14 mm or a depth of more than 4 mm on transverse view of the femoral condyle in T2-weighted MRI scans [69]. Sayyid et al. recently proposed that lesions greater than 26 mm combined, coronal and transverse, with individual measurements of >16.5 mm and

>10.5 mm, respectively, as factors with a poor prognosis [70]. Other factors to consider are the presence of articular damage that can be seen in MRI as a fracture band over an articular edge [7]; contour deformity or depression and subchondral low intensity over the fracture site, indicating further detachment of the lesion [64, 69, 71].

13.4.5 Bone Markers

There are multiple studies that have found an association with the levels of bone marker turnover, and the persistence or severity of fractures [72]. Bone-specific alkaline phosphatase, osteocalcin, collagen type *I* cross-linked C-telopeptide (CTX-1), and N-terminal telopeptide of collagen type I (NTX-1) have been related with joint space narrowing in patients with SIF. Serum N-terminal propeptide of type 1 collagen (P1NP) and tartrate-resistant acid phosphatase (TRACP-5b) have also been associated with joint space narrowing, as well as with length of fracture band, suggesting that the levels of bone metabolic markers might be predictors for failure and reflect the severity among patients with subchondral fractures [7]. In DM patients, bone remodeling biomarkers FGF-2 and IGF1 have been shown to be significantly decreased making them a potential method for diagnosis of fracture persistence and progression [73].

Therefore, further research and development of available bone turnover markers, with known levels, would be of significant interest for diagnostic and prognosis establishment of subchondral fractures.

13.5 Conclusions

Delayed healing of subchondral fractures is a novel concept in which clinical findings are often nonspecific, and initial radiographs may seem irrelevant. Therefore, delayed diagnosis is common and is one of the main causes of fracture failure and joint collapse. A high level of suspicion must be maintained when persistent pain and bone marrow edema like lesions are present,

and this must be associated within the clinical context. When properly diagnosed, it is crucial to consider a global assessment of the patient risk factors for delayed healing and to rule out systemic diseases in order to maintain homeostasis and properly assess modifiable factors. Finally, it is important to consider the joint as a system, to address the state of the meniscus, the alignment, and possible causes of joint overload before any treatment decision is arrived at in treating injury to the osteochondral unit.

References

1. United States Food and Drug Administration. Guidance document for the preparation of investigational device exemptions and pre-marked approval application for bone growth stimulator devices. Rockville, MD: United States Food and Drug Administration; 1988. Federal Register Volume 63, Issue 81
2. Müller ME, Allgöwer M, Schneider R, Willenegger H, Müller ME, Allgöwer M, et al. Introduction. In: Manual of internal fixation. Berlin/Heidelberg: Springer; 1979.
3. Graf BK, Cook DA, de Smet AA, Keene JS. "Bone bruises" on magnetic resonance imaging evaluation of anterior cruciate ligament injuries. Am J Sports Med. 1993;21(2):220–3.
4. Zura R, Xiong Z, Einhorn T, Watson JT, Ostrum RF, Prayson MJ, et al. Epidemiology of fracture nonunion in 18 human bones. JAMA Surg. 2016;151(11):e162775.
5. Pape D, Seil R, Fritsch E, Rupp S, Kohn D. Prevalence of spontaneous osteonecrosis of the medial femoral condyle in elderly patients. Knee Surg Sports Traumatol Arthrosc. 2002;10(4):233–40.
6. Clark D, Nakamura M, Miclau T, Marcucio R. Effects of aging on fracture healing. Curr Osteoporos Rep. 2017;15(6):601–8.
7. Shimizu T, Yokota S, Kimura Y, Asano T, Shimizu H, Ishizu H, et al. Predictors of cartilage degeneration in patients with subchondral insufficiency fracture of the femoral head: a retrospective study. Arthritis Res Ther. 2020;22(1):150.
8. Deng Z, Gao X, Sun X, Cui Y, Amra S, Huard J. Gender differences in tibial fractures healing in normal and muscular dystrophic mice. Am J Transl Res. 2020;12(6):2640–51.
9. Yamamoto T, Karasuyama K, Iwasaki K, Doi T, Iwamoto Y. Subchondral insufficiency fracture of the femoral head in males. Arch Orthop Trauma Surg. 2014;134(9):1199–203.
10. Gaston MS, AHRW S. Inhibition of fracture healing. J Bone Joint Surg Br. 2007;89(12):1553–60.

11. Tabrizi A. Effect of metabolic syndrome on union rate of fractures. J Anal Res Clin Med. 2015;3(1):37–42.

12. Sun K, Liu J, Lu N, Sun H, Ning G. Association between metabolic syndrome and bone fractures: a meta-analysis of observational studies. BMC Endocr Disord. 2014;14:13.

13. Dickson BM, Roelofs AJ, Rochford JJ, Wilson HM, De Bari C. The burden of metabolic syndrome on osteoarthritic joints. Arthritis Res Ther. 2019;21(1):289.

14. Marin C, Luyten FP, Van der Schueren B, Kerckhofs G, Vandamme K. The impact of type 2 diabetes on bone fracture healing. Front Endocrinol (Lausanne). 2018;9:6.

15. Tyndall WA, Beam HA, Zarro C, O'Connor JP, Lin SS. Decreased platelet derived growth factor expression during fracture healing in diabetic animals. Clin Orthop Relat Res. 2003;408:319–30.

16. Funk JR, Hale JE, Carmines D, Gooch HL, Hurwitz SR. Biomechanical evaluation of early fracture healing in normal and diabetic rats. J Orthop Res. 2000;18(1):126–32.

17. Chen Y, Huang YC, Yan CH, Chiu KY, Wei Q, Zhao J, et al. Abnormal subchondral bone remodeling and its association with articular cartilage degradation in knees of type 2 diabetes patients. Bone Res. 2017;5(1):17034.

18. Beam HA, Russell Parsons J, Lin SS. The effects of blood glucose control upon fracture healing in the BB Wistar rat with diabetes mellitus. J Orthop Res. 2002;20(6):1210–6.

19. Gandhi A, Beam HA, O'Connor JP, Parsons JR, Lin SS. The effects of local insulin delivery on diabetic fracture healing. Bone. 2005;37(4):482–90.

20. Meesters DM, Wijnands KAP, Brink PRG, Poeze M. Malnutrition and fracture healing: are specific deficiencies in amino acids important in nonunion development? Nutrients. 2018;10(11):1597.

21. Lavet C, Ammann P. Osteoarthritis like alteration of cartilage and subchondral bone induced by protein malnutrition is treated by nutritional essential amino acids supplements. Osteoarthr Cartil. 2017;25:S293.

22. Meesters DM, Neubert S, Wijnands KAP, Heyer FL, Zeiter S, Ito K, et al. Deficiency of inducible and endothelial nitric oxide synthase results in diminished bone formation and delayed union and nonunion development. Bone. 2016;83:111–8.

23. Hughes MS, Kazmier P, Burd TA, Anglen J, Stoker AM, Kuroki K, et al. Enhanced fracture and soft-tissue healing by means of anabolic dietary supplementation. J Bone Joint Surg Am. 2006;88(11):2386–94.

24. Zura R, Mehta S, Della Rocca GJ, Steen RG. Biological risk factors for nonunion of bone fracture. JBJS Rev. 2016;4(1):1–12.

25. Gorter EA, Hamdy NAT, Appelman-Dijkstra NM, Schipper IB. The role of vitamin D in human fracture healing: a systematic review of the literature. Bone. 2014;64:288–97.

26. Dao D, Sodhi S, Tabasinejad R, Peterson D, Ayeni OR, Bhandari M, et al. Serum 25-hydroxyvitamin D levels and stress fractures in military personnel: a systematic review and meta-analysis. Am J Sports Med. 2015;43(8):2064–72.

27. Childs BR, Andres BA, Vallier HA. Economic benefit of calcium and vitamin D supplementation: does it outweigh the cost of nonunions? J Orthop Trauma. 2016;30(8):e285–8.

28. Kapania EM, Reif TJ, Tsumura A, Eby JM, Callaci JJ. Alcohol-induced Wnt signaling inhibition during bone fracture healing is normalized by intermittent parathyroid hormone treatment. Anim Model Exp Med. 2020;7:3(2).

29. Akamatsu Y, Mitsugi N, Hayashi T, Kobayashi H, Saito T. Low bone mineral density is associated with the onset of spontaneous osteonecrosis of the knee. Acta Orthop. 2012;83(3):249–55.

30. Nelson FR, Craig J, Francois H, Azuh O, Oyetakin-White P, King B. Subchondral insufficiency fractures and spontaneous osteonecrosis of the knee may not be related to osteoporosis. Arch Osteoporos. 2014;9(1):1–7.

31. Berger CE, Kröner AH, Minai-Pour MB, Ogris E, Engel A. Biochemical markers of bone metabolism in bone marrow edema syndrome of the hip. Bone. 2003;33(3):346–51.

32. Habib GS. Systemic effects of intra-articular corticosteroids. Clin Rheumatol. 2009;28(7):749–56.

33. Liu YZ, Akhter MP, Gao X, Wang XY, Wang XB, Zhao G, et al. Glucocorticoid-induced delayed fracture healing and impaired bone biomechanical properties in mice. Clin Interv Aging. 2018;13:1465–74.

34. Zura R, Kaste SC, Heffernan MJ, Accousti WK, Gargiulo D, Wang Z, et al. Risk factors for nonunion of bone fracture in pediatric patients. Medicine (Baltimore). 2018;97(31):e11691.

35. Kompel AJ, Roemer FW, Murakami AM, Diaz LE, Crema MD, Guermazi A. Intra-articular corticosteroid injections in the hip and knee: perhaps not as safe as we thought? Radiology. 2019;293(3):656–63.

36. Song YW, Zhang T, Wang WB. Gluocorticoid could in influence extracellular matrix synthesis through Sox9 via p38 MAPK pathway. Rheumatol Int. 2012;32(11):3669–73.

37. Wernecke C, Braun HJ, Dragoo JL. The effect of intra-articular corticosteroids on articular cartilage: a systematic review. Orthop J Sport Med. 2015;3(5):1–7.

38. Bednar DA. Teriparatide treatment of a glucocorticoid-associated resorbing nonunion of a type III odontoid process fracture: A case report. J Spinal Disord Tech. 2013;26(8):E319–22.

39. Mitani Y. Effective treatment of a steroid-induced femoral neck fracture nonunion with a once-weekly administration of teriparatide in a rheumatoid patient: a case report. Arch Osteoporos. 2013;8(1):131.

40. Riew KD, Long J, Rhee J, Lewis S, Kuklo T, Kim YJ, et al. Time-dependent inhibitory effects of indomethacin on spinal fusion. J Bone Joint Surg Am. 2003;85(4):632–4.

41. Dodwell ER, Latorre JG, Parisini E, Zwettler E, Chandra D, Mulpuri K, et al. NSAID exposure and

risk of nonunion: a meta-analysis of case-control and cohort studies. Calcif Tissue Int. 2010;87(3):193–202.

42. Schemitsch EH, Bhandari M, Guyatt G, Sanders DW, Swiontkowski M, Tornetta P, et al. Prognostic factors for predicting outcomes after intramedullary nailing of the tibia. J Bone Joint Surg Am. 2012;94(19):1786–93.

43. Goodman S, Ma T, Trindade M, Ikenoue T, Matsuura I, Wong N, et al. COX-2 selective NSAID decreases bone ingrowth in vivo. J Orthop Res. 2002;20(6):1164–9.

44. George MD, Baker JF, Leonard CE, Mehta S, Miano TA, Hennessy S. Risk of nonunion with nonselective NSAIDs, COX-2 inhibitors, and opioids. J Bone Joint Surg Am. 2020;102(14):1230–8.

45. Liu B, Ji C, Shao Y, Liang T, He J, Jiang H, et al. Etoricoxib decreases subchondral bone mass and attenuates biomechanical properties at the early stage of osteoarthritis in a mouse model. Biomed Pharmacother. 2020;127:110144.

46. Chakkalakal DA, Novak JR, Fritz ED, Mollner TJ, McVicker DL, Garvin KL, et al. Inhibition of bone repair in a rat model for chronic and excessive alcohol consumption. Alcohol. 2005;36(3):201–14.

47. Kakar S, Einhorn TA, Vora S, Miara LJ, Hon G, Wigner NA, et al. Enhanced chondrogenesis and Wnt signaling in PTH-treated fractures. J Bone Miner Res. 2007;22(12):1903–12.

48. Castillo RC, Bosse MJ, MacKenzie EJ, Patterson BM, Burgess AR, Jones AL, et al. Impact of smoking on fracture healing and risk of complications in limb-threatening open tibia fractures. J Orthop Trauma. 2005;19(3):151–7.

49. McKee MD, DiPasquale DJ, Wild LM, Stephen DJG, Kreder HJ, Schemitsch EH. The effect of smoking on clinical outcome and complication rates following Ilizarov reconstruction. J Orthop Trauma. 2003;17(10):663–7.

50. Pearson RG, Clement RGE, Edwards KL, Scammell BE. Do smokers have greater risk of delayed and non-union after fracture, osteotomy and arthrodesis? A systematic review with meta-analysis. BMJ Open. 2016;6(11):e010303.

51. Scolaro JA, Schenker ML, Yannascoli S, Baldwin K, Mehta S, Ahn J. Cigarette smoking increases complications following fracture: a systematic review. J Bone Joint Surg Am. 2014;96(8):674–81.

52. Ru J-Y, Chen L-X, Hu F-Y, Shi D, Xu R, Du J-W, et al. Factors associated with development of re-nonunion after primary revision in femoral shaft nonunion subsequent to failed intramedullary nailing. J Orthop Surg Res. 2018;13(1):180.

53. Chen Y, Guo Q, Pan X, Qin L, Zhang P. Smoking and impaired bone healing: Will activation of cholinergic anti-inflammatory pathway be the bridge? Int Orthop. 2011;35(9):1267–70.

54. Sloan A, Hussain I, Maqsood M, Eremin O, El-Sheemy M. The effects of smoking on fracture healing. Surgeon. 2010;8(2):111–6.

55. Song Y, Greve JM, Carter DR, Koo S, Giori NJ. Articular cartilage MR imaging and thickness mapping of a loaded knee joint before and after meniscectomy. Osteoarthr Cartil. 2006;14(8):728–37.

56. Yamagami R, Taketomi S, Inui H, Tahara K, Tanaka S. The role of medial meniscus posterior root tear and proximal tibial morphology in the development of spontaneous osteonecrosis and osteoarthritis of the knee. Knee. 2017;24(2):390–5.

57. Klontzas ME, Vassalou EE, Zibis AH, Bintoudi AS, Karantanas AH. MR imaging of transient osteoporosis of the hip: an update on 155 hip joints. Eur J Radiol. 2015;84(3):431–6.

58. Akamatsu Y, Kobayashi H, Kusayama Y, Aratake M, Kumagai K, Saito T. Predictive factors for the progression of spontaneous osteonecrosis of the knee. Knee Surg Sports Traumatol Arthrosc. 2017;25(2):477–84.

59. Tsukamoto H, Saito H, Saito K, Yoshikawa T, Oba M, Sasaki K, et al. Radiographic deformities of the lower extremity in patients with spontaneous osteonecrosis of the knee. Knee. 2020;27(3):838–45.

60. Marti CB, Rodriguez M, Zanetti M, Romero J. Spontaneous osteonecrosis of the medial compartment of the knee: a MRI follow-up after conservative and operative treatment, preliminary results. Knee Surg Sports Traumatol Arthrosc. 2000;8(2):83–8.

61. Aglietti P, Insall JN, Buzzi R, Deschamps G. Idiopathic osteonecrosis of the knee. Aetiology, prognosis and treatment. J Bone Joint Surg Br. 1983;65(5):588–97.

62. Roemer FW, Bohndorf K. Long-term osseous sequelae after acute trauma of the knee joint evaluated by MRI. Skelet Radiol. 2002;31(11):615–23.

63. Mantyh PW. The neurobiology of skeletal pain. Eur J Neurosci. 2014;39(3):508–19.

64. Costa-Paz M, Muscolo DL, Ayerza M, Makino A, Aponte-Tinao L. Magnetic resonance imaging follow-up study of bone bruises associated with anterior cruciate ligament ruptures. Arthroscopy. 2001;17(5):445–9.

65. Ergun T. The relationship between MRI findings and duration of symptoms in transient osteoporosis of the hip. Acta Orthop Traumatol Turc. 2008;42(1):10–5.

66. Davies NH, Niall D, King LJ, Lavelle J, Healy JC. Magnetic resonance imaging of bone bruising in the acutely injured knee—short-term outcome. Clin Radiol. 2004;59(5):439–45.

67. Muheim G, Bohne WH. Prognosis in spontaneous osteonecrosis of the knee. Investigation by radionuclide scintimetry and radiography. J Bone Joint Surg Br. 1970;52(4):605–12.

68. Lotke PA, Nelson CL, Lonner JH. Spontaneous osteonecrosis of the knee: tibial plateaus. Orthop Clin North Am. 2004;35(3):365–70.

69. Lecouvet FE, Vande Berg BC, Maldague BE, Lebon CJ, Jamart J, Saleh M, et al. Early irreversible osteonecrosis versus transient lesions of the femoral condyles: prognostic value of subchondral bone and marrow changes on MR imaging. Am J Roentgenol. 1998;170(1):71–7.

70. Sayyid S, Younan Y, Sharma G, Singer A, Morrison W, Zoga A, et al. Subchondral insufficiency fracture of the knee: grading, risk factors, and outcome. Skelet Radiol. 2019;48(12):1961–74.
71. Sonoda K, Yamamoto T, Motomura G, Karasuyama K, Kubo Y, Iwamoto Y. Fat-suppressed T2-weighted MRI appearance of subchondral insufficiency fracture of the femoral head. Skelet Radiol. 2016;45(11):1515–21.
72. Cox G, Einhorn TA, Tzioupis C, Giannoudis PV. Bone-turnover markers in fracture healing. J Bone Joint Surg Br. 2010;92(3):329–34.
73. Chen QQ, Wang WM. Expression of FGF-2 and IGF-1 in diabetic rats with fracture. Asian Pac J Trop Med. 2014;7(1):71–5.

Avascular Necrosis

14

Katarzyna Herman, Przemysław Pękala,
Dawid Szwedowski, Radosław Grabowski,
and Jerzy Cholewiński

14.1 Introduction

Avascular necrosis (AVN), also known as osteonecrosis (ON), is progressive process of bone destruction characterized by cell death subsequent to an incident of bone ischemia. It is a common condition affecting mostly the femoral head, less frequently the shoulder and knee and rarely observed in other locations [1, 2]. The natural development of ON mostly leads to collapse and deformation of the affected joint surface, inevitably leading to joint destruction, causing secondary arthritis. If three or more anatomical sites are affected by osteonecrosis, it is defined as multifocal osteonecrosis [3]. On the cellular level, the pathology can be characterized as a local apoptosis of the bone cells, which can affect patients at any age and has a complicated multifactorial etiology [4]. It is believed that it is directly caused by impaired local blood distribution (in atraumatic cases, the anatomy of the bone's vascular network is untouched, but physiology of the blood distribution is significantly altered). The subchondral area of bone is mainly affected, and progress of this disease leads to irreversible joint cartilage and subchondral bone damage. The prognosis is better in children than in adults, due to their capability for bone growth and remodeling [5].

K. Herman (✉)
Department of Orthopedics and Traumatology,
Brothers Hospitallers Hospital, Katowice, Poland

Department of Medical Rehabilitation, Medical
University of Silesia, Katowice, Poland

Orthopaedic Arthroscopic Surgery International
(O.A.S.I.) Bioresearch Foundation, Gobbi N.P.O.,
Milan, Italy

P. Pękala
International Evidence-Based Anatomy Working
Group, Department of Anatomy, Jagiellonian
University Medical College, Krakow, Poland

Faculty of Medicine and Health Sciences, Andrzej
Frycz Modrzewski Krakow University,
Krakow, Poland

Lesser Poland Orthopedic and Rehabilitation
Hospital, Krakow, Poland

D. Szwedowski
Orthopaedic Arthroscopic Surgery International
(O.A.S.I.) Bioresearch Foundation, Gobbi N.P.O.,
Milan, Italy

Department of Orthopaedics and Trauma Surgery,
Provincial Polyclinical Hospital, Torun, Poland

R. Grabowski
Medical University of Lodz, Lodz, Poland

SPORTO-Sport Injuries and Rehabilitation Clinic,
Lodz, Poland

Widzew Lodz Football Club, Lodz, Poland

J. Cholewiński
Department of Orthopedics and Traumatology,
Brothers Hospitallers Hospital, Katowice, Poland

Department of Medical Rehabilitation, Medical
University of Silesia, Katowice, Poland

© ISAKOS 2022
A. Gobbi et al. (eds.), *Joint Function Preservation*, https://doi.org/10.1007/978-3-030-82958-2_14

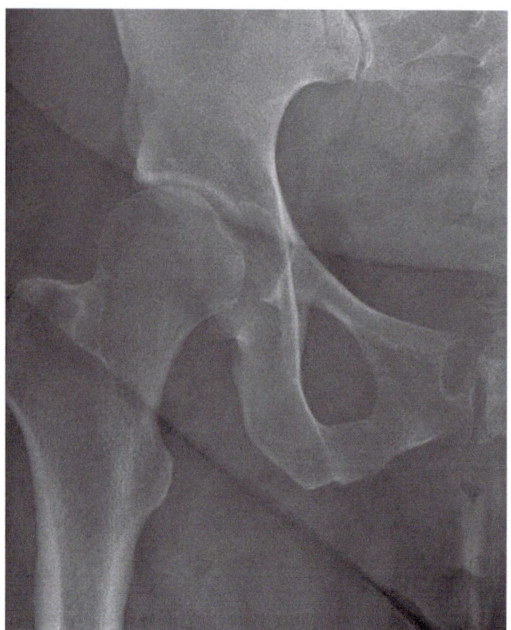

Fig. 14.1 A case of ON of the femoral head in a 52-year-old female, a former smoker, patient reported severe hip pain without traumatic incident. An AP view of the hip taken at the first visit, showing no particular pathological changes

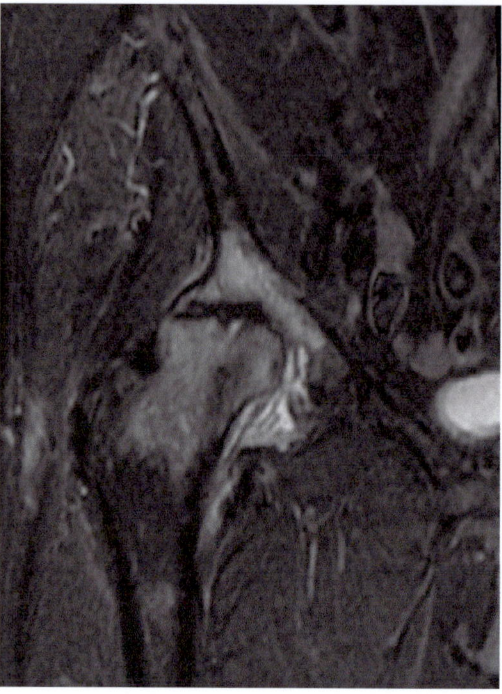

Fig. 14.2 MRI (coronal cut) of the same patient 2 months later showing ON of the femoral head, which is deformed, articular surface in the upper part clearly flattened and collapsed, bone marrow edema is visible in the femoral head with neck penetration and also within the acetabulum

Patients' symptoms in advanced stages usually consists of pain in the area of the joint affected by AVN, restricted range of motion, and swelling. It is believed that early stages of AVN are asymptomatic, especially in non-weightbearing joints, and therefore many patients are referred to the orthopedic surgeon with an advanced presentation and irreversible damage to the bone and articular surface (Figs. 14.1, 14.2, 14.3, and 14.4).

14.2 Epidemiology

It is estimated that the incidence of AVN is 1.4–3.0 per 100,000 worldwide and the most commonly affected site is the hip, which accounts for three-fourth of all AVN cases [6]. However, in long-term study following up patients with multifocal osteonecrosis, hips were affected in all cases and bilateral hip involvement was seen in

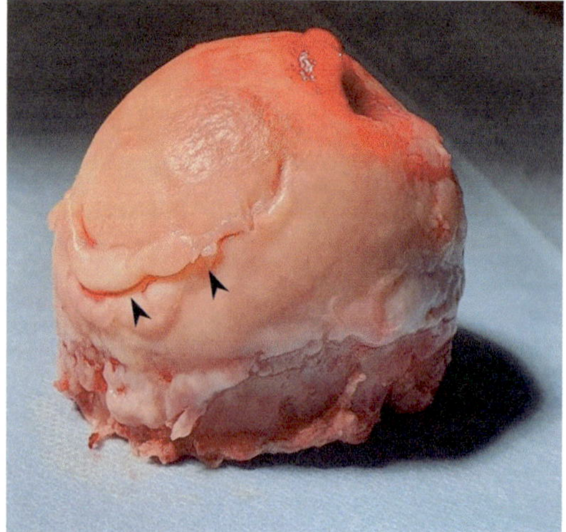

Fig. 14.3 Femoral head removed during hip arthroplasty performed due to ON. The arrow is indicating a line where the subchondral bone has collapsed

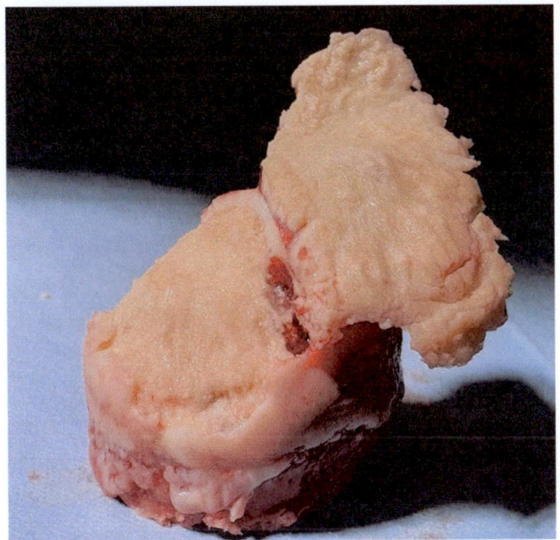

Fig. 14.4 The collapsed region has been elevated showing underlying necrotic bone

70% [7]. About 75% of patients are between 30 and 60 years of age, males are most commonly affected (7:3 male-female ratio) [1, 8]. The second most commonly affected site is the humeral head. It accounts for approximately 7% of AVN in the body, and generally patients with humeral head ON have concomitant femoral head ON [9]. Joint destruction due to humeral head ON accounts for 5% of all shoulder arthroplasties [10].

The knee is involved 90% less frequently than the hip, and typically the contralateral knee is affected in >80% of all cases [11]. ON in the knee can be divided into three categories: spontaneous osteonecrosis of the knee (SONK), secondary ON, and post-arthroscopic ON. Secondary ON occurs due to ischemia and is mostly seen in patients younger than 45 years of age, usually exposed to corticosteroids or alcohol. Lesions are usually multiple and affect more than one joint [12]. Noteworthy is the fact that SONK, historically classified as osteonecrosis, is in actually a subchondral insufficiency fracture in an osteopenic bone rather than a spontaneous necrosis as shown in histological studies [13]. In a recent review, authors have stressed the fact that SONK has multifactorial pathogenesis but the increase in joint contact pressure seen after meniscectomy and in meniscal tears, especially of the medial

posterior root, may result in insufficiency fractures [14]. Postarthroscopic ON occurs rarely and has been associated with meniscectomy and chondroplasty though some authors have suggested that it may be an actual subchondral fracture following arthroscopic procedure rather that an ischemic event [11, 15].

Other locations are less common, in the case-control study of ON in the United Kingdom authors found that just 3.6% of all ON cases were found in the foot [6]. The data on the epidemiology of foot and ankle atraumatic lesions is very limited though, so the true prevalence is unknown. ON of the talus accounts for a small percentage (approx. 2%) of all symptomatic cases, and it is characteristically localized in posterolateral region of the talar dome, where the vascular supply is the poorest [16]. Other described locations include first and second metatarsals and the navicular, but then again ON has been documented for nearly every bone in the foot [17]. In extreme cases, ON could also affect even sesamoids, causing significant pain symptoms leading to a decrease in quality of life of patients [18].

14.3 Pathophysiology

Osteonecrosis can be described as a combination of cell destruction due to ischemia, resorption, and complex regeneration of the affected site. The natural history of ON is variable, thus it mainly depends on the size of the ischemic region and the affected site. Pathophysiology of the disease is similar in adults and children beginning with an ischemic event, but due to different cartilage maturity and regeneration capability outcome in adolescents is much better than in adults [4, 19].

There are two main phases of ON; in first ischemic phase, impaired blood flow leads to necrosis of mesenchymal stem cells. Then, the changes in biology of the local environment cause apoptosis of the osteoblasts and lastly osteocytes to die [4, 20]. Necrotic changes in osteocytes begin after 2–3 h of anoxia; however, histological changes can be seen after 24–72 h after the ischemic incident [21]. In the second phase, capillary revascularization arises in the periphery of the necrotic tissue, commencing the continuous process of parallel bone resorption and bone production when the new bone tissue is laminated onto the dead bone. In the subchondral layer, resorption exceeds new bone production, consequently mechanical properties of the tissue dramatically decrease. This may lead to subchondral bone damage and irreversible change in the bone architecture caused by normal forces acting on an impaired bone [20]. It is mostly the loss of subchondral trabeculae that leads to following subchondral fracture [22].

Studies imply a pathophysiological path concerning disturbed subchondral microcirculation, which may occur in three possible mechanisms: mechanical vascular interruption, intravascular occlusion, and extravascular occlusion [20, 23]. Posttraumatic ON is caused by an interruption in vascular supply, which may occur in femoral neck, humeral neck, talar neck, and scaphoid waist due to either a fracture or a dislocation [24]. However, in the non-traumatic situation pathological changes leading to ON vary depending on the ischemic etiology that may be either intra- or extravascular. An intravascular obstruction may be caused by various abnormalities such as: clots, lipid thrombi or sickle cell aggregations [25]. In contrast, occlusion may occur due to extravascular pressure collapsing the vessels and therefore impairing the blood supply. The femoral head can be compared to a rigid wall chamber, with vessels passing through it. Fluid movement depends on the amount of pressure applied on them and as the vessels are compressible; a rise in extravascular pressure within the bone leads to collapse of these vessels. This hypothesized mechanism may occur due to lipid (or other substance) deposition within bone marrow or adipocyte hypertrophy that is seen in corticosteroid treatment, alcohol abuse or Gaucher disease [20].

14.4 Etiology

The causes of ON can be classified as traumatic and non-traumatic. It was long believed that cases not related to trauma are idiopathic, but recent studies suggest that there is a metabolic cause behind almost every case of ON.

14.4.1 Corticosteroids

The most prevalent and confirmed cause by many studies of ON not related to trauma is elevated corticosteroids level, reported to be 10–30% of all cases in retrospective studies [26]. The higher the dose and the longer the duration, the greater chance that ON will develop in the particular patient. Corticoids play a significant role in the development of the majority of ON cases. Noteworthy is the fact that not only in patients treated with corticosteroids therapy can ON occur, but in all patients with elevated levels of adrenal cortex hormones, even in those suffering from adrenocortical carcinoma [27]. Many hypothesized mechanisms have been proposed, and it has been accepted that compromised blood flow leads to subsequent cell death, but the pathophysiological pathway behind this process remains unclear. It should be emphasized that corticosteroids have numerous and extremely complex effects on immune and bone homeostasis [28]. Interestingly, many studies suggested the role of altered phospholipids metabolism in ON. In a mice model, ultrastructural changes in osteocytes were observed after high doses of methylprednisolone intramuscular injections. Lipid droplets accumulated in osteocytes and their gradual enlargement lead to compression of the nucleus, discontinuity of cell membrane, and finally cell disintegration [29]. Weinstein et al. found an increase in apoptosis of both osteocytes and osteoblasts after prednisolone administration in mice [30]. Moreover, animal models also show that corticosteroids prolong a life span of osteoclasts and this direct interaction is the cause of early and rapid bone loss [31]. Other systemic effects of corticosteroids also promote bone necrosis through suppression of angiogenesis, hyperlipidemia, increase in intracortical pressure and coagulation pathways alterations [28]. Koo et al. reported a total dose of prednisone used until ON was detected on MRI oscillated from 1.8 to 15.505 g and mean time from the beginning of steroid therapy until diagnosis of ON was 5.3 months [32]. Intriguingly, Liberman et al. found symptomatic ON only in 6 of 204 patients treated with steroids for immunosuppression after cardiac transplantation [33]. On the contrary, there are reports that even a single corticosteroid injection may cause an iatrogenic ON [34]. Therefore, physicians should be aware about this dramatic complication and always consider risk when prescribing corticoid injection therapy.

14.4.2 Alcohol Abuse

Alcohol effect on bone metabolism is dose dependent. Studies have shown that negative effects are seen in heavy and chronically drinking patients [35]. The final mechanism of the blood supply interruption and the death of bone cells in alcohol-related ON is unknown. It is hypothesized that an ischemic incident occurs due to adipocyte hypertrophy and a rise in intraosseous pressure [20]. In an animal model, chronic alcohol consumption has been proven to increase adipogenesis in bone marrow, cause cell hypertrophy, reduce hematopoiesis in the subchondral bone, and increase empty osteocyte lacunae [36]. What is more, cells obtained from bone marrow of the proximal femur during hip replacements of patients with alcohol-induced femoral head ON have shown decreased ability to differentiate into the osteogenic lineage compared with cells obtained from patients with femoral neck fractures [37]. A meta-analysis by Byung-Ho et al. has shown that the risk of femoral head ON increased by 35.3% for every 100 g of ethanol consumed per week [38]. Hirota et al. reported a dose-dependent relationship of weekly ethanol intake and a risk of ON, the odds ratio for > or =800 g/week was 14.8 [39].

14.4.3 Smoking

Although an exact mechanism is still unknown, a multifactorial cause has been hypothesized. Smoking modifies nitric oxide (NO) availability triggering oxidative stress and endothelial dys-

function, therefore altering the coagulation processes. Moreover, direct nicotine effect through sympathetic system effect and a subsequent vasoconstriction of microcirculation may also affect the bone [40, 41]. Studies have shown an increased risk of non-traumatic ON in the cigarette smoking population [40, 42, 43]. Matsuo et al. reported a significantly increased probability of ON for current smokers in their study group [43]. Interestingly, in a systematic review Wen et al. reported an elevated risk also for former smokers [40].

14.4.4 Sickle Cell Disease and Coagulation Abnormalities

Bone and other organs are gradually damaged by repeated incidents of vaso occlusion and inflammation that occur in coagulation abnormalities. Sickle cell anemia is an autosomal recessive hemoglobinopathy that leads to abnormal hemoglobin S polymerization in the deoxygenated state [44]. Affected erythrocytes are crescent shaped and become less deformable than the normal ones, which inevitably leads to vascular occlusion and subsequent ischemia [45]. Therefore, ON is a common complication of sickle cell disease reported mainly in the femoral head [46] and unlike ON caused by other diseases it affects the entire epiphysis [44].

Glueck et al. stated that thrombophilia and hypofibrinolysis are a risk factor for development of femoral head ON [47]. Also, Zalavras et al. reported that a significantly higher number of patients with non-traumatic ON had factor V Leiden mutation compared with controls [48]. It has also been reported that Hemophilia can cause femoral head ON due to repeated intraosseous hemorrhage that leads to intravascular occlusion [49].

14.4.5 Genetic

In the last few decades, several genetic studies on etiology of the AVN have been performed. There have been reports regarding links between the polymorphisms in several genes and the development of this disease. Among possible genetic predispositions to AVN, variations in VEGFA, IL-1B, IL-1R1, IL-1R2, IL-4, MMP-14, and many others have been identified [50–54]. Although there is a plethora of genetic studies on the genetic links in development of ON, most of the published articles are cross-sectional studies, with no statistical power to make a final conclusion on causative relationships. Many valuable genome-wide association studies have been performed. However, only further studies with well-designed methodology can give valuable answers, which will help develop new prophylactic and therapeutic approaches.

14.4.6 Dysbarism

Dysbarism is a known complication of diving caused by rapid changes in ambient pressure. Throughout diving, the nitrogen from the respirator tank is inhaled along with oxygen and is dissolved in blood and diffused into the tissues under pressure. If divers ascend too rapidly, the nitrogen will not be able to be excreted from the blood as the ambient pressure decreases and will form bubbles in the blood and soft tissues. Dysbaric ON occurs when the osteo vasculature is constricted due to an intraosseous pressure rise, caused by gaseous bubbles formation in the medullary cavity of long bones [55]. Though it is not a common cause of ON in the general population, a high incidence was reported among divers, 65% in a group of Hawaiian coral divers, and 85% in a group of Japanese sponge divers [56, 57].

14.4.7 Other

The prevalence of ON in patients infected with human immunodeficiency virus (HIV) was reported by Fessel et al. to be 0.9% [58], but some studies have stated that overall incidence may be 100 times higher than in the general population [59]. On the other hand, it has been debated that patients infected with HIV also have other risk factors of ON different to the HIV or the antiviral therapy [60–62]. The bone homeostasis may be disturbed due to immune-mediated consequences of HIV, changes in parathormone (PTH) and calcitonin levels that stimulate pro-inflammatory cytokines and osteoclastogenesis, but the influence of antiviral treatment has also been debated [63].

Gaucher disease (GD) is a disorder leading to accumulation of glucocerebrosides in the lysosomes of histiocytes (Gaucher cells) due to lack of beta-glucocerebrosidase. In GD, normal adipocytes in bone marrow are displaced by the Gaucher cells [64]. Though etiology is not obvious, it is possibly a rise in intraosseous extravascular pressure due to Gaucher cells deposition in bone marrow that leads to vessel occlusion and therefore necrosis [20, 65]. ON in GD is frequently multifocal and may be related to puberty or pregnancy, but new lesions can develop at any time [66, 67]. Studies have also reported ON in Fabry disease, another lysosomal storage disorder caused by deficiency of alpha-galactosidase A [68, 69].

Patients with systemic diseases related to phospholipids such as lupus erythematosus (SLE) are at particular risk as their primary condition makes them prone to the ON and the treatment (corticosteroids) further increases the chance of bone necrotic changes [70]. Herningou et al. described an increased probability of femoral head ON development during pregnancy and in postpartum period [71]. An elevated risk of ON has also been described for metabolic diseases such as hyperlipidemia, hypertriglyceridemia, and hyperuricemia [20]. ON has also been reported in pancreatitis, leukemia or lymphoma, patients undergoing radiation, bone marrow transplant and graft-versus-host disease, and patients with metastatic or disseminated malignancies [20, 72].

14.5 Diagnostic Imaging

The radiological tests which are used in detecting the ON are magnetic resonance imaging (MRI), X-ray, and positron emission tomography (PET). Changes in MRI scans precede manifestation of the ON in standard radiographs because MRI enables visualization of the metabolic changes in the bone marrow cells, while the radiographs show only the characteristics of mineralized structures of the bone, which are affected in later stages of the ON. Bone marrow edema is always present in ON; however, it is not specific as many other conditions (including trauma) can cause such changes in bone structure.

MRI is an essential diagnostic tool for early-stage (precollapse) lesions without subchondral fracture. Additionally, it is critical for early diagnosis, staging, and prognosis. A basic MR imaging examination protocol for suspected AVN includes coronal Tl-weighted spin-echo (SE) and coronal short tau inversion recovery (STIR) or fat-suppressed proton-density (PD)/T2-weighted fast (turbo spin-echo [TSE]) sequences with large fields of view. The use of intravenous contrast shows decreased enhancement in the necrotic bone and increased enhancement at the reparative interface and has been suggested as a means to differentiate viable from necrotic tissue [73]. Noteworthy is the fact that strong signal in T2 sequence in MRI can be also observed in young healthy athletes, who do not need any treatment [74]. Thus, physicians should always use a holistic approach and treat patient's symptoms, not only the results of the scans.

Differential diagnosis includes stress fractures, regional migratory osteoporosis, transient osteoporosis, complex regional pain syndrome (Sudeck's syndrome), primary bone tumors, metastases, infectious, reactive or rheumatoid arthritis, osteomyelitis, and leukemia. Differentiation between avascular necrosis and transient osteoporosis poses the greatest diagnostic challenge. Transient osteoporosis can only be recognized after more aggressive and irreversible diseases with similar clinical symptoms have been excluded. All of these disease entities may be accompanied by edema of the bone marrow presenting on an MRI as low signal intensity on T1-weighted imaging, T1-weighted TSE and high intensity on T2-weighted, turbo inversion recovery magnitude (TIRM) and STIR sequences, and increased intensity after administration of a

paramagnetic contrast [75, 76]. The absence of additional focal lesions in the subchondral bone is a very sensitive and specific sign of transient osteoporosis that differentiates it from chronic conditions. Additionally, T2-weighted imaging may reveal joint effusion which commonly accompanies this disorder [77]. In the case of differentiation with complex regional pain syndrome, the focus should be shifted to additional changes, such as skin atrophy, sensorimotor impairment, and contractures [78]. In the early stages, symptoms of both diseases are identical and affect a similar age group of patients. T1-weighted imaging shows very early changes involving the surrounding area of the necrotic bone ring of the fibrous connective tissue with a weak signal [75]. In contrast to transient osteoporosis, bone marrow edema does not occur in the early stages of avascular necrosis. However, it presents in more advanced stages of necrosis and correlates with fracture and progression of pain, according to some authors it is a weak prognostic factor [79–81].

Positron emission tomography (PET) scanning is very sensitive for AVN identification. Reactive hyperemic changes, hypervascularity, and higher turnover in bone metabolism are mirrored by higher uptake in PET. However, specificity is poor and access and costs present additional barriers [82].

Although radiographs are not the gold standard for the diagnosis of AVN, they are especially useful to identify subchondral collapse [83]. Classifications and scoring systems, regardless of the assessed joint, are generally based on the presence or absence of abnormalities such as sclerosis, subchondral fracture, head collapse, destruction of the underlying trabecular pattern and acetabular/glenoid involvement [84–86]. As previously mentioned, if an early phase of a subchondral fracture is suspected and is not clearly delineated on plain radiographs, computed tomography (CT) should be performed [87].

Studies have attempted to use genetic and biochemical markers from serum or joint tissue for early diagnosis; however, these methods are still in their infancy and their role is unclear [88, 89].

14.6 Treatment Perspectives

The natural history of ON leads to secondary osteoarthritis and in many cases, it results in the necessity of joint replacement surgery. Additionally, the hip hemiresurfacing showed to be an effective procedure, which can give a patient many years of a quality life before conversion to a total hip replacement [90]. However, in recent years the rapid development of the biological regenerative therapies resulted in a better prognosis for those patients, in which well-designed and performed treatments have been conducted with overall survivorship of 10 years as high as 80% [91]. In particular, autologous bone marrow transplant has been demonstrated to be an effective and safe therapeutic option [92]. Noteworthy is the fact that recent studies show possible complications of subchondroplasty. It is now believed that approach to treat impairment in local blood supply with bone cement is not particularly adequate, and more attention should be paid to therapeutic options related to the bone marrow [93, 94].

In recently published case series, authors reported successful treatment of spontaneous ON of the knee joint with administration of daily teriparatide. There was a significant reduction in pain and no further progression observed in 6 months of follow-up. However, the study was conducted only on the group of three patients [95]. There are also promising results of in vitro studies, suggesting possible positive effect of vitamin C and magnesium supplementation, which may play a significant role in the development new approaches of ON prevention [96].

References

1. Assouline-Dayan Y, Chang C, Greenspan A, Shoenfeld Y, Gershwin ME. Pathogenesis and natural history of osteonecrosis. Semin Arthritis Rheum. 2002;32(2):94–124.
2. Kang JS, Park S, Song JH, Jung YY, Cho MR, Rhyu KH. Prevalence of osteonecrosis of the femoral head: a nationwide epidemiologic analysis in Korea. J Arthroplast. 2009;24(8):1178–83.
3. Collaborative Osteonecrosis Group. Symptomatic multifocal osteonecrosis. A multicenter study. Clin Orthop Relat Res. 1999;369:312–26.
4. Guerado E, Caso E. The physiopathology of avascular necrosis of the femoral head: an update. Injury. 2016;47:S16–26.
5. Leroux J, Abu Amara S, Lechevallier J. Legg-Calvé-Perthes disease. Orthop Traumatol Surg Res. 2018;104(1S):S107–12.
6. Cooper C, Steinbuch M, Stevenson R, Miday R, Watts NB. The epidemiology of osteonecrosis: findings from the GPRD and THIN databases in the UK. Osteoporos Int. 2010;21(4):569–77.
7. Flouzat-Lachaniette C-H, Roubineau F, Heyberger C, Bouthors C, Hernigou P. Multifocal osteonecrosis related to corticosteroid: ten years later, risk of progression and observation of subsequent new osteonecroses. Int Orthop. 2016;40(4):669–72.
8. Liu F, Wang W, Yang L, Wang B, Wang J, Chai W, et al. An epidemiological study of etiology and clinical characteristics in patients with nontraumatic osteonecrosis of the femoral head. J Res Med Sci. 2017;22:15.
9. Mont MA, Payman RK, Laporte DM, Petri M, Jones LC, Hungerford DS. Atraumatic osteonecrosis of the humeral head. J Rheumatol. 2000;27(7):1766–73.
10. Denard PJ, Wirth MA, Orfaly RM. Management of glenohumeral arthritis in the young adult. J Bone Joint Surg Am. 2011;93(9):885–92.
11. Karim AR, Cherian JJ, Jauregui JJ, Pierce T, Mont MA. Osteonecrosis of the knee: review. Ann Transl Med. 2015;3(1):6.
12. Lerebours F, ElAttrache NS, Mandelbaum B. Diseases of subchondral bone 2. Sports Med Arthrosc Rev. 2016;24(2):50–5.
13. Yamamoto T, Bullough PG. Spontaneous osteonecrosis of the knee: the result of subchondral insufficiency fracture. J Bone Joint Surg Am. 2000;82(6):858–66.
14. Hussain ZB, Chahla J, Mandelbaum BR, Gomoll AH, LaPrade RF. The role of meniscal tears in spontaneous osteonecrosis of the knee: a systematic review of suspected etiology and a call to revisit nomenclature. Am J Sports Med. 2019;47(2):501–7.
15. MacDessi SJ, Brophy RH, Bullough PG, Windsor RE, Sculco TP. Subchondral fracture following arthroscopic knee surgery. A series of eight cases. J Bone Joint Surg Am. 2008;90(5):1007–12.
16. Delanois RE, Mont MA, Yoon TR, Mizell M, Hungerford DS. Atraumatic osteonecrosis of the talus. J Bone Joint Surg Am. 1998;80(4):529–36.
17. DiGiovanni CW, Patel A, Calfee R, Nickisch F. Osteonecrosis in the foot. J Am Acad Orthop Surg. 2007;15(4):208–17.
18. Bartosiak K, McCormick JJ. Avascular necrosis of the sesamoids. Foot Ankle Clin. 2019;24(1):57–67.
19. Leroux J, Abu Amara S, Lechevallier J. Legg-Calvé-Perthes disease. Orthop Traumatol Surg Res. 2018;104(1S):S107–12.

20. Shah KN, Racine J, Jones LC, Aaron RK. Pathophysiology and risk factors for osteonecrosis. Curr Rev Musculoskelet Med. 2015;8(3):201–9.

21. James J, Steijn-Myagkaya GL. Death of osteocytes. Electron microscopy after in vitro ischaemia. J Bone Joint Surg Br. 1986;68(4):620–4.

22. Brown TD, Baker KJ, Brand RA. Structural consequences of subchondral bone involvement in segmental osteonecrosis of the femoral head. J Orthop Res. 1992;10(1):79–87.

23. Cohen-Rosenblum A, Cui Q. Osteonecrosis of the femoral head. Orthop Clin North Am. 2019;50(2):139–49.

24. Large TM, Adams MR, Loeffler BJ, Gardner MJ. Posttraumatic avascular necrosis after proximal femur, proximal humerus, talar neck, and scaphoid fractures. J Am Acad Orthop Surg. 2019;27(21):794–805.

25. Mont MA, Cherian JJ, Sierra RJ, Jones LC, Lieberman JR. Nontraumatic osteonecrosis of the femoral head: where do we stand today? A ten-year update. J Bone Joint Surg Am. 2015;97(19):1604–27.

26. Aaron R, Gray R. Osteonecrosis: etiology, natural history, pathophysiology, and diagnosis. In: The adult hip. Philadelphia: Lippincott Williams & Wilkins; 2007. p. 465–76.

27. Siddiqui N, Khan BA. A case of avascular necrosis in a patient with adrenocortical carcinoma and disseminated metastasis. J Pak Med Assoc. 2018;68(8):1267–9.

28. Powell C, Chang C, Gershwin ME. Current concepts on the pathogenesis and natural history of steroid-induced osteonecrosis. Clin Rev Allergy Immunol. 2011;41(1):102–13.

29. Kawai K, Tamaki A, Hirohata K. Steroid-induced accumulation of lipid in the osteocytes of the rabbit femoral head. A histochemical and electron microscopic study. J Bone Joint Surg Am. 1985;67(5):755–63.

30. Weinstein RS, Nicholas RW, Manolagas SC. Apoptosis of osteocytes in glucocorticoid-induced osteonecrosis of the hip. J Clin Endocrinol Metab. 2000;85(8):2907–12.

31. Jia D, O'Brien CA, Stewart SA, Manolagas SC, Weinstein RS. Glucocorticoids act directly on osteoclasts to increase their life span and reduce bone density. Endocrinology. 2006;147(12):5592–9.

32. Koo K-H, Kim R, Kim Y-S, Ahn I-O, Cho S-H, Song H-R, et al. Risk period for developing osteonecrosis of the femoral head in patients on steroid treatment. Clin Rheumatol. 2002;21(4):299–303.

33. Lieberman JR, Roth KM, Elsissy P, Dorey FJ, Kobashigawa JA. Symptomatic osteonecrosis of the hip and knee after cardiac transplantation. J Arthroplast. 2008;23(1):90–6.

34. Al-Omari AA, Aleshawi AJ, Marei OA, Younes HMB, Alawneh KZ, ALQuran E, et al. Avascular necrosis of the femoral head after single steroid intra-articular injection. Eur J Orthop Surg Traumatol. 2020;30(2):193–7.

35. Maurel DB, Boisseau N, Benhamou CL, Jaffre C. Alcohol and bone: review of dose effects and mechanisms. Osteoporos Int. 2012;23(1):1–16.

36. Wang Y, Li Y, Mao K, Li J, Cui Q, Wang G-J. Alcohol-induced adipogenesis in bone and marrow: a possible mechanism for osteonecrosis. Clin Orthop Relat Res. 2003;410:213–24.

37. Suh KT, Kim SW, Roh HL, Youn MS, Jung JS. Decreased osteogenic differentiation of mesenchymal stem cells in alcohol-induced osteonecrosis. Clin Orthop Relat Res. 2005;431:220–5.

38. Yoon B-H, Kim T-Y, Shin I-S, Lee HY, Lee YJ, Koo K-H. Alcohol intake and the risk of osteonecrosis of the femoral head in Japanese populations: a dose-response meta-analysis of case-control studies. Clin Rheumatol. 2017;36(11):2517–24.

39. Hirota Y, Hirohata T, Fukuda K, Mori M, Yanagawa H, Ohno Y, et al. Association of alcohol intake, cigarette smoking, and occupational status with the risk of idiopathic osteonecrosis of the femoral head. Am J Epidemiol. 1993;137(5):530–8.

40. Wen Z, Lin Z, Yan W, Zhang J. Influence of cigarette smoking on osteonecrosis of the femoral head (ONFH): a systematic review and meta-analysis. Hip Int. 2017;27(5):425–35.

41. Bevan JA, Su C. Uptake of nicotine by the sympathetic nerve terminals in the blood vessel. J Pharmacol Exp Ther. 1972;182(3):419–26.

42. Takahashi S, Fukushima W, Kubo T, Iwamoto Y, Hirota Y, Nakamura H. Pronounced risk of non-traumatic osteonecrosis of the femoral head among cigarette smokers who have never used oral corticosteroids: a multicenter case-control study in Japan. J Orthop Sci. 2012;17(6):730–6.

43. Matsuo K, Hirohata T, Sugioka Y, Ikeda M, Fukuda A. Influence of alcohol intake, cigarette smoking, and occupational status on idiopathic osteonecrosis of the femoral head. Clin Orthop Relat Res. 1988;234:115–23.

44. Vanderhave KL, Perkins CA, Scannell B, Brighton BK. Orthopaedic manifestations of sickle cell disease. J Am Acad Orthop Surg. 2018;26(3):94–101.

45. Driscoll MC. Sickle cell disease. Pediatr Rev. 2007;28(7):259–68.

46. Hernigou P, Bachir D, Galacteros F. The natural history of symptomatic osteonecrosis in adults with sickle-cell disease. J Bone Joint Surg Am. 2003;85(3):500–4.

47. Glueck CJ, Freiberg RA, Wang P. Heritable thrombophilia-Hypofibrinolysis and osteonecrosis of the femoral head. Clin Orthop Relat Res. 2008;466(5):1034–40.

48. Zalavras CG, Vartholomatos G, Dokou E, Malizos KN. Genetic background of osteonecrosis: associated with thrombophilic mutations? Clin Orthop Relat Res. 2004;422:251–5.

49. Kilcoyne RF, Nuss R. Femoral head osteonecrosis in a child with hemophilia. Arthritis Rheum. 1999;42(7):1550–1.

50. An F, Wang J, Gao H, Liu C, Tian Y, Jin T, et al. Impact of IL1R1 and IL1R2 gene polymorphisms on risk of osteonecrosis of the femoral head from a case-control study. Mol Genet Genomic Med. 2019;7(3):e00557.

51. Jin T, Zhang Y, Sun Y, Wu J, Xiong Z, Yang Z. IL-4 gene polymorphisms and their relation to steroid-induced osteonecrosis of the femoral head in Chinese population. Mol Genet Genomic Med. 2019;7(3):e563.

52. Ma W, Xin K, Chen K, Tang H, Chen H, Zhi L, et al. Relationship of common variants in VEGFA gene with osteonecrosis of the femoral head: a Han Chinese population based association study. Sci Rep. 2018;8(1):16221.

53. Qi Y, Wang J, Sun M, Ma C, Jin T, Liu Y, et al. MMP-14 single-nucleotide polymorphisms are related to steroid-induced osteonecrosis of the femoral head in the population of northern China. Mol Genet Genomic Med. 2019;7(2):e00519.

54. Yu Y, Zhang Y, Wu J, Sun Y, Xiong Z, Niu F, et al. Genetic polymorphisms in IL1B predict susceptibility to steroid-induced osteonecrosis of the femoral head in Chinese Han population. Osteoporos Int. 2019;30(4):871–7.

55. Sharareh B, Schwarzkopf R. Dysbaric osteonecrosis: a literature review of pathophysiology, clinical presentation, and management. Clin J Sport Med. 2015;25(2):153–61.

56. Shinoda S, Hasegawa Y, Kawasaki S, Tagawa N, Iwata H. Magnetic resonance imaging of osteonecrosis in divers: comparison with plain radiographs. Skelet Radiol. 1997;26(6):354–9.

57. Wade CE, Hayashi EM, Cashman TM, Beckman EL. Incidence of dysbaric osteonecrosis in Hawaii's diving fishermen. Undersea Biomed Res. 1978;5(2):137–47.

58. Fessel WJ, Chau Q, Leong D. Association of osteonecrosis and osteoporosis in HIV-1-infected patients. AIDS. 2011;25(15):1877–80.

59. Morse CG, Mican JM, Jones EC, Joe GO, Rick ME, Formentini E, et al. The incidence and natural history of osteonecrosis in HIV-infected adults. Clin Infect Dis. 2007;44(5):739–48.

60. Morse CG, Kovacs JA. Metabolic and skeletal complications of HIV infection: the price of success. JAMA. 2006;296(7):844–54.

61. García-Alvarez García I, García-Alvarez García F, Crusels MJ, Cuesta Muñoz J, Letona Carbajo S, Amiguet García JA. Osteonecrosis and HIV. An Med Interna. 2003;20(4):187–90.

62. Munhoz Lima ALL, Oliveira PR, Carvalho VC, Godoy-Santos AL, Ejnisman L, Oliveira CR, et al. Osteonecrosis of the femoral head in people living with HIV: anatomopathological description and p24 antigen test. HIV AIDS (Auckl). 2018;10:83–90.

63. Lafferty MK, Fantry L, Bryant J, Jones O, Hammoud D, Weitzmann MN, et al. Elevated suppressor of cytokine signaling-1 (SOCS-1): a mechanism for dysregulated osteoclastogenesis in HIV transgenic rats. Pathog Dis. 2014;71(1):81–9.

64. Miller SP, Zirzow GC, Doppelt SH, Brady RO, Barton NW. Analysis of the lipids of normal and Gaucher bone marrow. J Lab Clin Med. 1996;127(4):353–8.

65. Deegan PB, Pavlova E, Tindall J, Stein PE, Bearcroft P, Mehta A, et al. Osseous manifestations of adult Gaucher disease in the era of enzyme replacement therapy. Medicine (Baltimore). 2011;90(1):52–60.

66. Zimran A, Morris E, Mengel E, Kaplan P, Belmatoug N, Hughes DA, et al. The female Gaucher patient: the impact of enzyme replacement therapy around key reproductive events (menstruation, pregnancy and menopause). Blood Cells Mol Dis. 2009;43(3):264–88.

67. Hughes D, Mikosch P, Belmatoug N, Carubbi F, Cox T, Goker-Alpan O, et al. Gaucher disease in bone: from pathophysiology to practice. J Bone Miner Res. 2019;34(6):996–1013.

68. Lien Y-HH, Lai L-W. Bilateral femoral head and distal tibial osteonecrosis in a patient with Fabry disease. Am J Orthop. 2005;34(4):192–4.

69. Lidove O, Zeller V, Chicheportiche V, Meyssonnier V, Sené T, Godot S, et al. Musculoskeletal manifestations of Fabry disease: a retrospective study. Joint Bone Spine. 2016;83(4):421–6.

70. Gladman DD, Dhillon N, Su J, Urowitz MB. Osteonecrosis in SLE: prevalence, patterns, outcomes and predictors. Lupus. 2018;27(1):76–81.

71. Hernigou P, Jammal S, Pariat J, Flouzat-Lachaniette CH, Dubory A. Hip osteonecrosis and pregnancy in healthy women. Int Orthop. 2018;42(6):1203–11.

72. Kaushik AP, Das A, Cui Q. Osteonecrosis of the femoral head: An update in year 2012. World J Orthop. 2012;3(5):49–57.

73. Karantanas AH, Drakonaki EE. The role of MR imaging in avascular necrosis of the femoral head. Semin Musculoskelet Radiol. 2011;15(3):281–300.

74. Kornaat PR, Van de Velde SK. Bone marrow edema lesions in the professional runner. Am J Sports Med. 2014;42(5):1242–6.

75. Asadipooya K, Graves L, Greene LW. Transient osteoporosis of the hip: review of the literature. Osteoporos Int. 2017;28(6):1805–16.

76. Szwedowski D, Nitek Z, Walecki J. Evaluation of transient osteoporosis of the hip in magnetic resonance imaging. Pol J Radiol. 2014;79:36–8.

77. Vande Berg BE, Malghem JJ, Labaisse MA, Noel HM, Maldague BE. MR imaging of avascular necrosis and transient marrow edema of the femoral head. Radiographics. 1993;13(3):501–20.

78. Gemmel F, Van Der Veen HC, Van Schelven WD, Collins JMP, Vanneuville I, Rijk PC. Multi-modality imaging of transient osteoporosis of the hip. Acta Orthop Belg. 2012;78(5):619–27.

79. Ito H, Matsuno T, Minami A. Relationship between bone marrow edema and development of symptoms in patients with osteonecrosis of the femoral head. AJR Am J Roentgenol. 2006;186(6):1761–70.

80. Fernandez-Canton G. From bone marrow edema to osteonecrosis. New concepts. Reumatol Clin. 2009;5(5):223–7.

81. Kim YM, Oh HC, Kim HJ. The pattern of bone marrow oedema on MRI in osteonecrosis of the femoral head. J Bone Joint Surg Br. 2000;82(6):837–41.
82. Ullmark G, Sundgren K, Milbrink J, Nilsson O. Femoral head viability following resurfacing arthroplasty. A clinical positron emission tomography study. Hip Int. 2011;21(1):66–70.
83. Nelson FRT, Bhandarkar VS, Woods TA. Using hip measures to avoid misdiagnosing early rapid onset osteoarthritis for osteonecrosis. J Arthroplast. 2014;29(6):1243–7.
84. Hasan SS, Romeo AA. Nontraumatic osteonecrosis of the humeral head. J Shoulder Elb Surg. 2002;11(3):281–98.
85. Schmitt-Sody M, Kirchhoff C, Mayer W, Goebel M, Jansson V. Avascular necrosis of the femoral head: inter- and intraobserver variations of Ficat and ARCO classifications. Int Orthop. 2008;32(3):283–7.
86. Lee G-C, Steinberg ME. Are we evaluating osteonecrosis adequately? Int Orthop. 2012;36(12):2433–9.
87. Yeh L-R, Chen CKH, Huang Y-L, Pan H-B, Yang C-F. Diagnostic performance of MR imaging in the assessment of subchondral fractures in avascular necrosis of the femoral head. Skelet Radiol. 2009;38(6):559–64.
88. Li T, Zhang Y, Wang R, Xue Z, Li S, Cao Y, et al. Discovery and validation an eight-biomarker serum gene signature for the diagnosis of steroid-induced osteonecrosis of the femoral head. Bone. 2019;122:199–208.
89. Floerkemeier T, Hirsch S, Budde S, Radtke K, Thorey F, Windhagen H, et al. Bone turnover markers failed to predict the occurrence of osteonecrosis of the femoral head—a preliminary study. J Clin Lab Anal. 2012;26(2):55–60.
90. Chatterjee D, Harb MA, Yiachos C, Tangl S, Vigorita VJ, Tischler HM, et al. A histologic analysis of a retrieved specimen 24 years after Hemiresurfacing for avascular necrosis. J Long-Term Eff Med Implants. 2019;29(1):19–27.
91. Andriolo L, Merli G, Tobar C, Altamura SA, Kon E, Filardo G. Regenerative therapies increase survivorship of avascular necrosis of the femoral head: a systematic review and meta-analysis. Int Orthop. 2018;42(7):1689–704.
92. Hernigou P, Poignard A, Manicom O, Mathieu G, Rouard H. The use of percutaneous autologous bone marrow transplantation in nonunion and avascular necrosis of bone. J Bone Joint Surg Br. 2005;87(7):896–902.
93. Chatterjee D, McGee A, Strauss E, Youm T, Jazrawi L. Subchondral calcium phosphate is ineffective for bone marrow edema lesions in adults with advanced osteoarthritis. Clin Orthop Relat Res. 2015;473(7):2334–42.
94. Kasik CS, Martinkovich S, Mosier B, Akhavan S. Short-term outcomes for the biologic treatment of bone marrow Edema of the knee using bone marrow aspirate concentrate and injectable demineralized bone matrix. Arthrosc Sports Med Rehabil. 2019;1(1):e7–14.
95. Horikawa A, Miyakoshi N, Hongo M, Kasukawa Y, Shimada Y, Kodama H, et al. Treatment of spontaneous osteonecrosis of the knee by daily teriparatide: a report of 3 cases. Medicine (Baltimore). 2020;99(5):e18989.
96. Zheng L-Z, Wang J-L, Xu J-K, Zhang X-T, Liu B-Y, Huang L, et al. Magnesium and vitamin C supplementation attenuates steroid-associated osteonecrosis in a rat model. Biomaterials. 2020;238:119828.

Current Concepts in Subchondral Bone Pathology

15

Alberto Gobbi, Ramiro Alvarez, Eleonora Irlandini, and Ignacio Dallo

15.1 Introduction

Articular cartilage and subchondral bone act as a functional unit, the osteochondral unit (OCU), to maintain joint homeostasis. Numerous research efforts have focused on articular cartilage damage and their pathophysiology. Still, few are focused on subchondral bone pathology, which should be viewed as a critical element of the osteochondral unit and a key player in joint health. Focal changes in the subchondral bone, termed bone marrow lesions (BMLs), are features detected by magnetic resonance imaging (MRI). BMLs describe an alteration of bone marrow signal intensity, with high signal on fluid-sensitive sequences (T2/proton density with fat suppression and short tau inversion recovery (STIR)) with or without low T1WI signal. BMLs are present in a wide range of pathologies, including traumatic contusion and fractures, post-cartilage surgery, osteoarthritis (OA), transient BML syndromes, spontaneous insufficiency fracture of the knee (SIFK), osteonecrosis (ON), and

conditions included in complex regional pain syndrome (CRPS). These MRI alterations may correspond histologically to true edema, but also trabecular necrosis, cysts, fibrosis, and cartilage fragments. Therefore, instead of the commonly used term "bone marrow edema," the expressions "bone marrow edema-like signal," or "BMLs" are more appropriate [1]. MRI plays a fundamental role in guiding the diagnosis based on recognizable typical patterns even at the early stages. There is a growing interest in the study of the subchondral bone in several pathological conditions. However, these BMLs remain controversial for their still unidentified role in etiopathological processes, clinical impact, and treatment. This chapter will focus on the current understanding of subchondral bone pathologies.

15.2 Classification of Subchondral Bone Pathology

A classification of BMLs according to cause into ischemic, mechanical, and reactive has been proposed. However, as the etiology and pathogenesis are at best poorly understood for many of the lesions, such differentiation might be misleading. For these reasons, subchondral bone marrow edema-like lesions around the knee can be classified into traumatic or non-traumatic and into reversible or irreversible [1, 2]. The reversibility of the BML depends on its cause and whether

A. Gobbi (✉) · E. Irlandini · I. Dallo
Orthopaedic Arthroscopic Surgery International (O.A.S.I.) Bioresearch Foundation, Gobbi N.P.O., Milan, Italy
e-mail: gobbi@cartilagedoctor.it;
e.irlandini@oasiortopedia.it

R. Alvarez
Sports Medicine and Rehabilitation,
Buenos Aires, Argentina

© ISAKOS 2022
A. Gobbi et al. (eds.), *Joint Function Preservation*, https://doi.org/10.1007/978-3-030-82958-2_15

there is an alteration to the structure of the osteochondral unit. There are distinctive features on MRI that can help to predict if the BML is reversible. Prognostic criteria that appear to indicate a benign course are no changes on plain radiographs, the lack of additional subchondral changes other than BML, and the absence of focal epiphyseal contour depression. Conversely, the presence on MRI of low signal intensity lines deep in the condyles [3, 4], or a subchondral area of low signal thicker than 4 mm strongly predicts irreversibility [5, 6]. The differential diagnosis includes a wide range of conditions based on typical patterns that rely on location, age, coexisting pathologies, clinical history, and MRI findings. The following paragraphs will focus on the characteristic imaging findings, the pathology, the prognosis, and possible complications of the most common types of BMLs.

15.2.1 Traumatic Subchondral Bone Lesions

Trauma-induced BMLs can be associated with acute direct or indirect trauma such as bone contusions, or with subacute lesions as a result of overload, such as stress fractures and repetitive microtrauma occurring during physical activity [2, 5]. These types of BMLs are mostly related to traumatic episodes and are often associated with knee ligament tears [7]. However, there is a spectrum of asymptomatic patients subject to repetitive microtrauma, with a subchondral edema-like signal on MRI [8, 9], that has been shown in 41% of collegiate basketball players [10]. Osseous injury can be caused by a direct blow, applied shear forces, bones impacting each other, or from traction forces in the context of avulsion injury [11]. These mechanisms and the associated soft-tissue injuries can be revealed through the study of the distribution of marrow edema-like signal [11, 12]. The pivot shift injuries, often related to ACL tears, are the most common cause of subchondral contusions [7]. This can be a result of valgus stress on the knee with the femur in external rotation relative to a fixed tibia or an internal tibial rotation during cut and jump activities [13].

These mechanisms could be the explanation of why the lateral compartment is more involved than the medial one [14].

Other location-specific patterns are those related to hyperextension, dashboard injury, clip injury, and lateral patellar dislocation [11]. Hyperextension and dashboard [15] injuries may lead to posterior cruciate ligament (PCL) tear and cause a subchondral contusion in the anterior tibia or femur [12, 16]. During clip injury, bone marrow edema is usually most prominent in the lateral femoral condyle produced by a direct blow. In contrast, a second smaller area of edema may be present in the medial condyle secondary to avulsive stress to the medial collateral ligament (MCL) [11]. Injuries related to spontaneously reduced lateral patellar dislocation in teenagers are characterized by one or a combination of kissing impaction in the medial patellar facet or the median ridge, a medial patellar traction contusion of the medial retinaculum, impaction contusion in the anterior lateral femoral condyle, and an osteochondral grade 4–5 defect in the lateral femoral condyle [5, 17]. Another age-location-specific pattern in children with open physis is the ACL tear, which typically avulses from the tibia, resulting in a traction contusion around the insertion site [1].

Histopathological findings following a single direct impact or resulting from repetitive microtrauma are characterized by microfractures of the subarticular spongiosa with osteocyte necrosis and empty lacunae, hemorrhage, and edema [18]. The natural history of post-traumatic bone contusions has been poorly investigated, especially in the long term. BML evolution is influenced by several factors like location of the lesion, the mechanism, the severity, osteochondral unit damage, and the association with other injuries. Bone contusions typically present themselves as reversible edema-like lesions, with indistinct margins that resolve within 2–4 months [1], like the one associated with isolated MCL tear. However, it has been reported that BML in a complex ligament injury may have a slower resolution [19].

The damage to the OCU plays a major role in the evolution of BMLs. While edema-like signal in ACL lesions without cortical involvement tend to resolve spontaneously in 95% of cases [20], BMLs are still present at 3 years follow-up in lesions associated with a disruption of the femoral cortical surface. According to the mechanism of the ACL tear, it has been noted that the non-contact lesion appears to cause more severe BML in both the medial and lateral compartments, than the contact ones [7]. The location of the lesion is another factor that may affect the evolution of BML. During an ACL tear, the BMLs located on the femoral condyle tend to resolution at 3 months compared to lateral tibial BMLs at 6 months [5, 12]. Moreover, 67% of lateral femoral condyle ACL injury-associated bone bruises have been related to osteochondral damage, whereas no cartilage defects were found in cases of BML of the posterolateral tibial plateau [21].

If the traumatic impact is severe, a subchondral fracture may cause a local depression and collapse of the cartilage surface. Osteoarthritis signs may appear in the course of bone remodeling as the subchondral bone increases in thickness and becomes stiffer. Biopsy samples of the articular cartilage overlying the bone bruise lesions showed degeneration or necrosis of chondrocytes and a loss of proteoglycan. These data support the suggestion that a severe bone bruise is a precursor of early degenerative changes [22]. There is no agreement in the literature regarding a correlation at short-term follow-up between BMLs and functional status, even though it has been reported that patients with an ACL tear and BML had increased pain scores and longer rehabilitation time [23], mainly if the alteration is still detectable 3 months after the injury [24]. It is still under debate if the initial joint injury and BML are directly correlated to long-term function and OA development [7].

15.2.2 Atraumatic Subchondral Bone Lesions

15.2.2.1 Transient Bone Marrow Lesion Syndromes

Transient conditions include regional migratory osteoporosis (RMO), transient osteoporosis (TOP), and complex regional pain syndrome (CRPS). All transient conditions have a similar MRI presentation of diffuse subchondral bone marrow high-signal intensity with indistinct margins, reaching but preserving the joint surface. Gender, age, and clinical history help to differentiate between the three diagnoses [1].

Complex regional pain syndrome (CRPS) was previously referred to as reflex sympathetic dystrophy or algodystrophy. A diagnosis of CRPS is made on a clinical basis and is divided into CRPS type I and II accordingly without or with identifiable peripheral nerve injury. CRPS Type 1 is initiated by a major or a minor traumatic injury and is often associated with pain, swelling, vasomotor instability, contracture, and osteoporosis [25]. Transient osteoporosis (TO) is a rare condition that is characterized by sudden onset localized pain, usually in a weight-bearing joint, most commonly the hip, that usually resolves with conservative management, hence the term "transient." It is also characterized by the demonstration of localized demineralization on plain radiography or computed tomography (CT) scan; therefore, the term "osteoporosis" [26]. The populations mainly affected are middle-aged men and pregnant women during the last trimester or the immediate postpartum period [27].

Regional migratory osteoporosis (RMO) is a disorder manifested by arthralgia migrating to other joints or within the same joint. RMO is limited to the weight-bearing joints of the lower appendicular skeleton. Clinical findings will usually include tenderness, joint effusion, and swelling with no significant restriction in the range of motion. In most cases, plain radiographs and bone densitometry will reveal localized demineralization in the juxta-articular bone. The pattern of symptoms migration has been reported as typically sequential, proximal to distal with a migratory interval of up to 9 months [28].

15.2.2.2 Subchondral Insufficiency Fractures

Subchondral insufficiency fractures of the knee (SIFK) are non-traumatic fractures with no histological evidence of necrosis, usually occurring in overweight, elderly female patients. SIFK involves a physiologic force applied to weakened trabeculae, often in association with osteopenia and diminished protective function of the articular cartilage and meniscus, which leads to a fracture along the subchondral area of the bone [29]. SIFK can be reversible but also can progress to a collapse of the articular surface and rapidly destructive OA [5].

In MRI evaluation, SIFK is best shown on T2-weighted and proton density-weighted images and is associated with marked bone marrow edema. The findings in SIFK include a hypointense line that is irregular, sometimes discontinuous, in the subarticular marrow, and an area of low signal intensity immediately subjacent to and creating the appearance of a thickened subchondral bone plate. These localized abnormalities represent the fracture line and the granulation tissue [29]. The low signal intensity area has prognostic relevance. If it is thicker than 4 mm or longer than 14 mm, the lesion may be irreversible and evolve into irreparable epiphyseal collapse and articular destruction [1, 6]. Edema-like signals present in SIFK extents from the subchondral region over large areas, often involving the entire femoral condyle and reaching the metaphysis [30]. It differs from the more localized BMLs subjacent to cartilage loss in osteoarthritis. However, the extent of the lesion has no prognostic significance [29]. SIFK typically is observed along the central weight-bearing aspect of the femoral condyle (60–90%) and is commonly associated with medial meniscus tears [31–33]. It has been proposed that more than 50% of patients demonstrate radial or posterior root tears [34]. These findings support the proposed role of mechanical stress in the development of SIFK and emphasize the rationale for meniscal conservation.

The clinical course and earliest stage of SIFK can be unpredictable and does not necessarily progress in every patient [5]. Typically, the initial phase consists of severe pain with functional impairment for at least 3 to 6 months, followed by spontaneous resolution with functional and radiographic improvement [35]. While subtle contour deformities occasionally can be observed in self-resolving lesions, prominent contour deformity, and the collapse of the subchondral bone plate are poor prognostic factors [29]. On the contrary, the lack of additional subchondral changes other than BML is 100% predictive of reversibility [6]. Markers of high-grade lesions include medial meniscus posterior root tears with associated moderate to severe extrusion, high-grade chondrosis, larger lesion sizes, and articular surface collapse [33].

15.2.2.3 Osteonecrosis

Ahlback first described osteonecrosis (ON) of the knee in 1968 [36]. Since then, the improvement of knowledge in this field has led to the identification of three distinct categories of ON: spontaneous osteonecrosis of the knee (SONK), avascular osteonecrosis (AVN), and postarthroscopic ON. SONK was recognized early as a distinct form of epiphyseal osteonecrosis. This condition typically is seen in patients after the sixth decade of life and more frequently in women. Patients usually report a sudden onset of knee joint pain related to minimal or no trauma and often recall a precise moment when the symptoms started [29]. SONK is the most common form of osteonecrosis of the knee [37].

The etiology of SONK is still not completely understood, but two hypotheses have been proposed. Avascular origin was initially suggested. However, the evidence in favor of this theory is limited. More recently, SONK has been associated with subchondral insufficiency fractures of the knee (SIFK). A study by Yamamoto and Bullough [38], which was supported by results of later studies [27, 39], showed that the first event is an SIFK, that has progressed into collapse followed by secondary necrosis limited to the area between the fracture line and the subchondral bone plate. Moreover, the MRI features of this lesion also are profoundly different from those of AVN studies [29].

Avascular necrosis (also called atraumatic, ischemic, or idiopathic osteonecrosis) is a degenerative bone condition characterized by the death of cellular components of the bone secondary to an interruption of the subchondral blood supply [40], usually affecting the epiphysis of long bones in patients below 45 years of age. It can be secondary to systemic diseases, radiation, chemotherapy, or substance consumption of alcohol, corticosteroids, and tobacco. These underlying systemic conditions and bone infarcts at other locations can narrow the differential diagnosis between SONK and AVN [1, 37]. In most cases, a "double-line sign," an inner high-signal-intensity band (vascularized granulation tissue), and an outer low-signal-intensity band (sclerotic appositional new bone) are visible on T2-weighted [29]. Advanced disease may result in the subchondral collapse, which threatens the viability of the joint involved. Lesions involving more than one-third of the condyle on midcoronal MR images or the middle and posterior one-third of the condyle on midsagittal MR images are at higher risk of collapse [41].

Post-arthroscopic ON or "osteonecrosis in the postoperative knee" (ONPK) [42] was first described by Brahme et al. in 1991 [43]. ONPK is the least common form of ON in the knee, and it is not related to age or sex predominance. The etiology of ONPK is debated; it may occur after meniscectomy, cartilage debridement, and radiofrequency surgery. Altered knee biomechanics after meniscectomy may be a predisposing factor for osteonecrosis. It has been proposed that increased tibiofemoral contact pressure might result in insufficiency fracture of the cartilage and subchondral bone with an intraosseous leak of synovial fluid and subsequent osteonecrosis [44, 45]. However, other authors have described the lesion as being, in fact, a subchondral fracture and not pure osteonecrosis as traditionally described [46, 47]. MRI obtained in the early stages of ONPK will demonstrate a nonspecific, large area of bone marrow edema (BME) in the femoral condyle, always coinciding with the site of the arthroscopic procedure [42].

15.2.2.4 Bone Marrow Lesions in Osteoarthritis

Subchondral BMLs are a common finding in patients with both early and advanced OA. These are often associated with meniscus damage, thinning, or focal cartilage defects and subchondral cyst-like lesions [5]. The most common histologic findings in bone marrow edema-like lesions in OA include bone necrosis, fibrosis, hemorrhage, and trabecular abnormalities, while edema is infrequent. These findings might be seen as well in SIFK. However, the bone marrow edema-like pattern is typically localized in osteoarthritis and extensive in SIFK, and the articular cartilage may be preserved in early SIFK, while significant cartilage loss typically accompanies eburnation in osteoarthritis. Once SIFK progresses to collapse and articular surface destruction, distinguishing it from primary osteoarthritis at imaging may be impossible [29]. The evolution of BML in the setting of OA is extremely variable. Subchondral lesions may regress or resolve completely within 30 months follow-up [48], but some studies showed the persistence of BML in the majority of patients [49, 50]. The clinical correlation of BMLs in the setting of OA is still under debate; moderate evidence supports that the severity and enlargement of BML are predictors of pain, the progression of cartilage damage, and subchondral bone attrition [35, 51–53].

15.2.2.5 Bone Marrow Lesions after Surgery for Cartilage Repair

The increasing awareness of the role played by the subchondral bone in cartilage lesions has recently led to investigations into the meaning of such MRI findings in patients with cartilage treatments. BML is a common finding after cartilage surgery, ranging from about 40% to 80% of both chondral and osteochondral procedures [54, 55]. The evolution of post-surgical BML is still unclear, with evidence of both a reduction and an increase in its incidence over time. From a histological point of view, the entire osteochondral unit can be altered, either as a short-term maturation result, or as long-term tissue evolution showing changes in bone mineral density, bone volume, and trabecular thickness [56]. In a review of 10 years of follow-up imaging of subchondral bone edema in knees after matrix-assisted autologous chondrocyte transplantation, BML was present in the 50% of cases during the first postoperative phases, markedly reduced to 30% at 2 and 3 years, and then again increased to 60% [57]. These changes might be related to the maturation phase marked by the decreasing of the

BML followed by an increase at long-term follow-up explained by the fact that the tissue obtained as a result of cartilage procedures maybe not sufficient to protect the subchondral bone from mechanical forces, thus leading to progressive abnormal subchondral bone stimulation. However, no correlation has been found between BML pre- and postoperative and clinical outcomes [58], making its significance questionable and difficult to rate [54, 55, 57].

15.3 Conclusions

There is a growing interest in the study of subchondral bone pathology and its pathogenesis. However, many aspects remain unsolved. BMLs are present in a wide range of conditions, including traumatic and non-traumatic. A key factor to address when studying a patient with subchondral BMLs is the distinction between reversible and irreversible lesions. MRI images play a significant role for a correct diagnosis, together with the clinical presentation, patient demographics, and history of trauma. A fundamental step to predict the evolution of the lesion will be the comprehension of what determines the different kinds of BMLs, as well as what turns a reversible lesion into an irreversible one. A better understanding of BMLs will be mandatory in the future to enable accurate differential diagnoses and to develop appropriately targeted treatments.

References

1. Kon E, Ronga M, Filardo G, et al. Bone marrow lesions and subchondral bone pathology of the knee. Knee Surg Sports Traumatol Arthrosc. 2016;24:1797–814.
2. Roemer FW, Frobell R, Hunter DJ, et al. MRI-detected subchondral bone marrow signal alterations of the knee joint: terminology, imaging appearance, relevance and radiological differential diagnosis. Osteoarthr Cartil. 2009;17:1115–31.
3. Yates PJ, Calder JD, Stranks GJ, Early MRI. Diagnosis and non-surgical management of spontaneous osteonecrosis of the knee. Knee. 2007;14(2):112–6.
4. Mont M, Marker DR, Zywiel MG, et al. Osteonecrosis of the knee and related conditions. J Am Acad Orthop Surg. 2011;19:482–94.
5. Marcacci M, Andriolo L, Kon E, et al. Etiology and pathogenesis of bone marrow lesions and osteonecrosis of the knee. EFORT Open Rev. 2016;1:219–24.
6. Lecouvet FE, Van de Berg BC, Be M, et al. Early irreversible osteonecrosis versus transient lesions of the femoral condyles: prognostic value of subchondral bone and marrow changes on MR imaging. AJR Am J Roentgenol. 1998;170:71–7.
7. Viskontas DG, Giuffre BM, Duggal N, et al. Bone bruises associated with ACL rupture: correlation with injury mechanism. Am J Sports Med. 2008;36:927–33.
8. Soder RB, Simões JD, Soder JB, et al. MRI of the knee joint in asymptomatic adolescent soccer players: a controlled study. AJR Am J Roentgenol. 2011;196(1):W61–5.
9. Pappas GP, Vogelsong MA, Staroswiecki E, et al. Magnetic resonance imaging of asymptomatic knees in collegiate basketball players: the effect of one season of play. Clin J Sport Med. 2016;26(6):483–9.
10. Major NM, Helms CA. MR imaging of the knee: findings in asymptomatic collegiate basketball players. AJR Am J Roentgenol. 2002;179:641–4.
11. Sanders T, Medynski MA, Feller JF, et al. Bone contusion patterns of the knee at MR imaging: footprint of the mechanism of injury. Radiographics. 2000;20 Spec No:S135–51.
12. Berger N, Andreisek G, Karer AT, et al. Association between traumatic bone marrow abnormalities of the knee, the trauma mechanism and associated soft-tissue knee injuries. Eur Radiol. 2017;27(1):393–403.
13. Koga H, Nakamae A, Shima Y, et al. Mechanisms for noncontact anterior cruciate ligament injuries. Am J Sports Med. 2010;38(11):2218–25.
14. Bretlau T, Tuxøe J, Larsen L, et al. Bone bruise in the acutely injured knee. Knee Surg Sports Traumatol Arthros. 2002;10:96–101.
15. Lee BK, Nam SW. Rupture of posterior cruciate ligament: diagnosis and treatment principles. Knee Surg Relat Res. 2011;23(3):135–41.
16. Sonin AH, Fitzgerald SW, Hoff FL, et al. MR imaging of the posterior cruciate ligament: nor- mal, abnormal, and associated injury patterns. Radiographics. 1995;15(3):551–61.
17. Sanders TG, Paruchuri NB, Zlatkin MB. MRI of osteochondral defects of the lateral femoral condyle: incidence and pattern of injury after transient lateral dislocation of the patella. AJR Am J Roentgenol. 2006;187:1332–7.
18. Rangger C, Kathrein A, Freund MC, et al. Bone bruise of the knee: histology and cryosections in 5 cases. Acta Orthop Scand. 1998;69:291–4.
19. Miller M, Osborne JR, Gordon WT, et al. The natural history of bone bruises. A prospective study of magnetic resonance imaging-detected trabecular microfractures in patients with isolated medial collateral ligament injuries. Am J Sports Med. 1998;26:15–9.
20. Costa-Paz M, Muscolo D, Ayerza M, et al. Magnetic resonance imaging follow-up study of bone bruises

associated with anterior cruciate ligament ruptures. Arthroscopy. 2001;17:445–9.

21. Vellet A, Marks P, Fowler P. Occult posttraumatic osteochondral lesions of the knee: prevalence, classification, and short-term sequelae evaluated with MR imaging. Radiology. 1991;178:271–6.

22. Nakamae A, Engebretsen L, Bahr R, et al. Natural history of bone bruises after acute knee injury: clinical outcome and histopathological findings. Knee Surg Sports Traumatol Arthrosc. 2006;14:1252–8.

23. Johnson DL, Bealle DP, Brand JC Jr, et al. The effect of a geographic lateral bone bruise on knee inflammation after acute anterior cruciate ligament rupture. Am J Sports Med. 2000;28(2):152–5.

24. Filardo G, Kon E, Tentoni F, et al. Anterior cruciate ligament injury: post-traumatic bone marrow edema correlates with long-term prognosis. Int Orthop. 2016;40(1):183–90.

25. Lee JW, Lee SK, Choy WS. Complex regional pain syndrome type 1: diagnosis and management. J Hand Surg Asian Pac Vol. 2018;23(1):1–10.

26. Kotwal A, Hurtado MD, Sfeir JG. Transient osteoporosis: clinical spectrum in adults and associated risk factors. Endocr Pract. 2019;25(7):648–56.

27. Takeda M, Higuchi H, Kimura M, et al. Spontaneous osteonecrosis of the knee: histopathological differences between early and progressive cases. J Bone Joint Surg Br. 2008;90(3):324–9.

28. Karantanas AH, Drakonaki E, Karachalios T, et al. Acute non-traumatic marrow Edema syndrome in the knee: MRI findings at presentation, correlation with spinal DEXA and outcome. Eur J Radiol. 2008;67(1):22–33.

29. Gorbachova T, Melenevsky Y, Cohen M, et al. Osteochondral lesions of the knee: differentiating the Most common entities at MRI. Radiographics. 2018;38(5):1478–95.

30. Vidoni Λ, Shah R, Mak D, et al. Metaphyseal hurst sign: a secondary sign on MRI of subchondral insufficiency fracture of the knee. J Med Imaging Radiat Oncol. 2018;62(6):764–8.

31. Pareek A, Parkes CW, Bernard C, et al. Spontaneous osteonecrosis/subchondral insufficiency fractures of the knee: high rates of conversion to surgical treatment and arthroplasty. J Bone Joint Surg Am. 2020;102(9):821–9.

32. Plett SK, Hackney LA, Heilmeier U, et al. Femoral condyle insufficiency fractures: associated clinical and morphologi- cal findings and impact on outcome. Skelet Radiol. 2015;44(12):1785–94.

33. Sayyid S, Younan Y. Sharma Gulshan. Subchondral insufficiency fracture of the knee: grading, risk factors, and outcome. Skelet Radiol. 2019;48(12):1961–74.

34. Yao L, Stanczak J, Boutin RD. Presumptive subarticular stress reactions of the knee: MRI detection and association with meniscal tear patterns. Skelet Radiol. 2004;33(5):260–4.

35. Roemer FW, Neogi T, Nevitt MC, et al. Subchondral bone marrow lesions are highly associated with, and predict subchondral bone attrition longitudinally: the MOST study. Osteoarthr Cartil. 2010;18(1):47–53.

36. Ahlbäck S, Bauer GC, Bohne WH. Spontaneous osteonecrosis of the knee. Arthritis Rheum. 1968;11:705–33.

37. Karim A, Cherian JJ, Jauregui J, et al. Osteonecrosis of the knee: review. Ann Transl Med. 2015;3(1):6.

38. Yamamoto T, Bullough PG. Spontaneous osteonecrosis of the knee: the result of subchondral insufficiency fracture. J Bone Joint Surg Am. 2000;82(6):858–66.

39. Hussain ZB, Chahla J, Mandelbaum BR, et al. The role of meniscal tears in spontaneous osteonecrosis of the knee: a systematic review of suspected Etiology and a call to revisit nomenclature. Am J Sports Med. 2019;47(2):501–7.

40. Shah KN, Racine J, Jones LC, Aaron RK. Pathophysiology and risk factors for osteonecrosis. Curr Rev Musculoskelet Med. 2015 Sep;8(3):201–9.

41. Sakai T, Sugano N, Ohzono K, Matsui M, Hiroshima K, Ochi T. MRI evaluation of steroid- or alcohol-related osteonecrosis of the femoral condyle. Acta Orthop Scand. 1998;69(6):598.

42. Pape D, Seil R, Anagnostakos K, Kohn D. Postarthroscopic osteonecrosis of the knee. Arthroscopy. 2007;23(04):428–38.

43. Brahme SK, Fox JM, Ferkel RD, Friedman MJ, Flannigan BD, Resnick DL. Osteonecrosis of the knee after arthroscopic surgery: diagnosis with MR imaging. Radiology. 1991;178(03):851–3.

44. Fukuda Y, Takai S, Yoshino N, et al. Impact load transmission of the knee joint-influence of leg alignment and the role of meniscus and articular cartilage. Clin Biomech (Bristol, Avon). 2000;15(7):516–21.

45. Jones RS, Keene GC, Learmonth DJ, et al. Direct measurement of hoop strains in the intact and torn human medial meniscus. Clin Biomech (Bristol, Avon). 1996;11(5):295–300.

46. MacDessi SJ, Brophy RH, Bullough PG, Windsor RE, Sculco TP. Subchondral fracture following arthroscopic knee surgery. A series of eight cases. J Bone Joint Surg Am. 2008;90(5):1007–12.

47. Hall FM. Osteonecrosis in the postoperative knee. Radiology. 2005;236(01):370–1. author reply 371

48. Roemer FW, Guermazi A, Javaid MK, et al. MOST study investigators. Change in MRI-detected subchondral bone marrow lesions is associated with cartilage loss: the MOST study. A longitudinal multicentre study of knee osteoarthritis. Ann Rheum Dis. 2009;68:1461–5.

49. Hunter DJ, Zhang Y, Niu J, et al. Increase in bone marrow lesions associated with cartilage loss: a longitudinal magnetic resonance imaging study of knee osteoarthritis. Arthritis Rheum. 2006;54:1529–35.

50. PR K, Kloppenburg M, Sharma R, et al. Bone marrow edema-like lesions change in volume in the majority of patients with osteoarthritis; associations with clinical features. Eur Radiol. 2007;17:3073–8.

51. Xu L, Hayashi D, Roemer FW, et al. Magnetic resonance imaging of subchondral bone marrow lesions in association with osteoarthritis. Semin Arthritis Rheum. 2012;42(2):105–18.
52. Felson DT, Niu J, Guermazi A, et al. Correlation of the development of knee pain with enlarging bone marrow lesions on magnetic resonance imaging. Arthritis Rheum. 2007;56:2986–92.
53. Yusuf E, Kortekaas MC, Watt I, Huizinga TW, Kloppenburg M. Do knee abnormalities visualised on MRI explain knee pain in knee osteoarthritis? A systematic review. Ann Rheum Dis. 2011;70:60–7.
54. Gobbi A, Karnatzikos G, Kumar A. Long-term results after microfracture treatment for full-thickness knee chondral lesions in athletes. Knee Surg Sports Traumatol Arthrosc. 2014;22:1986–96.
55. Niethammer TR, Valentin S, Gülecyüz MF, et al. Bone marrow edema in the knee and its influence on clinical outcome after matrix-based autologous chondrocyte implantation: results after 3-year follow-up. Am J Sports Med. 2015;43:1172–9.
56. Orth P, Goebel L, Wolfram U, et al. Effect of subchondral drilling on the microarchitecture of subchondral bone: analysis in a large animal model at 6 months. Am J Sports Med. 2012;40:828–36.
57. Filardo G, Kon E, Di Martino A, et al. Is the clinical outcome after cartilage treatment affected by subchondral bone edema? Knee Surg Sports Traumatol Arthrosc. 2014;22:1337–44.
58. Ebert JR, Smith A, Fallon M, et al. Degree of preoperative subchondral bone Edema is not associated with pain and graft outcomes after matrix-induced autologous chondrocyte implantation. Am J Sports Med. 2014;42(11):2689–98.

Osteochondral Pathologies as Effect of General Diseases

16

Mary A. Ambach, Christopher J. Rogers, and John G. Lane

16.1 Rheumatologic-Associated Arthropathies

Rheumatologic conditions create an inflammatory type of arthritis through autoimmune and inflammatory processes that disrupt joint structure and function. T-lymphocytes induce monocytes to produce pro-inflammatory cytokines, stimulate B-lymphocytes and plasma cells while neutrophils subsequently release proteases and elastase which degrade joint and periarticular components resulting in synovitis with cartilaginous and osseous damage [1].

16.1.1 Rheumatoid Arthritis (RA)

Rheumatoid arthritis is a systemic autoimmune inflammatory disorder of unknown etiology affecting multiple organ systems. In the musculoskeletal system, it affects the synovial lining of the diarthrodial joints leading to chronic symmetric erosive synovitis and articular destruction. Pannus formation, granulation tissue that covers the articular cartilage at joint margins, is the most destructive element of RA [2].

The American College of Rheumatology (ACR) guidelines for the diagnosis of RA requires 4 of the 7 criteria to be present: morning joint stiffness, arthritis of three or more joints, arthritis of hand joints, symmetric arthritis, rheumatoid nodules, rheumatoid factor positive and characteristic radiographic findings of erosions, bony decalcification, and symmetric joint space narrowing on hand and wrist X-rays [2]. Characteristic hand and wrist deformities include Boutonniere Deformity (MCP hyperextension, PIP flexion, DIP hyperextension), Swan neck Deformity (MCP flexion contracture, PIP hyperextension, DIP flexion), and ulnar deviation of the fingers [2]. Rheumatoid arthritis can also cause atlantoaxial joint subluxation which can

M. A. Ambach
San Diego Orthobiologics Medical Group, Carlsbad, CA, USA

University of California San Diego School of Medicine, La Jolla, CA, USA
e-mail: ambach@sdomg.com

C. J. Rogers
San Diego Orthobiologics Medical Group, Carlsbad, CA, USA

American Board of Physical Medicine and Rehabilitation, Rochester, MN, USA

American College of Regenerative Medicine, Registered Musculoskeletal Sonography (RMSK), Rockville, MD, USA

Department of Orthopaedic Surgery, University of California, San Diego, La Jolla, CA, USA
e-mail: rogers@sdomg.com

J. G. Lane (✉)
Department of Orthopaedic Surgery, University of California, San Diego, La Jolla, CA, USA

Musculoskeletal and Joint Research Foundation, La Jolla, CA, USA
e-mail: jglane@san.rr.com

© ISAKOS 2022
A. Gobbi et al. (eds.), *Joint Function Preservation*, https://doi.org/10.1007/978-3-030-82958-2_16

cause instability of the C1-C2 articulation causing pain and myelopathy.

16.1.2 Seronegative Spondyloarthropathies: Ankylosing Spondylitis (AS), Reiter's Syndrome, Psoriatic Arthritis, Inflammatory Bowel Disease (IBD)

Seronegative spondyloarthropathies consist of a group of multisystem inflammatory disorders affecting various joints. Spine, peripheral joints, and periarticular structures are typically involved; however, the hallmark feature of spondyloarthropathies is sacroiliac involvement. Included in this group of disorders are ankylosing spondylitis, psoriatic arthritis, Reiter's disease, and arthritis of inflammatory bowel disease. Features include mucocutaneous lesions, aortitis, sacroiliitis, inflammation of tendon attachments (enthesopathy), and HLA- B27 positivity [3].

Ankylosing spondylitis is a chronic inflammatory disorder of the axial skeleton. It often has an insidious onset and often presents as morning stiffness in the low back. Clinical findings are succinctly described by Nucatolla et al. [2]. Findings on physical examination include a positive Schober's test, increased fingertip-to-floor distance on forward flexion, decreased chest expansion, pain on sacroiliac compression or stress, decreasing height, kyphosis. Symmetric SI joint narrowing, erosions and sclerosis can lead to fusion. There is subchondral bone resorption, erosion sclerosis, and calcification in the joints causing ankylosis. The characteristic "bamboo spine" on X-ray results from ossification of the spinal ligaments, syndesmophyte formation, and ankylosis of the facet joints.

Reiter's disease is characterized by conjunctivitis, arthritis, urethritis, and mucocutaneous lesions [3]. Hicks et al. describes the disease process and symptomatology in the Archives of Physical Medicine and Rehab article [3]. The onset often follows sexually acquired urethritis or enteric infection with *Salmonella, Shigella,* or *Yersinia.* There is a 1- to 3-week latent period

between the inciting event and the development of arthritis. The arthritis is often asymmetric, tendonitis, and fasciitis are common, and interphalangeal joint dactylia is characteristic. Although complete recovery may occur in up to 80%, recurrent arthritis is common.

Psoriatic arthritis has several peripheral joint arthritic presentations as described by Hicks et al. [3]: (a) distal interphalangeal joint arthritis (b) arthritis mutilans with severe osteolysis of the phalanges, (c) symmetric small joint polyarthritis resembling rheumatoid arthritis, (d) monoarticular or asymmetric oligoarticular arthritis. In 16%, the arthritis can precede skin lesions. Arthritis is more common in patients with severe skin lesions.

Arthritis associated with inflammatory bowel disease occurs in conjunction with regional enteritis and chronic ulcerative colitis. The peripheral arthritis is often asymmetric, affects the large joints, and subside with remission of the bowel disease. Synovitis affects the peripheral joints in addition to sacroiliitis [2].

16.1.3 Connective Tissue Disease: Systemic Lupus Erythematosus (SLE), Scleroderma, Polymyositis/ Dermatomyositis, Mixed Connective Tissue Disease

Systemic lupus erythematosus (SLE) is a multisystemic autoimmune disease that affects every organ of the body. The Diagnosis of SLE by American College of Rheumatology requires positive findings for any four of the 11 ACR Classification criteria [2]. Included in the criteria is a non-erosive arthritis involving two or more peripheral joints with tenderness, swelling, and effusion. This non-erosive deforming arthritis, called Jaccoud's arthritis, is symmetric, affects the small joints of the hands, wrist and knees, migratory and chronic.

Polymyositis and dermatomyositis are inflammatory myopathies hypothesized to be due to abnormal immunoregulating mechanisms and viral etiologies. Different types and clinical presentations of this condition have been described [2, 3]. They clinically present with profound symmetric weakness of the proximal muscles and have five types. Type IV presents in childhood and is associated with severe joint contractures. Type V is associated with collagen vascular diseases (SLE, RA, Progressive systemic sclerosis). With this type, the arthritis can be severe and deforming, and the clinical picture of the individual diseases predominates. Prognosis and functional problems depend on the type of PM/DM. Treatment involves corticosteroids, other immunosuppressant and other treatments to address problems associated with each type.

Mixed connective tissue disease combines clinical features of systemic sclerosis, SLE, and polymyositis, and has been found to be associated with high titers of an antinuclear antibody to ribonucleoprotein [3]. Clinically, patients with this disorder demonstrate Raynaud's phenomenon, swollen fingers, myositis, esophageal abnormalities, arthritis or arthralgia, and impairment of pulmonary diffusion capacity [3]. Many patients respond well to corticosteroids, and the prognosis is generally good.

16.2 Endocrine-Associated Arthropathies

16.2.1 Diabetes

Diabetes is a group of chronic diseases characterized by hyperglycemia. The injurious effects of hyperglycemia lead to macrovascular complications such as peripheral arterial disease and microvascular complications of peripheral neuropathy [4]. Due to loss of protective sensation and impaired vascular supply, these can lead to serious foot complications including deformity, diabetic foot ulcerations, Charcot neuroarthropathy (CN), and infection.

Charcot neuroarthropathy is a chronic progressively degenerative arthropathy that leads to joint instability and destruction. A review in Current Diabetes Reports describes the disease, its mechanisms and clinical presentations [4]. Incidence is estimated to be between 0.1% and 0.9%. Historically, Jean Marie Charcot described the condition as having a neurovascular etiology causing an alteration of bone and joint nutrition. Later, German scientists described the condition as a result of neurotraumatic influences. It is now learned that these theories are entangled and influenced by other systemic states like inflammation (pro-inflammatory state), bony regulation, neuropeptides like nitric oxide (NO), and calcintonin gene-related peptide (CGRP). Early clinical features include painless swelling, effusion, and joint destruction while late findings include destruction of cartilage and bones, intra-articular loose bodies and subtle fractures. Disorganization of the joint can lead to subluxation and dislocation. Histologically, the bone in diabetes with CN displayed inflammatory and myxoid infiltrates with a disorganized trabecular pattern.

16.2.2 Hypothyroid

Hypothyroidism is a disorder of the endocrine system where the thyroid gland does not produce enough thyroid hormones. Common causes of hypothyroidism include autoimmune (Hashimoto's thyroiditis), iodine deficiency, and thyroid gland removal. Hypothyroid individuals with antithyroid antibodies have been associated with musculoskeletal diseases such as osteoarthritis (OA) and inflammatory arthritis [5, 6]. It is theorized to be related to a thyroid-stimulating hormone (TSH)-dependent increase in hyaluronic acid and proteoglycan synthesis.

16.2.3 Hyperparathyroid

Hyperparathyroidism is the state of having excess parathyroid hormone (PTH) resulting in hypercalcemia. This disorder has been described by Pincus in the Rheumatology Advisor [7]. The primary disorder is most commonly due to an autonomously functioning solitary adenoma (80–85%), gland hyperplasia (10–15%) or multiple adenomas (5%). Secondary hyperparathyroidism, seen most commonly in renal failure, occurs when there is partial resistance to the metabolic actions of PTH, leading to excessive production of the hormone. Classically, the effects of excess parathyroid hormone affect the kidney and musculoskeletal system giving rise to the well-known moniker for the clinical manifestations "bones, stones, and groans."

Specific rheumatic disorders in hyperparathyroidism have been studied by Helliwell in a retrospective survey [8]. They include erosive arthritis, periarticular calcification with inflammation and gouty arthritis. The association of an erosive arthritis is recognized and may simulate rheumatoid arthritis. Furthermore, hyperparathyroidism and rheumatoid arthritis may coexist with an apparent worsening of the rheumatoid process suggesting a possible adverse effect of excessive PTH.

Osteitis fibrosa cystica and crystalline arthropathies have also been associated with hyperparathyridism and have been reported by Pincus [7]. Osteitis fibrosa cystica, also known as von Recklinghausen's disease of bone, applies to the characteristic subperiosteal erosions that were initially identified in primary hyperparathyroidism. Clinically, this is characterized by diffuse bone pain, bone tenderness, and skeletal deformities including bowing of the long bones and fractures. Pathologically, there is an increase in the giant multinucleated osteoclasts on the surface of bone and a replacement of the normal bone elements with fibrous tissue. Brown tumors are severe manifestations of this turnover with areas of necrosis and focal hemorrhage with hemosiderin deposition. They are lytic lesions that become sclerotic as they heal, mimicking blastic metastasis on imaging. Crystalline arthropathies, namely gout and pseudogout, have been associated with hyperparathyroidism. In the setting of parathyroid hormone excess, the elevated calcium is postulated to impair proteoglycans which act to inhibit crystallization in pseudogout. Additionally, an increase in enzymes which catalyze the production of pyrophosphate has been shown and may promote the formation of calcium pyrophosphate crystals leading to increased risk of a pseudogout flare. A higher incidence of gout attacks have been found in the setting of hyperuricemia, thus ultimately increasing the

risk for clinically significant gout in the primary hyperparathyroidism population. Although the pathophysiology is not known, it is suggested that PTH or calcium deposition may inhibit uric acid excretion in the proximal renal tubule.

16.2.4 Hypercortisolism (Cushing's Disease)

Hypercortisolism, also called Cushing Syndrome (CS) is an endocrine disorder characterized by increased levels of cortisol, a hormone produced by the adrenal gland. This occurs secondary to excessive exogenous glucocorticoids (like ingesting oral prednisone) or endogenous increase from increased hypothalamic-pituitary ACTH secretion. CS is associated with various catabolic effects as described by Chang et al. [9]. Loss of collagen can cause fragile skin and poor wound healing. Decreased bone resorption can cause osteoporosis with susceptibility to fractures. It can cause severe growth failure in children due to bone alterations and can prevent bone repair by loss of cortical osteocytes [10].

16.2.5 Growth Hormone Overproduction (Acromegaly) and Insufficiency

Acromegaly is a disorder caused by excessive production of growth hormone from the anterior pituitary gland, resulting in excessive growth of body tissues and other metabolic dysfunctions. Articular involvement in acromegaly was well illustrated by Killinger et al. [11]. It was noted to be one of the most frequent clinical complications and could present as the earliest symptom in a significant proportion of patients. The pathogenesis of arthropathy in acromegaly is described to be due to elevated growth hormone and IGF-I levels causing soft tissue and fibroblast proliferation promoting growth of the articular cartilage and thickening of periarticular ligaments and connective tissue. These changes cause limitations in the range of movement and instability of the joint.

Patients with Growth Hormone Deficiency (GHD) have decreased or absent growth hormone production as a result of hypothalamic or pituitary disorders resulting in underactive pituitary gland function (i.e., hypopituitarism). The most salient features of GHD in adults as described by Owens et al. [12] include decreased lean body mass, increased visceral fat and subcutaneous fat, hyperlipidemia and decreased bone mass. GHD has been linked to a higher risk of bone fractures. Reed et al. [13] also noted that cortical bone density and trabecular bone density measurements in GHD patients were below the mean for age- and sex-matched controls; and that GHD patients also have decreased muscle mass and strength.

16.3 Hematologic Illness Arthropathies

16.3.1 Hemophilia

Hemophilia is a hereditary disease associated with a defect on the X chromosome, leading to absence or deficiency of production of coagulation factors VIII, IX, and XI [14]. The symptomatology is generally secondary to bleeding with the musculoskeletal system being the most frequently involved.

Roosendaal characterized three stages of joint damage in hemophilia [15]: acute hemarthrosis, chronic proliferative synovitis, and chronic hemophilic arthropathy. When blood enters the joint space, the iron component of hemoglobin leads to the formation of destructive oxygen metabolites affecting the synovium and cartilage. An inflammatory reaction occurs, resulting in progressive hemosiderin deposition, synovial hypertrophy, and neovascularization, which causes fibrosis and joint destruction. Acute hemarthrosis results in a swollen, warm and painful joint, and mainly affects the large joints. Chronic hemophilic arthropathy is generally accompanied by severe contractures, angular deformity, and loss of bone tissue.

The incidence of septic arthritis among patients with hemophilia is increased up to 40 times higher than in the general population and

affects the knee joint most commonly [14]. The diagnosis in these patients is frequently delayed because early symptoms are often misdiagnosed as hemarthrosis.

16.3.2 Sickle Cell Disease

Sickle cell diseases (SCD) are a group of genetic hemoglobin disorders causing crescent sickle-shaped RBC, causing obstruction of the microvasculature. Clinical presentation is characterized by recurrent microvascular occlusion with subsequent tissue ischemia, leading to painful vaso-occlusive crises. Morais et al. described SCD's involvement of the musculoskeletal system [14]. Vaso-occlusive-related complications mainly included sickle cell dactylitis in children causing painful, non-pitting swelling of the hands and feet, avascular necrosis, or vertebral collapse. Non-inflammatory and secondary osteoarthritic change and synovial effusions were also described.

16.3.3 Hemochromatosis

Hereditary hemochromatosis (HH) comprises a group of inherited iron-storage diseases causing excessive iron stores and hemosiderin deposition.

Diagnosis is often delayed because early symptoms are nonspecific, notably fatigue, arthralgia, and abdominal pain. Disease progression leads to cirrhosis, hepatocellular carcinoma, diabetes mellitus, impotence, skin pigmentation, cardiomyopathy, and hypogonadism [16].

Morais et al. authored a comprehensive article in rheumatology as he discussed joint involvement in patients with HH [14]. Up to 80% of HH patients were shown to develop a chronic progressive non-inflammatory arthropathy, usually affecting the second and third MCP joints, causing pain during handshake. Other joints including the PIP joints, the knees, wrists and distal radio-ulnar joints, hips, shoulders, ankles, and elbows were also found to be often involved. Chondrocalcinosis is also reported commonly involving the meniscal and articular cartilage in the knee, wrist, symphysis pubis, and spine.

16.4 Infectious Arthritis

Septic arthritis usually presents with rapid onset of joint pain, erythema, and decreased ROM. Fever, chills, and leukocytosis may be present. Hicks and Sutin discussed this condition and its joint manifestations [3]. It generally follows hematogenous dissemination of the infecting agent but can occur by local extension or penetration. Causative organisms are bacterial (Neisseria gonorrhea, Staph aureus), viruses (rubella, infectious hepatitis, lyme), mycobacteria (tuberculosis), and fungal. Joint involvement is usually monoarticular; however, a major exception may be gonococcal arthritis, in which a migratory tenosynovitis and synovitis may precede monoarticular involvement. Lyme arthritis can also present with intermittent migratory episodes of polyarthritis. Mycobacterial infection can cause chronic indolent arthritis, tenosynovitis, or vertebral involvement.

16.5 Crystal-Induced Joint Disease

16.5.1 Gout

Gout is a condition that is extensively discussed in multiple texts and journals [3, 17]. It is described as a disease of purine metabolism characterized by an increase in serum urate concentration and deposits of urate crystals in joints, tendons, and subcutaneous tissues resulting in an acute inflammatory response. Gout attacks usually occur with an acute onset of monoarticular synovitis in joints of the lower extremities, classically in the first MTP joint (podagra) causing severe pain, swelling, redness, and tenderness. However, gout can occur in ankles, knees, hands, wrist, and elbows. Biundo et al. further describes these attacks and comorbidities [17]. An acute attack can be pre-

cipitated by an acute illness causing a relatively rapid rise in uric levels (e.g., myocardiac infarction, stroke), surgery, dehydration, and diuretics. Recurrent attacks can result to chronic tophaceous gout, where deposits of urate crystals or tophi form and cause structural damage to the articular cartilage and adjacent bone. Comorbidities that promote hyperuricemia include obesity, metabolic syndrome, type 2-diatetes mellitus, hypertension, and chronic kidney disease.

16.5.2 Pseudogout

Pseudogout or calcium pyrophosphate deposition (CPPD) disease is characterized by recurrent episodes of inflammation in large joints as a result of calcium pyrophosphate crystals. The disease may be familial, idiopathic, or associated with other diseases such as hyperparathyroidism, hemochromatosis, and hypothyroidism [3]. CPPD crystals deposit in articular cartilage, fibrocartilage, and sometimes in ligaments [17]. X-ray examination reveals deposits of calcium pyrophosphate in cartilage, known as "chondrocalcinosis." Patients with chronic pseudogout may present as pseudorheumatoid arthritis, pseudoosteoarthritis or pseudoneuropathic arthritis with pronounced destructive changes similar to those seen with Charcot's joints [3].

16.6 Deposition and Storage Disease

16.6.1 Alkaptonuria

Alkaptonuria is an autosomal recessive deficiency in the enzyme homogentisic acid oxidase leading to oxidation and alkalinization of tissues called ochronosis [2]. The accumulation of homogentisic acid causes bluish discoloration of the urine, cartilage, skin, and sclera and can cause progressive degenerative arthropathy of the large joints, chondrocalcinosis, joint effusions, and osteochondral bodies [2].

16.6.2 Wilson's Disease

Wilson's disease is an autosomal recessive genetic disorder characterized by excess copper stored in various body tissues, particularly the liver, brain, and corneas of the eyes [2]. Thirty-two patients hospitalized for Wilson's disease were studied by Golding and Walshe and their findings were described [18]. Patients with this condition have been found to have osteoporosis or skeletal demineralization and premature osteoarthrosis. Changes in the spine were common and included osteochondritis, reduction of intervertebral joint spaces, osteoarthrosis, and a tendency to squaring of vertebral bodies. Other bony changes included irregularity of femoral trochanters, osteochondritis dissecans, and osteophytes at joint margins. The symptoms associated with these findings include pain and stiffness in the spine and joints.

16.6.3 Gaucher Disease

Gaucher disease (GD) is the most common lysosomal storage disorder and is an autosomal recessive disorder due to a deficiency of the enzyme glucocerebrosidase leading to accumulation of glucocerebroside within macrophages [19].

The bone and joint complications related to Gaucher disease were extensively studied by Gregory et al. [19]. The most common types of GD, the non-neuropathic variant, were described presenting predominantly with anemia, thrombocytopenia, and hepatosplenomegaly. Skeletal involvement often coexists with these findings and bone complications were seen in up to 80% of patients with GD. There is a broad range of osseous GD manifestations described, and these were secondary to progressive marrow infiltration by Gaucher cells (lipid-engorged macrophage), pericellular fibrosis, increased deposition of reticulin, failure of bone remodeling, loss of trabecular bone, and osteonecrosis. Skeletal manifestations ranged from asymptomatic osteopenia to episodic bone pain, osteonecrosis with loss of mechanical integrity, and secondary joint disease in major joints such as the femoral or humeral

heads. These "bone crises" have been attributed to microcirculatory disease and signals an evolving bone infarction. MRI during the acute phase of the bone crises may show localized subchondral edema, joint effusion, and periosteal elevation. Due to heterogeneity of bone involvement found, radiographic and imaging modalities were variable and included focal areas of bone radiolucency, cortical thinning and radiographic signs of osteonecrosis (*i.e.*, osteosclerosis, subchondral crescent sign, epi-physeal cortical flattening, fragmentation, deformity, and secondary osteoarthritis).

16.7 Vasculitis

Systemic vasculitis is a heterogeneous group of rare diseases characterized by inflammation and fibrinoid necrosis of blood vessel walls [20]. Vasculitis mechanisms and clinical manifestations were broadly discussed by Guillevin and Dörner [20]. It is defined as primary, with no identifiable cause or secondary to infection, malignancy, or autoimmune disease. Various pathogenic mechanisms have been implicated in the induction of vasculitis, including cell-mediated inflammation, immune complex-mediated inflammation and autoantibody-mediated inflammation. Vasculitis was shown to occur in many autoimmune diseases, including RA and SLE, Sjögren's syndrome, scleroderma, and sarcoidosis. Vasculitis accelerating atherosclerosis was found to be a complicating feature of most, possibly all, autoimmune diseases. Rheumatoid vasculitis is characterized by the occurrence of mononeuritis multiplex, purpura, and visceral involvement, with the latter sometimes being severe. The picture in SLE is similar to that in RA, except that patients with SLE exhibited more severe systemic symptoms. Primary Sjögren's syndrome is described to be associated with enhanced risk for B-cell lymphoma in which the presence of purpura or vasculitis are clinical risk indicators for future non-Hodgkin's lymphoma development. Vasculitis was shown to be rare in scleroderma and sarcoidosis.

Vasculitis may also occur in association with other types of inflammatory arthritis, including spondyloarthropathies and psoriatic arthritis [21].

16.8 Other Systemic Illness

16.8.1 Sarcoidosis

Sarcoidosis is a systemic, clinically heterogeneous disease characterized by the development of granulomas. Any organ system can be involved, and patients may present with any number of rheumatologic symptoms. The underlying etiology remains unclear though genetic, environmental, and infectious etiologies have all been suggested with varying levels of supportive evidence [22]. According to Sweis et al., up to 25% of patients with sarcoidosis have joint involvement; and sarcoid arthritis can be acute or chronic, and oligoarthritic or poly-arthritic (i.e., involving three or more joints) [22]. Chronic sarcoid arthritis typically occurs in the setting of systemic sarcoidosis and typically involves the knees, ankles, wrists, hands, and/or feet.

Joint effusions, synovitis, or even nodular proliferation of the synovium presenting as an intra-articular knee mass may also be present [23]. Patients with sarcoid arthritis often have periarticular inflammation but generally maintain normal range of motion [24]. Persistent inflammation can cause joint destruction or Jaccoud deformity [25].

16.8.2 Amyloidosis

Amyloidosis is a condition in which extracellular deposition of characteristic abnormal protein material occurs. This protein is predominantly in the form of fibrils which is relatively insoluble and resistant to proteolytic digestion accounting for their characteristic staining properties [26].

There are specific varieties of amyloidosis that present with osteoarticular manifestations as described by Rowe [26] and M'Bappé et al. [27]. Rheumatological manifestations of *AL immunoglobulin amyloidosis* are numerous and often

indicative of the disease. Deposits affect joint and periarticular structures. The most common presentation described is a progressively developing bilateral symmetric polyarthritis similar to RA. Joint involvement includes swelling especially on the back of the hands and wrists, a result of infiltration of the soft tissues next to the joints corresponding to amyloid deposits and responsible for painful limitation of motion for the shoulders, wrists, metacarpophalangeal and proximal interphalangeal joints and knees. This particular thickening is visible at the shoulders named the "shoulder pad" sign considered by some authors as pathognomonic of AL amyloidosis. There are associated cutaneous nodules consisting of amyloid deposits in about half of the cases which are firm, painless, and non-inflammatory, found mainly at the level of the olecranon, the wrists and the fingers. *b 2-Microglobulin amyloidosis* occurs in patients under chronic hemodialysis and can present with CTS, arthralgia, and a specific destructive spondyloarthropathy. Chronic inflammatory conditions like rheumatoid arthritis (RA) and ankylosing spondylitis have been associated with the development of *amyloid A amyloidosis* where amyloid deposits in the joints can be seen. Amyloid deposits in the joints have also been described as a rare complication of various forms of arthritis (AA amyloid).

16.8.3 Sjogren's Syndrome

Sjogren's syndrome (SS) is an autoimmune-mediated disorder of the exocrine glands. Patients present with dry eyes, dry mouth, skin lesions, and parotid involvement. Primary Sjogren's syndrome occurs in people with no other rheumatologic disorders. Secondary Sjogren's occurs in patients with other rheumatologic disorders like RA and SLE [2]. Pease et al. studied the clinical course of 48 patients with primary SS and their articular manifestations [28]. Symptoms suggestive of an arthralgia and/or arthritis have been reported as a presenting feature in up to 25% of patients. The arthropathy of SS had no single diagnostic feature. It presented as an intermittent polyarticular arthropathy affecting large and small joints. Some cases were symmetric while some were asymmetric. Arthralgia, synovitis, erosive joint disease, and joint space narrowing have been reported. On occasion, the arthropathy was shown to be associated with a purpuric vasculitis on the lower legs. Both et al. reviewed primary SS and found that arthritis is less common and occurs in about 16% of primary SS patients [29]. This predominantly consisted of symmetric, intermittent, non-erosive arthropathy mostly involving the proximal interphalangeal joints, metacarpal-phalangeal joints, and wrists.

16.9 Malignancy-Associated Arthropathies

16.9.1 Paraneoplastic Rheumatic Disease

Paraneoplastic rheumatic disorders are defined as rheumatic symptoms resulting from an underlying malignant disease, which is not directly related to a tumor or metastasis. The clinical course of the disease is in parallel with the underlying malignancy and most improve with successful treatment of the underlying malignancy [30].

These paraneoplastic symptoms are present at diagnosis in about 10% of patients with cancer, with up to 50% experiencing a paraneoplastic syndrome at some time during the course of their illness [31]. Fam extensively described this condition in a review article [31]. One-third of those with paraneoplastic syndrome were found to be endocrine in nature, while the remainder were hematological, rheumatic, and neuromuscular disorders. Fam classified paraneoplastic rheumatic syndromes into articular, neuromuscular, cutaneous, vascular, and miscellaneous. The articular syndromes include hypertrophic osteoarthropathy, carcinoma polyarthritis, amyloid arthritis, and secondary gout.

Hypertrophic osteoarthropathy (HOA) is characterized by Fam as the clinical triad of oligo- or polyarthritis, clubbing of fingers and toes, and periostitis of the distal ends of long bones [31]. The arthritis is described to be often symmetrical and painful and often affecting the knees, ankles, elbows, wrists, metacarpophalangeal and proximal interphalangeal joints. It is sometimes associated with tenderness of the adjacent bones and synovial effusions which are typically non-inflammatory. The etiology of HOA in patients with cancer is unknown, but humoral, vascular, immunological, and a vagal neural reflex mechanisms have been implicated [31].

Carcinoma polyarthritis is a seronegative inflammatory arthritis that may signal the onset of an occult malignancy [30]. Fam [31] and Faruk [30] had broad descriptions of this condition. The clinical presentation was found to be variable; however, certain features suggest the possibility of an underlying malignancy and serve to distinguish it from rheumatoid arthritis (RA). It differs from RA due to the initiation at advanced age, acute onset, predominantly asymmetrical lower extremity involvement, and sparing of wrist and hand joints. Also, unlike RA, the absence of erosions, deformities, rheumatoid factor, rheumatoid nodules, and family history is observed and joint radiographs tend to be normal. The pathogenesis of carcinoma polyarthritis is poorly understood however several autoimmune mechanisms have been suggested.

Amyloid arthritis occurs most commonly in patients with multiple myeloma and secondary gout arthritis and may be associated with leukemias, polycythemia rubra vera, essential thrombocythemia, lymphomas, myeloma, and rarely with carcinomas. These arthritides have been described in detail in the preceding sections [31].

References

1. Spires MC. Inflammatory arthritides. In: PM&R Knowledge NOW. 2012. https://now.aapmr.org/inflammatory-arthritides/. Accessed 8 Mar 2020
2. Nucatola TR, Freeman ED, Brown DP. Rheumatology. In: Cuccurullo SJ, editor. Physical medicine and rehabilitation board review. 3rd ed. New York: Demos Medical; 2015. p. 101–39.
3. Hicks JE, Sutin J. Rehabilitation in joint and connective tissue diseases. 3. Specific rheumatic diseases. Arch Phys Med Rehabil. 1998;69:S-84–96.

4. Schmidt BM, Holmes CM. Updates on diabetic foot and Charcot osteopathic arthropathy. Curr Diab Rep. 2018; https://doi.org/10.1007/s11892-018-1047-8.

5. Atzeni F, Doria A, Ghirardello A, Turiel M, Batticciotto A, Carrabba M, Sarzi-Puttini P. Antithyroid antibodies and thyroid dysfunction in rheumatoid arthritis: prevalence and clinical value. Autoimmunity. 2008;41:111–5. https://doi.org/10.1080/08916930701620100.

6. Waung JA, Bassett JH, Williams GR. Thyroid hormone metabolism in skeletal development and adult bone maintenance. Trends Endocrinol Metab. 2012;23:155–62. https://doi.org/10.1016/j.tem.2011.11.002.

7. Pincus M. Hyperparathyroidism. In: Rheumatology. Decision Support Medicine and Rheumatology Advisor. 2017. https://www.rheumatologyadvisor.com/home/decision-support-in-medicine/rheumatology/hyperparathyroidism/. Accessed 8 March 2020.

8. Helliwell M. Rheumatic symptoms in primary hyperparathyroidism. Postgrad Med J. 1983;59:236–40. https://doi.org/10.1136/pgmj.59.690.236.

9. Chang RG, Bazmi K, Beaufort A. Endocrine abnormalities affecting the musculoskeletal system. In: PM&R Knowledge NOW. 2018. https://now.aapmr.org/endocrine-abnormalities-affecting-the-musculoskeletal-system/ Accessed 8 Mar 2020.

10. Leong GM, Mercado-Asis LB, Reynolds JC, Hill SC, Oldfield EH, Chrousos GP. The effect of Cushing's disease on bone mineral density, body composition, growth, and puberty: a report of an identical adolescent twin pair. J Clin Endocrinol Metab. 1996;81:1905–191. https://doi.org/10.1210/jcem.81.5.8626856.

11. Killinger Z, Payer J, Lazúrová I, Imrich R, Homérová Z, Kužma M, Rovenský J. Arthropathy in acromegaly. Rheum Dis Clin N Am. 2010;36:713–20. https://doi.org/10.1016/j.rdc.2010.09.004.

12. Owens G, Balfour D, Biller BMK, Cohen J, Jacobs M, Lease M, et al. Clinical presentation and diagnosis: growth hormone deficiency in adults. Am J Manag Care. 2004. https://www.ajmc.com/journals/supplement/2004/2004-10-vol10-n13suppl/oct04-1929ps424-s430?p=2; Accessed 8 March 2020

13. Reed ML, Merriam GR, Kargi AY. Adult growth hormone deficiency—benefits, side effects, and risks of growth hormone replacement. Front Endocrinol. 2013;4:64. https://doi.org/10.3389/fendo.2013.00064.

14. Morais S, du Preez HE, Akhtar MR, Cross S, Isenberg DA. Musculoskeletal complications of haematological disease. Rheumatology. 2016;55:968–81. https://doi.org/10.1093/rheumatology/kev360.

15. Roosendaal G, Lafeber FP. Pathogenesis of haemophilic arthropathy. Haemophilia. 2006;12:117–21. https://doi.org/10.1111/j.1365-2516.2006.01268.x.

16. Frenzen K, Schafer C, Keyser G. Erosive and inflammatory joint changes in hereditary hemochromatosis arthropathy detected by low-field magnetic resonance imaging. Rheumatol Int. 2013;33:2061–7. https://doi.org/10.1007/s00296-013-2694-3.

17. Biundo J, Rush P. Gout and Pseudogout. In: PM&R Knowledge NOW. 2017. https://now.aapmr.org/gout-and-pseudogout/. Accessed 8 Mar 2020.

18. Golding DN, Walshe JM. Arthropathy of Wilson's disease. Study of clinical and radiological features in 32 patients. Ann Rheum Dis. 1977;36:99–111. https://doi.org/10.1136/ard.36.2.99.

19. Pastores GM, Patel MJ, Firooznia H. Bone and joint complications related to Gaucher disease. Curr Rheumatol Rep. 2000;2:175–80. https://doi.org/10.1007/s11926-000-0059-x.

20. Guillevin L, Dörner T. Vasculitis: mechanisms involved and clinical manifestations. Arthritis Res Ther. 2007;9:S9. https://doi.org/10.1186/ar2193.

21. Watts RA, Scott DGI. Vasculitis and inflammatory arthritis. Best Pract Res Clin Rheumatol. 2016;30:916–31. https://doi.org/10.1016/j.berh.2016.10.008.

22. Sweiss NJ, Patterson K, Sawaqed R, et al. Rheumatologic manifestations of sarcoidosis. Semin Respir Crit Care Med. 2010;31:463–73. https://doi.org/10.1055/s-0030-1262214.

23. Chu A, Ginat D, Terzakis J, Seneviratne A, Schneider KS. Chronic sarcoid arthritis presenting as an intra-articular knee mass. J Clin Rheumatol. 2009;15:190–2. https://doi.org/10.1097/RHU.0b013e3181a61c29.

24. Anandacoomarasamy A, Peduto A, Howe G, Manolios N, Spencer D. Magnetic resonance imaging in Löfgren's syndrome: demonstration of periarthritis. Clin Rheumatol. 2007;26:572–5. https://doi.org/10.1007/s10067-006-0360-9.

25. Spilberg I, Siltzbach LE, McEwen C. The arthritis of sarcoidosis. Arthritis Rheum. 1969;12:126–37. https://doi.org/10.1002/art.1780120209.

26. Rowe I. Editorial: amyloid arthropathy. Ann Rheum Dis. 1985;44:727–8. https://ard.bmj.com/content/annrheumdis/44/11/727.full.pdf. Accessed 26 Mar 2020

27. M'Bappé P, Grateau G. Osteo-articular manifestations of amyloidosis. Best Pract Res Clin Rheumatol. 2012;26:459–75. https://doi.org/10.1016/j.berh.2012.07.003.

28. Pease CT, Shattles W, Barrett NK, Maini RN. The arthropathy of sjögren's syndrome. Rheumatology. 1993;32:609–13. https://doi.org/10.1093/rheumatology/32.7.609.

29. Both T, et al. Reviewing primary Sjögren's syndrome: beyond the dryness—from pathophysiology to diagnosis and treatment. Int J Med Sci. 2017;14:191–200. https://doi.org/10.7150/ijms.17718.

30. Sendur OF. Paraneoplastic rheumatic disorders. Arch Rheumatol. 2012;27:18–23. https://doi.org/10.5606/tjr.2012.002.

31. Fam A. Paraneoplastic rheumatic syndromes. Baillieres Clin Rheumatol. 2000;14:515–33. https://doi.org/10.1053/berh.2000.0091.

Understanding Genetics in Osteochondral Pathologies

Dawid Szwedowski, Łukasz Paczesny,
Przemysław Pękala, Jan Zabrzyński,
and Joanna Szczepanek

17.1 Introduction

Articular cartilage is a highly specialized connective tissue. The most important components of this tissue are the extracellular matrix (ECM) with chondrocytes, collagen fibers (mainly type II and IX), small non-collagen proteins (aggrecan, high molecular weight proteoglycan and cartilage oligomeric matrix protein (COMP)), water, and a small volume of non-collagenous proteins and glycoproteins such as fibronectin [1]. This composition is regulated by chondrocytes in response to the changes in their chemical and mechanical environment.

The main collagen of cartilage tissue is type II, which together with few other types form a network of fibers, where aggregating and small non-aggregating proteoglycans are located. Aggrecans form aggregates with hyaluronic acid which are responsible for the mechanical properties of cartilage. Small non-aggregating proteoglycans (decorin and fibromodulin) limit the formation of collagen fibers. Moreover, other proteins—chondronectin, fibronectin, vitronectic, and thrombospondin are involved in the interaction between chondrocytes and the matrix [2–4]. The cell adhesion receptors—integrins play a major role in mediating the interactions between cells and the ECM. They create connections to a host of ECM proteins, most notably fibronectin and collagen types II and VI. Moreover, they provide signals regulating cells proliferation, survival, differentiation, and matrix remodeling [5]. Cartilage oligometric matrix protein prevents cartilage vascularization and probably is responsible for repair processes. In addition to structural elements, chondrocytes produce substances that perform only physiological functions—enzymes and cytokines. Those enzymes include metalloproteinases, adamalysins (a disintegrin and metalloproteinase domain or *ADAMs*), serine and cistern prostheses and their inhibitors [6, 7]. They participate in the reconstruction of the cartilaginous matrix. Cytokines (*IL-1β,TNFα, IL-6,IL-15, IL-17*, and *IL-18*) stimulate chondrocytes to increase the production of enzymes, thus enhancing matrix degradation, while *IL-4*, *IL-10*, and *IL-13* inhibit

D. Szwedowski (✉)
Orthopaedic Arthroscopic Surgery International
(O.A.S.I.) Bioresearch Foundation, Gobbi N.P.O.,
Milan, Italy

Ł. Paczesny · J. Zabrzyński
Orvit Clinic, Citomed Healthcare Center,
Torun, Poland

P. Pękala
Orthopaedic Arthroscopic Surgery International
(O.A.S.I.) Bioresearch Foundation, Gobbi N.P.O.,
Milan, Italy

Department of Anatomy, Jagiellonian University,
Kraków, Poland

J. Szczepanek
Centre for Modern Interdisciplinary Technologies,
Nicolaus Copernicus University, Torun, Poland

© ISAKOS 2022
A. Gobbi et al. (eds.), *Joint Function Preservation*, https://doi.org/10.1007/978-3-030-82958-2_17

this process [8, 9]. Human articular chondrocytes express constitutive complex of the major histocompatibility system (MHC) class I, molecules that regulate complement activation. After the activation, for example, under the influence of *IFN-α, IL-1, TNF-α,* or inflammatory joint diseases, chondrocytes express also MHC class II and *ICAM*-1 intercellular adhesion molecules [10]. In numerous studies, it was showed that chondrocytes also have tissue-specific antigens that induce the production of antibodies in patients with cartilage transplants, as well as in patients with RA and OA. The role of the genes coding structural components of cartilage and regulatory enzymes in the pathogenesis of osteochondral pathologies with respect to inter- and intracellular signaling pathways is still under investigation.

17.2 Genetic Basis of Osteochondral Pathology

Cartilage damages have a multifactorial nature, and genetic factors are their strong determinants. The molecular basis of degenerative changes in cartilage is at least partly understood, which is due to the results of numerous biochemical and genetic studies. The molecular markers associated with the development of cartilage pathology can be identified using molecular biology techniques (single gene analysis by PCR or genome-wide testing, for example, NGS, microarrays) or bioinformatics methods (e.g., meta-analyzes, association studies). Genetic changes underlying the etiopathogenesis of osteochondral pathologies include common DNA variants, mainly SNP type, differentiated gene expression, and epigenetic changes such as DNA methylation and microRNA regulation.

In the case of the characteristics of heritability of the disease, the most important data come from familial aggregation analyzes and studies of twins. Based on research on hereditary diseases (including familial cases of OA, chondrodysplasia, spondyloepiphyseal dysplasia), the genes collagen II and IX mutations [11, 12] and genes for structural proteins [13, 14] were selected. Heritability estimates ranging from 30% to 65%, depending on the joint site [15, 16]. Understanding the complex genetic background is useful primarily in the context of seeking new pathways leading to disease progression as well as in developing the new targeted therapies.

The potential biochemical and genetic markers in joint tissues which should be investigated are extracellular matrix components, such as precursor or degradation products of collagen and proteoglycans (Table 17.1), enzymes, cytokines (Tables 17.2 and 17.4), and transcription factors (Table 17.3). Metalloproteinases, inflammatory factors, signal molecules, and transcription factors are one of the best-described groups of genes involved in the pathogenesis of degenerative disease [17, 18]. Their expression level and concentration are related to tissue metabolism and can be measured in the blood, urine, or synovial fluid (Table 17.1). In clinical practice, inflammatory markers are considered to be well correlated with synovitis (Tables 17.2 and 17.4). Markers of cartilage degeneration have a moderate or good correlation with clinical and radiological findings in the course of degenerative joint diseases, especially osteoarthritis [19].

Changes in the structural components of cartilage, detected by biochemical and genetic methods in serum, synovial fluids, and urine, mainly include:

1. ECM components:
   - Increased level of type II collagen in serum and urine and increased level of procollagen type IIA N-terminal propeptide (*PIIANP*) in serum for cartilage synthesis
   - Increased level of C-telopeptide fragment of collagen type II (*CTX-II*) in urine, procollagen type II N-terminal propeptide (*PIINP*), cartilage oligomeric matrix protein (*COMP*), and binding proteins in serum and synovial fluid for cartilage degradation
2. matrix degrading enzymes:
   - proteolytic enzymes: metalloproteinases (MMPs), like: *MMP-3, MMP-9,* and *MMP-13*

Table 17.1 Selected biomarkers of cartilage, bone, and synovium metabolism (based on Nguyen et al. [20])

Molecule type origination	Markers of synthesis	Markers of degradation	Sample type
Tissue origination—cartilage			
Type II collagen	*PIICP*		Synovial fluid
	PIIANP		Serum
		CTX-II	Urine, synovial fluid
		HELIX-II	Urine
		C2C	Urine, serum, synovial fluid
		CIIM, Coll 2-1, NO2	Serum
Type X collagen	*C-Col10*		Serum
Aggrecan	*Epitope 846*		Synovial fluid
		ARGS	Synovial fluid
Non-collagenous and non-aggrecan proteins		*COMP*	Serum
		Pentosidine	Serum, synovial fluid
		FSTL1	Serum, synovial fluid
		Fib3-1, Fib3-2	Serum
Proteolytic enzymes		*MMP-3, MMP-9*	Serum
		MMP-1, MMP-13	Synovial fluid
		ADAMTS-4	Serum
Proteolytic enzyme inhibitors		*TIMP-1, TIMP -2*	Synovial fluid
Tissueorigination–bone			
Type I collagen	*PINP*		Serum
Non-collagenous protein	*OC*		Serum
		MidOC, CTX-I, NTX-I,PYD, DPD	Urine
Tissue origination–synovium			
Non-collagenous proteins	*HA*		Serum
	YKL-40		Serum, synovial fluid
Type III collagen		*Glc-Gal-PYD*	Urine

Table 17.2 Proinflammatory cytokines involved in OA (based on Mobasheri et al. [30])

Cytokine	Expression	Function
TNF-α	Synoviocytes Chondrocytes	Increase cartilage degradation and bone resorption Inhibit glycoprotein and collagen synthesis Upregulate MMP expression Stimulate other cells to produce proinflammatory cytokines and growth factors Stimulate proangiogenic factor release Stimulate other cells to produce chemotactic cytokines Stimulate nitric oxide production Induce chondrocyte apoptosis
IL-1β	Synoviocytes ChondrocytesMacrophages	Increase cartilage degradation and bone resorption Inhibit proteoglycan synthesis Upregulate MMP expression Production of proteolytic enzymes Stimulate other cells to produce proinflammatory cytokines Stimulate other cells to produce chemotactic cytokines Stimulate proangiogenic factor release Stimulate NO production Induce chondrocyte apoptosis
IL-6	Synoviocytes Chondrocytes Osteoblasts	Inhibit proteoglycan synthesis Reduce chondrocyte proliferation Increase MMP-2 activity Increase aggrecanase-mediated proteoglycan catabolism

Table 17.4 Inflammatory mediators and epigenetic modifications in an osteoarthritic joint (based on Raman et al. [44])

Protein category	Mediator	Epigenetic effect
Cytokines	*TNF-α, IL-1β, IL-6*	DNA methylation Histone modification miRNA regulation
Inducible nitric acid oxide synthetase	*iNOS*	DNA methylation Histone modification
Proteinase	*MMP-3, 9, 13, ADAMTS-4,5*	DNA methylation Histone modification miRNA regulation
ECM proteins	*COL2A1, COL99A1, ACAN*	DNA methylation Histone modification miRNA regulation
Chondrocyte growth gene	*GDF-5*	DNA methylation
Transcription factors	*SOX9, NFAT1, RUNX2*	DNA methylation Histone modification miRNA regulation

- aggrecanases such as disintegrin and metalloproteinase with thrombospondin-like motif (*ADAMTS*): *ADAMTS*-4 or *ADAMTS*-5 [20].

The homeostasis of cartilage is maintained by the balance of catabolic and anabolic processes, except for a pathological condition when degeneration exceeds regeneration and the loss of cartilage matrix occurs (Fig. 17.1). Changes in the genome and transcriptome levels lead to alterations in the protein level. Several homeostasis abnormalities within cartilaginous tissue have been found, including various structural changes, differentiated gene expression, as well as epigenetic regulation.

Although every cartilage damage or disease probably have a significant genetic background, cellular pathways leading to OA and RA are best described. Changes at the genetic level determine the progression of the disease and are present at every stage of its progression. They start from progressive cartilage loss, osteophyte formation, subchondral bone thickening to develop of synovial inflammation [21]. One of the most important families of enzymes are metalloproteinases, which are responsible for the irreversible proteolytic destruction of cartilage, especially of type II collagen. Seven matrix metalloproteinases have been shown to be expressed under varying circumstances in articular cartilage [22]. Only *MMP-1, MMP-2,MMP-13*, and *MMP-14* are constitutively expressed in adult cartilage. Their physiological function is tissue turnover, and the level of their expression increases significantly in disease states. The presence of the *MMP-3, MMP-8*, and *MMP-9* in cartilage appears to be characteristic for pathologic circumstances only [22]. The soluble collagenases *MMP-1, MMP-8*, and *MMP-13* play a key role in cartilage destruction. The collagenolytic activity of other MMPs (such as:*MMP-2* and *MMP-14*) is likely minor. In addition, it was experimentally demonstrated that *MMP-3, MMP-9*, and *MMP-10* degrade other ECM components, but in vivo they are unable to cleave native type II collagen [21–23]. The proper regulation of expression of the metalloproteinase family depends on many factors and trigger several intracellular signaling pathways. The expression patterns of MMPs in cartilage depend on proinflammatory and pleiotropic cytokines and growth factors [24, 25]. Overexpression of MMPs is an important marker of progression of osteochondral diseases regardless of etiology. It indicates this family of genes (especially MMP-13) as the main biomarkers of bone and cartilage damage as well as mediators of joint destruction. There is a relationship between the increase in MMPs expression and the rapid rate of joint destruction. The correlation of *MMP-13* expression with cartilage damage makes this gene an interesting candidate for pharmacological intervention.

Table 17.3 Classes of gene products that aid cartilage repair (based on Steinert et al. [33])

Therapeutic mechanism		Gene product
Stimulation of chondrogenic differentiation	*Anabolic growth factors*	*TGF-β 1, 2, 3*
		BMP-2, -4, -7
		CDMP-1, -2, -3 (GDF-5, -6, -7)
		Wnts
	Signal transduction molecules	*Smad-4, -5*
	Transcription factors	*SOX9, -5, -6*
		Brachyury
Stimulation of cartilage matrix synthesis and/or cell proliferation	*Anabolic growth factors*	*TGF-β 1, 2, 3*
		BMP-2, -4, -7
		CDMP-1, -2, -3 (GDF-5, -6, -7)
		IGF-1
		PDGF, EGF, HGF
	ECM component	*Type II collagen minigene*
		COMP
	Enzymes for GAG synthesis	*GlcAT-1*
Inhibition of osteogenesis/hypertrophy	*Inhibiting TGF-β/BMP action (growth factors)*	*Noggin, chordin*
	Inhibiting terminal differentiation (growth factors)	*PTHrP*
		IHH, SHH, DHH
	Signal transduction molecules	*Smad 6, 7*
		mLAP-1
Anti-inflammatory	*IL-1 blockage (cytokine antagonist)*	*IL-1Ra, sIL-1R, ICE inhibitor*
	TNF-α inhibition (cytokine antagonist)	*sTNFR, anti-TNF-antibodies, TACE inhibitor*
	MMP inhibition (proteinase inhibitor)	*TIMP-1, -2, MMP inhibitors*
	Cytokines	*IL-4, -10, -11, -13*
	Enzymes for glucosamine derivates (IL-1 inhibition)	*GFAT*
Senescence inhibition	*Inhibition of telomere erosion*	*hTERT*
	Free radical antagonist	*NO-(iNOS) antagonists, SOD*
Apoptosis inhibition	*Caspase inhibition*	*Bcl-2, Bcl-XL*
	Fas-L blockage	*Anti-FasL*
	NO-induced apoptosis	*Akt, PI-3-kinase*
	TNF-α, TRAIL inhibition	*NFκB*

Cartilage diseases are often accompanied by synovitis. Symptoms of the inflammatory state are proliferation of synoviocytes and tissue hypertrophy. Synoviocytes release inflammatory mediators and matrix degenerating enzymes into the joint. Their activation occurs due to the action of inflammatory mediators and cartilage matrix molecules, initiating a feedback cycle within the synovium, which results in progressive degeneration of the joint. Based on biochemical and genetic findings [19, 26, 27], numerous markers of the inflammatory process have been selected. The strongest correlations with inflammation of the joints have:

- inflammatory mediators: cyclooxygenase (*COX*), prostaglandin E2 (*PGE2*), *PGD2*, *PGF2a*, thromboxane, and *PGI2*
- circulating or locally occurring cytokines (Table 17.2): interleukin-1 (*IL-1*), *IL-6,IL-17*, *IL*-18, *TNF*-α chemokines, such as C-C motif chemokine ligand 5 (*CCL5*) and *IL-8*
- nitric oxide (NO)
- synovial degradation products: hyaluronan or hyaluronic acid (HA)

Based on several studies it has been proved that proinflammatory cytokines and metalloproteinases are involved in matrix disruption.

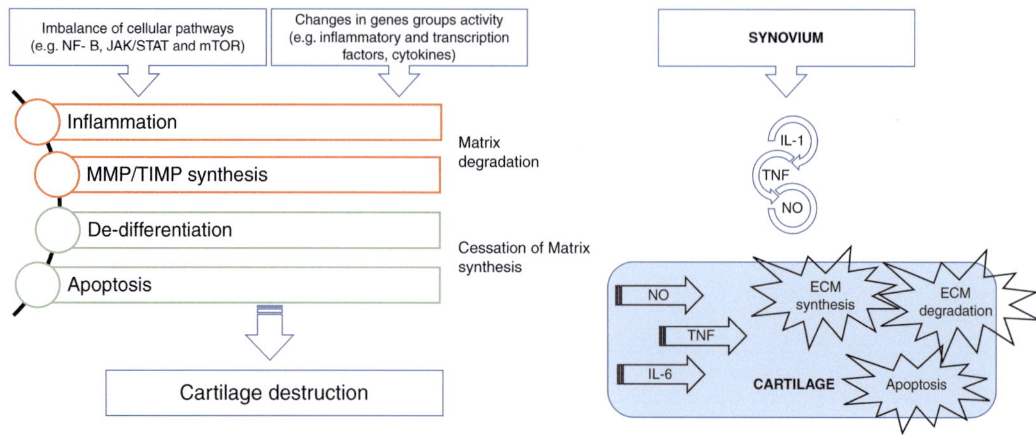

Fig. 17.1 Schematic summary of cartilage destruction

Targeted action on chondrocytes results in differentiated expression of catabolic and anabolic genes. In early and late OA the inflammation of the synovium, or synovitis is observed. It promotes acceleration of cartilage destruction and ultimately disease progression [28]. Synovial biopsy tests show differentiated expression of genetic markers of the inflammatory process, such as: chemokine ligand-5 (*CCL5*), *CCL7*, *CCL19*, and interleukin-8 (*IL-8*) [29]. Synovial inflammation is also correlated with secretion of proinflammatory cytokines (Table 17.2), like vascular endothelial growth factor (*VEGF*), blood vessel formation (factor VIII), intercellular adhesion molecule-1, and the proinflammatory cytokines (*TNF*-α, *IL-6* and IL-1ß) [29].

Although there are many joint biomarkers that may be potential diagnostic or prognostic tools for inflammation, their use in clinical practice is still difficult, and most of them are only relevant in clinical trials.

The genetic changes in cartilage are regulated directly and indirectly by genes associated with tissue metabolism. For several signaling pathways (like signal transduction, NF-κB, JAK/STAT, and mTOR) and transcription factors (e.g., *Runx2*, *C/EBPβ*, *HIF2α*, *Sox4*, and *Sox11*) important implication in these diseases were determined [17]. Several genes related to OA have been determinate.

Based on numerous studies, a strong correlation between OA and changes within chromosomes 2q, 9q, 11q, 16p, and 20q has been confirmed. Genes associated include *VDR*, *AGC1*, *IGF-1*, *ER* alpha, *TGF* beta, *CRTM* (cartilage matrix protein), *CRTL* (cartilage link protein), and collagen II, IX, and XI. It is worth mentioning that genes may be expressed differently according to sex and body site [31]. A powerful tool for identifying genes responsible for joint damage induction and progression is the combination of candidate gene approaches and quantitative trait loci (QTL). Such analysis with linkages to osteoarthritis were identified on chromosomes 2q (nodal OA, DIP OA, THR), 9q (hand OA), 11q (hand OA), and 16p (hip OA). Susceptibility to osteoarthritis has been associated with variation in several genes, including:

(a) variation in the *FRZB* gene on chromosome 2q32 (OA, osteoarthrosis, osteoarthritis of hip, female specific)
(b) variation in the *MATN3* gene on chromosome 2p24 (osteoarthritis of distal interphalangeal joints, OADIP, DIPOA, hand osteoarthritis; HOA)
(c) variation in the *ASPN* gene on chromosome 9q22 (osteoarthritis of knee/hip)
(d) variation in the *GDF5* gene on chromosome 20q11 (osteoarthritis of hip)

The identified chromosome fragments are also *loci* for such biomarkers as: fibronectin 9 (a glycoprotein present in the extracellular matrix of normal cartilage), collagen alpha-2(V) chain (alpha chain for one of the low abundance fibrillar collagens), interleukin 8 receptor (important in the regulation of neutrophil activation and chemotaxis), and matrix metalloproteinase gene cluster [31, 32].

An interesting research direction is also searching for single nucleotide polymorphisms (SNP) within the candidate. SNP, as the most common genetic varieties in the population, have been successfully correlated with the pathogenesis of many diseases of articular cartilage. For example, for OA, more than 50 SNPs have been identified in many genes that have been linked to the hip (*COL11A1*, *VEGF*, etc.), knee (*COL9A3*, *ASPN*, *GDF5*, etc.), or both (*IL-8*, *TGF-β1*, etc.) OA [34]. Multiple SNPs play different roles in the pathogenesis of OA and its subtypes. Wang et al. identified 56 SNPs from different genes that have been shown to be associated with either hip [23], knee [20], or both [13] OA [34]. SNPs in various genes appear to be associated with osteoarthritis [34–39]:

- *COL11A1* (rs1241164, rs4907986, rs2615977), hip OA
- *DVWA* (rs7639618, rs9864422, s11718863), knee OA
- *FRZB* (rs7775, rs288326), OA of the hip, knee, and hand
- *GDF5* (rs143383 (risk allele T)), OA of the hip, knee, and hand
- *CALM1* (rs12885713), hip OA
- *IL1RN* (rs9005, rs315952, rs419598), knee OA
- *MCF2L* (rs11842874 (risk allele G)), knee OA
- *ADAM12* (rs1871054)—knee OA

The phenotype of mature chondrocytes is stabilized by numerous epigenetic modifications), including DNA methylation (hypo- and hypermethylation mainly in promoter CpG sites of target genes, histone modification (methylation, ubiquitination, acetylation, and phosphorylation), and non-coding RNAs binding to the 3′-untranslated region of target gene [40–46]. Epigenetic changes have been described for many groups (transcription factors, proteinases, cytokines, chemokines, growth factors, and ECM proteins (Table 17.4) of genes relevant for cartilage destruction, hypertrophic chondrocyte formation, and synovitis. Modifications to the epigenetic pattern can lead to genetic disruptions that result in the overexpression of cartilage-degrading proteases and inflammatory process factors [44].

An interesting group of diagnostic markers are also microRNA molecules. Recent reports have demonstrated that microRNAs might play an important role in the development of joint disorders. As post-transcriptional regulators of target genes, participate in the modulation of cell signaling pathways. Differentiated miRNA expression has a significant effect on articular cartilage homeostasis (Table 17.5). The potential role of miRNAs in biological processes such as cartilage degeneration, chondrocyte proliferation, and differentiation are discussed [47]. Among microRNAs involved in cartilage-protective mechanisms, attention should be paid to miR-140 involved in cartilage development [48], miR-9 increased type II collagen [49], miR-27a prevented synovial fibroblast migration and invasion [48, 50], miR-221-3p prevented ECM degradation [51], anti-inflammatory miR-149 [52], or miR-125b prevented aggrecan loss [53]. Among microRNAs involved in cartilage-destructive mechanisms are, for example, miR-381 responsible for chondrocyte hypertrophy and cartilage degeneration [54], miR-216b inhibited chondrocyte proliferation [55], miR-302b promoted inflammation [56], miR-365 mediated mechanical stress and inflammatory pathway [57], miR-146a activated early OA [58] or miR-483-5p stimulated chondrocyte hypertrophy, ECM degradation and cartilage angiogenesis [59]. MicroRNAs are important post-transcriptional regulators of cell pathways, such as TGF-β/Smads and BMPs, MAPK, and NF-KB signaling, that are involved in the physiology of cartilage tissue [47]. As modulators of important miRNA genes, they are interesting candidates for targeted therapy.

Table 17.5 MicroRNAs involved in cartilage protection and degradation (based on [47])

miRNAs	Target Gene(s)	Function	Specimens
miR-140	*SMAD3*	Suppressing the Smad 2/3 pathway	Chondrocytes
	MMP13	Inhibition of the matrix metalloprotease	Cartilage C28/I2 cells
	RALA	Upregulates SOX9, ACAN, Col2a	Mesenchymal stem cells
	IL-1β *COLL3a1*	Inhibits (IL-1β)-induced signaling	Knee synovial fibroblasts
miR-145	*SMAD3* *SOX9*	Downregulates type II collagen and glycosaminoglycans concentration while up-regulating of MMP-13 expression	Knee OA cartilage
miR-29	*COL3a1* *osteonectin*	Promotes osteogenesis, inhibits osteoblast differentiation	hMSCs
miR-455	*SMAD2* *ACVR2B*	Promotes a degradative chondrocyte response	Hip articular cartilage
miR-302	*BMP 2R*	Shows pro-anabolic activities	hADSCs
miR-155	*MMP1* *MMP3*	Inhibits production of MMP1 and MMP13	Joint synovial tissue
miR-127	*MMP13*	Surpresses production of MMP1 and MMP13	Knee OA cartilage and chondrocytes
miR-148	*MMP13* *COL10A1* *ADAMTS5*	Shows pro-anabolic and anti-catabolic activities	OA articular cartilage and chondrocytes
miR-602	*SHH* *MMP13*	Negatively regulates the expression of SHH and MMP-13	OA articular cartilage and chondrocytes
miR-608	*SHH* *MMP13*	Negatively regulates the expression of SHH and MMP-13	OA articular cartilage and chondrocytes
miR-125	*ADAMTS4* *MMP13*	Negatively regulates the expression of ADAMTS4 and MMP-13	OA articular cartilage and chondrocytes
miR-27	*MMP13*	Negatively regulates the expression of MMP-13	OA articular cartilage and chondrocytes
miR-22	*PPARA* *BMP7*	Negatively regulates the expression of PPARA and BMP7, blocked inflammatory, and catabolic changes	Articular cartilage and chondrocytes
miR-9	*MMP13 (indirect)* *PRTG*	Negatively regulates the expression of MMP-13	Knee OA cartilage
miR-558	*COX-2* *MMP-1* *MMP-13*	Negatively regulates the expression of MMP-13, MMP-1, COX-2	Knee OA cartilage and chondrocytes
miR-488	*ZIP8* *MMP-13*	Reduces cartilage degradation	OA cartilage and chondrocytes
miR-320	*ADAMTS5*	Negatively regulates the expression of ADAMTS5	Knee OA cartilage and chondrocytes
miR-203	*MMP-1* *IL-6 (indirect)*	Increases secretion of MMP-1 and IL-6 via the NF-kB pathway	Articular synovial tissue and cell
miR-181	*MMP13 (indirect)*	Increases production of MMP13	Knee OA cartilage
miR-193	*COL2* *aggrecan* *SOX9*	Downregulates anabolic factors such as type 2 collagen, aggrecan, and SOX9	Knee OA cartilage

Epigenetic regulations of cytokine expression are well described in the literature. For example, *Il-1β* expression level is modulated by methylation and demethylation of CpG site of the promoter region [60], as well as the activities of miR-146a and miR-149 [46, 58]. In addition, it was found that use of histone deacetylase inhibitors in chondrocytes results in reduced secretion of inflammatory cytokines, like *Il-1β*, suppressing synovitis and preventing the re-differentiation of dedifferentiated chondrocytes [44, 61].

Genes and their polymorphisms play a key role in development of the osteochondral pathologies in the general population. The complexed background of the genetic heterogeneity of the cartilage diseases consists of a plethora of genes and their epigenetic modifications. It results in changes in their activity, and thus explains the background of the pathological process. Analyzes of allelic expression imbalances, genetic expression signatures, epigenetic regulatory mechanisms, and key cellular pathway disturbance provide comprehensive knowledge that brings us closer to understanding the normal tissue metabolism as well as induced by pathological processes. Therefore, the challenge is to find sensitive and specific biomarkers that will help in early diagnosis and targeted therapy of cartilage.

17.3 Gene Therapy Approaches

Osteochondral lesions are common, they are most likely responsible for the initiation of osteoarthritis development. As biochemical and genetic studies have added substantial knowledge in the past decade, the molecular basis of osteochondral pathologies has become clearer. Gene therapy is a technology utilizing the gene transfer to deliver therapeutic genes to the site of injury.

Bioactive proteins are difficult to administer effectively. However, gene transfer approaches are being developed to provide their sustained synthesis at sites of repair. The treatment of cartilage lesions is applied by transferring genes, encoding the specific growth factors, into chondrocytes or progenitor cells [62]. Gene delivery into the osteochondral unit is categorized as either in vivo (direct gene delivery into host tissue within the lesion site) or ex vivo (indirect gene delivery, for example, via stem cells or fibroblasts, following in vitro transfection or transduction). Viral gene vectors (e.g., adenovirus, retrovirus, adeno-associated virus (AAV), and herpes simplex virus) represent the efficient method for gene transfer [63]. Either the vectors can be injected directly into the host tissue (in vivo) or the cells from the injured tissue can be removed, genetically altered in vitro, and rein-

jected in the injury site (ex vivo). The direct method is less technically demanding, but indirect gene delivery is safer, because the gene manipulation takes place under controlled conditions in vitro. Moreover, safety tests can be performed in the genetically engineered cells before the implantation into the defects. Although the development of effective reagents for osteochondral defects remains complicated, gene transfer might improve regeneration at the site of injury by enabling the local, sustained, and potentially regulated expression of morphogens, growth factors, and anti-inflammatory proteins [64]. Understanding the molecular basis of cartilage and joint diseases is important and useful for the establishment of effective therapies.

Large scientific progress have been made in the last 30 years toward understanding the biology of articular cartilage healing and toward development of the new restoration techniques. Microfracture is one of several cartilage repair techniques that works by creating tiny fractures in the subchondral bone. It is a commonly used procedure to treat patients with small to moderate full-thickness chondral lesions [65]. Progenitor cells from the subchondral region enter the lesion site and become trapped in the fibrin clot, where some of them differentiate along a chondrogenic lineage to form repair tissue. Noteworthy is the fact that the newly formed tissue resulting from microfracture is a fibrocartilage, which is less durable compared to the articular cartilage. Additionally, there is a risk of an intralesional osteophyte formation after this procedure. For the larger lesions autologous chondrocyte implantation (ACI) is usually indicated [66]. The clinical trial design in emerging methods of cartilage repair requires that new cartilage repair methods are superior to microfracture [64].

The effectiveness of microfracture may be improved by two simple gene-based techniques. Pascher et al. presented approach to enhance natural repair mechanisms within cartilage lesions by targeting bone marrow-derived cells for genetic modification [67]. As an alternative medium for gene delivery, they investigated the feasibility of using coagulated bone marrow aspirates. Mixing an adenoviral suspension with the fluid phase of freshly aspirated bone marrow resulted in uniform vector dispersion. The rate of transgenic expression is in direct proportion to the density of nucleated cells in the corresponding clot. Sieker et al. presented good results with this method using cDNA that encodes bone morphogenetic protein *BMP-2* and Indian hedgehog protein. It was effective to improve cartilage repair in osteochondral defects in the trochlea of rabbit knees. However, the *BMP-2* treatment, carried the risk intralesional bone formation [68]. Ivkovic et al. used ovine autologous bone marrow transduced with adenoviral vectors containing cDNA for green fluorescent protein or transforming growth factor (*TGF-β1*). The marrow was allowed to clot forming a gene plug and then was implanted into partial-thickness defects, which were created on medial femoral condyle in sheep model [69]. This method improved the outcome and *TGF*-treated defects showed significantly higher amounts of collagen II in histologic examination. In the second approach, the recombinant adeno-associated virus is used directly to the exudate that is implanted into the osteochondral lesion. In a rabbit osteochondral defect model, fibroblast growth factor 2 (*FGF-2*), insulin-like growth factor 1 (*IGF-1*), and the transcription factor *SOX9*, have been delivered by transgene, with promising results [64, 70–72].

Over the years, ACI has been recognized as a good treatment option to deal with large full-thickness chondral lesions [73–75]. However, this approach requires two surgeries. Firstly, articular cartilage is harvested from a lesser-weight-bearing part of the joint. Then, autologous chondrocytes need to be expanded in culture and implanted into the defect. As the application of ACI has been limited by the high cost of autologous therapy and by the need for two surgeries, using genetically modified allografted chondrocytes could reduce complexity and improve cost-effectiveness. Kang et al. showed for the first time, that genetically modified allografted chondrocytes could persist and express transgenes in rabbit's osteochondral defects [76]. There is also a large body of evidence confirming that genetically modified allogenic or autogenous chondrocytes are effective in cartilage repair in animals

[77, 78]. Ortved et al. presented that transferring *IGF-1* by AAV to autologous chondrocytes improved repair outcomes of full-thickness chondral defects in equines [79]. Such genetically enhanced allograft chondrocytes were used in human clinical trials [80]. The transduced cells were surgically introduced into cartilage lesions using a fibrin scaffold. A line of human chondrocytes was obtained from a newborn with polydactyly. One cohort of cells was transduced with a retrovirus carrying *TGF-β1* cDNA.

Evans et al. demonstrated another therapy based on the remarkable potential of genetically modified, autologous skeletal muscle and fat grafts to heal large osseous and chondral defects. These tissues can be harvested, genetically modified, and then press-fit into the osteochondral lesions within the time frame of a single surgery. The theory behind the muscle-based tissue engineering is related to the unique biology of skeletal muscle-derived cells. Skeletal muscle contains satellite cells, which are capable of fusing to form post-mitotic, multinucleated myotubes and myofibers. The post-mitotic myofibers are stable cells and theoretically capable of long-lasting gene expression. Their potency is likely to reflect the presence of endogenous progenitor cells, the secretion morphogenetic signals by the genetically modified cells, and the scaffolding properties of the tissues themselves. When compared to ACI, the complexity of the procedure is reduced, which should lower the costs. Moreover, skeletal muscle and fat are easily accessible and available for biopsies. The results from pilot experiments with rabbits showed that the implanted tissues formed bone in the subchondral region and cartilage above, indicating the impact of the progenitor cell location on the process of the differentiation [81].

The main limitation of treatment of focal cartilage defects with non-scaffolds approach is that the genetically modified cells or gene vectors are diluted by the joint fluid and fail to maintain at the target lesion area. To avoid this obstacle, a promising approach is to deliver modified cells or gene vectors using different types of biomaterials. Scaffolds for cartilage repair present new options to structurally support cartilage repair

[82]. Gene therapy combined with scaffolds increases the efficiency and durability of transfected genes, forming an efficient system to promote osteochondral regeneration. When the scaffold is degraded, the contents are slowly released to the target area. These biomaterials can be implanted into the articular cartilage defects to provide gene transfer that enables the controlled release of vector over time. The regulated transmission of genetic material via biomaterials could enhance the properties of the gene products and protect these active agents against degradation [83]. Additionally, biomaterial-mediated gene delivery provides biomechanical environment in magnitudes similar to that of native articular cartilage, what is especially important for the repair of large defects. Scaffolds in gene therapy for osteochondral tissue repair can be applied using two methods: through the incorporation of the vector during scaffold preparation or by the connection of the vector to a formed scaffold. Among the numerous biologic and synthetic materials, biocompatible and biodegradable compounds matrices are thought to possess the most promising potential for repair of osteochondral defects [84–86]. Different gene delivery approaches use solid scaffolds, hydrogels, and micelles (alginate, poloxamer PF68, and poloxamine T908 polymeric micelles based on poly (ethyleneoxide)—PEO—and poly(propylene oxide)—PPO—triblock copolymers, self-assembling RAD16-I peptide hydrogels, polypseudorotaxane gels) to create vector-loaded biomaterials [63, 87, 88]. These biomaterials provide the environment for the formation of cartilage tissue with adapted mechanical properties and affording protection against tissue degradation in conditions that enable joint resurfacing.

The use of controlled gene delivery approaches to facilitate clinical cartilage repair is still a developing field. Emerging approaches include the use of progenitor cells, rather than chondrocytes, as agents of gene transfer for cartilage repair. Mesenchymal stem cells (MSCs) are the most widely studied due to their high availability and proliferative/differentiation ability. Genetically modified MSCs can be considered as a fundament for one of the existing cell-based cartilage repair methods. Bone marrow delivered MSCs (BMCs) and adipose-delivered MSCs (ASCs) are commonly employed for osteochondral therapy [88]. Leng et al. used transfected BMSCs with *hIGF-1* cDNA and mixed with calcium alginate gels for transplantation into osteochondral defects showing the improvement in the repair outcomes [89]. In another study, Venkatesan et al. designed 3D fibrin-polyurethane scaffolds in a hydrodynamic environment that provided a favorable growth environment for rAAV-infected *SOX9-* modified hBMCs and promoted their differentiation into the chondrocytes [90]. Yang et al. also transfected BMCs with adenoviruses expressing C-type natriuretic peptides and seeded the cells onto silk/chitosan scaffolds to promote chondrogenesis in rats [91].

Controlled tissue growth and biomimetic cartilage properties were maintained upon seeding the ASCs into large PCL-scaffolds immobilized with Dox-inducible lentiviruses expressing *IL-1Ra* [92]. Although, these techniques have not yet been confirmed in the clinical studies, they hold a great scientific promise for treating cartilage injuries in the near future. In addition, interest in improving the efficiency and targeting of non-viral vectors continues. For example, Pi et al. identified chondrocyte-affinity peptide that enhances cartilage-targeting transfection when attached to the polyethylenimine [93]. These data demonstrate that the potential for gene therapy of cartilage lesions is encouraging. However, this field is still developing. The assessments of possible associated toxicity are also lacking and it is essential for clinical translation.

17.4 Conclusion

Future research will continue to investigate the optimal combination of scaffolds, cells, peptides, and growth factors to repair articular cartilage defects. The combination of genetic engineering, gene transfer techniques, and tissue engineering is one of the potential new strategies for the treatment of osteochondral injuries. The advantage of gene transfer into the chondrogenic cells relies on sustained levels of growth factors, which can be reached through transgene expression in situ. Although, the field of genetic engineering is young, current research offers gene transfer approaches developed to provide sustained synthesis of bioactive reagents at the cartilage repair sites. To augment regeneration of articular cartilage, therapeutic genes can be delivered to the synovium, or directly to the cartilage lesion.

Because cartilage injuries are not life-threatening, the safety of gene transfer approaches for repair is of particular importance. To harness the potential of this technology for clinical use, it is crucial to use safe and efficient vectors, transgenes, and delivery systems. The major considerations for clinical translation are their biology, safety, ease of manufacture, and cost-effectiveness [64]. In this regard, combining gene therapy with tissue engineering concepts might overcome the various physiological barriers that impede the safe, effective, and long-term treatment of damaged articular surface [94]. Cartilage repair could become the domain of gene therapy because osteochondral pathologies are very common, and application provides local treatment with a relatively small amount of vector.

References

1. Sophia Fox AJ, Bedi A, Rodeo SA. The basic science of articular cartilage: structure, composition, and function. Sports Health. 2009;1(6):461–8.
2. Hewitt AT, Varner HH, Silver MH, Martin GR. The role of chondronectin and cartilage proteoglycan in the attachment of chondrocytes to collagen. Prog Clin Biol Res. 1982;110 Pt B:25–33.
3. Martin JA, Buckwalter JA. The role of chondrocyte-matrix interactions in maintaining and repairing articular cartilage. Biorheology. 2000;37(1–2):129–40.
4. Buckwalter JA, Mankin HJ. Articular cartilage: tissue design and chondrocyte-matrix interactions. Instr Course Lect. 1998;47:477–86.
5. Loeser RF. Integrins and chondrocyte-matrix interactions in articular cartilage. Matrix Biol. 2014;39:11–6.
6. Gill SE, Parks WC. Metalloproteinases and their inhibitors: regulators of wound healing. Int J Biochem Cell Biol. 2008;40(6–7):1334–47.
7. Cudic M, Fields GB. Extracellular proteases as targets for drug development. Curr Protein Pept Sci. 2009;10(4):297–307.
8. Wojdasiewicz P, Poniatowski LA, Szukiewicz D. The role of inflammatory and anti-inflammatory cytokines in the pathogenesis of osteoarthritis. Mediat Inflamm. 2014;2014:561459.
9. Kany S, Vollrath JT, Relja B. Cytokines in inflammatory disease. Int J Mol Sci. 2019;20(23)
10. Osiecka-Iwan A, Hyc A, Radomska-Lesniewska DM, Rymarczyk A, Skopinski P. Antigenic and immunogenic properties of chondrocytes. Implications for chondrocyte therapeutic transplantation and pathogenesis of inflammatory and degenerative joint diseases. Cent Eur J Immunol. 2018;43(2):209–19.
11. Olsen BR. Mutations in collagen genes resulting in metaphyseal and epiphyseal dysplasias. Bone. 1995;17(2 Suppl):45S–9S.
12. Cicuttini FM, Spector TD. Genetics of osteoarthritis. Ann Rheum Dis. 1996;55(9):665–7.
13. Palotie A, Vaisanen P, Ott J, Ryhanen L, Elima K, Vikkula M, et al. Predisposition to familial osteoarthrosis linked to type II collagen gene. Lancet. 1989;1(8644):924–7.
14. Chapman KL, Mortier GR, Chapman K, Loughlin J, Grant ME, Briggs MD. Mutations in the region encoding the von Willebrand factor a domain of matrilin-3 are associated with multiple epiphyseal dysplasia. Nat Genet. 2001;28(4):393–6.
15. Carr AJ. The genetic basis of severe osteoarthritis. Ann R Coll Surg Engl. 2003;85(4):263–8.
16. Kraus VB, Jordan JM, Doherty M, Wilson AG, Moskowitz R, Hochberg M, et al. The genetics of generalized osteoarthritis (GOGO) study: study design and evaluation of osteoarthritis phenotypes. Osteoarthr Cartil. 2007;15(2):120–7.
17. Nishimura R, Hata K, Takahata Y, Murakami T, Nakamura E, Ohkawa M, et al. Role of signal transduction pathways and transcription factors in cartilage and joint diseases. Int J Mol Sci. 2020;21(4)
18. Nishimura R, Hata K, Nakamura E, Murakami T, Takahata Y. Transcriptional network systems in cartilage development and disease. Histochem Cell Biol. 2018;149(4):353–63.
19. Lee AS, Ellman MB, Yan D, Kroin JS, Cole BJ, van Wijnen AJ, et al. A current review of molecular mechanisms regarding osteoarthritis and pain. Gene. 2013;527(2):440–7.
20. Nguyen LT, Sharma AR, Chakraborty C, Saibaba B, Ahn ME, Lee SS. Review of prospects of biological fluid biomarkers in osteoarthritis. Int J Mol Sci. 2017;18(3)
21. Young DA, Barter MJ, Wilkinson DJ. Recent advances in understanding the regulation of metalloproteinases. F1000Res. 2019;8.
22. Rose BJ, Kooyman DL. A tale of two joints: the role of matrix metalloproteases in cartilage biology. Dis Markers. 2016;2016:4895050.
23. Murphy G, Lee MH. What are the roles of metalloproteinases in cartilage and bone damage? Ann Rheum Dis. 2005;64 Suppl 4:iv44–7.
24. Liacini A, Sylvester J, Li WQ, Huang W, Dehnade F, Ahmad M, et al. Induction of matrix metalloproteinase-13 gene expression by TNF-alpha is mediated by MAP kinases, AP-1, and NF-kappaB transcrip-

tion factors in articular chondrocytes. Exp Cell Res. 2003;288(1):208–17.

25. Vincenti MP, Brinckerhoff CE. Transcriptional regulation of collagenase (MMP-1, MMP-13) genes in arthritis: integration of complex signaling pathways for the recruitment of gene-specific transcription factors. Arthritis Res. 2002;4(3):157–64.

26. Goldring MB, Berenbaum F. The regulation of chondrocyte function by proinflammatory mediators: prostaglandins and nitric oxide. Clin Orthop Relat Res. 2004;427 Suppl:S37–46.

27. Wang X, Hunter D, Xu J, Ding C. Metabolic triggered inflammation in osteoarthritis. Osteoarthr Cartil. 2015;23(1):22–30.

28. Egloff C, Hugle T, Valderrabano V. Biomechanics and pathomechanisms of osteoarthritis. Swiss Med Wkly. 2012;142:w13583.

29. Scanzello CR, McKeon B, Swaim BH, DiCarlo E, Asomugha EU, Kanda V, et al. Synovial inflammation in patients undergoing arthroscopic meniscectomy: molecular characterization and relationship to symptoms. Arthritis Rheum. 2011;63(2):391–400.

30. Mobasheri A, Henrotin Y, Biesalski HK, Shakibaei M. Scientific evidence and rationale for the development of curcumin and resveratrol as nutraceuticals for joint health. Int J Mol Sci. 2012;13(4):4202–32.

31. Spector TD, MacGregor AJ. Risk factors for osteoarthritis: genetics. Osteoarthr Cartil. 2004;12 Suppl A:S39–44.

32. Chapman K, Mustafa Z, Irven C, Carr AJ, Clipsham K, Smith A, et al. Osteoarthritis-susceptibility locus on chromosome 11q, detected by linkage. Am J Hum Genet. 1999;65(1):167–74.

33. Steinert AF, Noth U, Tuan RS. Concepts in gene therapy for cartilage repair. Injury. 2008;39(Suppl 1):S97–113.

34. Wang T, Liang Y, Li H, Li H, He Q, Xue Y, et al. Single nucleotide polymorphisms and osteoarthritis: an overview and a Meta-analysis. Medicine (Baltimore). 2016;95(7):e2811.

35. Evangelou E, Chapman K, Meulenbelt I, Karassa FB, Loughlin J, Carr A, et al. Large-scale analysis of association between GDF5 and FRZB variants and osteoarthritis of the hip, knee, and hand. Arthritis Rheum. 2009;60(6):1710–21.

36. Day-Williams AG, Southam L, Panoutsopoulou K, Rayner NW, Esko T, Estrada K, et al. A variant in MCF2L is associated with osteoarthritis. Am J Hum Genet. 2011;89(3):446–50.

37. Wang D, Zhou K, Chen Z, Yang F, Zhang C, Zhou Z, et al. The association between DVWA polymorphisms and osteoarthritis susceptibility: a genetic meta-analysis. Int J Clin Exp Med. 2015;8(8):12566–74.

38. Lv ZT, Liang S, Huang XJ, Cheng P, Zhu WT, Chen AM. Association between ADAM12 single-nucleotide polymorphisms and knee osteoarthritis: a Meta-analysis. Biomed Res Int. 2017;2017:5398181.

39. Valdes AM, Doherty M, Spector TD. The additive effect of individual genes in predicting risk of knee osteoarthritis. Ann Rheum Dis. 2008;67(1):124–7.

40. Im GI, Choi YJ. Epigenetics in osteoarthritis and its implication for future therapeutics. Expert Opin Biol Ther. 2013;13(5):713–21.

41. Kim H, Kang D, Cho Y, Kim JH. Epigenetic regulation of chondrocyte catabolism and anabolism in osteoarthritis. Mol Cells. 2015;38(8):677–84.

42. den Hollander W, Ramos YF, Bomer N, Elzinga S, van der Breggen R, Lakenberg N, et al. Transcriptional associations of osteoarthritis-mediated loss of epigenetic control in articular cartilage. Arthritis Rheumatol. 2015;67(8):2108–16.

43. Loughlin J, Reynard LN. Osteoarthritis: epigenetics of articular cartilage in knee and hip OA. Nat Rev Rheumatol. 2015;11(1):6–7.

44. Raman S, FitzGerald U, Murphy JM. Interplay of inflammatory mediators with epigenetics and cartilage modifications in osteoarthritis. Front Bioeng Biotechnol. 2018;6:22.

45. Reynard LN, Loughlin J. Genetics and epigenetics of osteoarthritis. Maturitas. 2012;71(3):200–4.

46. Zhang M, Wang J. Epigenetics and osteoarthritis. Genes Dis. 2015;2(1):69–75.

47. Li YP, Wei XC, Li PC, Chen CW, Wang XH, Jiao Q, et al. The role of miRNAs in cartilage homeostasis. Curr Genomics. 2015;16(6):393–404.

48. Tardif G, Hum D, Pelletier JP, Duval N, Martel-Pelletier J. Regulation of the IGFBP-5 and MMP-13 genes by the microRNAs miR-140 and miR-27a in human osteoarthritic chondrocytes. BMC Musculoskelet Disord. 2009;10:148.

49. Gu R, Liu N, Luo S, Huang W, Zha Z, Yang J. MicroRNA-9 regulates the development of knee osteoarthritis through the NF-kappaB1 pathway in chondrocytes. Medicine (Baltimore). 2016;95(36):e4315.

50. Zhou B, Li H, Shi J. miR27 inhibits the NF-kappaB signaling pathway by targeting leptin in osteoarthritic chondrocytes. Int J Mol Med. 2017;40(2):523–30.

51. Yang S, Yang Y. Downregulation of microRNA221 decreases migration and invasion in fibroblastlike synoviocytes in rheumatoid arthritis. Mol Med Rep. 2015;12(2):2395–401.

52. Santini P, Politi L, Vedova PD, Scandurra R, Scotto d'Abusco A. The inflammatory circuitry of miR-149 as a pathological mechanism in osteoarthritis. Rheumatol Int. 2014;34(5):711–6.

53. Ge FX, Li H, Yin X. Upregulation of microRNA-125b-5p is involved in the pathogenesis of osteoarthritis by downregulating SYVN1. Oncol Rep. 2017;37(4):2490–6.

54. Chen W, Sheng P, Huang Z, Meng F, Kang Y, Huang G, et al. MicroRNA-381 regulates chondrocyte hypertrophy by inhibiting histone deacetylase 4 expression. Int J Mol Sci. 2016;17(9)

55. He J, Zhang J, Wang D. Down-regulation of microRNA-216b inhibits IL-1beta-induced chondrocyte injury by up-regulation of Smad3. Biosci Rep. 2017;37(2)

56. Wang Y, Yu T, Jin H, Zhao C, Wang Y. Knockdown MiR-302b alleviates LPS-induced injury by target-

ing Smad3 in C28/I2 Chondrocytic cells. Cell Physiol Biochem. 2018;45(2):733–43.

57. Yang X, Guan Y, Tian S, Wang Y, Sun K, Chen Q. Mechanical and IL-1beta responsive miR-365 contributes to osteoarthritis development by targeting histone deacetylase 4. Int J Mol Sci. 2016;17(4):436.

58. Yamasaki K, Nakasa T, Miyaki S, Ishikawa M, Deie M, Adachi N, et al. Expression of MicroRNA-146a in osteoarthritis cartilage. Arthritis Rheum. 2009;60(4):1035–41.

59. Wang H, Zhang H, Sun Q, Wang Y, Yang J, Yang J, et al. Intra-articular delivery of Antago-miR-483-5p inhibits osteoarthritis by modulating Matrilin 3 and tissue inhibitor of metalloproteinase 2. Mol Ther. 2017;25(3):715–27.

60. Hashimoto K, Oreffo RO, Gibson MB, Goldring MB, Roach HI. DNA demethylation at specific CpG sites in the IL1B promoter in response to inflammatory cytokines in human articular chondrocytes. Arthritis Rheum. 2009;60(11):3303–13.

61. Huh YH, Ryu JH, Chun JS. Regulation of type II collagen expression by histone deacetylase in articular chondrocytes. J Biol Chem. 2007;282(23):17123–31.

62. Martinek V, Fu FH, Lee CW, Huard J. Treatment of osteochondral injuries. Genetic engineering. Clin Sports Med. 2001;20(2):403–16. viii

63. Venkatesan JK, Rey-Rico A, Cucchiarini M. Current trends in viral gene therapy for human orthopaedic regenerative medicine. Tissue Eng Regen Med. 2019;16(4):345–55.

64. Evans CH, Huard J. Gene therapy approaches to regenerating the musculoskeletal system. Nat Rev Rheumatol. 2015;11(4):234–42.

65. Gobbi A, Karnatzikos G, Kumar A. Long-term results after microfracture treatment for full-thickness knee chondral lesions in athletes. Knee Surg Sports Traumatol Arthrosc. 2014;22(9):1986–96.

66. Gobbi A, Berruto M, Kon E, Filardo G, Delcogliano M, Montaperto C, et al. Patellofemoral chondral defects treated with 2nd generation ACI: a clinical review at 6 years (SS-60). Arthroscopy. 2009;25(6):e33.

67. Pascher A, Palmer GD, Steinert A, Oligino T, Gouze E, Gouze JN, et al. Gene delivery to cartilage defects using coagulated bone marrow aspirate. Gene Ther. 2004;11(2):133–41.

68. Sieker JT, Kunz M, Weissenberger M, Gilbert F, Frey S, Rudert M, et al. Direct bone morphogenetic protein 2 and Indian hedgehog gene transfer for articular cartilage repair using bone marrow coagulates. Osteoarthr Cartil. 2015;23(3):433–42.

69. Ivkovic A, Pascher A, Hudetz D, Maticic D, Jelic M, Dickinson S, et al. Articular cartilage repair by genetically modified bone marrow aspirate in sheep. Gene Ther. 2010;17(6):779–89.

70. Cucchiarini M, Madry H. Overexpression of human IGF-I via direct rAAV-mediated gene transfer improves the early repair of articular cartilage defects in vivo. Gene Ther. 2014;21(9):811–9.

71. Cucchiarini M, Madry H, Ma C, Thurn T, Zurakowski D, Menger MD, et al. Improved tissue repair in articular cartilage defects in vivo by rAAV-mediated overexpression of human fibroblast growth factor 2. Mol Ther. 2005;12(2):229–38.

72. Cucchiarini M, Orth P, Madry H. Direct rAAV SOX9 administration for durable articular cartilage repair with delayed terminal differentiation and hypertrophy in vivo. J Mol Med (Berl). 2013;91(5):625–36.

73. Brittberg M, Lindahl A, Nilsson A, Ohlsson C, Isaksson O, Peterson L. Treatment of deep cartilage defects in the knee with autologous chondrocyte transplantation. N Engl J Med. 1994;331(14):889–95.

74. Gobbi A, Kon E, Filardo G, Delcogliano M, Montaperto C, Boldrini L, et al. 2nd Gen ACI vs microfracture in knee chondral defect treatment: comparative study at 5 years (SS-61). Arthroscopy. 2009;25(6):e33–e4.

75. Peterson L, Vasiliadis HS, Brittberg M, Lindahl A. Autologous chondrocyte implantation: a long-term follow-up. Am J Sports Med. 2010;38(6):1117–24.

76. Kang R, Marui T, Ghivizzani SC, Nita IM, Georgescu HI, Suh JK, et al. Ex vivo gene transfer to chondrocytes in full-thickness articular cartilage defects: a feasibility study. Osteoarthr Cartil. 1997;5(2):139–43.

77. Goodrich LR, Hidaka C, Robbins PD, Evans CH, Nixon AJ. Genetic modification of chondrocytes with insulin-like growth factor-1 enhances cartilage healing in an equine model. J Bone Joint Surg Br. 2007;89(5):672–85.

78. Orth P, Kaul G, Cucchiarini M, Zurakowski D, Menger MD, Kohn D, et al. Transplanted articular chondrocytes co-overexpressing IGF-I and FGF-2 stimulate cartilage repair in vivo. Knee Surg Sports Traumatol Arthrosc. 2011;19(12):2119–30.

79. Ortved KF, Begum L, Mohammed HO, Nixon AJ. Implantation of rAAV5-IGF-I transduced autologous chondrocytes improves cartilage repair in full-thickness defects in the equine model. Mol Ther. 2015;23(2):363–73.

80. Ha CW, Noh MJ, Choi KB, Lee KH. Initial phase I safety of retrovirally transduced human chondrocytes expressing transforming growth factor-beta-1 in degenerative arthritis patients. Cytotherapy. 2012;14(2):247–56.

81. Evans CH, Liu FJ, Glatt V, Hoyland JA, Kirker-Head C, Walsh A, et al. Use of genetically modified muscle and fat grafts to repair defects in bone and cartilage. Eur Cell Mater. 2009;18:96–111.

82. Zaslav K, McAdams T, Scopp J, Theosadakis J, Mahajan V, Gobbi A. New Frontiers for cartilage repair and protection. Cartilage. 2012;3(1 Suppl):77S–86S.

83. Cucchiarini M, Madry H. Biomaterial-guided delivery of gene vectors for targeted articular cartilage repair. Nat Rev Rheumatol. 2019;15(1):18–29.

84. Lam J, Lu S, Kasper FK, Mikos AG. Strategies for controlled delivery of biologics for cartilage repair. Adv Drug Deliv Rev. 2015;84:123–34.

85. Rey-Rico A, Frisch J, Venkatesan JK, Schmitt G, Rial-Hermida I, Taboada P, et al. PEO-PPO-PEO carriers for rAAV-mediated transduction of human articular chondrocytes in vitro and in a human osteochondral defect model. ACS Appl Mater Interfaces. 2016;8(32):20600–13.

86. Pannier AK, Shea LD. Controlled release systems for DNA delivery. Mol Ther. 2004;10(1):19–26.

87. Glass KA, Link JM, Brunger JM, Moutos FT, Gersbach CA, Guilak F. Tissue-engineered cartilage with inducible and tunable immunomodulatory properties. Biomaterials. 2014;35(22):5921–31.

88. Yan X, Chen YR, Song YF, Yang M, Ye J, Zhou G, et al. Scaffold-based gene therapeutics for osteochondral tissue engineering. Front Pharmacol. 2019;10:1534.

89. Leng P, Ding CR, Zhang HN, Wang YZ. Reconstruct large osteochondral defects of the knee with hIGF-1 gene enhanced mosaicplasty. Knee. 2012;19(6):804–11.

90. Venkatesan JK, Gardner O, Rey-Rico A, Eglin D, Alini M, Stoddart MJ, et al. Improved chondrogenic differentiation of rAAV SOX9-modified human MSCs seeded in fibrin-polyurethane scaffolds in a hydrodynamic environment. Int J Mol Sci. 2018;19(9)

91. Yang S, Qian Z, Liu D, Wen N, Xu J, Guo X. Integration of C-type natriuretic peptide gene-modified bone marrow mesenchymal stem cells with chitosan/silk fibroin scaffolds as a promising strategy for articular cartilage regeneration. Cell Tissue Bank. 2019;20(2):209–20.

92. Moutos FT, Glass KA, Compton SA, Ross AK, Gersbach CA, Guilak F, et al. Anatomically shaped tissue-engineered cartilage with tunable and inducible anticytokine delivery for biological joint resurfacing. Proc Natl Acad Sci U S A. 2016;113(31):E4513–22.

93. Pi Y, Zhang X, Shi J, Zhu J, Chen W, Zhang C, et al. Targeted delivery of non-viral vectors to cartilage in vivo using a chondrocyte-homing peptide identified by phage display. Biomaterials. 2011;32(26):6324–32.

94. Evans CH. The vicissitudes of gene therapy. Bone Joint Res. 2019;8(10):469–71.

Livia Roseti and Brunella Grigolo

18.1 Introduction

Osteochondral lesions affect both articular cartilage and subchondral bone, i.e., the osteochondral unit. In adolescents and young adults, particularly in athletes, they are mostly of traumatic origin, often in association with ligament and meniscal injuries. Osteochondral lesions typically occur at the knee, ankle, or elbow; severity ranges from a small crack to a piece of bone breaking off within the joint. The fragments, differing in both size and depth, can stay attached (stable) to the injured area or become loose (unstable) inside the joint [1, 2]. The causes of osteochondral injuries are not yet completely understood, but current theories include the lack of blood supply to the affected area, genetic factors, direct compressive trauma, or repetitive strains [3]. The repair process is initiated by undifferentiated mesenchymal stem/stromal cells (MSCs) from the bone marrow tissue of the subchondral bone. However, in general, it does not ensure neither the healing of the defect nor the symptom remission in the long-term period. Without proper treatments, Osteoarthritis may progress with aging, mostly leading to the

need for joint prosthetic replacement [4]. Osteochondral lesion's symptoms vary according to type, size, and site of the damage, and to the involvement of the surrounding structures. Pain may appear during efforts or simple movements; acute pain can arise limiting or even blocking the joint. Diagnosis is made through magnetic resonance imaging and arthroscopy. Proper treatment depends on the patient's age, type, severity, location, and size of the damage, as well as associated symptoms [1, 5].

The treatment of osteochondral lesions represents a challenge in the orthopedic field. The first approaches are usually conservative, based on drug administration (painkillers) or physical therapies. Hyaluronic acid infiltration in the joint may decrease symptoms due to its lubricating action [6]. If conservative treatment is not effective in reducing pain and/ or functional limitations, a surgical solution is needed. There are several surgical techniques that can be performed in arthroscopy (minimally invasive surgery) or with the traditional open-air technique, depending on lesion type. Reparative approaches generally foresee subchondral bone penetration, which induces bleeding. There is then migration of bone marrow MSCs to the site of injury along with blood clot formation. Although excellent short-term clinical outcomes have been demonstrated [7], the resulting repair tissue is mainly composed of fibrocartilage. Restorative

L. Roseti · B. Grigolo (✉)
Laboratorio RAMSES, Dipartimento RIT-Research,
Innovation & Technology, Istituto di Ricerca
Codivilla Putti, IRCCS Istituto Ortopedico Rizzoli,
Bologna, Italy
e-mail: livia.roseti@ior.it; brunella.grigolo@ior.it

© ISAKOS 2022
A. Gobbi et al. (eds.), *Joint Function Preservation*, https://doi.org/10.1007/978-3-030-82958-2_18

techniques' purpose is the reconstitution of joint structure and function: arthroplasty is a widely diffused technique recommended for elderly patients. Even if an invasive procedure, it has been shown to relieve pain and improve mobility. Complications may be stiffness, instability, aseptic loosening, infection, prosthesis failure, and mal-alignment [8].

Biological, regenerative approaches, including osteochondral grafting and autologous chondrocyte implantation (ACI) aim to reconstruct the native structure and function of the damaged tissue. Autologous osteochondral transplantation (mosaicplasty) has some drawbacks like donor site morbidity and graft failure. Limitations of allograft transplantation include graft availability, possible disease transmission, and short cell viability [9]. ACI, in its first generation, consisted of a cartilage biopsy harvesting from a non-load bearing area of the articular surface, an ex vivo expansion phase of the isolated chondrocytes, and finally the implantation of a chondrocyte suspension in the damaged area [10]. This procedure has also been utilized in conjunction with bone grafting to repair osteochondral lesions of varying depths and sizes. Despite good outcomes [11], issues were related to chondrocyte phenotype loss during cell expansion, cell leakage in the joint space, the need of two interventions, and the prohibitive costs [12]. Therefore, researches have been addresses toward the evaluation of cells alternative to chondrocytes and to the development of tissue engineering techniques.

Tissue engineering applies the principles and methods of engineering, biology, chemistry, and mechanics. The aim is to develop biological substitutes capable of restoring, maintaining, or improving tissue structure and function. In particular, tissue engineering combines cells, scaffolds, and growth factors to support and enhance regeneration [13]. The innovative feature of engineered tissues is that, if successful, they integrate within the patients and do not require expensive posttreatments drug therapies like traditional transplants (i.e., immunosuppressors) [14].

18.2　Scaffolds

Scaffolds can be defined as artificial structures used to support three-dimensional (3D) tissue formation [15, 16]. An ideal scaffold should be biocompatible, nontoxic, and non-immunogenic and should resorb/degrade in vivo in tandem with tissue formation. Another key feature is bioactivity, i.e., its ability to interact with the surrounding living tissues or organs. Over the last years, many scaffolds have been developed and tested for cartilage or bone regeneration purposes [5]. Scaffolds for cartilage tissue engineering can be fabricated starting from gelatin, chitosan, hyaluronic acid, collagen, alginate, glycosaminoglycan, starch, and bacteria. Biomimetic cartilage polymers of synthetic origin are poly (D, L-lactic-co-glycolic acid), poly (caprolactone), poly (ethyleneglycol), and poly (glycolic acid). Extracellular Matrix (ECM)-based materials are obtained after decellularization or devitalization of cartilage ECM. Such scaffolds facilitate cell differentiation, but the reduced content of glycosaminoglycan and the effect of residual cell components are still unknown. Cell-derived matrix (CDM) generated from cells grown in monolayer or 3D in vitro cultures may be an option. However, the clinical application is scarce due to complex, long-lasting, and expensive processes of production [17]. Bioceramics have been utilized for cartilage regeneration but revealed poor elasticity and high stiffness. Therefore, mixtures of polymers and bioceramics have been developed [18]. Scaffold for subchondral bone tissue engineering should guarantee compressive strength and are mostly metallic materials, bioglass, and bioceramics alone or in combination with polymers [5]. Metallic materials possess excellent mechanical properties but are inert. This drawback can be overcome by coating metallic materials with bioceramic nanoparticles. Bioceramics and bioglasses could promote biomineralization because of their excellent osteoconductivity. Bioglasses and glass-ceramics are bioactive ceramics able to interact with the surrounding tissues thus favoring the osteogenic process. However, there are some drawbacks, such as low elasticity, heavy brittleness, extremely high stiffness, and poor fracture toughness. The incorporation of ductile materials, such as gelatin, collagen, chondroitin sulfate, and polylactide acid, may enhance the mechanical properties of bioceramics [17].

Currently, scaffolds suitable for osteochondral tissue engineering can be divided into native biological and synthetic polymeric materials. Generally, natural polymers favor better cell proliferation and differentiation but display weaker biomechanical properties. Diversely, synthetic polymers can be developed into different shapes, possess controlled degradation kinetics and regulated biomechanical properties [19]. The development of osteochondral scaffolds able to address both cartilage and bone reconstruction thus restoring the properties of the entire osteochondral compartment is a challenge [20, 21]. An ideal scaffold suitable for osteochondral lesion treatment should possess specific structural and mechanical features to mimic all the components of the osteochondral unit: articular cartilage layer, tidemark, and subchondral bone. Single-phase scaffolds support cartilage and/or bone growth, but do not mimic the physical structure and composition of the osteochondral unit thus making regeneration and function restoration more difficult [14]. In this perspective, multilayered, stratified, or gradient scaffolds can better mimic such a complex interface structure [5, 14]. Cell viability, attachment, proliferation and homing, zonal chondrogenic and osteogenic differentiation, vascularization in the bony part, host integration, and load-bearing ability should be all guaranteed. Mechanical features should match with those found at the site of implantation [1]. In most of the biphasic and multilayered developed scaffolds, polymers such as gelatin, collagen, and polylactide acid serve as the cartilage layer, and bioceramics including hydroxyapatite and tricalcium phosphate (TCP) serve as the subchondral bone layer [5]. The group of Levingstone utilized type I collagen and hydroxyapatite for the subchondral bone layer, type I and II collagens and hydroxyapatite for the intermediate layer, and type I and II collagens and hyaluronic acid for the articular cartilage layer [22].

18.2.1 Nanostructured Scaffolds

Nanostructured scaffolds have recently emerged as promising biomimetic candidates since cartilage and bone are typical examples of nanomaterials [23, 24]. A Nanomaterial is defined as a natural, derived, or manufactured material containing particles in the free, aggregate, or agglomerated state, and in which, at least 50% of the particles in the numerical size distribution, are characterized by one or more external dimensions between 1 nm and 100 nm (Recommendation 2011/696/EU) [25]. A nanostructured, magnesium-hydroxyapatite composite, biomimetic porous, three-layer gradient scaffold reproducing cartilaginous layer, tidemark, and subchondral bone has been seeded with MSCs. Safranin O staining and collagen type II and proteoglycans immunostaining confirmed that chondrogenic differentiation was specifically induced only in the cartilaginous layer; von Kossa staining, osteocalcin, and osteopontin immunostaining positivity demonstrated that osteogenic differentiation occurred on both intermediate and lower layers [26].

18.2.2 Three-Dimensional Printed Scaffolds

Although multilayered scaffolds have good biomimetic ability, it is difficult to biologically and mechanically reconstruct the osteochondral unit features with the traditional biotechnologies. Furthermore, the adhesive strength between adjacent layers is often insufficient, leading to delamination. Hence, a smart single-phase scaffold that possesses bilineage functions for simultaneously regenerating both cartilage and subchondral may be the solution [17]. Additive manufacturing (AM) technologies (ISO/ASTM 52900:2015 Standard Terminology for Additive Manufacturing) [27] may respond to such a need. They fabricate objects by layer or drop-by-drop deposition, combining computer-assisted design (CAD) with computer-assisted manufacturing (CAM) [28]. Translated into the clinical practice, it should be possible to fabricate complex structures, "on-demand," starting from patient's medical images acquired with noninvasive techniques like Magnetic Resonance Imaging and Computerized Tomography. A more punctual control of porosity, pore size, and mechanical and chemical properties is then possible, permitting better mimic abilities. These methods allow also for variation of the composition of two or more materials across the surface, interface, or bulk of the scaffold. Mellor et al. printed scaffolds composed of polycaprolactone (PCL) with either β-TCP or dECM. The subsequent seeding of human ASCs allowed zonal cartilage- and bone-like ECM deposition [29] in both cases.

18.3 Bio-tissues

Cells and growth factors both act like adjuvant to the scaffolds in order to develop tissue-like structures able to enhance structure and function restoration.

An ideal cell component should be viable, easily available, non-immunogenic, non-tumorigenic, phenotypically stable, and responsive to bioactive factors [1]. Recently, stem/stromal cells have raised interest in osteochondral regeneration due to their ability to differentiate in culture into both chondrogenic and osteogenic phenotypes [30]. They can in fact be expanded ex vivo and then seeded onto a scaffold (eventually added with growth factors). The resulting construct is therefore ready to be implanted in the patient. MSCs from bone marrow were the first to be investigated for tissue engineering purposes. They have been demonstrated to possess a multilineage differentiation potential [31]. This implies that, when seeded in gradient scaffolds, MSCs have the possibility to differentiate toward the chondrogenic or osteogenic phenotype, depending on the layer structure, composition, and orientation. Moreover, MSCs display immunomodulatory properties allowing an allogenic utilize to solve issues due to autologous procedures like donor site morbidity and scarce availability. Finally, MSCs possess

the ability to migrate to sites of tissue injury. This is important because recent discoveries highlighted that exogenously supplied MSCs secrete in the sites of injury soluble growth factors and cytokines that exert immunomodulatory, anti-inflammatory, and trophic effects on the patient's own resident stem cells that form the new tissue [32, 33]. Recently, adipose-derived stem cells (ASCs) have garnered attention for their similarities with bone marrow cells, in relation to morphology and immunologic phenotype. With respect to bone marrow, adipose tissue is more abundant and easier to obtain; ASCs frequency and proliferation rate are higher. Moreover, ASCs possess a protective effect on MSCs [34]. Other sources of adult stem/stromal cells investigated include muscle, synovial membrane, trabecular bone, dermis, blood, and periosteum. Other types of stem cells enclose perinatal, embryonic, and induced pluripotent stem cells that have been demonstrated to be promising. Genetically modified stem cells are under investigation as well.

Despite several investigations, the cell expansion phase is generally expensive and requires two interventions to treat patients. An alternative option is the use of "concentrates" which carry a much lesser amount of MSCs but contains the "stem cell niche" rich in other cell types and growth factors favoring tissue regeneration. The concentrate can be obtained by minimal manipulation (centrifugation and/or enzymatic treatment) of the source tissue in the operating room and loaded to the scaffold just before implantation. In the orthopedic clinical practice, an often-utilized concentrate is bone marrow concentrate (BMC) [35] which has been demonstrated in vitro to differentiate toward chondrogenic and osteogenic lineages [36]. Recently, stromal vascular fraction (SVF) from adipose tissue has been characterized and applied [37]. Both concentrates can be utilized with adjuvants like Platelet-Rich Plasma (PRP), which was derived from the patient's own blood throughout a centrifugation process. PRP contains a high concentration of platelets that, when activated, secrete growth factors and other proteins that regulate cell division, stimulate tissue regeneration, and promote healing [38]. A drawback is the difficulty to standardize concentrate and PRP preparations, due to patient variability.

Synthetic growth factors are recombinant molecules available on the market. Some of the mostly investigated for cartilage and bone healing are: Transforming Growth Factor-beta-1 (TGF-®1), Bone Morphogenetic Proteins (BMPs), Insulin-like Growth Factor-1 (IGF-1), Platelet-derived Growth Factor (PDGF), and Vascular Endothelial Growth factor (VEGF) [39]. Administration can be performed directly by infiltration or mediated by a scaffold [1]. Scaffold loading may be direct, or indirect, by loading microspheres or microparticles, which are then incorporated into the scaffold. Alternatively, a growth factor's gene can be transfected into the cells inducing them to express the corresponding protein in the environment [39]. Differently from concentrates, synthetic growth factors can be standardized but may present important drawbacks like bone ectopic formation [40].

18.3.1 Bioprinted Tissues

Scaffold-based 3D bioprinting is a further innovative development of AM techniques. It consists of the fabrication of living tissue/organ-like structures throughout the bottom-up deposition of either cell-laden droplets or cells embedding in a hydrogel, in both cases termed as "bioink" [41]. Such technology makes it possible to overcome issues present in more conventional methods like static (manual) seeding onto scaffold or dynamic seeding using bioreactors [5]. In these cases, problems are due to cell accumulation at the surface of the scaffold and to the low density in the inner part where cells tend to die because of the scarcity of nutrients. This may lead to inaccurate experimental results and consequent speculations. Differently, 3D bioprinting offers the advantage of fine control of cell spatial distribution in terms of homogeneity. Gao and colleagues bioprinted a construct composed of poly(ethylene glycol) dimethacrylate, gelatin methacrylate (PEG-GelMA), and bone marrow-derived human MSCs. The printing process

resulted in 80% cell survival. The resulting construct was mechanically stable, with a uniform cell distribution. When the scaffold was cultured with chondrogenic or osteogenic medium, cartilage and bone tissues were produced, respectively, as determined by specific gene and protein expression [42].

Further benefits from 3D bioprinting technique include the reduced production times, an increased versatility, and the possibility to work under room temperature and "solvent-free" conditions, taking advantage of the features of water-based gels such as bioinks [5]. Even 3D bioprinting allows the fabrication of custom-made products based on patient's medical images. Such options improve the match between implant and defect size thus shortening the time required for surgery and for patient recovery, and positively affecting the success of treatment. To date, the most used scaffold-based 3D bioprinting technologies are based on jetting, extrusion, and laser technology, each with advantages and disadvantages. To overcome existing issues and to obtain improved performance, hybrid cell-printing techniques have been developed [5].

18.4   Preclinical Investigation

In the literature, there are many preclinical studies aimed at investigating the feasibility of scaffold-based procedures for the treatment of osteochondral defects. In vitro testing is performed to characterize cell activity, material toxicity and immunogenicity, growth factor dose and release, and to evaluate the interaction between cells and biomaterials and ECM production. Animal models are required to better reflect the complexity of the osteochondral unit. The most common anatomical site for osteochondral lesion creation is the stifle joints (both medial and lateral condyles and trochlea). Histological, histochemical, immunohistochemical, histomorphometric, and mechanical investigations are generally performed to evaluate clinical outcomes [43].

Some studies investigated the scaffold alone, as the above described nanostructured three-layer gradient one based on magnesium-hydroxyapatite [26]. Results of the study in sheep highlighted that the composite is safe and easy to use and may represent a suitable matrix to direct the process of osteochondral regeneration [44, 45]. A further horse study evaluated the mechanical behavior of the scaffold when tested at the maximum challenge of weight loading and motion. Results obtained 2 months after surgery demonstrated good defect filling without any inflammatory reactions [44, 45]. A study in rabbit investigated a 3D bioprinted single-phase Mn-TCP-based scaffold with bilineage functions. After implantation, both cartilage and subchondral bone were reconstructed in a rabbit osteochondral defect model [17].

Other researches evaluated cell-seeded scaffolds. For example, a biphasic composite sponge scaffold based on collagen and poly(DL-lactic-co-glycolic acid) (PLGA) was seeded with canine bone marrow MSCs and implanted in the knee osteochondral defect of a 1-year-old beagle. An osteochondral-like tissue was regenerated 4 months after implantation. An alginate-based hydrogel bioprinted by a distinct encapsulation first with human osteogenic progenitors and then with chondrocytes was implanted in mice. Results evidenced that the cells stayed in their compartment remaining viable and separately producing cartilage or bone [46].

Other investigations concerned the effect of growth factors. In a study, critical size osteochondral defects (10×6 mm) were created in medial femoral condyles of Goettingen minipigs. Animals were randomized into four groups: the first, control group, treated with the scaffold (a blend of poly DL-lactide-co-glycolide, calcium sulfate, polyglycolide fibers) alone, the second with the scaffold added with PRP, the third with scaffold with BMC, and the fourth with scaffold in combination with BMC and PRP. After 26 weeks the authors found that the addiction of BMC or PRP led to a significant improvement of the histological score compared to the control group, but the combination of BMC and PRP did not further enhance the histological score [47].

The diversity of tested protocols (scaffold alone, cells, concentrates, etc.), animal models, experimental times, and evaluation methods makes a comparison between studies difficult. The progress of histological and image analysis techniques has increased the level of investigations, but, at the same time, has increased the number of evaluation types. In general, it can be assessed that the use of comparable methods and protocols for biomechanical tests will be essential to make the obtained results more comparable and reliable. Importantly, the peculiarity of the osteochondral unit makes evaluations quite complex. In particular, post-explant biomechanical tests are generally conducted on one tissue rather than another, instead of on both [43].

18.5 Clinical Applications

Among the many different scaffolds developed to reproduce the osteochondral unit, only a few have been investigated in clinical studies [48].

The clinical trials utilizing cell-free scaffolds highlighted, in general, good outcomes, but further confirmation is needed. In a multicenter randomized controlled clinical trial, patients affected by symptomatic chondral and osteochondral knee lesions were treated with a nanostructured multilayer scaffold constituted by collagen and magnesium-enriched hydroxyapatite and evaluated for up to 2 years. Bone marrow stimulation was used as a reference intervention. A statistically significant improvement of all clinical scores was obtained in both groups, although no overall statistically significant differences were detected between the two treatments [49]. The same scaffold was utilized in a population affected by early OA of the knee. Clinical outcomes improved; MRI analysis showed integration of the scaffold only in 47% of the patients, with partial regeneration of the subchondral bone. No correlation between clinics and radiological images was found [50]. In another study, two different shapes (cylindrical and tapered) of a crystalline aragonite biphasic scaffold were tested in patients affected by focal chondral-osteochondral knee lesions of the condyle and trochlea. A statistically significant improvement in all clinical scores was documented in both groups, without any differences, except for revision rate, which was lower in the tapered implant group [51]. A three-layered scaffold mimicking the entire osteochondral unit was tested in patients affected by isolated large osteochondral knee lesions. An improvement was observed in all the assessed scores [52]. A retrospective therapeutic study was conducted in order to assess the effectiveness of the combination of microfracture and cell-free hyaluronic acid-based scaffold in the treatment of talus osteochondral defects smaller than 1.5 cm^2 and deeper than 7 mm. Score results were positive highlighting that the investigated combination appears to be a safe and efficient technique [53]. Another retrospective study was conducted to evaluate the long-term outcomes of a biphasic scaffold for the treatment of osteochondral lesions of the talus. The authors concluded that postoperative scores were good, but randomized controlled clinical trials comparing established treatment methods were needed [54].

Other clinical trials utilized stem cells and growth factors. Hyaluronic acid and bone marrow aspirate concentrate was administered to treat full-thickness cartilage lesions of the knee associated with significant subchondral bone loss. The technique, a novel one-step procedure [55] gave good results. Another population of patients who had osteochondral lesions of the knee was treated with a hyaluronic acid-based scaffold filled with bone marrow concentrate and covered by a layer of platelet-rich fibrin. The clinical imaging and the histological results were satisfactory in terms of tissue healing. Giannini et al. applied a similar treatment to repair talar osteochondral lesions, obtaining comparable results [56, 57]. Young, active patients with knee full-thickness cartilage or osteochondral defect, including traumatic or atraumatic full-thickness knee cartilage defects or osteochondritis dissecans, were treated with a novel autologous-made matrix consisting of hyaline cartilage chips combined with mixed plasma poor rich in platelets clot and plasma rich in growth factors. Excellent clinical, functional, and MRI-based outcomes were observed.

Recently, an international consensus group composed of a panel of expert orthopedic clinicians and researchers met to identify a series of indications on the use of scaffold-based procedures for the treatment of chondral and osteochondral knee defects. As a conclusion of their work, they considered the use of scaffold-based procedures as appropriate in almost all cases of chondral or osteochondral lesions in non-OA knees. All experts agreed on the contraindication for the use of scaffolds in advanced OA, and on the importance to define potential and limitations within its earlier phases [58]. The consensus group highlighted that the difficulty to find evidence on the superiority of one scaffold-based procedure over the others and they did not provide indications regarding the best product or process. The use of osteochondral scaffold was considered uncertain in case of small lesions and appropriate for bigger lesions, especially in younger patients.

18.6 Future Perspectives

Advancements in the complex and multidisciplinary strategies of tissue engineering and the development of 3D tailored tissue-like substitutes have proved to be particularly interesting and promising for the progress and exploitation of therapies specialized in the treatment of osteochondral defects. Personalized tissue engineering strategies are envisioned to evolve into more effective and successful 3D templates with regenerative action. Advances in 3D bioprinting technologies will allow moving from time- and labor-intensive fabrication technologies into mass production of patient-tailored tissue and organ substitutes, which will lead to a cost reduction and easier access to these technologies in medical institutions. It is also expected that 3D bioprinting advances will influence the progress of imaging technologies. This will result in improved equipment and software to translate tissue scans with a higher level of detail and information into a virtual 3D model.

The possibility to combine multiple biomaterials, cells, or growth factors within the same 3D template will allow the regeneration of complex biological systems such as the osteochondral unit that requires specific structures that fuse with the nearby tissue with complementary functionality. The combination of all these developments with the possibility to cryobank cells and more complex systems as tissue-engineered constructs assists in the translation of effective off-the-shelf strategies involving custom-made products available upon request. Thus, the advances described are paving the way for enhanced personalized treatments searching for innovative and effective solutions to promote real tissue regeneration meeting individual patient requirements and needs.

Within the next years, it is expected that orthopedics will evolve toward the use of products customized to the individual characteristics of each patient. Ongoing technological advances will exponentially increase the level of detail and information, bringing new knowledge and the need to provide more precise diagnosis and pathology management. Scientific developments and clinical trials will help to understand and guide personalized strategies toward a successful clinical scenario. It is envisioned that, in a relatively short period, this kind of technology may become available in clean rooms close to the surgical theater of some leading-edge hospitals to assist grafting or replacement surgeries with customized 3D scaffolds. Undoubtedly, in the next years, several hurdles will remain, but surely, several of these approaches and technologies will be one step closer to meeting the enthusiastic challenges of personalized medicine and revolutionize the therapeutic field with custom-made therapies and effective tailored treatments for a wide range of pathologies.

18.7　Conclusions

This chapter focuses on the developments of osteochondral tissue engineering, scaffolds, and bio-tissues and their application in orthopedics. It outlines the essence of preclinical and clinical research conducted to find new solutions, create, and establish new specific treatments. Scaffolds, cell sources, and growth factors have extensively grown in the last two decades, having a positive effect on the expansion of truly functional engineered tissue.

The importance of personalized medicine in tissue engineering has been recognized for leading to the customization of scaffold architectures that should perfectly fit individual tissue defects. Tailoring the treatment to the patient improves outcome and recovery time and decreases the health care and social co-lateral costs associated with ineffective or inadequate approaches.

The combination of various scientific disciplines is necessary to further develop appropriate new materials or blending the properties of synthetic and biologic matrices in an ideal way.

References

1. Roseti L, Grigolo B. Chapter 7: Host environment: scaffolds and signaling (tissue engineering) articular cartilage regeneration: cells, scaffolds, and growth factors. In: Gobbi A, Espregueira-Mendes J, Lane J, Karahan M, editors. Bio-orthopaedics. Berlin: Springer; 2017. p. 87–103. ISBN 978-3-662-54180-7.
2. Seo SG, Kim JS, Seo DK, et al. Osteochondral lesions of the talus. Acta Orthop. 2018;89:462–7.
3. Durur-Subasi I, Durur-Karakaya A, Yildirim OS. Osteochondral lesions of major joints. Eurasian J Med. 2015;47:138–44.
4. Lane NE, Shidara K, Wise BL. Osteoarthritis year in review 2016: clinical. Osteoarthr Cartil. 2017;25:209–15.
5. Roseti L, Parisi V, Petretta M, et al. Scaffolds for bone tissue engineering: state of the art and new perspectives. Mater Sci Eng C. 2017;78:1246–62.
6. Ahmed TA, Hincke MT. Strategies for articular cartilage lesion repair and functional restoration. Tissue Eng Part B Rev. 2010;16:305–29.
7. Gobbi A, Karnatzikos G, Kumar A. Long-term results after microfracture treatment for full-thickness knee chondral lesions in athletes. Knee Surg Sports Traumatol Arthrosc. 2014;22(9):1986–96.
8. Harris JD, Siston RA, Pan X, et al. Autologous chondrocyte implantation: a systematic review. J Bone Joint Surg Am. 2010;92:2220–33.
9. Giannini S, Buda R, Ruffilli A, et al. Failures in bipolar fresh osteochondral allograft for the treatment of end-stage knee osteoarthritis. Knee Surg Sports Traumatol Arthrosc. 2015;23:2081–9.

10. Brittberg M, Lindahl A, Nilsson A, et al. Treatment of deep cartilage defects in the knee with autologous chondrocyte transplantation. N Engl J Med. 1994;331:889–95.

11. Krill M, Early N, Everhart J, et al. Autologous chondrocyte implantation (ACI) for knee cartilage defects: a review of indications, technique, and outcomes. JBJS Rev. 2018;6:e5.

12. Andriolo L, Merli G, Filardo G, et al. Failure of autologous chondrocyte implantation. Sports Med Arthrosc Rev. 2017;25:10–8.

13. Atala A, Kasper FK, Mikos AG. Engineering complex tissues. Sci Transl Med. 2012;4:160rv12.

14. Nukavarapu SP, Dorcemus DL. Osteochondral tissue engineering: current strategies and challenges. Biotechnol Adv. 2013;31:706–21.

15. Bouet G, Marchat D, Cruel M, et al. In vitro three-dimensional bone tissue models: from cells to controlled and dynamic environment. Tissue Eng B Rev. 2015;21:133–56.

16. Navarro M, Michiardi A, Castaño O, et al. Biomaterials in orthopaedics. J R Soc Interface. 2008;5:1137–58.

17. Deng C, Yao Q, Feng C, Li J, et al. 3D printing of bilineage constructive biomaterials for bone and cartilage regeneration. Adv Funct Mater. 2017;27:1703117.

18. Xue D, Zheng Q, Zong C, et al. Osteochondral repair using porous poly(lactide-co-glycolide)/nanohydroxyapatite hybrid scaffolds with undifferentiated mesenchymal stem cells in a rat model. J Biomed Mater Res. 2010;94A:259e70.

19. Deng C, Chang J, Wu C. Bioactive scaffolds for osteochondral regeneration. J Orthop Transl. 2019;17:15e25.

20. Madry H, van Dijk CN, Mueller-Gerbl M. The basic science of the subchondral bone. Knee Surg Sports Traumatol Arthrosc. 2010;18:419–33.

21. Schinhan M, Gruber M, Vavken P, et al. Critical-size defect induces unicompartmental osteoarthritis in a stable ovine knee. J Orthop Res. 2012;30:214–20.

22. Levingstone TJ, Matsiko A, Dickson GR, et al. A biomimetic multi-layered collagen-based scaffold for osteochondral repair. Acta Biomater. 2014;10:1996–2004.

23. Florencio-Silva R, Rodrigues da Silva Sasso G, Sasso-Cerri E, et al. Biology of bone tissue: structure, function, and factors that influence bone cells. Biomed Res Int. 2015;2015:421746.

24. Moran CJ, Pascual-Garrido C, Chubinskaya S, et al. Restoration of articular cartilage. J Bone Joint Surg. 2014;96:336–44.

25. European commission. Recommendation on the definition of a nanomaterial (2011/696/EU); 2011

26. Manferdini C, Cavallo C, Grigolo B, et al. Specific inductive potential of a novel nanocomposite biomimetic biomaterial for osteochondral tissue regeneration. J Tissue Eng Regen Med. 2013;10:374–91.

27. ISO/ASTM 52900:2015 Standard Terminology for Additive Manufacturing -General Principles- Terminology-Section 2.2 Process categories.

28. Ventola CL. Medical applications for 3D printing: current and projected uses. Pharmacol Ther. 2014;39:704–11.

29. Mellor F, Nordberg RC, Huebner P, et al. Investigation of multiphasic 3D-bioplotted scaffolds for site-specific chondrogenic and osteogenic differentiation of human adipose-derived stem cells for osteochondral tissue engineering. J Biomed Mater Res B Appl Biomater. 2019; https://doi.org/10.1002/jbm.b.34542.

30. Li X, Wang M, Jing X, et al. Bone marrow- and adipose tissue-derived mesenchymal stem cells: characterization, differentiation, and applications in cartilage tissue engineering. Crit Rev Eukaryot Gene Expr. 2018;28:285–310.

31. Caplan AI. Mesenchymal stem cells. J Orthop Res. 1991;9:641–50.

32. Caplan AI. Mesenchymal stem cells: time to change the name! Stem Cells Transl Med. 2017;6:1445–51.

33. Caplan AI. There is no "stem cell mess". Tissue Eng B Rev. 2019;25:291–3.

34. Rehman J, Traktuev D, Li J, et al. Secretion of angiogenic and antiapoptotic factors by human adipose stromal cells. Circulation. 2004;109:1292–8.

35. Grigolo B, Cavallo C, Desando G, et al. Novel nanocomposite biomimetic biomaterial allows chondrogenic and osteogenic differentiation of bone marrow concentrate derived cells. J Mater Sci Mater Med. 2015;26:173.

36. Cavallo C, Desando G, Cattini L, et al. Bone marrow concentrated cell transplantation: rationale for its use in the treatment of human osteochondral lesions. J Biol Regul Homeost Agents. 2013;27:165–75.

37. Bianchi F, Maioli M, Leonardi E, et al. A new non-enzymatic method and device to obtain a fat tissue derivative highly enriched in Pericyte-like elements by mild mechanical forces from human Lipoaspirates. Cell Transplant. 2013;22:2063–77.

38. Shahid M, Kundra R. Platelet-rich plasma (PRP) for knee disorders. EFORT Open Rev. 2017;2:28–34.

39. Fortier LA, Barker JU, Strauss EJ, et al. The role of growth factors in cartilage repair. Clin Orthop Relat Res. 2011;469:2706–15.

40. Tachi K, Takami M, Sato H, et al. Enhancement of bone morphogenetic protein-2-induced ectopic bone formation by transforming growth factor-β1. Tissue Eng Part A. 2011;17:597–606.

41. Schubert C, van Langeveld MC, Donoso LA. Innovations in 3D printing: a 3D overview from optics to organs. Br J Ophthalmol. 2014;98:159–61.

42. Gao G, Schilling AF, Hubbell K, et al. Improved properties of bone and cartilage tissue from 3D inkjet-bioprinted human mesenchymal stem cells by simultaneous deposition and photocrosslinking in PEG-GelMA. Biotechnol Lett. 2015;37:2349–55.

43. Maglio M, Brogini S, Pagani S, et al. Current trends in the evaluation of osteochondral lesion treatments: histology, histomorphometry, and biomechanics in preclinical models. BioMed Res Int. 2019;4040236:27.

44. Kon E, Delcogliano M, Filardo G, et al. Orderly osteochondral regeneration in a sheep model using

a novel nano-composite multilayered biomaterial. Orthop Res. 2010a;28:116–24.

45. Kon E, Mutini A, Arcangeli E, et al. Novel nano-structured scaffold for osteochondral regeneration: pilot study in horses. J Tissue Eng Regen Med. 2010b;4:300–8.

46. Fedorovich NE, Schuurman W, Wijnberg HM, et al. Biofabrication of osteochondral tissue equivalents by printing topologically defined, cell-laden hydrogel scaffolds. Tissue Eng C Methods. 2012;18:33–44.

47. Betsch M, Schneppendahl J, Thuns S, et al. Bone marrow aspiration concentrate and platelet rich plasma for osteochondral repair in a porcine osteochondral defect model. PLoS One. 2013;8:e71602.

48. Kon E, Filardo G, Perdisa F, et al. Clinical results of multilayered biomaterials for osteochondral regeneration. J Exp Orthop. 2014;1:10.

49. Kon E, Filardo G, Brittberg M et al (2018) A multilayer biomaterial for osteochondral regeneration shows superiority vs microfractures for the treatment of osteochondral lesions in a multicentre randomized trial at 2 years. Knee Surg Sports Traumatol Arthrosc 26:2704–2715.

50. Condello V, Filardo G, Madonna V, et al. Use of a biomimetic scaffold for the treatment of osteochondral lesions in Early osteoarthritis. BioMed Res Int. 2018;2018:7937089.

51. Kon E, Robinson D, Verdonk P, et al. A novel aragonite-based scaffold for osteochondral regeneration: early experience on human implants and technical developments. Injury. 2016;47:S27–32.

52. Berruto M, Delcogliano M, de Caro F, et al. Treatment of large knee osteochondral lesions with a biomimetic scaffold: results of a multicenter study of 49 patients at 2-year follow-up. Am J Sports Med. 2014;42:1607–17.

53. Yontar NS, Aslan L, Can A, et al. One step treatment of talus osteochondral lesions with microfracture and cell free hyaluronic acid based scaffold combination. Acta Orthop Traumatol Turc. 2019;53:372e375.

54. Di Cavea E, Versaria P, Sciarretta F, et al. Biphasic bioresorbable scaffold (TruFit plug®) for the treatment of osteochondral lesions of talus: 6- to 8-year follow-up. Foot. 2017;33:48–52.

55. Sadlik B, Gobbi A, Puszkarz M, et al. Biologic inlay osteochondral reconstruction: arthroscopic one-step osteochondral lesion repair in the knee using Morselized bone grafting and hyaluronic acid-based scaffold embedded with bone marrow aspirate concentrate. Arthrosc Tech. 2017;6(2):e383–9.

56. Buda R, Vannini F, Cavallo M, et al. Osteochondral lesions of the knee: a new one-step repair technique with bone-marrow-derived cells. J Bone Joint Surg Am. 2010;92:2–11.

57. Giannini S, Buda R, Vannini F, et al. One-step bone marrow-derived cell transplantation in talar osteochondral lesions. Clin Orthop Relat Res. 2009;467(12):3307–20.

58. Filardo G, Andriolo L, Angele P, et al. Scaffolds for knee chondral and osteochondral defects: indications for different clinical scenarios. A Consensus Statement. Cartilage I-II; 2020.

3D Bioprinting of the Osteochondral Unit

19

Shanmugasundaram Saseendar
and Saseendar Samundeeswari

> "What is now proved was once only imagined"
> –William Blake, Poet and Painter

19.1 Introduction

A large osteochondral defect in a young patient represents one of the difficult and frustrating clinical scenarios for both the patient and the orthopedic surgeon. Articular cartilage carries poor healing potential and hence damages fail to heal spontaneously and lead to progressive impairment of joint structure and function [1].

Treatment options for cartilage injury have included palliative care with analgesics and anti-inflammatory drugs, lifestyle modifications, and reparative strategies that aim to fill the defect. But the newly formed repair tissue has mostly exhibited features of fibrous tissue that does not possess the unique characteristics of hyaline cartilage, namely resistance to shear, compression, and load.

Regenerative techniques including autologous or allogeneic osteochondral grafting and autologous chondrocyte implantation (ACI) aim at regenerating cartilage tissue with structural and functional features equivalent to the native cartilage [1, 2]. However, these procedures have certain limitations: donor site morbidity, graft failure, risk of disease transmission for allografts, limitations due to size/ volume of cartilage defect, need for two interventions, prolonged recovery, and increased costs.

Of late, tissue engineering has emerged as a promising strategy for cartilage regeneration [2]. The idea of tissue engineering emerged just over 30 years ago with the development of elastic cartilage in the shape of the human ear by seeding cow cartilage cells into a biodegradable ear-shaped mold [3]. Though tissue engineering has advanced since its introduction, this technology involves the use of uniform biomaterial scaffolds in order to replicate the parent tissue. Hence, this concept could never get close to replicating the vast heterogeneity and anisotropy in the anatomical structure and biomechanical properties of any known human tissue.

19.2 The Osteochondral Unit

Synovial joints are complex structures that permit near frictionless motion between bones. The bone ends are lined by hyaline cartilage, which has the unique ability to withstand high compressive load and shear forces. The compressive load is also shared by the subchondral bone—both functioning together to form the *osteochondral unit*. This dynamic relationship between carti-

S. Saseendar (✉)
Department of Orthopaedics, Apollo Hospital,
Muscat, Sultanate of Oman

S. Samundeeswari
Department of Orthopaedics, CARE Sports
Injury Clinic, Puducherry, India

© ISAKOS 2022
A. Gobbi et al. (eds.), *Joint Function Preservation*, https://doi.org/10.1007/978-3-030-82958-2_19

lage and bone helps to maintain a good joint function and integrity [4].

19.2.1 The Cartilage

Without blood vessels, nerves, and lymphatics and with only one cell-type, hyaline cartilage appears to be the easiest tissue to print in the laboratory. However, despite its simple appearance, cartilage is a tissue that exhibits great heterogeneity. Articular cartilage can be subdivided into three zones [5] (Fig. 19.1): the *superficial zone*, with the highest cell density, the lowest amount of glycosaminoglycans (GAGs), and the lowest biosynthetic activity; and the *middle* and the *deep zones* that show a progressive decrease in cell density and increase in the amount of GAGs with increasing depth, the greatest amount of GAGs, and the lowest cell density being in the deep zone. A high concentration of GAGs means a higher compressive modulus of the tissue, which is therefore highest in the deep zone. The cell morphology also varies in the different zones: the chondrocytes are smaller and flattened in the superficial zone, while they are larger and round in the deep zone.

Furthermore, the collagen fibers have a typical arcade-like alignment—originating vertically from the deep zone in a direction perpendicular to the articular surface and arching in the middle zone to become parallel to the articular surface in the superficial layer. The microstructural arrangement of the cells, GAGs, and the collagen fibrils in addition to proteoglycan aggregates between the fibrils contribute to the unique characteristics of the hyaline cartilage that include increased compressive stiffness, resilience, and shear resistance. There are also different types of proteins present in the articular cartilage (Fig. 19.1)—e.g., clusterin, proteoglycan-4 (PRG4), Lubricin, Del-1 in the superficial zone; cartilage intermediate layer protein (CILP) in the middle zone and cartilage oligomeric matrix protein (COMP) in the middle and deep zones. Thus, cartilage is a tissue that is much more complex than was initially thought.

19.2.2 The Subchondral Bone

The osteochondral unit is composed of hyaline cartilage connected through a zone of calcified cartilage to the subchondral cortical bone known as the subchondral plate, which gives way to metaphyseal trabecular bone [4, 6] (Fig. 19.1). The distinct histological boundary between hyaline and calcified cartilage is known as the *tidemark*. Duncan et al. [7] define the *subchondral plate* as a zone which separates the articular cartilage from the marrow cavity. It normally consists of two layers: the calcified region of the articular cartilage and a layer of lamellar bone, whereas the term *subchondral bone plate* or *subchondral zone* is used to refer to the layer of lamellar bone.

The line separating the calcified zone from the subchondral bone plate is the *cement line*. The subchondral bone plate varies in thickness and density depending on the joint and the region in the joint. The trabeculae beneath this layer is the *supporting trabeculae* and together with the subarticular spongiosa forms the *subchondral bone*.

The subchondral plate serves as an important support to the overlying articular cartilage [7]. It also absorbs most of the mechanical stress that is transmitted [8, 9]. Cartilage and bone act in concert, the former as the bearing and the latter as the structural girder and shock absorber [10]. Thus, each of these anatomically closely related tissues is affected by any alteration in the mechanical properties of the other.

Osteochondral defects arise in adults as a result of acute trauma to the cartilage and underlying bone or in association with meniscal/ligament tears or can be a result of osteochondritis dissecans in young, active children and adults [4].

The focal damage to the osteochondral unit initiates a cascade of repair and remodeling that turns out to be detrimental than protective to the health and function of the joints, leading to progressive osteoarthritis [11]. Thus, osteoarthritis is not a disease of just the cartilage but also the subchondral bone [7].

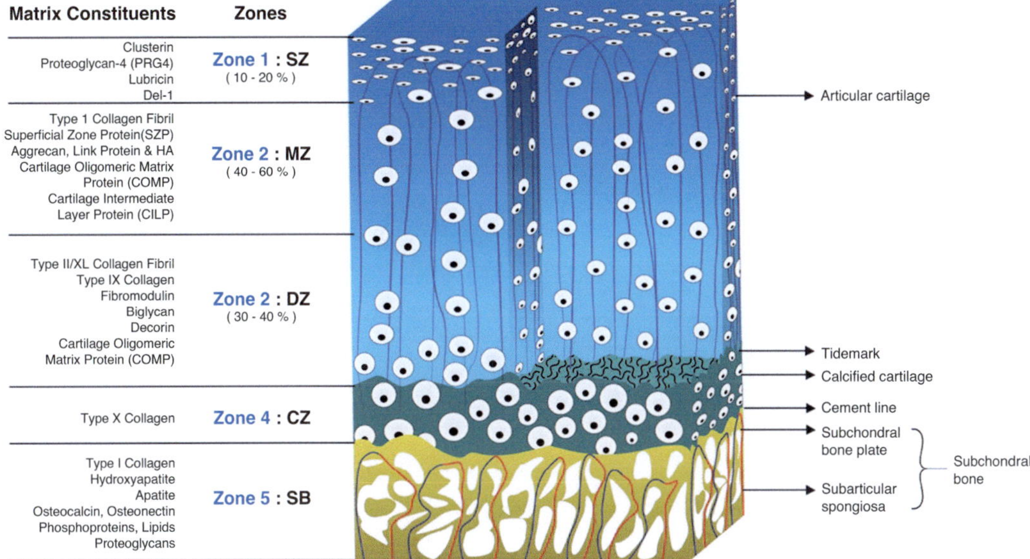

Fig. 19.1 Schematic illustration of the different zones of the osteochondral unit along with the matrix constituents in the different zones of the cartilage

19.3 Understanding the Principles of 3D Bioprinting

There are three types of **manufacturing** in industrial jargon (Fig. 19.2). An understanding of these is essential to understand 3D bioprinting.

1. **Formative manufacturing** is a process where a liquid material is injected or poured into a mold and allowed to cool. This method is mostly used in the creation of dental implants. The liquid material solidifies into the shape provided by the mold.

2. **Subtractive manufacturing** involves removing parts of a solid material in order to create an object of a particular size and shape and generally depend on a top-to-bottom approach. In tissue engineering (TE), this technique is used in creating scaffolds of a predetermined size and shape, following which particular cell types could be instilled for culture. The problem with this technique is the random cell distribution within the scaffold with lower density in the inner parts [12]. So also, it is not possible to replicate the heterogeneity and cells and extracellular substances in the created tissue, e.g., cartilage.

3. **Additive manufacturing**, on the other hand, involves creation of the final product from the bottom-up through layer-by-layer addition of material and offers the potential to fabricate tissue constructs with large heterogeneity.

3D bioprinting can be defined as a procedure of synchronous printing of biomaterials and living cells for biological applications. It works by additive manufacturing. The technique overcomes the limitations of scaffold-based tissue engineering technologies, such as restriction of structural complexity and spatial heterogeneities, by printing bioink, with or without cells, layer-by-layer, in a scaffold-free fashion to mimic the natural structure of the target tissue.

The time required to create the biological product is dramatically reduced by this technique, in comparison to scaffold-based bioengineering techniques. Hence, it is also called **rapid prototyping** [13].

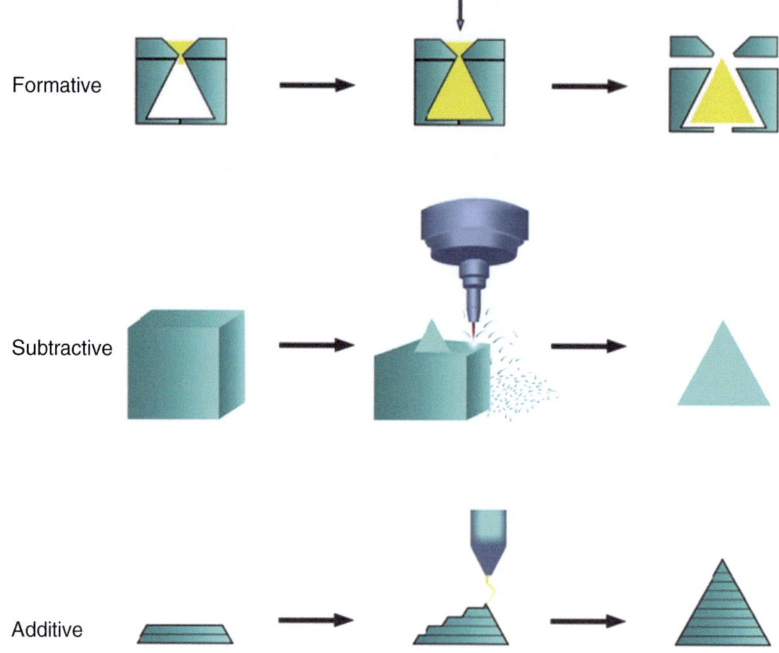

Fig. 19.2 Schematic illustration of the types of manufacturing—formative, subtractive, and additive

19.4 Steps of 3D Bioprinting

Although *bioprinting* appears to be a subfield of *3D printing*, the printing mechanisms are substantially different as cells cannot survive the rigorous processes of most 3D-printing procedures. (Advancements in digital control and highly precise positioning of cells and scaffold materials have made it feasible to fabricate engineered living organs and tissues through high throughput bioprinting technology.) It happens in three distinct phases [14]:

1. *The pre-processing phase* is the planning phase. 3D images are obtained through one of many ways—a 3D scanner, computed tomography, magnetic resonance, 3D ultrasound, or video system. The resolution is highest with a 3D MRI. These 3D images are then converted to 2D patterns using computer-aided designing (CAD) softwares to enable bioprinting through a layer-by-layer bottom-up fashion with precise deposition of biomaterials and cells.

2. *The processing phase* involves actual construction and manufacturing of the bioprinted tissue. This phase is influenced by the specific printing method used and the combination of materials—bioink, scaffold, additives used for printing.

3. *The post-processing phase* converts the bioprinted tissue into a fully mature tissue ready for in vivo usage, mostly with the help of a bioreactor.

Thus, bioprinting enables organ and tissue fabrication by precise deposition of specific cells and biomaterial to deliver customized living objects with complex structure.

19.5 Approaches to Bioprinting

There are three central approaches to bioprinting—biomimicry, autonomous self-assembly, and microtissue-based method. Their usage depends on the type of target tissue, user experience, and the printing technique used. The strategies may be used in combination for complex tissue types.

1. ***Biomimicry*** follows the principle of ***"function follows form"*** to engineer each individual component of the native tissue [14]. However, even the simplest of tissues has the most staggering complexity with numerous cell types, signaling molecules, and structural elements and external environmental factors including pressure, temperature, and electrical forces associated with the tissue. Choosing an appropriate scaffold material and the use of bioreactors minimizes the complexities by recreating the structural and mechanical properties of the target tissue.

2. ***Autonomous self-assembly*** makes use of embryonic elements that will self-organize and interact to develop into a normal tissue [15]. Hence, this method does not rely on the usage of a scaffold. This method can produce tissues of high cellular density, improved cellular interactions, accelerated growth, and improved long-term function.

3. ***Microtissue-based method*** relies on the fact that a typical complex tissue is composed of many simpler units. Tissue engineering techniques are used to form the smallest structural and functional units (minitissues) which are incorporated into the bioink and printed. Further consolidation to the final target tissue (macrotissue) occurs by the biomimetic or self-assembly strategies [15].

19.6   Techniques of 3D Bioprinting

The most commonly used scaffold-based 3D bioprinting technologies are inkjet bioprinting, extrusion bioprinting, and laser-assisted bioprinting [14, 16] (Fig. 19.4).

Inkjet bioprinting (Fig. 19.3), one of the oldest methods, uses a noncontact technique that uses thermal, piezoelectric, or electromagnetic forces to expel discrete droplets of bioink layer-by-layer onto a substrate to create a 3D structure. Advantages include high speed and low costs due to similarity with commercial printers.

Inkjet

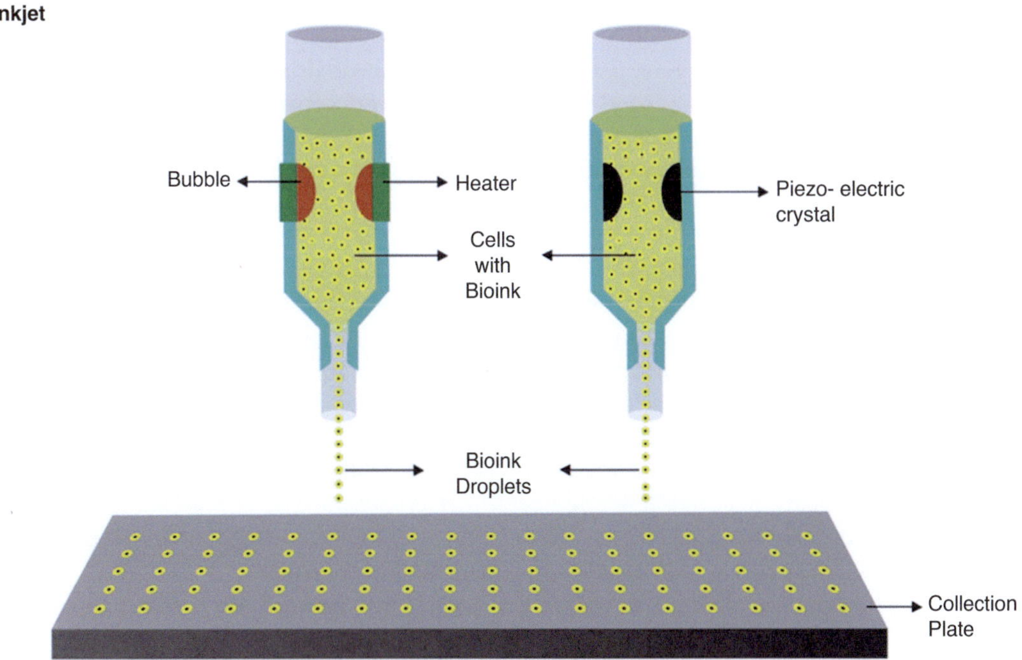

Fig. 19.3 Schematic illustration of Inkjet bioprinting technique—bubbles of bioink created and expelled through thermal heaters or piezoelectric crystals onto a collection plate

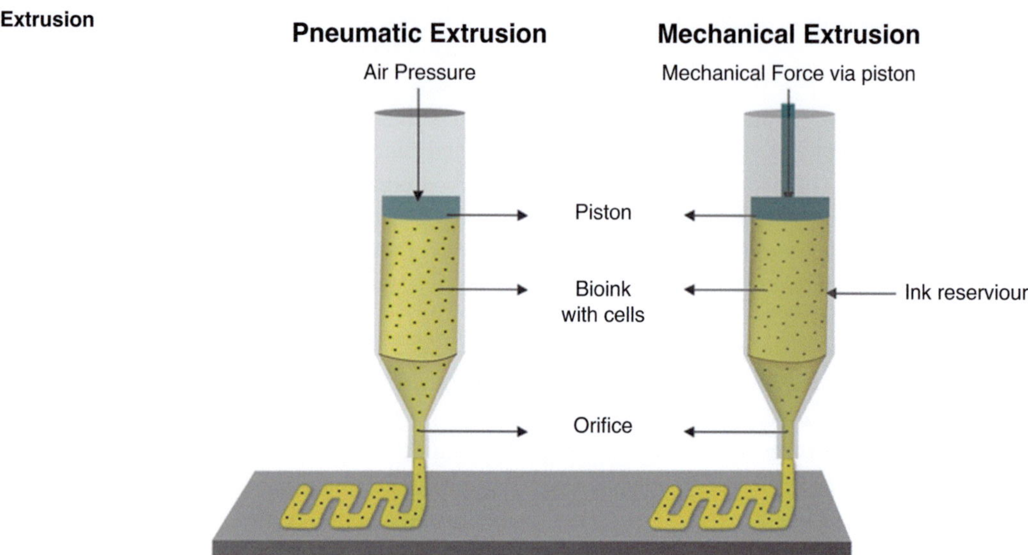

Fig. 19.4 Schematic illustration of Microextrusion bioprinting technique—pneumatic or mechanical force dispense bioink through a nozzle onto a collection plate

However, it lacks precision in droplet size and placement compared to other methods. Inkjet bioprinting is also limited to low viscosity biomaterials and low cell densities due to the risk of nozzle clogging and cellular distortion at higher densities [14, 16].

Microextrusion bioprinting (Fig. 19.4), the most commonly used method, has been utilized to fabricate heterogeneous scaffolds for osteochondral regeneration. It produces a continuous stream of bioink through a nozzle onto a stage through a pneumatic or mechanical extrusion system. It is suitable for most bioink of higher viscosities (e.g., complex polymers and cell spheroids) and for high cell densities. Limitations include low resolution and loss of cellular viability due to the deformation of cells due to the pressure of mechanical extrusion [16, 17].

Laser-assisted bioprinting (Fig. 19.5) is an expensive and complex, noncontact, nozzle-free printing technique where high-energy laser pulses are directed through a ribbon containing bioink to transfer materials to a receiving substrate.

Although this technique can produce tissues with higher resolution, the printing speed is slow, risking dehydration and lowering cellular viability [17].

19.7 Scaffolds, Hydrogels, and Bioinks

Scaffolds are materials that have been engineered to cause desirable cellular interactions to contribute to the formation of new functional tissues. Scaffolds mimic the ECM of the native tissue and create an in vivo-like microenvironment mimicking biological entities and stimulating cell-specific responses to lead to tissue regeneration and repair [18, 19]. Cells are often "seeded" into these structures in order to result in three-dimensional tissue formation.

Hydrogels, a type of scaffold, are moldable polymers that can absorb water thousands of times their dry weight.

Bioink is a material that is printed in layer-by-layer fashion. It consists of a particular combina-

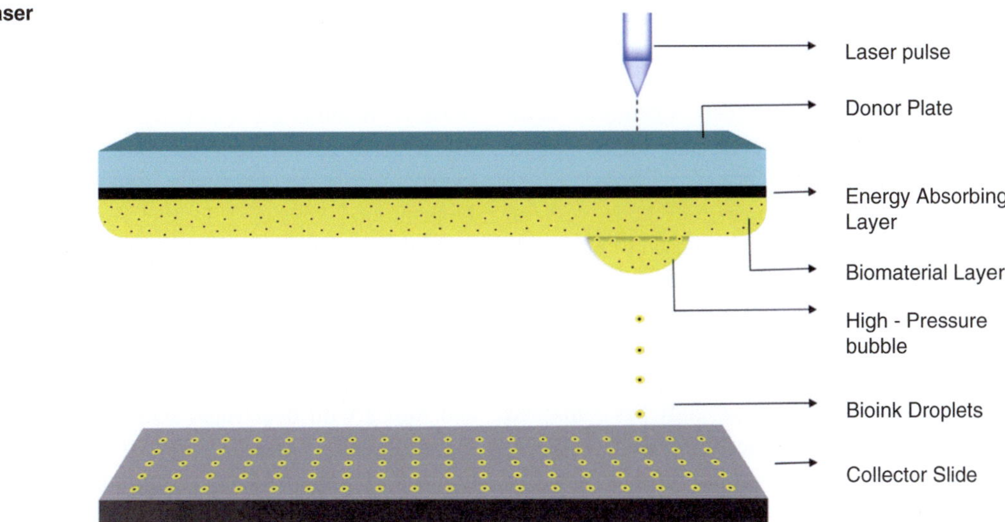

Fig. 19.5 Schematic illustration of Laser-assisted bioprinting technique—bioink with cells suspended on a ribbon get propelled to a collection plate when vaporized by a laser pulse

tion of cells or tissue spheroids, additives (e.g., growth factors, signaling molecules), and scaffold. The specific property of the bioink however depends on the printed tissue and the bioprinting approach or technique. There are two main bioink types:

Scaffold-based bioink is composed of cells and scaffold biomaterials which are released together in order to produce a construct. The scaffold creates an environment suitable for cell growth and differentiation [2, 16]. This is the older and most common method.

Scaffold-free bioink is characterized by aggregates such as cell pellets, tissue strands, or spheroids which secrete extracellular matrix (ECM)-like structures holding together the cell component [20].

Autonomous self-assembly and microtissue-based approaches can use bioink without scaffold. Inkjet bioprinting requires a bioink with low viscosity and low thermal conductivity to avoid nozzle clogging and heat damage, respectively. Microextrusion bioprinting can tolerate higher viscosities.

Both methods have advantages and limitations and at best complement each other to help cover the broad spectrum of tissue engineering/regenerative medicine applications [21].

19.8 3D Bioprinting the Osteochondral Unit

Initial attempts to create bilayered grafts for osteochondral tissue regeneration started with *3D printing* with conventional scaffold fabrication techniques, rather than with *3D bioprinting*. Polymeric scaffolds like polylactide (PLA) and polyethylene glycol have been used to mimic the cartilage tissue, whereas ceramic materials like hydroxyapatite and beta (β)-tricalcium phosphate were chosen to represent the subchondral bone. Culturing MSCs and chondrocytes on such scaffolds resulted in different tissue morphologies. Though 3D printing can recreate different mechanical and porosity properties, inferior cell-cell interactions and inhomogeneous cell growth and differentiation meant poor clinical outcomes [22].

One of the earliest attempts at *3D bioprinting* the osteochondral unit was performed by microextrusion of two different cell types: mesenchymal stem cells with osteoinductive calcium phosphate particles and chondrocytes on two sides of an alginate mesh scaffold [23]. Culture as well as in vivo experimentation demonstrated both osteogenic and chondrogenic differentiation.

Subsequently, bioprinting of cells with an appropriate hydrogel has been used to guide differentiation into the desired tissue. Collagen type-I and polycaprolactone (PCL) have been identified as suitable for bone tissue formation, and hyaluronic acid or alginate for cartilage tissue formation [24]. Another study has combined the use of bone morphogenetic protein-2 (BMP-2) with human mesenchymal stem cells (hMSCs) to commit towards osteoblast formation, and TGF-β1 with hMSCs to commit towards chondrocyte differentiation [25].

Articular cartilage is responsible for resisting compressive stress and enables proper distribution of mechanical loading on the subchondral bone. Deeper zones help articular cartilage to resist further compression force. The subchondral bone, on the part, is composed of concentric lamellar layers around the osteons and flat layers representing new bone formation. The peripheral bone is largely avascular, while the endosteal bone abuts directly on calcified cartilage [26].

The unique anisotropic arrangement is formed due to the external loads over time, which is transmitted through the matrix of the tissue and converted into a biochemical signal, alerting cells to either produce more or catabolize existing ECM [27].

Scaffold-based tissue engineering approaches interrupt this transmission as the scaffold material confines the cells and shields cells from this mechanotransductive signaling cascade [28]. This calls for novel scaffold-free tissue engineering approaches that can preserve the natural balance between external mechanical loading and the maintaining of zonal microenvironments for chondrocytes to develop into the heterogenic layers of cartilage.

Moreover, at the osteochondral interface, there is a transition between chondrocytes from the calcified cartilage zone and cells from subchondral bone differ in their differentiation sta-

tus and metabolic activities. This makes it challenging to recapitulate this interface. The subchondral bone has a suddenly higher compressive modulus and tensile modulus, demanding a more robust material like PCL, unlike hydrogels that are commonly used for cartilage bioprinting.

Multilayered osteochondral tissue constructs have been created by bioprinting mesenchymal stem cells (MSCs) and recombinant human bone morphogenetic protein (rhBMPs) on a PCL frame to mimic the subchondral bone [29, 30] and MSCs with hyaluronic acid and TGF-β on the subchondral bone structure to mimic cartilage tissue. The constructs showed promising results in rabbits [30].

19.9 Challenges in Clinical Application of 3D Bioprinted Osteochondral Tissues

Challenges to clinical application include both clinical and administrative [13, 31].

1. Clinical challenges:

- The bioprinted tissue will need to incorporate with the surrounding host chondral surfaces and host subchondral bone.
- Allogeneic tissues will have a similar potential incorporation challenges as viable allograft osteoarticular allografts.
- Autograft osteochondral composites will need tissue harvest, cell expansion, printing, and the maturation of the fabricated construct, making one-stage printing in situ challenging.
- The bilayered printed grafts will need sufficient biomechanical strength at implantation to sustain joint motion and immediate rehabilitation to avoid iatrogenic stiffness and pain.
- Expensive bioreactors will need to be designed and installed for cultivating bioprinted grafts to enhance biological and mechanical properties of the printed tissues.
- Postoperative rehabilitation will need to consider limited weight-bearing for a period of

time to offset time for incorporation and strengthening of the graft.

- Inflammatory conditions of the joint should be addressed earlier in order to avoid detrimental effects on the bioprinted graft.
- Bioprinting with cells/material from nonhuman sources can pose ethical problems.
- It is as yet unknown if the use of using pluripotent stem cells poses risks to the patient, e.g., formation of teratoma.
- There is a possibility of contracting zoonosis with the use of nonhuman materials for 3D bioprinting.
- It is possible that implantation of 3D bioprinted material can cause irreversible damage or loss of opportunity for future treatment.

2. Administrative challenges:

- Bioprinting of composite osteochondral tissues will need to demonstrate cost-efficacy. Like all new technologies, 3D Bioprinting can become more cost-effective over time, with more efficiency in production and equipment manufacturing.
- The technology has to go through daunting regulatory hurdles for clinical application. This can be a costly and time-consuming process.
- Certain religions or faiths do not allow the use of grafts/cellular material while others do not allow the use of animal products or human embryonic tissues. Such prohibitions can be at individual, institutional, or national levels. Such nonavailability can lead to 3D bioprinting-related medical tourism to access these materials and cell lines thereby running a shadow economy of hope.

19.10 Future Perspectives

Each tissue of the human body harbors a variety of highly specialized cells surrounded by a heterogeneous extracellular matrix that resulted from billions of years of evolution. While any amount of advancement in science cannot match natural evolution, an interdisciplinary effort combining the best international expertise in biology, engineering, chemistry, robotics, material science, medicine, noninvasive diagnostic imaging as well as computer-aided design is needed to take us ahead one step at a time towards the goal of bioprinting a completely viable and functional osteochondral unit with near normal biochemical, biomechanical, and heterogenic properties and good healing capabilities with adjacent parent tissue in vivo.

Future research should take into consideration:

1. development of advanced 3D imaging systems that better picture the structural and functional variations in different zones of various tissues,
2. harvest and storage of potential cell lineage for usage in bioprinting,
3. development of techniques for vascularization especially of the subchondral portion of the bilayered bioprinted tissues,
4. development of faster bioprinting systems that would not lose on the viability of cell-biomaterial suspensions as with the present slower systems,
5. development of automated robotic systems that could interact and work together to print different cell-biomaterial suspensions in order to recreate, with more accuracy, the anisotropic complex tissue structures,
6. further advancement of the scaffold-free bioprinting technique, which has shown promising results in manipulating both soft- as well as hard-matrix materials, is in order,
7. advances in designing constructs with gradient porosity, that are also capable of delivering growth factors/genes with precise spatiotemporal control, will further our attempts to reproduce the tissue heterogeneity,
8. devising smaller, less cumbersome bioprinting systems for intraoperative use.

"There is no innovation and creativity without failure. Period."

–Brene Brown, Researcher, Author, and Professor.

References

1. Moran CJ, Pascual-Garrido C, Chubinskaya S, Potter HG, Warren RF, Cole BJ, Rodeo SA. Restoration of articular cartilage. JBJS. 2014;96(4):336–44.
2. Roseti L, Grigolo B. Host environment: scaffolds and signaling (tissue engineering) articular cartilage regeneration: cells, scaffolds, and growth factors. In: Bio-orthopaedics. Berlin: Springer; 2017. p. 87–103.
3. Cao Y, Vacanti JP, Paige KT, Upton J, Vacanti CA. Transplantation of chondrocytes utilizing a polymer-cell construct to produce tissue-engineered cartilage in the shape of a human ear. Plast Reconstr Surg. 1997;100(2):297–302. https://doi.org/10.1097/00006534-199708000-00001.
4. Lepage SI, Robson N, Gilmore H, Davis O, Hooper A, St. John S, Kamesan V, Gelis P, Carvajal D, Hurtig M, Koch TG. Beyond cartilage repair: the role of the osteochondral unit in joint health and disease. Tissue Eng Part B Rev. 2019;25(2):114–25. https://doi.org/10.1089/ten.TEB.2018.0122.
5. Di Bella C, Fosang A, Donati DM, Wallace GG, Choong PF. 3D bioprinting of cartilage for orthopedic surgeons: reading between the lines. Frontiers in Surgery. 2015;2:39. https://doi.org/10.3389/fsurg.2015.00039.
6. Madry H, van Dijk CN, Mueller-Gerbl M. The basic science of the subchondral bone. Knee Surg Sports Traumatol Arthrosc. 2010;18(4):419–33. https://doi.org/10.1007/s00167-010-1054-z. Epub 2010 Jan 30
7. Duncan HO, Jundt J, Riddle JM, Pitchford W, Christopherson T. The tibial subchondral plate. A scanning electron microscopic study. The J Bone Joint Surg Am. 1987;69(8):1212–20.
8. Pugh JW, Rose RM, Radin EL. Elastic and viscoelastic properties of trabecular bone: dependence on structure. J Biomech. 1973;6(5):475–85.
9. Radin EL, Paul IL. Does cartilage compliance reduce skeletal impact loads?. The relative force-attenuating properties of articular cartilage, synovial fluid, periarticular soft tissues and bone. Arthritis Rheum. 1970;13(2):139–44.
10. Layton MW, Goldstein SA, Goulet RW, Feldkamp LA, Kubinski DJ, Bole GG. Examination of subchondral bone architecture in experimental osteoarthritis by microscopic computed axial tomography. Arthritis Rheum. 1988;31(11):1400–5.
11. Stiebel M, Miller LE, Block JE. Post-traumatic knee osteoarthritis in the young patient: therapeutic dilemmas and emerging technologies. Open Access J Sports Med. 2014;5:73.
12. Tuan RS, Chen AF, Klatt BA. Cartilage regeneration. J Am Acad Orthop Surg. 2013;21(5):303.
13. Saseendar S. 3D Bioprinting. In: Vaishya R, Haleem A, Maini L, editors. An update on medical 3D printing. New Delhi: Salubris; 2020. p. 183–201.
14. Bishop ES, Mostafa S, Pakvasa M, Luu HH, Lee MJ, Wolf JM, Ameer GA, He TC, Reid RR. 3-D bioprinting technologies in tissue engineering and regenerative medicine: current and future trends. Genes Dis. 2017;4(4):185–95. https://doi.org/10.1016/j.gendis.2017.10.002. Epub 2017 Nov 22
15. Murphy SV, Atala A. 3D bioprinting of tissues and organs. Nat Biotechnol. 2014;32(8):773.
16. Roseti L, Parisi V, Petretta M, Cavallo C, Desando G, Bartolotti I, Grigolo B. Scaffolds for bone tissue engineering: state of the art and new perspectives. Mater Sci Eng C. 2017;78:1246–62.
17. Irvine SA, Venkatraman SS. Bioprinting and differentiation of stem cells. Molecules. 2016;21(9):1188.
18. Sultana N. Scaffolds for tissue engineering. MRS Bull. 2003;28(4):301–6.
19. Aldana AA, Abraham GA. Current advances in electrospun gelatin-based scaffolds for tissue engineering applications. Int J Pharm. 2017;523(2):441–53.
20. Gruene M, Deiwick A, Koch L, Schlie S, Unger C, Hofmann N, Bernemann I, Glasmacher B, Chichkov B. Laser printing of stem cells for biofabrication of scaffold-free autologous grafts. Tissue Eng Part C Methods. 2011;17(1):79–87.
21. Ozbolat IT. Scaffold-based or scaffold-free bioprinting: competing or complementing approaches? J Nanotechnol Eng Med. 2015;6(2):024701.
22. Datta P, Dhawan A, Yu Y, Hayes D, Gudapati H, Ozbolat IT. Bioprinting of osteochondral tissues: a perspective on current gaps and future trends. Int J Bioprinting. 2017;3(2):109–20. https://doi.org/10.18063/IJB.2017.02.007.
23. Fedorovich NE, Schuurman W, Wijnberg HM, Prins HJ, Van Weeren PR, Malda J, Alblas J, Dhert WJ. Biofabrication of osteochondral tissue equivalents by printing topologically defined, cell-laden hydrogel scaffolds. Tissue Eng Part C Methods. 2012;18(1):33–44. https://doi.org/10.1089/ten.tec.2011.0060.
24. Park JY, Choi JC, Shim JH, Lee JS, Park H, Kim SW, Doh J, Cho DW. A comparative study on collagen type I and hyaluronic acid dependent cell behavior for osteochondral tissue bioprinting. Biofabrication. 2014;6(3):035004. https://doi.org/10.1088/1758-5082/6/3/035004.
25. Gurkan UA, El Assal R, Yildiz SE, Sung Y, Trachtenberg AJ, Kuo WP, Demirci U. Engineering anisotropic biomimetic fibrocartilage microenvironment by bioprinting mesenchymal stem cells in nanoliter gel droplets. Mol Pharm. 2014;11(7):2151–9. https://doi.org/10.1021/mp400573g.
26. Clark JM, Huber JD. The structure of the human subchondral plate. The J Bone Joint Surg Br. 1990;72(5):866–73.
27. Makris EA, Gomoll AH, Malizos KN, Hu JC, Athanasiou KA. Repair and tissue engineering techniques for articular cartilage. Nat Rev Rheumatol. 2015;11(1):21.
28. MacBarb RF, Chen AL, Hu JC, Athanasiou KA. Engineering functional anisotropy in fibrocartilage neotissues. Biomaterials. 2013;34(38):9980–9. https://doi.org/10.1016/j.biomaterials.2013.09.026.

29. Shim JH, Lee JS, Kim JY, Cho DW. Bioprinting of a mechanically enhanced three-dimensional dual cell-laden construct for osteochondral tissue engineering using a multi-head tissue/organ building system. J Micromech Microeng. 2012;22(8):085014. https://doi.org/10.1088/0960-1317/22/8/085014.

30. Shim JH, Jang KM, Hahn SK, Park JY, Jung H, Oh K, Park KM, Yeom J, Park SH, Kim SW, Wang JH. Three-dimensional bioprinting of multilayered constructs containing human mesenchymal stromal cells for osteochondral tissue regeneration in the rabbit knee joint. Biofabrication. 2016;8(1):014102. https://doi.org/10.1088/1758-5090/8/1/014102.

31. Ravnic DJ, Leberfinger AN, Koduru SV, Hospodiuk M, Moncal KK, Datta P, Dey M, Rizk E, Ozbolat IT. Transplantation of bioprinted tissues and organs: technical and clinical challenges and future perspectives. Ann Surg. 2017;266(1):48–58. https://doi.org/10.1097/SLA.0000000000002141.

Biphasic Osteochondral Restoration Techniques Using Synovial Stem Cells and Artificial Bone

George Jacob, Kazunori Shimomura, Wataru Ando, David A. Hart, and Norimasa Nakamura

20.1 Introduction

Osteochondral lesions (OL) pose a complex clinical scenario to manage, and effective solutions for osteochondral restoration remain elusive. OLs are commonly a result of trauma, where shear forces create a stress fracture in the chondral matrix extending into the subchondral bone [1]. They are also noted in association with osteonecrosis (ON), osteochondritis dissecans (OD), subchondral insufficiency fractures, and osteoarthritis (OA) [2]. Subchondral bone stiffening can lead to inefficient shock absorbance and chondral breakdown [3]. These lesions can increase in size over time and lead to overall joint degeneration, with progression to advanced OA. When viewing the chondral and subchondral components of the joint together, it can be termed as an osteochondral unit where each layer possesses different biologic and biomechanical characteristics. This is the main reason why regenerative techniques have focused on biphasic constructs which emulate both the characteristics of cartilage and bone. It has been an especially difficult task to develop an implant that can be mechanically strong, bioactive, biomimetic, and maintain a stable interface between the two biphasic material surfaces. Various preclinical and clinical studies have employed scaffolds seeded with cells, as well as acellular implants to address the problem. In this chapter, we focus on a synovial mesenchymal stem cell-derived tissue-engineered construct (TEC) combined with hydroxyapatite for the repair of osteochondral lesions.

20.2 Synovium-Derived Mesenchymal Stem Cells

Synovium-derived mesenchymal stem cells (SDMSCs) have been of particular interest in cartilage regeneration following the publication of literature demonstrating them to have superior chondrogenic and osteogenic differentiation capacity [4–7], along with a higher proliferation

G. Jacob · K. Shimomura · W. Ando
Department of Orthopaedic Surgery, Osaka University Graduate School of Medicine, Osaka, Japan
e-mail: kazunori-shimomura@umin.net; w-ando@umin.ac.jp

D. A. Hart
McCaig Institute for Bone and Joint Health, University of Calgary, Calgary, AB, Canada
e-mail: hartd@ucalgary.ca

N. Nakamura (✉)
Institute for Medical Science in Sports, Osaka Health Science University, Osaka, Japan

Global Centre for Medical Engineering and Informatics, Osaka University, Osaka, Japan
e-mail: norimasa.nakamura@ohsu.ac.jp

© ISAKOS 2022
A. Gobbi et al. (eds.), *Joint Function Preservation*, https://doi.org/10.1007/978-3-030-82958-2_20

potential [8]. SDMSCs are also readily obtained during arthroscopy harvest of synovial tissue. This procedure is relatively painless and has the advantage of minimal donor site morbidity. Also, SDMSCs have displayed less senescence and minimal variability in multi-potency when comparing subjects of different ages [4, 9]. Interestingly, it has also been noted that SDMSCs from OA and rheumatoid patients exhibited a similar capacity for regeneration and chondral repair as did normal donors [10]. Along with these advantages, SDMSCs also overcome the problems of limited cell numbers associated with bone marrow aspiration and cartilage harvest. There is, however, when used in an autologous manner, a need for staged surgeries as harvested synovial tissue requires in vitro expansion with this approach (synovial explant surgery and then subsequent surgical implantation into the defect).

20.3 Tissue-Engineered Construct

Tissue-engineered constructs (TEC) have been developed for the repair of chondral defects using autologous or allogeneic SDMSCs in three-dimensional culture. TEC is a scaffold-free construct generated as a result of the extracellular matrices (ECM) produced by the SDMSCs themselves [11]. TECs are manufactured using a high density culture (4.0×10^5 cells/cm^2) of SDMSCs in growth medium containing 0.2 mM of ascorbate-2-phosphate [11, 12]. Ascorbic acid aids in collagen synthesis and secretion, and thus improves the production of the collagen matrices [13]. With regular media changes, a complex ECM is synthesized by the SDMSCs after ~14 days which can be detached from the culture dish by applying shear stress along the borders of the dish. This results in a detached monolayer complex which when left suspended spontaneously forms a three-dimensional tissue-like structure by active contraction. Being a three-dimensional culture, the cells can effectively maintain their cellular phenotype and not undergo dedifferentiation commonly encountered in chondrocyte and bone marrow-derived mesenchymal stem cell two-dimensional cultures [14–18]. With the addition of a chondrogenic medium

to such TEC, the cells express increased GAG synthesis and expression of type II collagen, aggrecan, and SOX9 [12] proving it to be an appropriate 3D microenvironment for chondrogenic culture. TEC has also proven to have excellent adhesive properties for fixation to a chondral surface due to the presence of fibronectin and vitronectin throughout the whole tissue [12]. Therefore, no fixation methods employing sutures or fibrin glues are required for the delivery of a TEC to a chondral defect. TEC has proven to be a valuable scaffold-free tissue construct that can be easily handled surgically and implanted into chondral defects [12]. The use of TEC for chondral repair has been reported in pre-clinical studies, as well as one in a human pilot study with five subjects [12, 19, 20]. In the clinical pilot study, at 24 months follow-up, magnetic resonance imaging, and second look arthroscopy confirmed the presence of defect fill, and biopsy indicated development of a tissue resembling hyaline cartilage [19]. The TEC protocol is currently in further transition from bench to bedside as larger clinical trials are currently underway.

20.4 Biphasic Osteochondral Implants

As an osteochondral solution has many functions to satisfy, the research to date has been focused on combining different materials to fulfill all the necessary criteria. Zonal restoration of the chondral and subchondral layer in a layer-by-layer fashion is the goal of these biphasic implants and several exist to date [1, 21, 22]. Current concerns when employing polymers and synthetics for osteochondral restoration are the biocompatibility of the materials and their long-term safety implications. Effective cell delivery and integration of the implants have also been challenging with selected implants showing some promise [23–25]. Of note, only a few biphasic scaffolds have been utilized in clinical trials using materials such as collagen, hydroxyapatite (HA), various synthetics, and a novel material aragonite [26]. The most studied implant has been Maioregen™ (Fin-Ceramica, Faenza, Italy) which is composed of collagen and HA organized in three layers. Clinical trials have shown good integration and filling of the chondral defect and demonstrated the implant to be an efficacious option with a low complication rate [27–31]. A recent systematic review also reported similar findings but did highlight that the available literature was not of high-level evidence and thus this implant at present cannot be deemed superior to any other techniques. Another reported implant has been a synthetic biphasic scaffold consisting of poly-lactic, poly-glycolic acid, and calcium phosphate marketed as TruFit™ (Smith and Nephew, Andover. MA). A few studies reported good results [32, 33]; however, literature has reported delayed integration [32–34] and longer follow-up noted subchondral cysts and bone edema [35–38]. A more recent bilayer scaffold under study has been a crystalline coral aragonite-hyaluronic acid implant (Agili-C™, Cartiheal (2009) Ltd., Israel) which showed good integration and clinical improvement in a study up to 12 months follow-up [39] and at further 24 months in a doctoral dissertation by Di Matteo [40]. At present, there is not enough evidence to deem any of these implants superior or more efficient than the other except for Trufit™ being less favorable due to adverse reactions.

20.5 Biphasic TEC Osteochondral Implants

With the encouraging potential that TEC has demonstrated for in vivo chondral repair [12, 20], it was then decided to focus on the development of a biphasic osteochondral solution utilizing TEC for lesions that extend beyond the chondral component into the subchondral region of the joint. It was postulated that TEC had some notable advantages over other currently available constructs in that it is bioactive, leading to the production of a high-quality ECM for the chondral component of the defect. In addition, the strong presence of SDMSCs within the TEC allows for the continued regeneration of the chondral layers [41]. With the TEC being entirely natural and lacking any artificial components, it mitigates many of the safety concerns regarding synthetic polymers. The adhesive properties of TEC also allow for it to be easily combined with a subchondral material component

towards developing a biphasic implant to address the complexity of osteochondral lesions.

For the initial combined biphasic implant, a porous synthetic HA plug was prepared with dimensions 5 mm in diameter and 4 mm in height with 75% porosity (NEOBONE®; MMT Co.LTD., Osaka, Japan). HA displays adequate mechanical strength and high porosity, allowing for the cell penetration required for integration with the neighboring bone tissues. HA has proven to be both osteoinductive and allows for bony ingrowth [42–44], making it a reasonable choice for use as the subchondral component of the biphasic implant with a TEC (Fig. 20.1).

Another studied option for the subchondral component was beta-tricalcium phosphate (β-TCP) with same size and porosity (OSferion®; Olympus Terumo Biomaterials, Tokyo, Japan) as the previously described HA component. Beta-TCP is a suitable candidate as it is equally biocompatible to HA, but undergoes resorption faster than HA in vivo [45], a feature that theoretically may make it more efficient regarding integration after implantation. TEC was created using the standard protocol, and then without any additional adhesive measures, placed onto the artificial bone plug which immediately bonded to form a stable biphasic implant (Fig. 20.1).

20.5.1 Results of TEC Biphasic Osteochondral Implantation

To date, only two preclinical studies have been conducted employing TEC as part of a biphasic osteochondral implant, with both studies conducted in rabbit models. The first was conducted using TEC + HA in artificially created knee osteochondral defects, with biological and biomechanical comparison to a control group of HA implanted defects, as well as a group of normal knees [41]. The TEC + HA group demonstrated superior repair in all objective outcome measures when compared to the control and normal groups. There was an earlier restoration of both the chondral and subchondral bone components, with excellent integration. The quality of the chondral tissue was superior to that of the control group where the repair tissue was cracked and fissuring.

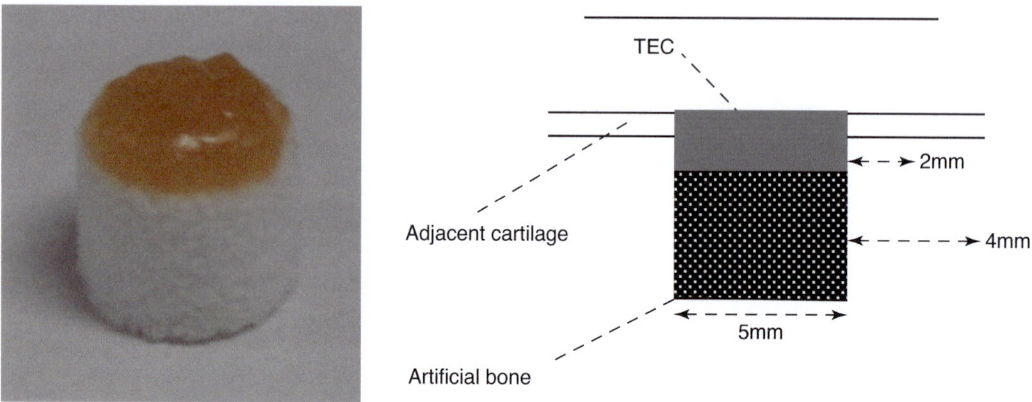

Fig. 20.1 Biphasic Osteochondral TEC Implant (this figure was quoted and modified from Shimomura et al. Tissue Engineering Part A. 2014 Sep 1;20(17–18):2291–304.)

TEC/HA

TEC/βTCP

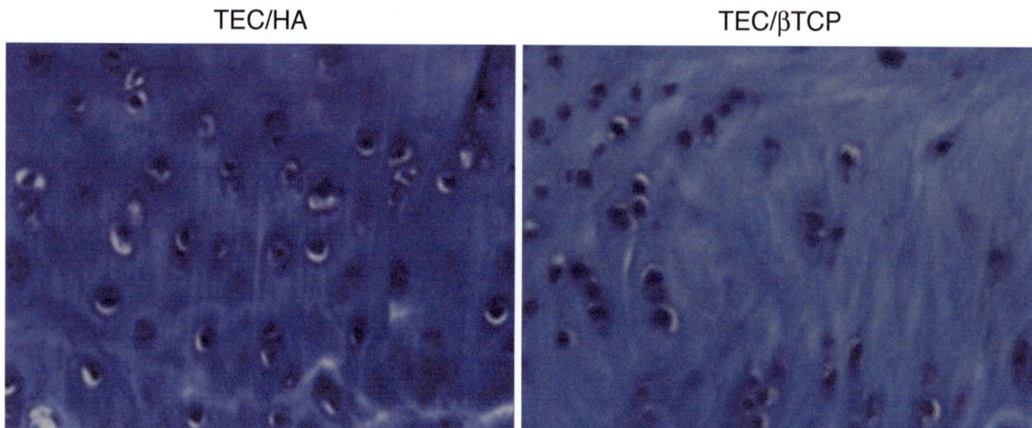

Fig. 20.2 High magnification positive Toluidine blue (TB) staining of cartilage repair tissue exhibiting hyaline cartilage-like repair with chondrocytes arranged in longitudinal columns in the TEC/HA group. The TEC/β-TCP group resembled a more fibrocartilage repair, demonstrating a more disorganized structure. (this figure was quoted and modified from Shimomura et al. The American Journal of Sports Medicine, 45(3), 666–675)

Biomechanical testing revealed the TEC + HA repaired defects displayed similar stiffness to normal osteochondral joint surfaces.

The second study compared two different combinations of the biphasic implant, one being TEC + HA and the second TEC + β-TCP [46]. When comparing these two combinations, TEC + HA demonstrated better and more efficient healing at 6 months (Fig. 20.2). The TEC + β-TCP group showed earlier degradation of the β-TCP when compared to the HA of the TEC + HA, although the subchondral bone repair was inferior with higher porosity at 6 months. Biomechanical testing also revealed the stiffness of the TEC + β-TCP group was significantly lower than that of normal tissue (Fig. 20.3). Thus,

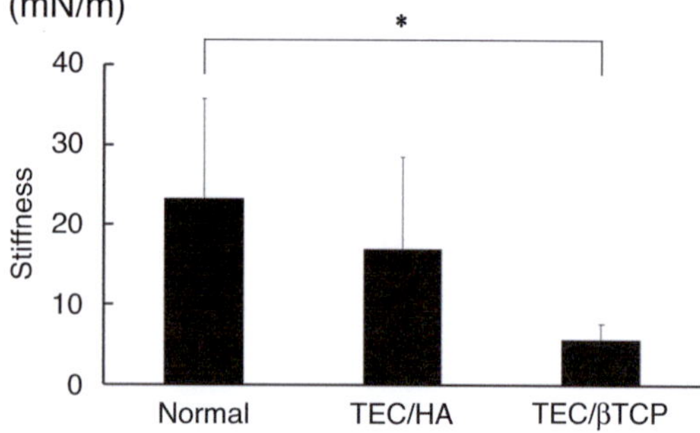

Fig. 20.3 The stiffness of biphasic implants at 6 months postimplantation. The TEC/HA implant restored stiffness to a mean of 73% that of healthy osteochondral tissue. TEC/β-TCP implants exhibited a significantly lower stiffness than that of normal tissue (P < 0.05) (this figure was quoted and modified from Shimomura et al. The American Journal of Sports Medicine, 45(3), 666–675)

Table 20.1 Summary of preclinical studies employing TEC as an osteochondral implant

Author/year	Shimomura et al. 2014	Shimomura et al. 2016
Study design/animal	Controlled laboratory study/rabbit	Controlled laboratory study/rabbit
Implant material	TEC + HA	TEC+ HA vs TEC + β-TCP
Defect Details	HA + TEC: 23 HA alone:18 Control: 5	TEC + HA: 35 TEC + β-TCP:16 Control: 5
Follow-up	6 months	6 months
Evaluation	Macro, Histo, Biomech	Macro, Histo, Biomech
Results	*Macro*: TEC group had better fill at all time points, control had less fill and cracking of the repair tissue over time. *Histo*: 6 months osteochondral repair noted in both groups, but TEC demonstrated better integration and hyaline-like cartilage compared to control. *Histo score:* Significantly higher in TEC group for chondral and subchondral tissue *Biomech testing:* TEC group: Restoration of stiffness values close to normal tissue	*Macro*: No OA changes noted, good defect fill, and no significant difference b/w groups. *Histo*: 6 months complete osteochondral repair in both groups. *Histo score*: Chondral region TEC + HA showed better scores. Subchondral region by 2 months both groups had identical scores. *Biomech testing:* TEC + β-TCP were sig weaker while TEC + HA resorted stiffness to 73% of healthy tissue.

Abbreviations: *TEC* Tissue-engineered construct; *HA* Hydroxyapatite; *Macro* Macroscopy; *Histo* Histology; *Biomech* Biomechanical; *β-TCP* Beta-Tricalcium phosphate

from these studies, it was concluded that HA was a better material than β-TCP for combination with a TEC. However, further investigation is warranted as to which material is the best choice in the long term. Table 20.1 summarizes the studies with TEC biphasic osteochondral implants.

20.6 Conclusion

TEC has proven to be an effective repair strategy for chondral injuries, showing superior results in both preclinical and human trials. In preclinical studies, a biphasic TEC/HA construct has demonstrated good potential to also repair osteochondral defects. HA has shown to be superior to β-TCP when combined with a TEC in currently available preclinical studies in rabbits, but further long-term testing is required to strengthen this conclusion. The addition of other materials and/ or addition of growth factors may further improve the efficacy of these biphasic implants and future research may determine a more optimal construct. However, combining a TEC with a HA component appears to be an attractive choice for a biphasic osteochondral repair strategy, owing in part to its ease of manufacture and the positive results to date. Larger animal model studies will help confirm the positive results of TEC and artificial bone as a biphasic osteochondral implant obtained thus far, as well as offer the ability to assess the biological and biomechanical functionality of the implant.

References

1. Gomoll AH, Madry H, Knutsen G, van Dijk N, Seil R, Brittberg M, et al. The subchondral bone in articular cartilage repair: current problems in the surgical management. Knee Surg Sport Traumatol Arthrosc. 2010;18(4):434–47.
2. Gorbachova T, Melenevsky Y, Cohen M, Cerniglia BW. Osteochondral lesions of the knee: differentiating the most common entities at MRI. Radiographics. 2018;38(5):1478–95.
3. Willers C, Wood DJ, Zheng MH. A current review on the biology and treatment of articular cartilage defects (Part I & Part II). J Musculoskelet Res. 2003;7:157–81.
4. Sakaguchi Y, Sekiya I, Yagishita K, Muneta T. Comparison of human stem cells derived from various mesenchymal tissues: superiority of synovium as a cell source. Arthritis Rheum. 2005;52:2521–9.

5. Mochizuki T, Muneta T, Sakaguchi Y, Nimura A, Yokoyama A, Koga H, et al. Higher chondrogenic potential of fibrous synovium- and adipose synovium-derived cells compared with subcutaneous fat-derived cells: distinguishing properties of mesenchymal stem cells in humans. Arthritis Rheum. 2006;54(3):843–53.

6. Fan J, Varshney RR, Ren L, Cai D, Wang DA. Synovium-derived mesenchymal stem cells: a new cell source for musculoskeletal regeneration. Tissue Eng B Rev. 2009;15:75–86.

7. Shirasawa S, Sekiya I, Sakaguchi Y, Yagishita K, Ichinose S, Muneta T. In vitro chondrogenesis of human synovium-derived mesenchymal stem cells: optimal condition and comparison with bone marrow-derived cells. J Cell Biochem. 2006;97(1):84–97.

8. Fernandes TL, Kimura HA, Pinheiro CCG, Shimomura K, Nakamura N, Ferreira JR, et al. Human synovial mesenchymal stem cells good manufacturing practices for articular cartilage regeneration. Tissue Eng C Methods. 2018;24(12):709–16.

9. De Bari C, Dell'Accio F, Tylzanowski P, Luyten FP. Multipotent mesenchymal stem cells from adult human synovial membrane. Arthritis Rheum. 2001;44:1928–42.

10. Koizumi K, Ebina K, Hart DA, Hirao M, Noguchi T, Sugita N, et al. Synovial mesenchymal stem cells from osteo- or rheumatoid arthritis joints exhibit good potential for cartilage repair using a scaffold-free tissue engineering approach. Osteoarthr Cartil. 2016;24(8):1413–22.

11. Ando W, Tateishi K, Katakai D, Hart DA, Higuchi C, Nakata K, et al. In vitro generation of a scaffold-free tissue-engineered construct (TEC) derived from human synovial mesenchymal stem cells: biological and mechanical properties and further chondrogenic potential. Tissue Eng A. 2008;14(12):2041–9.

12. Ando W, Tateishi K, Hart DA, Katakai D, Tanaka Y, Nakata K, et al. Cartilage repair using an in vitro generated scaffold-free tissue-engineered construct derived from porcine synovial mesenchymal stem cells. Biomaterials. 2007;28(36):5462–70.

13. Yasui Y, Ando W, Shimomura K, Koizumi K, Ryota C, Hamamoto S, et al. Scaffold-free, stem cell-based cartilage repair. J Clin Orthop Trauma. 2016;7:157–63.

14. Schnabel M, Marlovits S, Eckhoff G, Fichtel I, Gotzen L, Vécsei V, et al. Dedifferentiation-associated changes in morphology and gene expression in primary human articular chondrocytes in cell culture. Osteoarthr Cartil. 2002;10(1):62–70.

15. Darling EM, Athanasiou KA. Rapid phenotypic changes in passaged articular chondrocyte subpopulations. J Orthop Res. 2005;23(2):425–32.

16. Takahashi K, Yamanaka S. Induction of pluripotent stem cells from mouse embryonic and adult fibroblast cultures by defined factors. Cell. 2006;126(4):663–76.

17. Duan L, Ma B, Liang Y, Chen J, Zhu W, Li M, et al. Cytokine networking of chondrocyte dedifferentiation in vitro and its implications for cell-based cartilage therapy. Am J Transl Res. 2015;7:194–208.

18. Takahashi T, Ogasawara T, Asawa Y, Mori Y, Uchinuma E, Takato T, et al. Three-dimensional microenvironments retain chondrocyte phenotypes during proliferation culture. Tissue Eng. 2007;13(7):1583–92.

19. Shimomura K, Yasui Y, Koizumi K, Chijimatsu R, Hart DA, Yonetani Y, et al. First-in-human pilot study of implantation of a scaffold-free tissue-engineered construct generated from autologous synovial mesenchymal stem cells for repair of knee chondral lesions. Am J Sports Med. 2018;46(10):2384–93.

20. Shimomura K, Ando W, Moriguchi Y, Sugita N, Yasui Y, Koizumi K, et al. Next generation mesenchymal stem cell (MSC)–based cartilage repair using scaffold-free tissue engineered constructs generated with synovial mesenchymal stem cells. Cartilage. 2015;6:13–29.

21. Kon E, Delcogliano M, Filardo G, Busacca M, Di Martino A, Marcacci M. Novel nano-composite multilayered biomaterial for osteochondral regeneration: a pilot clinical trial. Am J Sports Med. 2011;39(6):1180–90.

22. Shimomura K, Moriguchi Y, Murawski CD, Yoshikawa H, Nakamura N. Osteochondral tissue engineering with biphasic scaffold: current strategies and techniques. Tissue Eng B Rev. 2014;20:468–76.

23. Ahn JH, Lee TH, Oh JS, Kim SY, Kim HJ, Park IK, et al. Novel hyaluronate-atelocollagen/beta-TCP-hydroxyapatite biphasic scaffold for the repair of osteochondral defects in rabbits. Tissue Eng Part A. 2009;15(9):2595–604.

24. Hung CT, Lima EG, Mauck RL, Taki E, LeRoux MA, Lu HH, et al. Anatomically shaped osteochondral constructs for articular cartilage repair. J Biomech. 2003;36(12):1853–64.

25. Alhadlaq A, Mao JJ. Tissue-engineered osteochondral constructs in the shape of an articular condyle. J Bone Joint Surg Am. 2005;87(5):936–44.

26. Kon E, Filardo G, Perdisa F, Venieri G, Marcacci M. Clinical results of multilayered biomaterials for osteochondral regeneration. J Exp Orthop. 2014;1(1):10.

27. Filardo G, Kon E, Di Martino A, Busacca M, Altadonna G, Marcacci M. Treatment of knee osteochondritis dissecans with a cell-free biomimetic osteochondral scaffold: clinical and imaging evaluation at 2-year follow-up. Am J Sports Med. 2013;41(8):1786–93.

28. Kon E, Filardo G, Di Martino A, Busacca M, Moio A, Perdisa F, et al. Clinical results and MRI evolution of a nano-composite multilayered biomaterial for osteochondral regeneration at 5 years. Am J Sports Med. 2014;42(1):158–65.

29. Delcogliano M, de Caro F, Scaravella E, Ziveri G, De Biase CF, Marotta D, et al. Use of innovative biomimetic scaffold in the treatment for large osteochondral lesions of the knee. Knee Surg Sport Traumatol Arthrosc. 2014;22(6):1260–9.

30. Brix M, Kaipel M, Kellner R, Schreiner M, Apprich S, Boszotta H, et al. Successful osteoconduction but limited cartilage tissue quality following osteochon-

dral repair by a cell-free multilayered nano-composite scaffold at the knee. Int Orthop. 2016;40:625–32.

31. Di Martino A, Kon E, Perdisa F, Sessa A, Filardo G, Neri MP, et al. Surgical treatment of early knee osteoarthritis with a cell-free osteochondral scaffold: results at 24 months of follow-up. Injury. 2015;46 Suppl 8:S33–8.

32. Carmont MR, Carey-Smith R, Saithna A, Dhillon M, Thompson P, Spalding T. Delayed incorporation of a TruFit plug: perseverance is recommended. Arthroscopy. 2009;25:810–4.

33. Bedi A, Foo LF, Williams RJ, Potter HG. The maturation of synthetic scaffolds for osteochondral donor sites of the knee: an MRI and T2-mapping analysis. Cartilage. 2010;1:20–8.

34. Bugelli G, Ascione F, Dell'Osso G, Zampa V, Giannotti S. Biphasic bioresorbable scaffold (TruFit®) in knee osteochondral defects: 3-T MRI evaluation of osteointegration in patients with a 5-year minimum follow-up. Musculoskelet Surg. 2018;102:191–9.

35. Joshi N, Reverte-Vinaixa M, Díaz-Ferreiro EW, Domínguez-Oronoz R. Synthetic resorbable scaffolds for the treatment of isolated patellofemoral cartilage defects in young patients: magnetic resonance imaging and clinical evaluation. Am J Sports Med. 2012;40(6):1289–95.

36. Dhollander AAM, Liekens K, Almqvist KF, Verdonk R, Lambrecht S, Elewaut D, et al. A pilot study of the use of an osteochondral scaffold plug for cartilage repair in the knee and how to deal with early clinical failures. Arthroscopy. 2012;28(2):225–33.

37. Barber FA, Dockery WD. A computed tomography scan assessment of synthetic multiphase polymer scaffolds used for osteochondral defect repair. Arthroscopy. 2011;27(1):60–4.

38. Verhaegen J, Clockaerts S, Van Osch GJVM, Somville J, Verdonk P, Mertens P. TruFit plug for repair of osteochondral defects—where is the evidence? Systematic review of literature. Cartilage. 2015;6:12–9.

39. Kon E, Robinson D, Verdonk P, Drobnic M, Patrascu JM, Dulic O, et al. A novel aragonite-based scaffold for osteochondral regeneration: early experience on human implants and technical developments. Injury. 2016;47 Suppl 6:S27–32.

40. A DMB. Aragonite-based scaffold for the treatment of knee osteochondral defects: results of a prospective multi-center trial at 2 years' follow-up. Bologna University; 2019.

41. Shimomura K, Moriguchi Y, Ando W, Nansai R, Fujie H, Hart DA, et al. Osteochondral repair using a scaffold-free tissue-engineered construct derived from synovial mesenchymal stem cells and a hydroxyapatite-based artificial bone. Tissue Eng A. 2014;20(17–18):2291–304.

42. Gleeson JP, Plunkett NA, O'Brien FJ. Addition of hydroxyapatite improves stiffness, interconnectivity and osteogenic potential of a highly porous collagen-based scaffold for bone tissue regeneration. Eur Cells Mater. 2010;20:218–30.

43. Shen C, Ma J, Chen XD, Dai LY. The use of β-TCP in the surgical treatment of tibial plateau fractures. Knee Surg Sport Traumatol Arthrosc. 2009;17(12):1406–11.

44. Tamai N, Myoui A, Tomita T, Nakase T, Tanaka J, Ochi T, et al. Novel hydroxyapatite ceramics with an interconnective porous structure exhibit superior osteoconduction in vivo. J Biomed Mater Res. 2002;59(1):110–7.

45. Yamasaki N, Hirao M, Nanno K, Sugiyasu K, Tamai N, Hashimoto N, et al. A comparative assessment of synthetic ceramic bone substitutes with different composition and microstructure in rabbit femoral condyle model. J Biomed Mater Res B Appl Biomater. 2009;91(2):788–98.

46. Shimomura K, Moriguchi Y, Nansai R, Fujie H, Ando W, Horibe S, et al. Comparison of 2 different formulations of artificial bone for a hybrid implant with a tissue-engineered construct derived from synovial mesenchymal stem cells. Am J Sports Med. 2017;45(3):666–75.

Proximal Tibial Subchondral Cystic Lesion Treatment with Osteo-Core-Plasty

Alberto Gobbi, Arvin Jonathan Arbas, and Ignacio Dallo

21.1 Subchondral Cyst

Subchondral bone cysts are widely observed but poorly understood [1]. Patients with proximal tibial subchondral cyst had lower tibial cartilage volume and have more changes structurally compared to patients without subchondral cyst [2]. Subchondral cyst commonly coexists with bone marrow lesion (BML) especially those with Grade 3 BML or higher [3].

21.2 Epidemiology

Almost 50% of knee osteoarthritis (OA) patients have subchondral bone cysts [2, 4, 5] while only 13.6% of healthy individuals have subchondral bone cyst [2, 6]. Females had more and larger subchondral cysts in the lateral compartment than males. This finding is perhaps due to the loading effect as females also showed more valgus loading knee [7–9].

21.3 Formation

Subchondral cyst formation often occurs in osteoarthritis of the knee, more commonly in the advanced stage of osteoarthritis [2, 10, 11]. Two theories are proposed as the mechanism of cyst formation. One is the synovial breach theory [2, 12, 13] and the other one is the bony contusion theory [2, 10, 14]. Bony contusion theory explains that excessive loading or trauma can lead to trabecular microfractures, bone necrosis, and focal bone resorption, eventually resulting in cyst formation [1, 10, 15]. Synovial breach theory states that the calcified barrier between cartilage and subchondral bone is injured, allowing for fluid to seep into the subchondral bone. This eventually creates a fluid-filled cyst lesion [1, 16, 17].

The relationship between subchondral cyst and structural change of the knee is examined by one study that shows there is a correlation between alteration of the subchondral cyst size and the cartilage loss in the medial femoral condyle for a period of 2 years [2, 4].

The relationship of the BML and subchondral cyst is unclear but some studies show that it was the bone marrow lesion that was directly involved in developing subchondral cyst [2, 18–20]. Repetitive compressive load-bearing or shear loading leads to bone marrow lesion due to subchondral damage [11, 21, 22]. Some studies show that subchondral bone cysts are related to higher

A. Gobbi (✉) · A. J. Arbas · I. Dallo
Orthopaedic Arthroscopic Surgery International (O.A.S.I.) Bioresearch Foundation, Milan, Italy
e-mail: gobbi@cartilagedoctor.it

© ISAKOS 2022

A. Gobbi et al. (eds.), *Joint Function Preservation*, https://doi.org/10.1007/978-3-030-82958-2_21

localized stress which could stimulate bone alterations or bone remodeling [1, 23].

Most likely, subchondral cyst formation is a response to altered loading distribution through the proximal tibia and also possible through joint space narrowing with disease progression [1, 24].

Subchondral cysts are usually ellipsoidal or spherical within the subchondral bone cavity and are associated with subchondral bone and cartilage degeneration of osteoarthritic knees [1, 19, 25, 26].

21.4 Location

The ratio of cyst volume to tibial volume range may be as high as 14.8% over the total proximal tibia, and up to 24.5% in the medial compartment but only up to 5.3% in the lateral compartments [1] Fig. 21.1.

The subchondral cyst volume and number were associated with the bone mineral density of medial and lateral compartments of the knee [1]. Lateral compartment subchondral cyst volume and number were associated with osteoarthritis severity, joint alignment, joint space narrowing, and gender [1]. On the other hand, in the medial region, higher medial bone mineral density (BMD) was associated with greater cyst incidence and volume [1].

Over the total tibial region, there was a strong association that was observed between subchondral cyst incidence and alignment. This study suggests that cartilage degeneration can be associated with proportionally larger and more numerous cysts [1, 26].

21.5 Classification

There are different ways of classifying subchondral bone cysts. They can be classified according to the following parameters, number of cysts, cyst number per total volume, cyst volume per total volume, total cyst volume, maximum cyst volume, and average cyst volume [1].

Subchondral cysts are assessed as Grade 0, without lesion; Grade 1, mild to moderate lesion; and Grade 2, severe (large) lesion [2].

Twenty-three percent to 35.7% of the patients with subchondral cysts could progress and 13% of those that have progression develop at least one or more subchondral cysts. On the other hand, subchondral cyst regression was observed

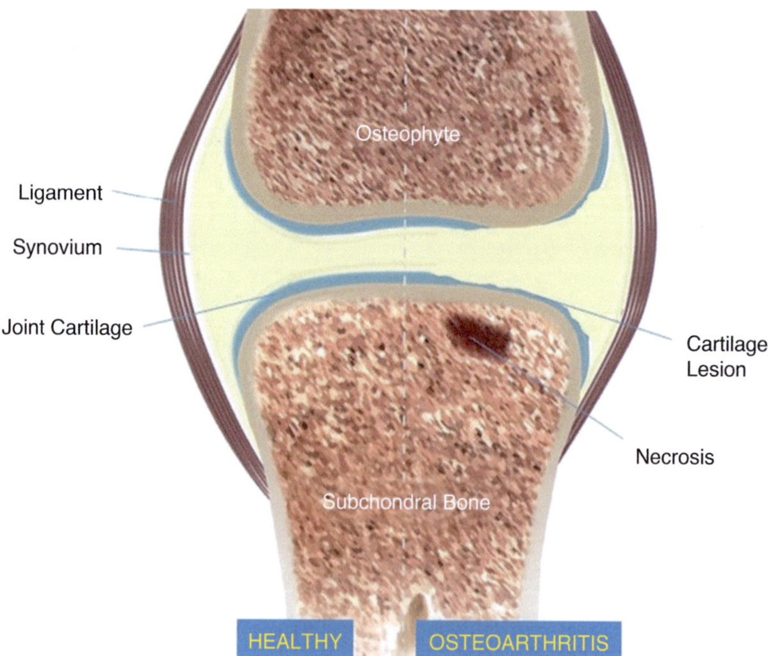

Fig. 21.1 Comparison of Healthy and Osteoarthritic Subchondral Bone; 14.8% can be seen over the proximal tibia. Courtesy of Aspire-Medical [27]

in 23.8% of the patients, of which, 14.3% experienced complete regression resolution [2].

Those with subchondral cysts had a cartilage loss rate of 9.3%. Lateral compartment regression typically has a significant lateral tibial cartilage reduction. However, greater loss of medial cartilage was noted with patients that have subchondral cyst progression [2].

21.6 Presentation

Knee Osteoarthritis is a debilitating disease, which is painful and illustrates cartilage deterioration and altered subchondral bone [1]. Recent studies show that subchondral bone has its role in progression of osteoarthritis, how it influences knee pain, and how it influences mechanical behavior of subchondral bone [1, 7, 28, 29].

Pain severity can be measured at the affected knee joint using the pain subsection of the Western Ontario McMasters Osteoarthritis Index (WOMAC) [1, 30–32], Visual Analog Scale (VAS) Scores, and Knee Injury and Arthritis Outcome Scores (KOOS) [11]. There is no correlation between cyst parameters and total WOMAC pain or nocturnal pain [1].

Osteoarthritis knee pain severity was associated with bone marrow lesions, subchondral bone attrition, effusion or synovitis, and meniscal tears, but not with subchondral bone cysts [33]. Although rare, the ganglion of the underlying subchondral bone cyst may exert pressure on the soft tissue causing it to swell and increase pain development [34, 35].

Patients with valgus alignment may be inclined to higher cyst numbers before clinical signs of osteoarthritis, such as pain [1].

21.7 Imaging

Magnetic Resonance Imaging (MRI) can be used to distinguish a cyst from BMLs [1, 36, 37]. However, MRI cannot quantify the BMD of the patient [1]. MRI slice that yielded the greatest lesion were the cuts used to measure the subchondral cysts [2]. The extent of the subchondral cyst must be assessed on the medial and lateral tibiofemoral compartments [1].

It is unclear which specific cyst parameters, such as the number or the size, are associated with clinical symptoms, and which parameters are associated with BMD [1]. However, both medial and lateral region, cyst number, and volume were related to BMD. High cyst number per volume was also associated with high bone volume per total volume (BV/TV) [1, 25] and high trabecular thickness [1, 26]. High BMD is likely a response to higher stress, whereby local bone remodeling is affected and bone structure near the subchondral surface is changed [1, 26, 38].

A subchondral cyst was defined as a well-demarcated hypersignal, Fig. 21.2 whereas a BML could be seen as an ill-defined hypersignal.

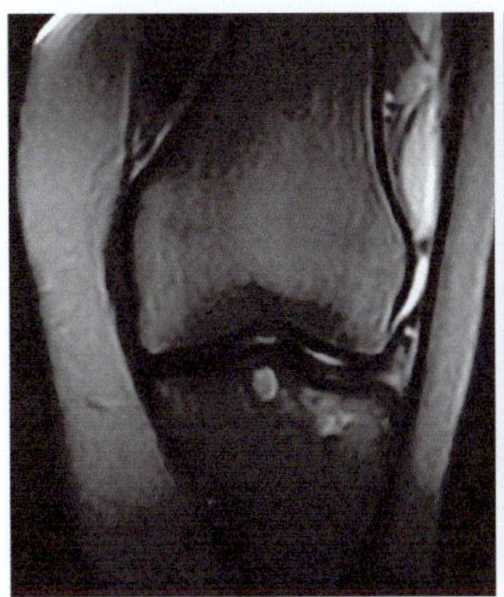
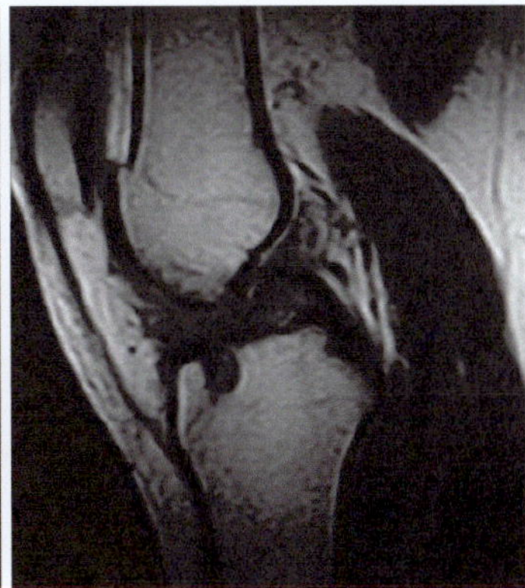
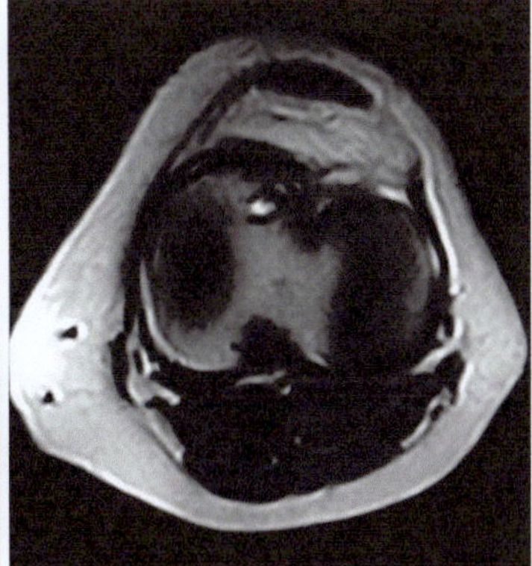

Fig. 21.2 MRI of proximal tibial subchondral cyst, left knee. Courtesy of Gobbi A

BMLs were present in 91.2% of the subregions where subchondral cysts were found [2]. Patient with a subchondral cyst had less lateral tibial cartilage volume but with greater tibial plateau bone area compared to those without subchondral bone cyst. A subchondral cyst was more likely to have large BMLs (Grade 3). On the other hand, those with BML without cyst tend to be small BML (Grade 1) [2].

Clinical quantitative computed tomography (QCT) can characterize the cyst but it is still unclear if the QCT findings are correlational with the severity of the knee pain [1]. It is reported that there are changes in bone mineral density to regions adjacent to subchondral bone cyst [1, 26]. Both MRI and QCT can offer a three-dimensional character of the cyst [1]. Kellgren-Lawrence scoring can be used to classify the osteoarthritis [1, 39].

Medial and lateral joint space widths were assessed at equal distances from the tibial spine allowing an estimation of alignment between the femoral and tibial axes. Neutral alignment was defined as $178° \pm 2°$ [1, 29]. Total subchondral bone cyst number and lateral cysts were associated with valgus alignment [1]. Lateral compartment is subjected to higher tibial loads in patient with valgus alignment [1, 40, 41].

21.8 Treatment Options

Treating subchondral bone defects and cartilage comprises both the biological as well as the structural component. Biological aspects of treatment include marrow stimulation techniques like K-wire drilling, microfracturing, nanofracturing, and core decompression. This treatment also includes additive therapies like autologous Platelet Rich Plasma (PRP) injections, adipose derivatives treatment, and bone marrow cell injections [27]. Structural component includes the subchondroplasty aspects such as cement injections, ACI procedures, allograft transplantation, bone marrow cell graft injections, and iliac crest bone grafting options [27]. Fig. 21.3.

Bone marrow stimulation technique such as microfracture led to the formation of subchondral cyst (63% of cases). Drilling prompted significant changes in almost all parameters of the architecture of the subchondral bone. It weakens the micro-architecture of the subchondral bone plate and the subarticular spongiosa. Entire osteochondral unit is altered after drilling [42].

There are a lot of treatment options in the market depending on the patient's condition. This chapter will focus on Osteo-Core-Plasty as a viable option in treating subchondral bone cyst.

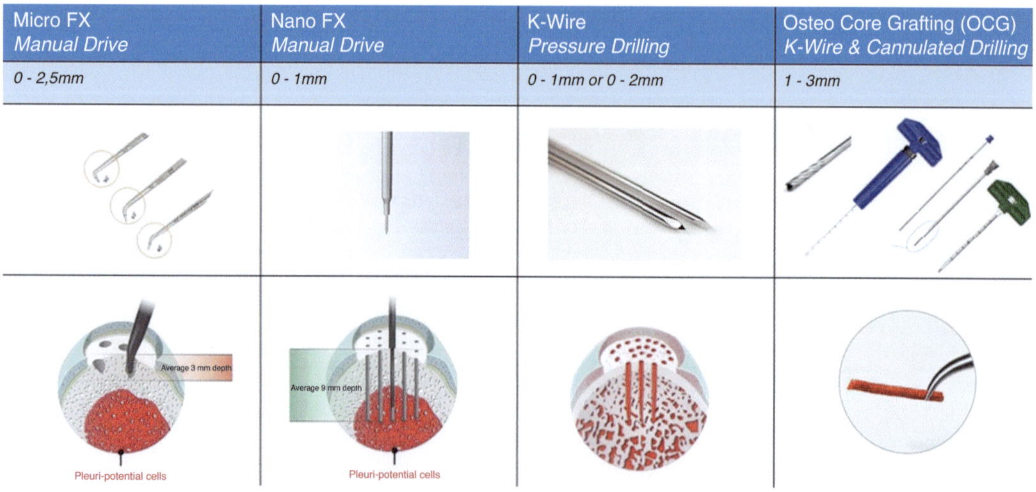

Micro FX *Manual Drive*	Nano FX *Manual Drive*	K-Wire *Pressure Drilling*	Osteo Core Grafting (OCG) *K-Wire & Cannulated Drilling*
0 - 2,5mm	*0 - 1mm*	*0 - 1mm or 0 - 2mm*	*1 - 3mm*

Fig. 21.3 Comparison of Different Subchondral Bone Treatment with their corresponding hole size in drilling and instruments used. Courtesy by Aspire-Medical [27]

21.9 Osteo-Core-Plasty

Osteo-Core-Plasty (Marrow Cellution™) is a minimally invasive subchondral bone augmentation procedure that provides both biologic and structural components to provide an optimized environment for regeneration. It is a fluoroscopic guided, minimally invasive, autologous, biologic procedure that allows necrotic bone segment resection and transplant living, live, intact bone segments that have the capabilities to reincorporate naturally without foreign body implantation [27].

It is an approach that could potentially overcome the issue of centrifugation techniques wherein there is an increased level of peripheral blood nucleated cells which contain very few stem or progenitor cells [43]. It uses multiple small volume draws (1 mL) from a single puncture that utilizes lateral flow from multiple sites near the inner cortical bone space in bone marrow (SSLM method) [43]. It is identified that this anatomical location contains a high number of bone marrow stem or progenitor cells [43–46].

Osteo-Core-Plasty starts with bone marrow aspiration process. Figure 21.4 All the materials and instruments are prepared. Aseptic technique is applied over the iliac crest and operative site. First is to heparinize all kit components using 2.000 units/mL heparin. Then, Introducer Needle with sharp stylet is inserted just past cortex into the medullary space. Sharp stylet is then removed. Syringe is attached and 1 mL marrow is aspirated to ensure proper positioning of the needle tip. Then, the syringe is removed. Blunt Stylet is inserted and locked. Introducer Needle may now be advanced to the desired depth. Then, Guide Grip is now rotated to skin level. Blunt Stylet is then removed. Next, the Aspiration Cannula is inserted and secured. Then, syringe is attached and 1 ml marrow is aspirated. Then, Guide Grip is held at the handle and rotated 360° counterclockwise then another 1 mL is aspirated. Guide Grip could be rotated as needed and could be reassembled for additional puncture sites [27].

Application could be done arthroscopically or open access method. Arthroscopic method is done with fluoroscopic guidance. Necrotic Tissue Zone is identified. K-wire is then inserted to the target zone and cannulated drill is inserted over the K-Wire. K-Wire and necrotic bone core are then removed. Extraction/Delivery Tool contain-

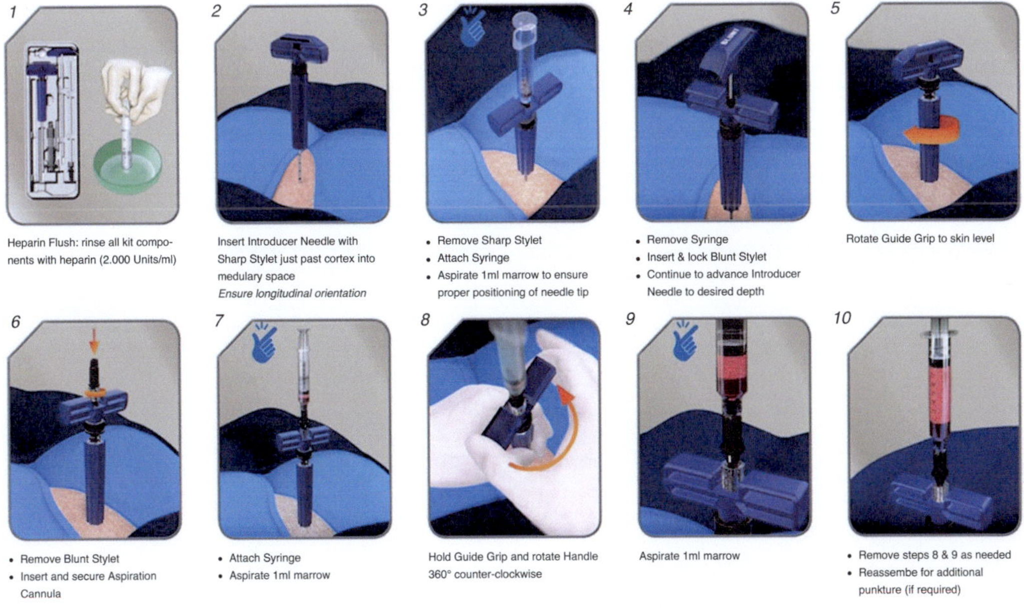

Fig. 21.4 Steps in Marrow Cellution™ Bone Marrow Aspiration Process. Courtesy by Aspire-Medical [27]

ing Marrow Cellution Bone Core Graft. Next, Probe is inserted to push bone core graft to target zone position. Lastly, Marrow Cellution™ is injected as liquid bone graft [27] Fig. 21.5.

Open technique is also done with fluoroscopic guidance wherein the necrotic tissue zone is identified. Then, cartilage bed is now debrided. After debridement, cannulated drill is inserted to required depth. Necrotic core is removed. Extraction/Delivery Tool containing Marrow Cellution™ Bone Core Graft is then inserted. Then Probe is used to push Bone Core Graft to Distal Position. Then, Marrow Cellution™ Liquid Bone Graft is injected. Then the Marrow Cellution™ Saturated Matrix Scaffold Membrane is applied. Finally, Fibrin Glue is applied to seal the membrane [27] Fig. 21.6.

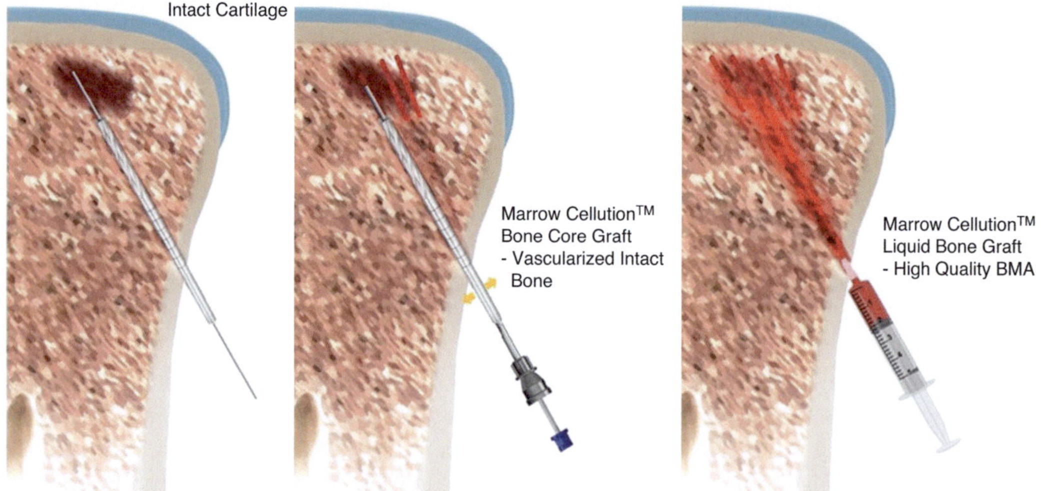

Fig. 21.5 Osteo-Core-Plasty for Intact Cartilage: Arthroscopic Application Steps. Courtesy by Aspire-Medical [27]

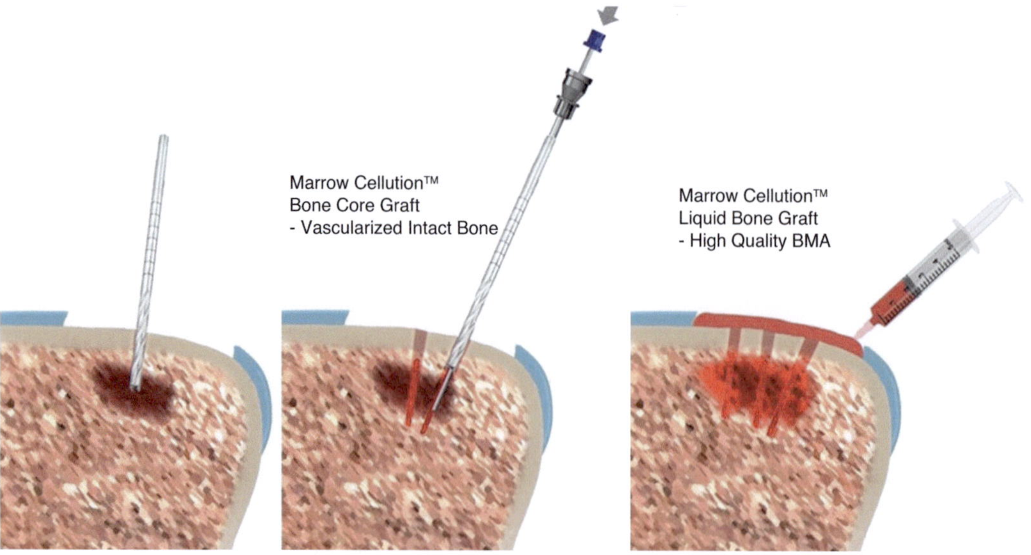

Fig. 21.6 Osteo-Core-Plasty: Open Access Surgical Procedure Steps. Courtesy by Aspire-Medical [27]

	Marrow Cellution™	Harvest BMAC®
Aspiration Volume	7-10 mL	60 mL
Final Volume	7-10 mL (No change)	7 mL
Aspiration Sites	1	3
Aspiration Time	1-2 Minutes	3-5 Minutes
Manipulated off Sterile Field	NO	YES
Processing Time	0 Minutes	17 Minutes
CFU-f/million TNC	51.89	12.37
Avg. CFU-f Concentration	1697.8 per mL	835 per mL

Fig. 21.7 Table of Comparison of Osteo-Core-Plasty using Marrow Cellution™ Technique versus Centrifugation Technique using Harvest BMAC®. Courtesy by Aspire-Medical [47]

Studies show that bone marrow samples containing a relatively high CFU-fs/mL and CD34+/mL can be attained without the need for centrifugation using the Marrow Cellution™ system. The level of CFU-fs/mL was significantly higher in the Osteo-Core-Plasty compared to BMACs in side-by-side comparison from the same patients using the contralateral iliac crest [43]. Another study showed that the Osteo-Core-Plasty had over twice as many fibroblast-like colony forming units (CFU-f) and only half as many nucleated cells compared to centrifugation techniques [47] Fig. 21.7. Moreover, the Osteo-Core-Plasty showed the same numbers of CD34+ and CD117+ cells compared to centrifugation techniques [43].

There are several benefits of Osteo-Core-Plasty. It allows the clinician to retain the product entirely on the sterile area rather than necessitating the product to leave the sterile area for centrifugation and reenter the sterile area for administration to the patient, decreases procedural expenses, and maintains all the cells and growth factors obtained during aspiration [48]. Users of this technique reported that another advantage is the ability to advance into and retreat from the marrow area in both precise and controlled manner [48]. This technique produced a higher quality aspirate with the necessity to aspirate only the volume needed for regeneration treatment [48].

21.10 Take Home Message

There is still no gold standard treatment protocol in treating subchondral bone cysts. Different treatment modalities have been tested in the hope that they might reduce pain and stop the progression of the disease.

Subchondral cysts may not directly cause the pain [33] but they are associated with subchondral bone and cartilage degeneration which furtherly causes painful osteoarthritic knees [1, 19, 25, 26].

Advancement in MRI and early diagnosis of osteoarthritis has opened a broader knowledge about the significance of subchondral bone. Long-term results using bone marrow aspirate concentrate showed promising clinical outcomes

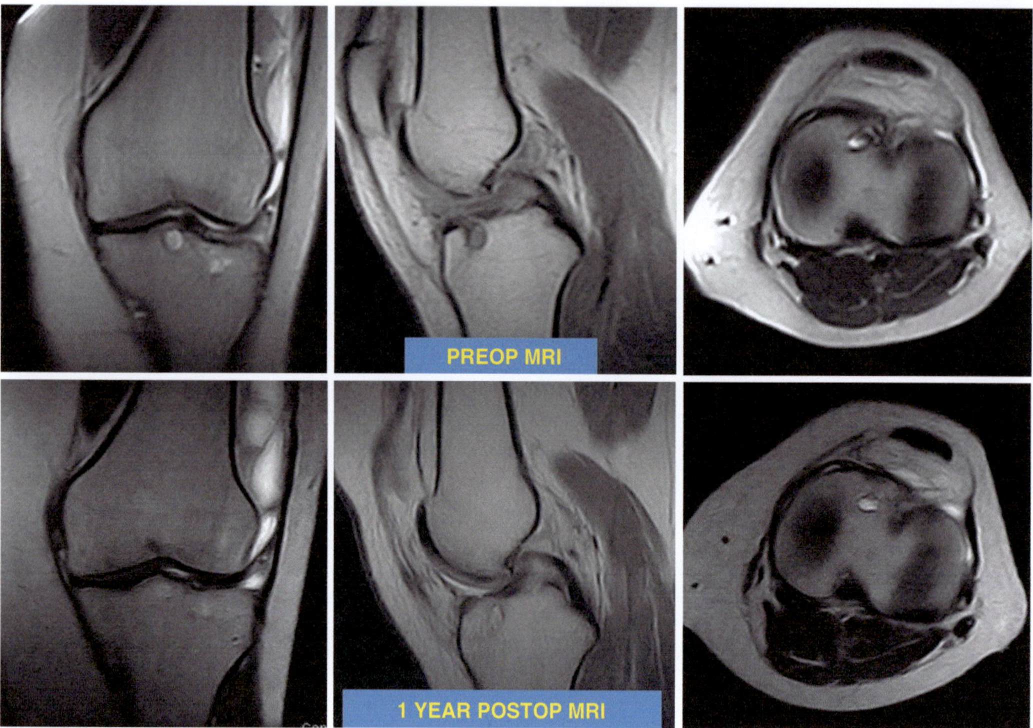

Fig. 21.8 Comparison of preoperative and postoperative Osteo-Core-Plasty MRI showing improvement manifested as increased hypointensity over the proximal tibial subchondral cyst. Courtesy of Gobbi A

in repair of cartilage lesions [49, 50]. Similar biologic treatment for subchondral cyst can aid in the healing response of such lesions.

Osteo-Core-Plasty is a viable option in treating proximal tibia subchondral cyst by reducing pain over the affected area, returning to activity early [11, 51] and improved MRI imaging showing increased hypointensity over the subchondral cyst [52, 53]. Fig. 21.8

There is still a need for high-quality RCTs studies and systematic reviews in the future to further improve treatment strategy in preventing or treating subchondral bone cyst manifested as a late stage of osteoarthritis.

References

1. Burnett WD, Kontulainen SA, McLennan CE, Hazel D, Talmo C, Wilson DR, et al. Knee osteoarthritis patients with more subchondral cysts have altered tibial subchondral bone mineral density. BMC Musculoskelet Disord. 2019;20(1):1–9.

2. Tanamas SK, Wluka AE, Pelletier J-P, Martel-Pelletier J, Abram F, Wang Y, et al. The association between subchondral bone cysts and tibial cartilage volume and risk of joint replacement in people with knee osteoarthritis: a longitudinal study [Internet]. Arthritis Res Ther. 2010;12. [cited 2020 Feb 18]. Available from: http://arthritis-research.com/content/12/2/R58

3. Tanamas SK, Wluka AE, Pelletier JP, Martel-Pelletier J, Abram F, Wang Y, et al. The association between subchondral bone cysts and tibial cartilage volume and risk of joint replacement in people with knee osteoarthritis: a longitudinal study. Arthritis Res Ther. 2010;12(2):1–7.

4. Raynauld JP, Martel-Pelletier J, Berthiaume MJ, Abram F, Choquette D, Haraoui B, et al. Correlation between bone lesion changes and cartilage volume loss in patients with osteoarthritis of the knee as assessed by quantitative magnetic resonance imaging over a 24-month period. Ann Rheum Dis [Internet]. 2008;67(5):683–8. https://doi.org/10.1136/ard.2007.073023.

5. Wu H, Webber C, Fuentes CO, Bensen R, Beattie K, Adachi JD, et al. Prevalence of knee abnormalities in patients with osteoarthritis and anterior cruciate ligament injury identified with peripheral magnetic resonance imaging: A pilot study. Can Assoc Radiol J. 2007;58(3):167–75.

6. Beattie KA, Boullos P, Pui M, O'Neill J, Inglis D, Webber CE, et al. Abnormalities identified in the knees of asymptomatic volunteers using peripheral magnetic resonance imaging. Osteoarthr Cartil [Internet]. 2005;13(3):181–6. Available from: https://linkinghub.elsevier.com/retrieve/pii/S1063458404002547

7. Burnett WD, Kontulainen SA, McLennan CE, Hazel D, Talmo C, Wilson DR, et al. Proximal tibial trabecular bone mineral density is related to pain in patients with osteoarthritis. Arthritis Res Ther [Internet]. 2017;19(1):200. Available from: http://arthritis-research.biomedcentral.com/articles/10.1186/s13075-017-1415-9

8. Sharma L, Song J, Dunlop D, Felson D, Lewis CE, Segal N, et al. Varus and valgus alignment and incident and progressive knee osteoarthritis. Ann Rheum Dis. 2010;69(11):1940–5.

9. Hvid I, Andersen LI. Acta Orthopaedica scandinavica the quadriceps angle and its relation to femoral Torsion. 2009 [cited 2020 Mar 2]; Available from: https://www.tandfonline.com/action/journalInformation?journalCode=iort20

10. Ondrouch AS. Cyst Formation in osteoarthritis. J Bone Joint Surg Br [Internet]. 1963;45(4):755–60.

11. Chua K, Kang JYB, Ng FDJ, Pang HN, Lie DTT, Silva A, et al. Subchondroplasty for bone marrow lesions in the arthritic knee results in pain relief and improvement in function. J Knee Surg [Internet]. 2019; https://doi.org/10.1055/s-0039-1700568.

12. Freund E. The pathological significance of intra-articular pressure. Edinb Med J [Internet]. 1940;47(3):192–203.

13. Landells JW. The bone cysts of osteoarthritis. J Bone Joint Surg Br. 1953;35B(4):643–9.

14. Rhaney K, Lamb DW. The cysts of osteoarthritis of the hip; a radiological and pathological study. J Bone Joint Surg Br. 1955;37–B(4):663–75.

15. Dürr HR, Martin H, Pellengahr C, Schlemmer M, Maier M, Jansson V. The cause of subchondral bone cysts in osteoarthrosis: A finite element analysis. Acta Orthop Scand [Internet]. 2004;75(5):554–8.

16. Resnick D, Niwayama G, Coutts RD. Subchondral cysts (geodes) in arthritic disorders: pathologic and radiographic appearance of the hip joint. Am J Roentgenol. 1977;128(5):799–806.

17. Freund E. The pathological significance of intra-articular pressure. Edinb Med J. 1940;47(3):192–203.

18. Carrino JA, Blum J, Parellada JA, Schweitzer ME, Morrison WB. MRI of bone marrow edema-like signal in the pathogenesis of subchondral cysts. Osteoarthr Cartil. 2006;14(10):1081–5.

19. Crema MD, Roemer FW, Marra MD, Niu J, Lynch JA, Felson DT, et al. Contrast-enhanced MRI of subchondral cysts in patients with or at risk for knee osteoarthritis: The MOST study. Eur J Radiol. 2010;75(1):e92–6.

20. Roemer FW, Guermazi A, Javaid MK, Lynch JA, Niu J, Zhang Y, Felson DT, Lewis CE, Torner J, Nevitt MC; MOST Study investigators. Change in MRI-detected subchondral bone marrow lesions is associated with cartilage loss: the MOST Study. A longitudinal multicentre study of knee osteoarthritis. Ann Rheum Dis. 2009;68(9):1461–5. https://doi.org/10.1136/ard.2008.096834. Epub 2008 Oct 1. PMID: 18829615; PMCID: PMC2905622.

21. Pathria MN, Chung CB, Resnick DL. Acute and stress-related injuries of bone and cartilage: pertinent anatomy, basic biomechanics, and imaging perspective [Internet]. Radiology. 2016;280:21–38. https://doi.org/10.1148/radiol.16142305.

22. Sanchez-Adams J, Leddy HA, McNulty AL, O'Conor CJ, Guilak F. The mechanobiology of articular cartilage: bearing the burden of osteoarthritis [Internet]. Curr Rheumatol Rep. 2014;16:1–9. https://doi.org/10.1007/s11926-014-0451-6.

23. McErlain DD, Milner JS, Ivanov TG, Jencikova-Celerin L, Pollmann SI, Holdsworth DW. Subchondral cysts create increased intra-osseous stress in early knee OA: A finite element analysis using simulated lesions. Bone [Internet]. 2011;48(3):639–46.

24. Machado M, Flores P, Ambrosio J, Completo A. Influence of the contact model on the dynamic response of the human knee joint.

25. Chiba K, Burghardt AJ, Osaki M, Majumdar S. Three-dimensional analysis of subchondral cysts in hip osteoarthritis: an ex vivo HR-pQCT study. Bone [Internet]. 2014;66:140–5.

26. Chen Y, Wang T, Guan M, Zhao W, Leung FKL, Pan H, et al. Bone turnover and articular cartilage differences localized to subchondral cysts in knees with advanced osteoarthritis. Osteoarthr Cartil. 2015;23(12):2174–83.

27. Osteo-Core-Plasty™ with Marrow Cellution™ | Aspire Medical Innovation [Internet]. [cited 2020 Mar 3]. Available from: https://aspire-medical.eu/marrowcellution/osteo-core-plasty-with-marrow-cellution/

28. Javaid MK, Kiran A, Guermazi A, Kwoh CK, Zaim S, Carbone L, et al. Individual magnetic resonance imaging and radiographic features of knee osteoarthritis in subjects with unilateral knee pain: The health, aging, and body composition study. Arthritis Rheum [Internet]. 2012;64(10):3246–55. https://doi.org/10.1002/art.34594.

29. Burnett WD, Kontulainen YZSA, Mclennan CE, Hazel D, Talmo C, Hunter DJ, et al. Knee osteoarthritis patients with severe nocturnal pain have altered proximal tibial subchondral bone mineral density. Osteoarthr Cartil [Internet]. 2015;23:1483–90. https://doi.org/10.1016/j.joca.2015.04.012.

30. Bellamy N. Outcome measurement in osteoarthritis clinical trials. J Rheumatol Suppl [Internet]. 1995;43:49–51.

31. Samaroo KJ, Tan M, Putnam D, Bonassar LJ. Binding and lubrication of biomimetic boundary lubricants on articular cartilage. J Orthop Res [Internet]. 2017;35(3):548–57.

32. Bellamy N, Buchanan WW, Goldsmith CH, Campbell J, Stitt LW. Validation study of WOMAC: A health status instrument for measuring clinically impor-

tant patient relevant outcomes to antirheumatic drug therapy in patients with osteoarthritis of the hip or knee. J Rheumatol. 1988;15(12):1833–40.

33. Torres L, Dunlop DD, Peterfy C, Guermazi A, Prasad P, Hayes KW, et al. The relationship between specific tissue lesions and pain severity in persons with knee osteoarthritis. Osteoarthr Cartil [Internet]. 2006;14(10):1033–40.

34. Audrey HX, Bin ARHR, THC A. The truth behind Subchondral cysts in osteoarthritis of the knee. Open Orthop J. 2014;8(1):7–10.

35. Schajowicz F, Clavel Sainz M, Slullitel JA. Juxta-articular bone cysts (intra-osseous ganglia). A clinico-pathological study of eighty-eight cases. J Bone Jointt Surg Br. 1979;61(1):107–16.

36. Kornaat PR, Bloem JL, Ceulemans RYT, Riyazi N, Rosendaal FR, Nelissen RG, et al. Osteoarthritis of the knee: association between clinical features and MR imaging findings. Radiology. 2006;239:811–7.

37. Roemer FW, Eckstein F, Guermazi A. Magnetic resonance imaging-based semiquantitative and quantitative assessment in osteoarthritis [Internet]. Rheum Dis Clin N Am. 2009;35:521–55. Available from: https://linkinghub.elsevier.com/retrieve/pii/S0889857X09000647

38. Frazer LL, Santschi EM, Fischer KJ. The impact of subchondral bone cysts on local bone stresses in the medial femoral condyle of the equine stifle joint. Med Eng Phys [Internet]. 2017;48:158–67.

39. KELLGREN JH, LAWRENCE JS. Radiological assessment of osteo-arthrosis. Ann Rheum Dis. 1957;16(4):494–502.

40. Werner FW, Ayers DC, Maletsky LP, Rullkoetter PJ. The effect of valgus/varus malalignment on load distribution in total knee replacements. J Biomech [Internet]. 2005;38(2):349–55.

41. Hsu HP, Garg A, Walker PS, Spector M, Ewald FC. Effect of knee component alignment on tibial load distribution with clinical correlation. In: Clinical Orthopaedics and Related Research; 1989. p. 135–44.

42. Orth P, Goebel L, Wolfram U, Ong MF, Gräber S, Kohn D, et al. Effect of subchondral drilling on the microarchitecture of subchondral bone: Analysis in a large animal model at 6 months. Am J Sports Med [Internet]. 2012;40(4):828–36.

43. Scarpone M, Kuebler D, Chambers A, De Filippo CM, Amatuzio M, Ichim TE, et al. Isolation of clinically relevant concentrations of bone marrow mesenchymal stem cells without centrifugation. J Transl Med [Internet]. 2019;17:10. https://doi.org/10.1186/s12967-018-1750-x.

44. Jones E, English A, Churchman SM, Kouroupis D, Boxall SA, Kinsey S, et al. Large-scale extraction and characterization of CD271+ multipotential stromal cells from trabecular bone in health and osteoarthritis: Implications for bone regeneration strategies based on uncultured or minimally cultured multipotential stromal cells. Arthritis Rheum [Internet]. 2010;62(7):1944–54. https://doi.org/10.1002/art.27451.

45. Jones E, Schäfer R. Where is the common ground between bone marrow mesenchymal stem/stromal cells from different donors and species? [Internet]. Stem Cell Res Ther. 2015;6:143. Available from: http://stemcellres.com/content/6/1/143

46. Birbrair A, Frenette PS. Niche heterogeneity in the bone marrow. Ann N Y Acad Sci. 2016;1370:82–96.

47. Harrell DB. Comparative analysis of marrow cellution TM and the BMAC® HARVEST®/TERUMO® SYSTEM; 2016.

48. Harrell DB, Purita JR, Gonzalez R, Liriano L. Novel technology to increase concentrations of stem and progenitor cells in marrow aspiration.

49. Gobbi A, Karnatzikos G, Scotti C, Mahajan V, Mazzucco L, Grigolo B. One-step cartilage repair with bone marrow aspirate concentrated cells and collagen matrix in full-thickness knee cartilage lesions: results at 2-year follow-up. Cartilage [Internet]. 2011;2(3):286–99.

50. Gobbi A, Karnatzikos G, Sankineani SR. One-step surgery with multipotent stem cells for the treatment of large full-thickness chondral defects of the knee. Am J Sports Med [Internet]. 2014;42(3):648–57.

51. Pietropaoli MA. Victory Sports Medicine & Orthopedics | Subchondroplasty: Knee repair, not knee replacement [Internet]. Victory Sports Medicine. 2019 [cited 2020 Mar 4]. Available from: https://victorysportsmedicine.com/services/knee-repair/

52. Calderone ME, Patel M, Brant MK. MRI findings after a subchondroplasty procedure of the ankle: a MRI findings after a subchondroplasty procedure of the ankle: a case report case report. 2015 [cited 2020 Mar 4]; Available from: https://rdw.rowan.edu/stratford_research_day/2019/may2/13

53. Bonadio MB, Ormond Filho AG, Helito CP, Stump XM, Demange MK. Bone marrow lesion: image, clinical presentation, and treatment. Magn Reson Insights. 2017;10:1178623X1770338.

Inflammatory Environment and Cartilage Repair

Fabio Valerio Sciarretta

Articular cartilage is the highly specialized connective tissue that covers and protects the joints surfaces of diarthrodial joints. It normally appears as a rubber-like, smooth tissue that offers an elastic resistance and rebounces when probed during arthroscopic procedures and its principal function is to provide a smooth, lubricated surface for articulation and to facilitate the transmission of loads with a low frictional coefficient.

The historical identification and comprehension of cartilage as an isolated tissue represents a very particular and fascinating attractive story, [1] starting from the description of the joint environment made from Galen of Pergamon (Pergamon, 129 A.D.—Rome, 210 A.D.) in his treatise "On the usefulness of various parts of the body," written between A.D. 165 and 175 and known for centuries just in Greek, Latin, or Arabic, when referring to joint protection, the famous Greek physician and surgeon affirms "Nature has again searched out a double remedy, first covering each member of the joint with cartilage and then pouring over the cartilages themselves a sort of oily substance, a greasy, glutinous fluid, which gives every joint an easy movement and protection against wear" so that cartilage serves as grease for the joints" [2]. Also one of the most significant physicians, astronomers, and writers of the Islamic Golden Age, considered the father of modern medicine, the Persian Avicenna (Bukhara, 980–Hamadan, 1037), after eight centuries, was confirming the particular shock-absorbing ability of cartilage "it was made for the purpose of providing a cushion between hard bone and the soft members, so that the latter should not be injured when exposed to a blow or fall, or compression" [3].

Passed five centuries, the Flemish anatomist and physician Andreas Vesalius (Brussels, 1514–1564 Zante), father of the modern human anatomy, following Galen, stated that cartilage "has no sensation and no marrow," but deepened the description of cartilage evolution during life decades "In younger people cartilages are soft, but with age they harden and resemble the fragility and friability of bone… In older people epiphysis are no longer joined to their bones by the intervention of cartilage which plays the part of glue, but have lost the cartilage and are joined in such a way… it is difficult to see the point of union" [4].

The first scientific description of the articular cartilage structure dates to the eighteenth century, exactly in 1743, in "Of the structure and diseases of articulating cartilages" treatise written by the famous London surgeon and anatomist William Hunter (East Kilbride, 1718–London, 1783), where, in the wonderful incipit of the book He states "The Fabric of the Joints in the human body is a subject as much the more entertaining, as it must strike everyone that considers it

F. V. Sciarretta (✉)
Clinica Nostra Signora della Mercede, Rome, Italy

© ISAKOS 2022
A. Gobbi et al. (eds.), *Joint Function Preservation*, https://doi.org/10.1007/978-3-030-82958-2_22

attentively with an idea of fine mechanical composition. Wherever the motion of one bone upon another is requisite, there we find an excellent apparatus for rendering that motion safe and free." Following He refers that "An articulating Cartilage is an elastic Substance uniformly compact, of a white Colour, and somewhat diaphanous, having a smooth polished Surface covered with a Membrane; harder and more brittle than a Ligament, softer and more pliable than a Bone… When an articulating Cartilage is well prepared, it feels soft, yields to the touch, but restores itself to its former Equality of Surface when the pressure is taken off. This Surface, when viewed through a Glass, appears like a Piece of Velvet… a mass of short and nearly parallel Fibres rising up from the Bone as the silky Threads of that rise from the woven Cloth or Basis… Now these perpendicular Fibres make the greatest Part of the cartilaginous Substance; but without Doubt there are likewise transverse Fibrils which connect them, and make the Whole a solid Body, though these last are not easily seen, because being very tender, they are destroyed in preparing the Cartilage" [5].

In the same Century, the famous Italian anatomist Giovanni Battista Morgagni (Forli, 1682–Padua 1771) gave us the first description of the possible damage to the cartilaginous coating of bones describing the osteoarthritic cartilage changes in the report of an autopsy performed in 1741 of a woman who was "frequently afflicted with ischiadic pains"… "The head of the right os femoris was not rounded into a globular form: and was depressed, and not covered by a smooth and white cartilage, but by one of a pale ash-colour: and, indeed, this cartilage was totally deficient in the posterior part of the head; so that the bone appeared naked in that part, and formed into many roundish and protuberant particles" [6].

We had to wait almost another century before the rudimentary chemistry determined that cartilage had two principal components: collagen fibers and chondrin and that chondrin was shown to be the "chondroitin sulphuric acid"; but still in 1944 Comroe in "Arthritis and Allied Conditions" stated that "Articular cartilage contains glycogen, collagen, chondroitin sulfuric acid, lactic acid, and calcium salts; but its exact chemical composition is not accurately known" [7].

In 1925, light microscope analysis revealed three layers in the articular cartilage according to the distribution of chondrocytes and collagen fibers orientation. In 1969, Mankin [8] showed that chondrocytes have metabolic activity and in 1971 Strawich [9] that type II is the main collagenous component of articular cartilage, completed in 1978 by Rhodes [10] who identified the physicochemical composition of articular collagen. Finally, in the following years, electronic microscopy made it possible to reach modern informations and opened the way to the modern era of cartilage repair.

Osteoarthritis (OA) is a chronic, long-term debilitating disease characterized by the deterioration of articular cartilage covering bone surfaces of the joints and most of the surrounding tissues, which creates various discomforting symptoms as pain, scrosci articolari, swelling, stiffness, and limited range of motion and ambulation. The burden of OA is physical, psychological, and socioeconomic. OA can be associated with significant disability, such as a reduction in mobility and activities of daily living. Psychological sequelae include distress, devalued self-worth, and loneliness. Given the high frequency of OA in the population, its economic burden is large [11]. OA most commonly affects the joints in the knees, hands, feet, and spine and is also relatively common in shoulder and hip joints. Everybody knows it is a very common disease, in fact, OA has a global impact that represents one of the major challenges for national health systems all over the world in the twenty-first century. In 2005, it was estimated that 26.9 million US adults have clinical OA defined on the basis of symptoms and physical findings and that up to 8.5 million people in the UK are affected by joint pain that can be attributed to osteoarthritis [12–14]. Prevalence of OA increases with age: 13.9% of adults age 25 and older have clinical OA of at least one joint, while 33.6% of adults age 65 and older have OA. The global prevalence of hip and knee OA is approaching 5%, as already confirmed in 1998 when the highest prevalence of knee pain was found in women over 75 years (35%) and is projected to increase as the population ages [15]. The most recent update of the Global Burden of Disease figures (GBD 2013) estimated that 242 million people were living in the world with symptomatic and activity limiting OA of the hip and/or knee, accounting for 13 million years lived with disability. These figures are likely to be an underestimate of the true global burden of OA, as these rates only consider hip and knee OA and not OA at other sites.

Apart from the incredible epidemiological importance for our society, for the goal of this chapter, OA has to be understood and focused in relation to its origin. OA is a heterogeneous, common, very complex disorder, which presents in its background not just one but many and many risk factors and causes. Various specific risk factors have been identified including obesity, metabolic diseases, age, sex, ethnicity and race, occupation, smoking, bone density, and muscle function. The identified risk factors can mainly be distinguished in genetic, constitutional, environmental, and local factors, each one of them being more or less determinant in the different joints [16]. The most studied and well-known OA risk factor is obesity, whose association with hip OA is weaker than with knee OA [17].Among the genetic predisposition, it is certainly to be emphasized that hand, hip, and knee OA are hereditary in 40–60% of cases.

Obese constitution represents a risk factor, that, as some local factors as recreational trauma or joint hypermobility or muscle weakness, can be reversed. Some constitutional risk factors may differ for developing OA, as high bone density, or for OA progression and poor clinical results, as low bone density represents a risk factor both in knee and hip osteoarthritis.

The precise recognition and identification of the various osteoarthritic features have been studied in order to determine a radiographic classification in different stages of OA. The Kellgren-Lawrence classification is the most widely used, especially in clinical researches. This classification evaluates the appearance of osteophytes and cysts, joint space loss, and sclerosis, and it grades the severity from 0 to 5 points.

The radiological features found in OA joints have been graded as follows: (1) formation of osteophytes on joint margins or on tibia spines for knee OA; (2) periarticular ossicles in relation to distal and proximal interphalangeal joints; (3) narrowing of joint cartilage and sclerosis of subchondral bone; (4) pseudocystic areas with sclerotic walls in the subchondral bone; and (5) altered shape of the bone ends [18]. Some of these criteria were adopted by the World Health Organization (WHO) as the standard for studies on OA.

At present time there is no medical treatment that can prevent, limit, or stop the progression of OA [19]. Pain is the main symptom that renders OA a functionally devastating condition, being able to determine important joint loss of function and disability. OA-derived disability has a higher incidence in women, especially in those with lower educational levels and the socially disadvantaged and in people relying on manual labor, weight-bearing, or positions that require knee bending, long-time standing, and walking during their daily activities.

Current medical modalities to reduce pain in the treatment of OA are nonsteroidal anti-inflammatory drugs (NSAIDs) and joint visco-supplementation by intra-articular injections of hyaluronic acid. Main downsides of these treatments are the short-term effect and risks connected to the chronic use of NSAIDs, mainly toxicity and risk of thromboembolism [20, 21]. In the severity progression of the cases, surgical procedures, from biological repair and partial resurfacing procedures to mostly joint replacement surgeries, as total hips or total knees, become suggested [11]. During the last years, orthobiology has emerged looking to anticipate tissue degradation and promote tissue regeneration. Clinical trials using orthobiologics, such as platelet-rich plasma (PRP), bone marrow aspirate concentrate (BMAC), fat graft, and mesenchymal stem cells, have shown promising results for the treatments of OA.

One of the possible revolutions in the treatment of OA may be derived from the use of Mesenchymal Stem Cells (MSCs). MSCs are perivascular cells and are known to be able to accumulate in damaged tissues where they develop their function of promoting tissue regeneration by replacing cells or by empowering the regenerative capacity of in situ quiescent cells and through their immunomodulatory activities in order to reduce inflammation in cartilage breakdown and OA. In fact, MSCs answer to inflammation by producing and releasing several growth factors that promote angiogenesis, remodel the extracellular matrix, and differentiate the progenitor cells thus promoting tissue repair, but at the same time are able to modulate the immune cells in the inflamed tissue microenvironment, basing their action on the type and intensity of inflammation [22]. All innate immune cells precedently described (macrophages, mast cells, natural killer, and others) present in the site of inflammation can be regulated by MSCs. This MSCs action on the innate immune system will provoke indirect effects also on adaptive immune system cells, including T cells. For these cells, it has been proven that MSCs are able to inhibit their proliferation through the expression of chemokines and inducible NO synthase (iNOS; in the case of rodent MSCs) or IDO (in the case of human and other mammalian MSCs) [23, 24]. MSCs secrete contains, in fact, multiple growth factors, cytokines, anti-inflammatory mediators (PGE2, TSG6, HO1, and galectins), and exosomes, all contributing to the immunomodulatory and immunosuppressive action of MSCs, properties that have been used and proven to be effective in resolving inflammation in systemic lupus erythematosus, multiple sclerosis, kidney injury, fibrosis, and arthritis [25]. MSCs may be isolated by various sources, mainly bone marrow and adipose tissue, have been demonstrated to possess chondrogenic differentiation capacity and their use relies on their capacity to increase the population and the function of the pluripotent cells present in cartilage defects. For this reason, articular cartilage damage and OA have been considered primary areas of MSCs-based therapies [26]. Since OA may be the result of dysfunction in the MSCs population, giving rise to degenerative changes in the absence of repair [27], MSCs may represent and effective cartilage and OA degraded tissues repair. Recently, MSCs action and effects have been greatly correlated to MSCs paracrine effects, that can be distinguished into various separate actions, mainly trophic (antiapoptosis, angiogenesis, and support of growth and differentiation of stem and progenitor cells) immunomodulation, antiscarring, and chemoattraction through the secretion of a myriad of bioactive molecules, including bFGF (basic fibroblast growth factor), CCL (chemokine ligand), CXCL = chemokine (C-X-C ligand), GM-CSF (granulocyte–macrophage colony-stimulating factor), HGF (hepatocyte growth factor), IGF-1 (insulin growth factor-1), LIF (leukemia inhibitory factor), M-CSF (macrophage colony-stimulating factor), mDC (macrophage-derived chemokine), NK (natural killer), PGE2 (prostaglandin E2), SCF (stem cell factor), SDF-1 (stromal cell-derived factor 1), TGF-β (transforming growth factor-β), and VEGF (vascular endothelial growth factor) that represent the agents of MSCs revolution. Preclinical animal works have confirmed MSCs' success in repairing cartilage and preventing OA by intra-articular injections of bone marrow-derived MSCs (BM-MSCs) or adipose tissue-derived MSCs (AD-MSCs). In the literature, we can actually find also several studies reporting the effects and results of intra-articular injection of autologous MSCs in humans for the treatment of knee OA [28–33]. Centeno et al. [28] reported significant cartilage and meniscus growth, and reduced pain and increased joint mobility in patients with degenerative joint disease at 24 weeks after autologous bone marrow-derived MSC injection [35]. Emadedin et al. [29] observed satisfactory effects, in terms of pain, functional status, and walking ability, of just one intra-articular injection of bone marrow-derived MSCs in patients with knee, ankle, or hip OA, at 30 months after injection of MSCs. They observed no severe adverse events such as pulmonary embolism, death, or systemic complications and no tumor growth. A limited number of patients had very minor localized adverse effects such as rash and erythema. Positive clinical scores results were confirmed by MRI. The authors concluded that injection of MSCs in different OA-affected joints is safe and therapeutically beneficial. The same group, in 2019, [30] conducted a triple-blind, randomized controlled trial (RCT) with a placebo control in knee OA patients. Forty-three patients (Kellgren-Lawrence grades 2, 3, and 4) were assigned to either the MSCs ($n = 19$) or placebo ($n = 24$) group. The study demonstrated the safety and efficacy of single intra-articular implantation of 40 × 106 autologous MSCs in patients with knee OA. Regarding adipose tissue-derived MSCs, Jo et al. [31] evaluated the potential of intra-articular injection of adipose tissue-derived MSCs for the treatment of knee OA at low (1.0×10^7 cells), mid-dose (5.0×10^7), and high-dose (1.0×10^8). "The WOMAC score improved at 6 months after injection in the high-dose group. The size of cartilage defect decreased while the volume of cartilage increased in the medial femoral and tibial condyles of the high-dose group. Arthroscopy showed that the size of cartilage defect decreased in the medial femoral and medial tibial condyles of the high-dose group. Histology demonstrated thick, hyaline-like cartilage regeneration. These results showed that intra-articular injection of $1.0 \times 10(8)$ AD-MSCs into the osteoarthritic knee improved function and pain of the knee joint without causing adverse events, and reduced cartilage defects by regeneration of hyaline-like articular cartilage," authors concluded. Hong et al. [32] in 2019 evaluated intra-articular injection of autologous stromal vascular cells in a double blind randomized study with HA injection in opposite knees of same patients and concluded that autologous adipose-derived SVF cells treatment is safe and can effectively relieve pain, improve function, and repair cartilage defects in patients with knee osteoarthritis at 12 months follow-up and on MRI controls WORMS and MOCART measurements revealed a significant improvement of articular cartilage repair in SVF-treated knees compared with HA-treated knees. Another group of authors some months ago conducted a prospective double-blinded, randomized controlled, phase IIb clinical trial [33]. Where a group of 12 patients who underwent injection with autologous adipose-derived mesenchymal stem cells (AD-MSCs) was compared with 12 knees with injection of normal saline up to 6 months. The Authors found statistical improvement of VAS, WOMAC, and KOOS scores in AD-MSCs patients treated group. At MRIs follow-ups, WORMS and MOCART measurements revealed a significant improvement of articular cartilage repair in SVF-treated knees compared with saline-treated knees [33].

But, while no definite treatment has been demonstrated able to cure the disease, great efforts have been done in trying to better understand the etiopathogenesis of OA with the goal that better understanding would have forwarded the development of new cartilage and OA repair treatments. Until the end of the last century, OA was still retained due to a mechanical derangement of the joint environment, that, in consequence of the overloading and the modifications of the joint pressures inside the joint envelope, or the chronic damage from prior mechanical derangements such as a meniscal tear, hypermobility, or anatomic malalignment would have brought to the progressive degeneration of the cartilage sheet that covers the bony surfaces of the joints, damaging, in particular, its extracellular matrix of the cartilage, especially if associated to a genetic alteration of its components [34]. This "wear and tear paradigm" was deeply connected to the fact that cartilage, without blood vessels and nerves terminations was unable, when damaged, to react, as it should normally, by an inflammatory answer and was not able to repair itself due to the low metabolic capacity of its cells, the chondrocytes. In fact, under a loading that exceeds the capacity of the tissue, degradation of matrix macromolecules exceeds their synthesis, causing joint tissue degeneration, possibly leading to OA [35]. Over the last 20 years, molecular biology found that some mediators, as the cytokines, could increase the production of matrix metalloproteinases (MMPs) by chondrocytes, and this brought, during the following years, to develop a new "inflammatory" theory at the basis of OA pathogenesis, retaining synovitis one of most important features of OA, considered now as an inflammation-associated multifactorial disorder.

The Osteoarthritis Research Society International (OARSI) having emphasized the highly heterogenous phenotypical origin of OA, in 2015, has led to create a new shared and updated definition of OA: "Osteoarthritis is a disorder involving movable joints characterized by cell stress and extracellular matrix degradation initiated by micro- and macro-injury that activates maladaptive repair responses including proinflammatory pathways of innate immunity. The disease manifests first as a molecular derangement (abnormal joint tissue metabolism) followed by anatomic, and/or physiologic derangements (characterized by cartilage degradation, bone remodeling, osteophyte formation, joint inflammation and loss of normal joint function), that can culminate in illness." As per the definition, OA-specific manifestations of illness will likely differ according to OA phenotype. OA may be manifested by a prolonged period of musculoskeletal tissue abnormalities at a molecular but clinically silent level that can precede the anatomic organ system disease and illness manifestations by years or even decades [36].

Scientists have concluded that inflammation in OA is different from that encountered in rheumatoid arthritis and other autoimmune diseases: it is chronic, comparatively low-grade, and mediated primarily by the innate immune system and in lower size by the adaptive immune system [37]. So the actual frame of OA development is retained to be:

1. Modification of joint loading due to a single important trauma or repeated microtrauma or aging and genetic abnormalities that cause damage to joint tissues and initiate cartilage breakdown [38].
2. Chondrocytes release more catabolic enzymes as MMPs, which in turn fourthly cause cartilage damage [39].
3. Release of matrix components which elicits inflammation [40].
4. Activation of immune, innate, and adaptive response. This is the most important step that has modified the interpretation of OA etiopathogenesis based on the results of many studies that have confirmed a relationship between synovitis and cartilage deterioration and pain, [41, 42] infiltrate of T cells, B cells, and macrophages in OA patients synovial

membrane, [43–45] retrieval of immunoglobulins in OA patients cartilage, [39] synovium and plasma, and the confirmation of the fact that the complement results importantly activated in OA synovitis [46].

In fact, inflammation is encountered in the early stages of OA and has several specific characteristics that make this form different from the one, high-grade, encountered during rheumatoid arthritis. Specifically, OA inflammation is caused by joint damage occurring more often in patients with the various risk factors identified as obesity, advanced age, prior trauma, joint overuse, disruption of circadian rhythms, or genetic abnormalities. The body reacts to the initial damage by the activation of several molecular, immune and mechanical pathways, transducing joint trauma, chronic injury, or overuse damage into inflammatory processes [47]. The innate immune system recognizes the endogenous damage-associated molecular patterns (DAMPs), that are molecules produced during tissue damage, through its innate immune cells like mast cells (regulators of vascular permeability that seem to play a crucial role in OA joint inflammation since they facilitate leukocyte infiltration) and macrophages and initiates a protective or reparative immune response normally guided by various inflammatory mediators. In OA there is a prolonged or dysregulated activation of DAMPs, through an exacerbated cytokine release, that causes a destructive inflammation [48]. DAMPs have been identified in several sources: in cartilage ECM components (Biglycan, Fibronectin, Tenascin C, and LMW hyaluronic acid), in plasma proteins (α1 microglobulin, α2 microglobulin, fibrinogen), in crystals (basic calcium phosphate, calcium pyrophosphate dehydrate, uric acid) and in other sources, all of them contributing to the chronic OA inflammation. Also, the complement system malfunction has been involved in OA etiology. Normally, the complement system enhances the ability of antibodies and phagocytic cells to clear pathogens from an organism through chemotaxis, exudation of plasma proteins at inflammatory sites, and opsonization of damaged cells. Interestingly, several products of tissue breakdown, and especially ECM components, in the joint are capable of activating both DAMPs and complement, as cartilage oligomeric matrix protein, osteoadherin, and chondroadherin [49, 50].

Also, different soluble mediators of inflammation have been demonstrated to be implicated in OA etiopathogenesis, contributing to cartilage degradation and synovial cell activation. Among these need to emphasized several cytokines, chemokines, growth factors, adipokines, prostaglandins, and leukotrienes that are produced by different cell types within the joint, including fibroblast-like synoviocytes and chondrocytes. Among the cytokines, IL-6, IL-8, and IL-15, have been detected in elevated levels in OA joints and are retained to have a catabolic role inside the joints (In OA joints, IL-1β and TNF amplify the arthritic condition by inducing the production of proinflammatory cytokines, such as IL-6, IL-8, and monocyte chemoattractant protein 1); many cytokines can also promote OA progression by inhibiting anabolic processes critical to cartilage homeostasis [51, 52]. Many chemokines (IL-8, CCL5, CCL19, CCR1, CCR2, CCR3, and CCR5), that represent a subset of cytokines that induce the recruitment and trafficking of inflammatory cells and mesenchymal progenitors, produced in the joint affected by OA, might facilitate the onset and progression of the disease [53], while others (stromal cell-derived factor-1—CXCL12) seem to promote tissue repair [54]. TGF-β growth factors that work to maintain cartilage homeostasis have been associated with the development of osteophytosis and synovial fibrosis in OA joints [55]. The vascular endothelial growth factor (VEGF) generated by the inflamed synovium can promote angiogenesis and thereby facilitate infiltration of the joint by immune cells [56]. Also, the family of cytokines mainly derived from adipose tissue, namely adipokines, have been associated with OA [57], because part of them, including leptin, adiponectin, visfatin, and resistin, have been demonstrated to be able to induce the production of inflammatory mediators and cartilage-degrading factors, leading to chondrocyte degradation and the development of OA [57, 58]. Interesting and to be emphasized is also the relationship between the behavior of the infrapatellar fat pad as a local mediator of pain and inflammation in OA. Clockaerts et al. in 2010 concluded that the infrapatellar fat pads, derived from the knees of patients with OA, contain an increased number of macrophages, lymphocytes, and granulocytes [59]. Studies from fat pad explants have demonstrated the ability of this tissue to produce and secrete large amounts of leptin and adiponectin and various inflammatory mediators such as VEGF, TNFα, and IL-6 [60, 61]. Additionally, the OA fat pad is highly innervated by small C-fiber neurons containing the neuroinflammatory mediator and vasodilator Substance P that mediates not only pain sensation, but also directly acts on a variety of immune cells and the vascular system to induce proinflammatory cytokine (IL-1β and TNFα) production and vascular leak, respectively [62]. In conclusion, fat pads can represent another source of inflammatory mediators, such as adipokines and neuropeptides, and of soluble mediators of inflammation such as IL-1β, TNFα, and IL-6.

Prostaglandins and leukotrienes are generated from arachidonic acid via distinct enzymatic cascades that can be induced by inflammation or trauma. The cyclooxygenase-2 (COX-2) enzyme is upregulated in inflamed joint tissues and is responsible for elevated production of lipid mediators including prostaglandins, such as PGE2, which promotes inflammation, apoptosis, and angiogenesis in the OA joint. Leukotriene B4 (LTB4) and its metabolite LTC4, are produced by OA synovium and to a lesser extent OA bone and cartilage and acts as leukocyte chemoattractant, possibly stimulating the production of IL-1β and TNF by synovial tissues [63, 64].

As stated earlier, OA is a greatly complex disease in which the inflammatory mediators are released not only by cartilage, but also by bone and synovium and their source may differ by OA phenotype [65]. The OA symptom joint swelling is clearly related to synovitis and some studies have evidenced that the presence of synovitis observed during arthroscopies and on MRIs may be a surrogate marker of severity and associated with increased risk of radiographic evidence of disease progression [41, 66]. In fact, the paper by

Ayral et al. [41] showed, by serial arthroscopies performed on knees with symptomatic but preradiographic OA stages, the association between the presence of synovitis and the future development of medial cartilage loss. Other papers have confirmed the onset of synovitis in the early stages of OA. In the study by Haywood et al. [66] synovial inflammation was present in many patients with minimal OA radiographic signs and in the study by Scanzello et al. [67] in patients who underwent arthroscopic meniscectomy without evidence of radiographic OA, synovial inflammation was noted in 43% of cases.

In the literature, no clear explication of synovitis origin is found [68]. It is thought that the detached cartilage fragments get in contact with the synovium membrane which reacts by inflammatory mediators production retaining the foreign bodies. These mediators can activate the chondrocytes to synthesize metalloproteinase enzymes and, eventually, increase cartilage degradation. The same mediators can also increase the synthesis of inflammatory cytokines and MMPs by synovial cells themselves and OA synovitis perpetuates the cartilage degradation.

Another theory retains, instead, that synovial Inflammation may drive synovial angiogenesis, linked to OA pain, through macrophage activation [56].

As stated by Sellam and Berembaum [69], although synovium is not the only tissue involved in OA inflammation it represents a major site of gross and microscopic inflammatory change, with a hyperplasia of the tissue layers and presence of various inflammatory cells, including macrophages, T and B cells, and natural killer cells, normally absent. But we now know that OA disease affects the entire joint structure, because it really is a whole-organ disease of the joint [47, 65]. Cartilage is involved and undergoes fibrillation and degradation, single or multiple cartilage defects with a possible detachment of fragments that cause synovium activation, bone reacts by subchondral bone thickening and formation of osteophytes; menisci present damage with tears; the capsule undergoes process of inflammation, thicknening, and hypertrophy and also ligaments and tendons degenerate. All these OA changes

follow several cellular and molecular processes: an increase in cartilage catabolism and a concomitant decrease in cartilage anabolism and repair; hypertrophy and death of chondrocytes; impairment or dysregulation of autophagy; osteoclast-mediated remodeling of bone; and infiltration and activation of immune cells [65].

Imaging is greatly improving our revelation and identification of all tissues modifications involved in OA, especially in the early stages of OA and particularly in weight-bearing joints as knees and hips. In particular, MRI represents the main synovium examining the instrument, but enables also to identify the main bone early involvement and sign in OA as the bone marrow edema, that can be supposed to be the sign of excessive stresses delivered at the bone-cartilage interface [70].

Specifically looking at the knee, in 2012, Luyten et al. [71] have proposed a classification of early OA that can be defined based on clinical and imaging findings, and should meet three criteria: (I) knee pain, (II) Kellgren-Lawrence (KL) (6) grade 0, I, or II (osteophytes only) using plain radiographs, and (III) cartilage lesion confirmed by arthroscopy and/or OA-related MRI findings such as degenerations of cartilage and meniscus, and/or subchondral bone marrow lesions (BMLs).

MRI features of degenerative changes of the cartilage, BMLs, and/or meniscus are based on the Boston Leeds Osteoarthritis Knee Score (BLOKS), the Whole Organ Magnetic Resonance Imaging Score (WORMS), and their comparisons. Specifically, to confirm an early OA at least two of the four following items need to be fulfilled:

- Cartilage morphology scores: at least grade 3 (WORMS grade 3–6); Cartilage Score 1: at least grade 2 (BLOKS grade 2 and 3); Meniscal tears: at least grade 3 (BLOKS grade 3 and 4);
- BML: at least WORMS grade 2.

After this classification, interest has grown in diagnosing early OA in order to prevent or slow the progression of this pathology and improve cartilage defects and osteochondral degenerative changes treatments. This improved interest has brought to develop and evaluate cartilage compositional MRI techniques such as T1rho (T1ρ), T2 and delayed gadolinium-enhanced MRI of cartilage (dGEMRIC) that have proven to be sensitive to the alteration in cartilage extracellular matrix (ECM) composition therefore possibly allowing to detect molecular changes before appearance of gross morphological changes.

MRI is more sensitive to identify bone structure changes than radiographic findings. Schiphof et al. compared the findings between MRI and plain radiograph in a population-based study and concluded the definition of knee OA based on MRI more sensitively detected structural knee OA than the definition based on plain radiograph [72]. Zhu et al. [73] have performed a prospective cohort study of 895 participants, showing that 85% had MRI-detected osteophytes at baseline while only 10% were detected by radiographs. This has led Nagai et al. in their review to conclude that "MRI is a tool that provides useful information for early OA diagnosis. Although MRI is not recommended for now to diagnose early knee OA in daily clinical practice, because of lack of validated consensus criteria and the frequent prevalence of structural knee joint changes with MRI, the literature suggests that such MRI-detected lesions may represent early knee OA, and add support for the investigation of intervention effectiveness at the early stage of OA, including several advantages, including the absence of radiation exposure" [74].

As seen, many are the pathways by which inflammation intervenes and represents one of the major determining causes of cartilage breakdown and OA. The inflammatory mediators change the normal rules of cellular biology and through these processes, chondrocytes modify their function and their differentiation rates, catabolism prevails on anabolism: time passes, OA inflammation and process progress and end in the increase of oxidative stresses and in the entire joint degeneration. Additional studies have also shown that it is not just a matter of local inflammation, but also of systemic inflammation, as may be in obesity and certain chronic diseases, may contribute to OA pathogenesis.

Nowadays, have these new informations and theories been able to change our therapeutical approach to cartilage repair? The present authorized treatment protocols rely only on symptom-modifying agents, such as analgesics, NSAIDs, steroids, and hyaluronic acid until the OA progression requires surgical procedures, from orthobiologics, biological repair, and partial resurfacing procedures to total joint arthroplasties. Several promising findings showing disease-modifying effects of inhibitors of low-grade inflammation, such as growth factors, cytokines, MMP inhibitors, strontium ranelate, infliximab, pralacasan, FGF18, etoricoxib, flavocoxid and licofelone, and many others, in animal models of OA warrant follow-up in human clinical trials hoping to be able to establish new treatment paradigms that may modify the disease development and progression.

In 2007, Felson and Kim [75] already concluded their work affirming that OA is a final common pathway following many predisposing factors and thus therapeutics may just have limited efficacy in all those patients with preexisting joint damage, biomechanical predisposition, or obesity, especially in those with relatively advanced OA. This means that the more advanced is the OA process less chance will the patients have to be treated by new recent advanced molecular or orthobiological anti-inflammatory and regenerative interventions described earlier. So all our efforts need to be devoted to anticipate the diagnosis, identify early stages of cartilage breakdown, and OA in order to be able to develop and then prescribe specific anti-inflammatory interventions with greater hope of success, and promote and adopt new orthobiological tissue treatments favoring both inflammation reduction and cartilage damaged tissue regrowth and repair. Among these efforts need to be included the need for an increasing appreciation of clinical risk factors, an augmented ability to identify and quantify synovial inflammation, degeneration, and hypertrophy, increased use and continuously adjourned validation of modern and new highly sensitive imaging modalities capable of visualizing cartilage damage at the earlier ICRS classification stages, before the onset of irreversible joint failure.

References

1. Benedek TG. A history of the understanding of cartilage. Osteoarthr Cartil. 2006;14(3):203–9.
2. Galen on the usefulness of various parts of the body. Cornell University Press, Ithaca, NY, 1968 Trans. by M.T. May, (a) Book 16, ii (1:683); (b) Book 12, ii (1:552).
3. Avicenna the canon of medicine. AM Kelley, New York 1970; Trans. by O.C. Gruner, Chap. 16, p. 94.
4. A. Vesalius. On the fabric of the human body, cartilage, its nature, function and differentiation. San Francisco, CA: Norman Publishing, 1998. Trans. by Richardson WF and Carman JB, Chap. 2, vol. 1, pp. 8–11, 39–40.
5. Hunter W. Of the structure and diseases of articulating cartilages. Philos Trans R Soc Lond. 1743;42:514–21.
6. Morgagni JB. The seats and causes of diseases investigated by anatomy. London: Millar and Cadell; 1769. p. 3.
7. Comroe BI. Arthritis and allied conditions. 3rd ed. Philadelphia: Lea & Febiger; 1944. p. 93.
8. Mankin HJ, Lipiello L. The turnover of adult rabbit articular cartilage. J Bone Joint Surg. 1969;51A:1591–600.
9. Strawich E, Nimni ME. Properties of a collagen molecule containing three identical components extracted from bovine articular cartilage. Biochemistry. 1971;10:3905–11.
10. Rhodes RK, Miller EJ. Physicochemical characterization and molecular organization of the collagen a and B chains. Biochemistry. 1978;17:3442–8.
11. Litwic A, Edwards MH, Dennison EM, Cooper C. Epidemiology and burden of osteoarthritis. Br Med Bull. 2013;105:185–99.
12. Helmick CG, Felson DT, Lawrence RC, Gabriel S, Hirsch R, Kwoh CK, et al. Estimates of the prevalence of arthritis and other rheumatic conditions in the United States. Part I. Arthritis Rheum. 2008;58(1):15–25.
13. Lawrence RC, Felson DT, Helmick CG, et al. Estimates of the prevalence of arthritis and other rheumatic conditions in the United States Part II. Arthritis Rheum. 2008;58:26–35.
14. National Collaborating Centre for Chronic Conditions. Osteoarthritis: national clinical guideline for care and management in adults. London: Royal College of Physicians; 2008.
15. Urwin M, Symmons D, Allison T, Brammah T, Busby H, Roxby M, Simmons A, Williams G. Estimating the burden of musculoskeletal disorders in the community: the comparative prevalence of symptoms at different anatomical sites, and the relation to social deprivation. Ann Rheum Dis. 1998;57(11):649–55.
16. Sharma L, Cooke TD, Guermazi A, Roemer FW, Nevitt MC. Valgus malalignment is a risk factor for lateral knee osteoarthritis incidence and progression: findings from the multicenter osteoarthritis study and the osteoarthritis initiative. Arthithis Rheum. 2014;65(2):355–62.
17. Grotle M, Hagen KB, Natvig B, et al. Obesity and osteoarthritis in knee, hip and/or hand: an epidemiological study in the general population with 10 years follow-up. BMC Musculoskelet Disord. 2008;9:132.
18. Kellgren JH, Lawrence JS. Radiological assessment of osteo-arthrosis. Ann Rheum Dis. 1957;16(4):494–502.
19. Martel-Pelletier J, Barr AJ, Cicuttini FM, Conaghan PG, Cooper C, Goldring MB, et al. Osteoarthritis. Nat. Rev. Dis. 2016;2:16072.
20. Solomon DH, Husni ME, Libby PA, Yeomans ND, Lincoff AM, Lüscher TF, et al. The risk of major NSAID toxicity with celecoxib, ibuprofen, or naproxen: a secondary analysis of the PRECISION trial. Am J Med. 2017;130(12):1415–22.
21. Lee T, Lu N, Felson DT, Choi HK, Dalal DS, Zhang Y, et al. Use of non-steroidal anti-inflammatory drugs correlates with the risk of venous thromboembolism in knee osteoarthritis patients: a UK population-based case-control study. Rheumatology. 2016;55(6):1099–105.
22. Wang Y, Chen X, Cao W, Shi Y. Plasticity of mesenchymal stem cells in immunomodulation: pathological and therapeutic implications. Nat Immunol. 2014;15:1099–16.
23. Ren G, et al. Mesenchymal stem cell-mediated immunosuppression occurs via concerted action of chemokines and nitric oxide. Cell Stem Cell. 2008;2:141–50.
24. Su J, et al. Phylogenetic distinction of iNOS and IDO function in mesenchymal stem cell-mediated immunosuppression in mammalian species. Cell Death Differ. 2014;21:388–96.
25. Shi Y, Wang Y, Li Q, et al. Immunoregulatory mechanisms of mesenchymal stem and stromal cells in inflammatory diseases. Nat Rev Nephrol. 2018;14(8):493–507.
26. Kong L, Zheng LZ, Qin L, Ho KKW. Role of mesenchymal stem cells in osteoarthritis treatment. J Orthop Transl. 2017;9:89–103.
27. Barry FM. Murphy Mesenchymal stem cells in joint disease and repair. Nat Rev Rheumatol. 2013;910:584–94.
28. Centeno CJ, Busse D, Kisiday J, Keohan K, Freeman K, Karli D. Increased knee cartilage volume in degenerative joint disease using percutaneously implanted, autologous mesenchymal stem cells. Pain Physician. 2008;113:343–53.
29. Emadedin M, Ghorbani Liastani M, Fazeli R, Mohseni F, Moghadasali R, Mardpour S, Hosseini SE, Niknejadi M, Moeininia F, Aghahossein Fanni A, Baghban Eslaminejhad R, Vosough Dizaji A,

Labibzadeh N, Mirazimi Bafghi A, Baharvand H, Aghdami N. Long-term follow-up of intra-articular injection of autologous mesenchymal stem cells in patients with knee, ankle, or hip osteoarthritis. Arch Iran Med. 2015;18(6):336–44.

30. Emadedin M, Labibzadeh N, Liastani MG, Karimi A, Jaroughi N, Bolurieh T, Hosseini SE, Baharvand H, Aghdami N. Intra-articular implantation of autologous bone marrow-derived mesenchymal stromal cells to treat knee osteoarthritis: a randomized, triple-blind, placebo-controlled phase 1/2 clinical trial. Cytotherapy. 2018;20(10):1238–46.

31. Jo CH, Lee YG, Shin WH, Kim H, Chai JW, Jeong EC, et al. Intra-articular injection of mesenchymal stem cells for the treatment of osteoarthritis of the knee: a proof- of-concept clinical trial. Stem Cells. 2014;325:1254–66.

32. Hong Z, Chen J, Zhang S, Zhao C, Bi M, Chen X, Bi Q. Intra-articular injection of autologous adipose-derived stromal vascular fractions for knee osteoarthritis: a double-blind randomized self-controlled trial. Int Orthop. 2019;43:1123–34.

33. Lee WS, Hwan JK, Kim K-IL, Kim GB, Jin W. Intra-articular injection of autologous adipose tissue-derived mesenchymal stem cells for the treatment of knee osteoarthritis: a phase IIb, randomized, placebo-controlled clinical trial. Stem Cells Transl Med. 2019;00:1–8.

34. Andriacchi T, Mundermann A, Smith R, Alexander E, Dyrby C, Koo S. A framework for the in vivo pathomechanics of osteoarthritis at the knee. Ann Biomed Eng. 2004;32:447–57.

35. Chen C, Tambe DT, Deng L, Yang L. Biomechanical properties and mechanobiology of the articular chondrocyte. Am J Physiol Cell Physiol. 2013;305:C1202–8.

36. Kraus VB, Blanco FJ, Englund M, Karsdal MA, Lohmander LS. Call for standardized definitions of osteoarthritis and risk stratification for clinical trials and clinical use. Osteoarthr Cartil. 2015;23(8):1233–41.

37. Haseeb A, Haqqui TM. Immunopathogenesis of osteoarthritis. Clin Immunol. 2013;146(3):185–96.

38. Brandt KD, Dieppe P, Radin E. Etiopathogenesis of osteoarthritis. Med Clin North Am. 2009;93:1–24.

39. Lane Smith R, Trindade MC, Ikenoue T, et al. Effects of shear stress on articular chondrocyte metabolism. Biorheology. 2000;37:95–107.

40. Jasin HE. Immune mediated cartilage destruction. Scand J Rheumatol Suppl. 1988;76:111–6.

41. Ayral X, Pickering EH, Woodworth TG, et al. Synovitis: a potential predictive factor of structural progression of medial tibiofemoral knee osteoarthritis - results of a 1 year longitudinal arthroscopic study in 422 patients. Osteoarthr Cartil. 2005;13:361–7.

42. Hill CL, Hunter DJ, Niu J, et al. Synovitis detected on magnetic resonance imaging and its relation to pain and cartilage loss in knee osteoarthritis. Ann Rheum Dis. 2007;66:1599–603.

43. Revell PA, Mayston V, Lalor P, Mapp P. The synovial membrane in osteoarthritis: a histological study including the characterisation of the cellular infiltrate present in inflammatory osteoarthritis using monoclonal antibodies. Ann Rheum Dis. 1988;47:300–7.

44. Sakkas LI, Scanzello C, Johanson N, et al. T cells and T-cell cytokine transcripts in the synovial membrane in patients with osteoarthritis. Clin Diagn Lab Immunol. 1998;5:430–7.

45. Nakamura H, Yoshino S, Kato T, et al. T-cell mediated inflammatory pathway in osteoarthritis. Osteoarthr Cartil. 1999;7:401–2.

46. Wang Q, Rozelle AL, Lepus CM, et al. Identification of a central role for complement in osteoarthritis. Nat Med. 2011;17:1674–9.

47. Robinson WH, Lepus CM, Wang Q, Raghu H, Mao R, Lindstrom TM, Sokolove J. Low-grade inflammation as a key mediator of the pathogenesis of osteoarthritis. Nat Rev Rheumatol. 2016;12(10):580–92.

48. Orlowsky EW, Kraus VB. The role of innate immunity in osteoarthritis: when our first line of defense goes on the offensive. J Rheumatol. 2015;42:363–71.

49. Happonen K, Saxne T, Aspberg A, Morgelin M, Heinegard D, Blom A. Regulation of complement by cartilage oligomeric matrix protein allows for a novel molecular diagnostic principle in rheumatoid arthritis. Arthritis Rheum. 2010;62:3574–83.

50. Sjoberg A, Manderson G, Morgelin M, Day A, Heinegard D, Blom A. Short leucine-rich glycoproteins of the extracellular matrix display diverse patterns of complement interaction and activation. Mol Immunol. 2009;46:830–9.

51. Gabay C. Interleukin-6 and chronic inflammation. Arthritis Res Ther. 2006;8(Suppl. 2):S3.

52. Goldring M, Fukuo K, Birkhead J, Dudek E, Sandell L. Transcriptional suppression by interleukin-1 and interferon-gamma of type II collagen gene expression in human chondrocytes. J Cell Biochem. 1994;54:85–99.

53. Scanzello CR, Goldring SR. The role of synovitis in osteoarthritis pathogenesis. Bone. 2012;51:249–57.

54. Miller RJ, Banisadr G, Bhattacharyya BJ. CXCR4 signaling in the regulation of stem cell migration and development. J Neuroimmunol. 2008;198:31–8.

55. van Lent PL, et al. Crucial role of synovial lining macrophages in the promotion of transforming growth factor β-mediated osteophyte formation. Arthritis Rheum. 2004;50:103–11.

56. Haywood L, McWilliams DF, Pearson CI, Gill SE, Ganesan A, Wilson D, et al. Inflammation and angiogenesis in osteoarthritis. Arthritis Rheum. 2003;48:2173–7.

57. Conde J, Scotece M, Gomez R, Lopez V, Gomez-Reino J, Gualillo O. Adipokines and osteoarthritis: novel molecules involved in the pathogenesis and progression of disease. Arthritis. 2011;2011:203901.

58. Francin PJ, et al. Association between adiponectin and cartilage degradation in human osteoarthritis. Osteoarthr Cartil. 2014;22:519–26.

59. Clockaerts S, Bastiaansen-Jenniskens Y, Runhaar J, Van Osch G, Van Offel J, Verhaar J, et al. The infrapatellar fat pad should be considered as an active osteoarthritic joint tissue: a narrative review. Osteoarthr Cartil. 2010;18:876–82.
60. Hui W, Litherland G, Elias M, Kitson G, Cawston T, Rowan A, et al. Leptin produced by joint white adipose tissue induces cartilage degradation via upregulation and activation of matrix metalloproteinases. Ann Rheum Dis. 2012;71:455–62.
61. Ushiyama T, Chano T, Inoue K, Matsusue Y. Cytokine production in the infrapatellar fat pad: another source of cytokines in knee synovial fluids. Ann Rheum Dis. 2003;62:108–12.
62. Bohnsack M, Meier F, Walter G, Hurschler C, Schmolke S, Wirth C, et al. Distribution of substance-P nerves inside the infrapatellar fat pad and the adjacent synovial tissue: a neurohistological approach to anterior knee pain syndrome. Arch Orthop Trauma Surg. 2005;125:592–7.
63. Martel-Pelletier J, Pelletier JP, Fahmi H. Cyclooxygenase-2 and prostaglandins in articular tissues. Semin Arthritis Rheum. 2003;33:155–67.
64. Casale TB, Abbas MK, Carolan EJ. Degree of neutrophil chemotaxis is dependent upon the chemoattractant and barrier. Am J Respir Cell Mol Biol. 1992;7:112–7.
65. Loeser RF, Goldring SR, Scanzello CR, Goldring MB. Osteoarthritis: a disease of the joint as an organ. Arthritis Rheum. 2012;64:1697–707.
66. Roemer FW, Guermazi A, Felson DT, Niu J, Nevitt MC, Crema MD, et al. Presence of MRI- detected joint effusion and synovitis increases the risk of cartilage loss in knees without osteoarthritis at 30-month follow-up: the MOST study. Ann Rheum Dis. 2011;70:1804–9.
67. Scanzello C, McKeon B, Swaim B, DiCarlo E, Asomugha E, Kanda V, et al. Synovial inflammation in patients undergoing arthroscopic meniscectomy: molecular characterization and relationship to symptoms. Arthritis Rheum. 2011;63:391–400.
68. Berembaum F. Osteoarthritis as an inflammatory disease (osteoarthritis is not osteoarthrosis!). Osteoarthr Cartil. 2013;21(1):16–21.
69. Sellam J, Berenbaum F. The role of synovitis in pathophysiology and clinical symptoms of osteoarthritis. Nat Rev Rheumatol. 2010;6:625–35.
70. Lories R, Luyten F. The bone-cartilage unit in osteoarthritis. Nat Rev Rheumatol. 2011;7:43–9.
71. Luyten FP, Denti M, Filardo G, Kon E, Engebretsen L. Definition and classification of early osteoarthritis of the knee. Knee Surg Sports Traumatol Arthrosc. 2012;20(3):401–6.
72. Schiphof D, Oei EHG, Hofman A, et al. Sensitivity and associations with pain and body weight of an MRI definition of knee osteoarthritis compared with radiographic Kellgren and Lawrence criteria: a population-based study in middle-aged females. Osteoarthr Cartil. 2014;22:440–6.
73. Zhu Z, Laslett LL, Jin X, et al. Association between MRI-detected osteophytes and changes in knee structures and pain in older adults: a cohort study. Osteoarthr Cartil. 2017;25:1084–92.
74. Nagai K, Nakamura T. Fu FH the diagnosis of early osteoarthritis of the knee using magnetic resonance imaging. AOJ. 2018;3:110.
75. Felson D, Kim Y. The futility of current approaches to chondroprotection. Arthritis Rheum. 2007;56:1378–83.

Karun Amar, Anshuman Singh, and John G. Lane

23.1 Introduction

Articular cartilage lesions of the glenohumeral joint are thought to be rare and often diagnosed incidentally during arthroscopy for treatment of concomitant shoulder pathology. Recent advancements in imaging and increased awareness of the pathology among clinicians have resulted in an increase in the diagnosis of these lesions. Although there exists an abundance of high-quality data analyzing the various treatment options for management of chondral lesions in other joints, such as the knee and ankle, there is a surprising lack of similar data pertaining to the shoulder. As a result, clinicians are often forced to base treatment of the glenohumeral joint on outcomes from knee/ankle literature. Many treatment options, both nonoperative and surgical, have been described to address this issue [1–6]. In this chapter, the authors will discuss the evaluation and management of symptomatic, focal chondral, and osteochondral defects of the glenohumeral joint. The management of high-grade degenerative chondrosis and osteochondral defects for treatment of shoulder instability will not be covered as these topics represent a separate entity and merit their own discussion.

23.2 Anatomy

The glenohumeral joint is a non-weight-bearing ball and socket joint with the most mobility of any joint in the body. The articulating portions of the humeral head and glenoid fossa are lined with articular cartilage, which plays an important role in reducing the friction of the bones as they glide over each other [7]. An anatomic study which analyzed the articular geometry of the joint revealed that the humeral head and glenoid are exceedingly congruent with difference of radius of curvature less than 3 mm in all shoulders analyzed and less than 2 mm in 88% of shoulders analyzed [8]. Furthermore, the mean thickness of cartilage on the glenoid was found to be 2.16 mm versus 1.44 mm for the humeral head. The thickness of the cartilage on the glenoid was greater peripherally whereas the opposite was true for the humeral head [8]. The avascularity of articular cartilage results in a limited capacity for intrinsic healing and repair [9].

K. Amar · A. Singh
Department of Orthopaedics, Southern California
Permanente Medical Group, San Diego, CA, USA
e-mail: Karun.X.Amar@kp.org

J. G. Lane (✉)
Musculoskeletal and Joint Foundation, La Jolla,
California, USA

Department of Orthopedic Surgery, University of
California, San Diego, USA
e-mail: jglane@san.rr.com

© ISAKOS 2022
A. Gobbi et al. (eds.), *Joint Function Preservation*, https://doi.org/10.1007/978-3-030-82958-2_23

23.3   Etiology

Focal chondral defects of the glenohumeral joint are typically the result of an injury or pathology of the humerus or glenoid. Perhaps the most well-described cause is trauma due to acute or recurrent shoulder instability [10–13]. The Hill-Sachs lesion, or osteochondral defect of the posterolateral humeral head, has been reported to be present in 40–90% of acute dislocations and 100% of the time in patients with recurrent instability [14, 15]. In one study, Hinterman et al. performed arthroscopic examination of 212 patients with at least one documented shoulder dislocation. They found that 68% of patients had evidence of a Hill-Sachs lesion and 46% of patients had evidence of cartilage defects at other locations within the joint [12]. Another study reported Hill-Sachs lesions were found in 60.5% of shoulders and chondral lesions of the glenoid in 23% of shoulders during arthroscopic examination [13]. Although no definitive causal relationship has been established, focal chondral defects in the glenohumeral joint are often found when treating concomitant pathology such as rotator cuff tears or superior labrum anterior to posterior (SLAP) tears. Miller et al. reported on glenohumeral abnormalities associated with full-thickness rotator cuff tears and found Outerbridge grade III or higher lesions in 28 out of 100 shoulders [16]. This was supported by Gartsman et al. who found minor cartilage lesions in 8.5% and major lesions in 4.5% of patients undergoing surgical repair of rotator cuff tears [17]. During arthroscopic examination of patients with internal impingement syndrome, 17% of shoulders were found to also have osteochondral lesions of the humeral head near the insertion of the supraspinatus [18]. Similarly, chondral defects were found in 52% of patients with type II SLAP lesions at the time of arthroscopic examination. The defects were present adjacent to the biceps tendon on the humerus and along the anterior aspect of the glenoid [19]. Osteochondritis dissecans lesions have also been reported in the humeral head, albeit much less commonly than in the knee, elbow, and ankle [20–22]. These lesions tend to occur more often in middle-aged males

and are typically encountered in the anterosuperior or less commonly posterosuperior part of the humeral head [23]. The final major contributor to the development of glenohumeral chondral defects involves iatrogenic injury during prior surgical procedures [24–27]. These injuries can be the result of mechanical damage from surgical instruments or improperly placed implants or chondrolysis due to thermal ablation or continuous administration of local anesthetics through a pump.

23.4   Classification

There is not a classification system tailored specifically for glenohumeral joint chondral defects. Consequently, the Outerbridge and International Cartilage Repair Society(IRCS) Hyaline Cartilage Lesion grading scales are used. In the Outerbridge classification, grade 0 refers to normal cartilage, grade 1 has softening and swelling of the cartilage, grade 2 has fissuring up to half the depth of the cartilage, grade 3 has fissuring involving more than half the depth of the cartilage, and lastly grade 4 involves full-thickness loss of cartilage down to the level of subchondral bone [28]. The International Cartilage is a modification of the Outerbridge classification which uses arthroscopic findings to expand upon it with more subclassifications [29].

23.5   History and Physical Exam

A thorough history should be obtained as part of the workup of a patient presenting with shoulder pain. Details regarding the character, duration, and location of the pain are important. Any history of recent or remote shoulder trauma should also be ascertained. Special consideration should be provided for prior instability episodes given the high rate of osteochondral injury associated with instability of the shoulder. Oftentimes, it is difficult to differentiate a symptomatic chondral lesion of the glenohumeral joint from concomitant pathology. Patients may complain of mechanical symptoms such as clicking/catching or a vague, ill-defined pain deep in the shoulder. This is in contrast to the lateral-based pain that is classic for patients with rotator cuff pathology. Additionally, information regarding prior surgical procedures or injections to the ipsilateral shoulder should be obtained as chondral injuries can be iatrogenic. Lastly, it is important to understand what treatment modalities have already been attempted to address the current symptoms.

Given the lack of highly sensitive or specific examination tests for focal chondral defects, the goal of the physical exam in patients where chondral injury is suspected is often to rule out other causes of pain. A standard examination of the neck and shoulder should be performed including basic components such as inspection, palpation, range of motion, and strength. Special tests for impingement, rotator cuff, biceps, and instability should also be performed as needed based on previously obtained history. Although an audible crepitus with circumduction of the shoulder can be caused by a multitude of factors, irregularity of the glenohumeral joint surface due to a chondral defect should strongly be considered. The compression rotation test, as described by Ellman et al., can also be used to help identify lesions [30]. In this test, the patient is placed in the lateral decubitus position with the painful shoulder side up. A medially directed force is then applied to the shoulder to load the joint and the patient is asked to rotate the shoulder with the elbow bent. A positive test is when the maneuver reproduces pain or crepitus.

23.6 Imaging

A standard radiographic workup including Grashey, scapular-Y, and axillary views should be obtained during the initial assessment of the patient. These views can be used to assess for evidence of degenerative changes such as joint space narrowing, osteophytes, subchondral sclerosis/cysts. If there is suspicion of bone loss, Stryker notch and West Point views can also be obtained to assess for Hill-Sachs lesions or anterior glenoid bone loss, respectively. Additionally, if an axillary view is not feasible due to patient discomfort, a Velpeau view can be obtained instead.

CT scan imaging may also be a useful modality to assist in the evaluation of patients with shoulder pain and concern for potential osteochondral injuries. CT imaging can provide an accurate assessment of the degree of bony involvement in suspected osteochondral injuries which can aid in surgical planning. CT arthrography has also been validated as an effective means of detecting moderate to large chondral defects of the glenohumeral joint [31].

MRI is the gold standard of imaging chondral lesions as it can be used to evaluate for concomitant pathology as well. Although MRI allows for excellent visualization of the articular cartilage of the humeral head and glenoid, it is easy to overlook evidence of chondral injury unless the examiner is specifically looking for it [32]. One reason why the observer needs a high index of suspicion is that the thickness of the articular cartilage layer in the glenohumeral joint is much less than other joints such as the knee. Typical MRI findings of cartilage lesions include contour deformities with areas of abnormal signal intensity with possible underlying bony edema based on the chronicity of the injury [33]. Denti et al. [34] reported a 60% accuracy and 87% sensitivity for detecting osteochondral injuries of the humeral head with non-arthrographic MRI. Another study investigated the detection of glenohumeral cartilage lesions on MRI, reporting a sensitivity of 53%, specificity of 87%, and accuracy of 77% [35]. Intra-articular contrast may result in improved detection of chondral lesions. Additionally, new cartilage-specific sequences and protocols are being developed to further improve detection and assessment [32, 36–38].

23.7 Treatment Options

23.7.1 Nonoperative Management

Given the vast majority of chondral lesions are asymptomatic, a trial of nonoperative management is indicated as the first line of treatment for patients with these lesions. Conservative management includes NSAIDs, physical therapy, and intra-articular injections. Although there are no randomized controlled trials demonstrating efficacy of NSAIDs in the treatment of shoulder pain, there is data to suggest that upwards of 50% of patients will experience some improvement in pain [39, 40]. Physical therapy is an essential component of nonoperative management as it may address concomitant pathology in the shoulder contributing to the patient's symptomology. Therapy should focus on stretching to improve range of motion as well as strengthening of the periscapular, deltoid, and rotator cuff musculature.

A variety of different injections have been described for the management of articular cartilage defects, including corticosteroids, hyaluronic acid, platelet-rich plasma, and mesenchymal signaling cells. Much of the data behind these modalities is derived from other joints, but there is emerging evidence specific to the glenohumeral joint [41, 42]. Hyaluronic acid (HA) is a component of native articular cartilage that is thought to increase the viscosity and elasticity of synovial fluid [41]. Blaine et al. conducted a randomized controlled trial comparing the injection of HA to a placebo consisting of phosphate-buffered saline (PBS) in patients with persistent shoulder pain. Although they did not detect a statistically significant difference in outcomes at 13 weeks posttreatment, the HA groups did outperform the control group at 26 weeks posttreatment [43]. Of note, this study was not specific to patients with glenohumeral chondral defects and included all patients with shoulder pain. In a recent meta-analysis that analyzed the outcomes of HA for treatment of glenohumeral osteoarthritis, the authors failed to detect a difference in outcomes compared to injection of placebo [44]. Several studies have compared the efficacy of HA and corticosteroids. In one such study, Merolla et al. compared intra-articular injection of HA to methylprednisolone and assessed outcomes at multiple time points. They found that the HA group had decreased pain levels at 6 months whereas the corticosteroid group did not [45]. The data supporting use of platelet-rich plasma and mesenchymal signaling cells for management of glenohumeral chondral defects is even more sparse. There is one uncontrolled case report which describes the use of three intra-articular PRP injections (each spaced 1 week apart) for the treatment of glenohumeral OA. At the final follow-up of 42 weeks, the patient experienced a significant decrease in VAS and DASH scores [46]. The rest of the evidence for use of PRP and MSCs comes from data related to management of knee OA [47–50].

23.7.2 Surgical Management

When appropriate nonoperative treatment of symptomatic chondral/osteochondral glenohumeral lesions is not effective, surgical management may be considered. Some of the described surgical procedures to treat these lesions include debridement, microfracture, osteochondral autograft/allograft transplantation, autologous chondrocyte implantation (ACI), and resurfacing. When attempting to determine which procedure is most appropriate for a given patient, some important factors to consider include overall size and depth of the lesion, location of the lesion, containment, involvement of the subchondral bone, and previous surgical management. Positive prognostic indicators for any surgical procedure include size of the lesion <2 cm^2 and unipolar lesions involving the humeral head [2, 51]. In their review article, Elser et al. propose a rudimentary algorithm for surgical management of chondral lesions of the humeral head (Fig. 23.1) [5]. The algorithm is based on the size of the lesion and presence of bony involvement. For focal lesions, they advocate for treatment options such as debridement, microfracture, OATS, or ACI. For larger lesions with significant bone loss, they recommend osteochondral allografts or resurfacing.

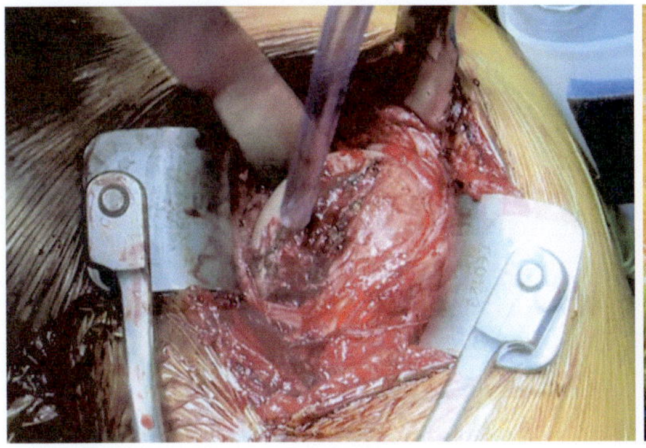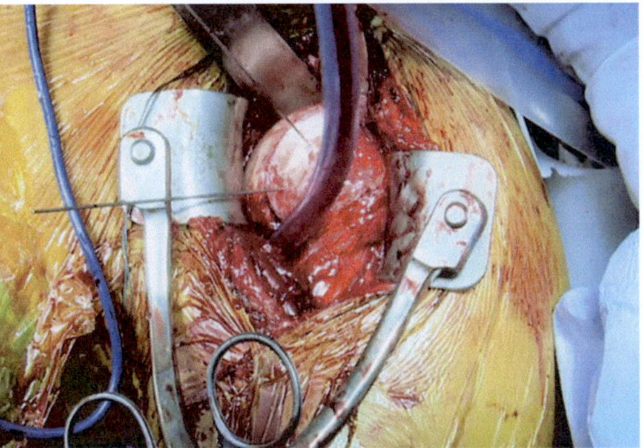
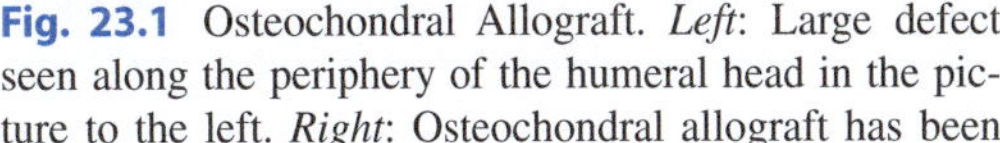

Fig. 23.1 Osteochondral Allograft. *Left*: Large defect seen along the periphery of the humeral head in the picture to the left. *Right*: Osteochondral allograft has been placed into the defect with provisional K-wire fixation. (Photos courtesy of Matthew Provencher, MD.)

23.8 Arthroscopic Debridement

Arthroscopic debridement is often employed as a first line of surgical management of symptomatic chondral lesions as it is not technically demanding and requires no extra equipment or implants. Debridement is typically performed with the use of a mechanical arthroscopic shaver and/or sharp curettes, which are generally part of shoulder arthroscopy instrumentation sets. The goal of this procedure is to remove any loose flaps or unstable edges of unhealthy cartilage that may be the source of mechanical symptoms and reduce the chance of propagation of the lesion. In cases where full-thickness chondral defects are encountered, debridement can be performed to create vertical shoulders of healthy cartilage around the defect. Biomechanical studies have shown that vertical shoulders around an area of cartilage loss allow a more normal transfer of load to the surrounding cartilage as compared to a tapered transition. This procedure is commonly performed for incidental chondral defects that are found when treating concomitant pathology within the shoulder.

Much of the reported data with respect to arthroscopic debridement of the glenohumeral joint involves patients with glenohumeral arthritis. Since this represents a distinct pathology from that of the focal chondral defect, results should be interpreted cautiously as outcomes may appear worse due to a higher degree of baseline dysfunction in patients with glenohumeral arthritis. In a case series of 33 patients, Skelley et al. reported on the outcomes of isolated arthroscopic debridement and capsular release for the management of glenohumeral osteoarthritis. Although there was an initial improvement in range of motion and pain scores, patients returned to their baseline levels by 3.8 months postoperatively. Furthermore, 60% of patients reported that they were unhappy with the outcome and 42% went onto subsequent arthroplasty [52].

In contrast to these findings, several other studies have shown good outcomes after arthroscopic debridement [53–56]. In one such study, the authors report good or excellent outcome in 80% of patients who underwent arthroscopic debridement to treat early glenohumeral osteoarthritis [54]. Kerr et al. also reported good outcomes after arthroscopic debridement. Interestingly, although bipolar lesions involving the glenoid and humeral head were associated with worse outcomes, high-grade unipolar chondral lesions were not [53]. In a retrospective study of 81 patients who underwent arthroscopic debridement for isolated degenerative joint disease of the shoulder, Van Thiel et al. reported generally good outcomes. The average ASES, Simple Shoulder Test, and VAS scores all improved at mean follow-up of 27 months post-procedure. Additionally, the range of motions improved significantly with respect to flexion, abduction, and external rotation. Lastly, they reported that risk factors for eventual conversion to arthroplasty included grade IV bipolar chondral loss, joint space of less than 2 mm, and presence of large osteophytes. It is important to note that in addition to debridement of the chondral lesions, other procedures such as capsular release, biceps tenotomy, subacromial decompression, and loose body removal were also performed in the same setting. Therefore, it is unclear whether the improved outcomes were the result of the debridement or other procedures [56].

There has been some conflicting data regarding the expected duration of symptomatic relief after arthroscopic debridement. In a recently published systematic review, Williams et al. investigated the outcomes and survivorship after arthroscopic treatment of glenohumeral arthritis [57]. While they found that symptomatic relief generally lasted a minimum of 6 months, the average and/or maximum duration of benefits was extremely variable. Mean time to conversion to arthroplasty ranged from 9 months to 56 months. There was significant heterogeneity among the studies in terms of the exact procedures performed and patient selection criteria, which made comparison difficult. The study that had the strictest inclusion criteria (excluded treatment of concomitant pathology in the same setting as the debridement) had among the highest failure rate and lowest mean duration of positive results postoperatively. Taken collectively, it appears that the benefit of arthroscopic debridement appears to wane after approximately 2 years.

23.9　Microfracture

Microfracture is a surgical technique that has been studied extensively for chondral lesions in the knee and is starting to gain acceptance for use in the glenohumeral joint [58, 59]. The technique

in the shoulder is identical to the knee and can also be performed arthroscopically. The chondral defect is first debrided down through the calcified layer at the base of the lesion. Vertical walls are then created to shoulder the lesion. This technique requires a well-shouldered lesion in order to contain the resulting fibrin clot and thus it cannot be performed in unconstrained lesions. Next, a tapered awl is used to create holes down into the subchondral bone and allow release of the marrow elements into the defect. The holes are typically created a minimum of 2–3 mm apart and to a depth of about 4–6 mm. The blood clot slowly remodels into fibrocartilaginous tissue as opposed to the hyaline cartilage that it is replacing. The fibrocartilaginous tissue has been found on histological analysis to be composed primarily of type I collagen as opposed to the type II collagen that predominates in articular cartilage. As a result, the wear properties and overall strength of the tissue are inferior to native articular cartilage. A variation of the technique involves drilling into the subchondral bone instead of using an awl. This can be performed utilizing a standard Kirschner wire or a motorized pick. The benefit of drilling is that it allows the release of the marrow elements without causing impaction of the surrounding subchondral bone as with the use of an awl. This impaction injury to the subchondral bone has been shown to stimulate a fracture-like healing response and can lead to the formation of heterotopic bone in the defect site. Although various postoperative rehabilitation protocols have been proposed, most allow full active and passive motion of the shoulder immediately after surgery [60]. Since the shoulder does not experience the same loading forces as the knee, weight-bearing through the shoulder is allowed.

In a case series of 31 shoulders, Millet et al. reported on their outcomes after performing microfracture for full-thickness articular cartilage defects in the glenohumeral joint [61]. Of the 31 shoulders, 6 underwent treatment for bipolar lesions, 13 for isolated glenoid lesions, and 12 for isolated humeral head lesions. Although 6 out of 31 shoulders (19%) treated were considered failures as they needed a subsequent surgical procedure, the rest did well. When analyzing the shoulders that did not require further surgery, there was a significant improvement in postoperative ASES score, pain score, and overall patient satisfaction level as compared to preoperative values. The greatest improvement in scores was achieved in patients with isolated lesions of the humeral head.

In another case series, Frank et al. reported on short-term outcomes after microfracture of glenohumeral chondral defects in 17 shoulders (16 patients) [62]. The average size of the defects was 5.07 cm² for humeral lesions and 1.66 cm² for glenoid lesions. Two patients were lost to follow-up. The mean follow-up for the remaining patients was 27.8 months. Three patients went on to subsequent surgery and were considered to have failed treatment. When analyzing outcomes for the remaining 12 shoulders, the authors found that there was a significant decrease in VAS after surgery from 5.6 to 1.9. There were also statistically significant improvements in ASES (44.3–86.3) and SST (5.7–10.3) scores. Although the authors claim that 92.3% of patients felt they were satisfied with the outcome and would do the surgery over again, this is misleading because they chose to exclude the three patients that failed treatment. It is likely that these patients would have reported dissatisfaction with the surgery, which would dramatically negatively impact the overall satisfaction rate. The authors of this study subsequently published a follow-up study reporting long-term outcomes of the same cohort of patients at an average of 10 years following surgery [63]. Although there were no significant changes in the values of the VAS, ASES, or SST scores between the short-term and long-term follow-up time points, the survivorship was only 66.7%. Of the patients who failed treatment, three went on to subsequent arthroplasty or resurfacing and two were unsatisfied with the surgery and were considering other surgical options.

Snow and Funk published a case series of eight patients who underwent arthroscopic microfracture for the treatment of full-thickness chondral lesions which were less than 4 cm² in size [64]. At mean follow-up of 15.4 months, the patients showed significant improvement in Constant and Oxford scores as compared to preoperative values. Two of the patients underwent reoperation for unrelated reasons and the lesions showed good filling of the defects with fibrocartilage.

When analyzing the results of the studies, it does appear that microfracture is a reasonable first-line option for the management of chondral defects in the glenohumeral joint. Although the results of the surgery tend to be durable in those patients who experience improvement in symptoms, there is a potentially large cohort of patients who do not seem to respond to microfracture.

23.10 Osteochondral Autograft Transfer

Osteochondral autograft transfer is a surgical technique in which one or more cylindrical osteochondral plugs are harvested from the patient and transferred into a separate chondral defect. Typically, the plugs are harvested from a less weight-bearing portion of the knee such as the periphery of the trochlea and intercondylar notch. The donor plugs are slightly larger in diameter than the defect size, which allows press-fit fixation without the need for supplementary internal fixation with hardware. This can be accomplished with the use of specialized instrumentation that allows for the precise formation of cylindrical plugs of specific diameter and depth. It is critically important that the plugs be harvested and placed in a manner that is orthogonal to the articular cartilage. This prevents graft obliquity and mismatch of contour between the graft and surrounding cartilage. Due to the inherent constraints of donor site morbidity when harvesting plugs, typically no more than three plugs can safely be obtained. Therefore, this technique is best utilized on smaller lesions between 0.5 cm² and 2 cm², although it can be attempted for slightly larger lesions. The primary advantages of the technique involve transfer of viable chondrocytes into the defect site, which results in formation of hyaline cartilage as opposed to fibrocartilage. Additionally, the procedure can be performed as a single stage. The main disadvantage is the donor site morbidity and the need to harvest plugs from an uninvolved knee. Also, in order to gain full access to the joint, this procedure typically requires an open approach as opposed to arthroscopic.

In a case series of eight patients, Scheibel et al. described the outcomes after osteochondral autograft transplant of full-thickness articular cartilage defects in the shoulder [65]. The lesions, which had an average size of 1.5 cm², were present on the humeral head in seven cases and on the glenoid in one. The grafts were harvested from the outer edge of the lateral femoral condyle and open arthrotomies were performed on both the knee and shoulder. The authors reported a significant improvement in Constant score at the time of final follow-up. MRIs were obtained postoperatively for all patients and osteointegration of the graft was noted in all but one shoulder. Additionally, second-look arthroscopy was performed in two cases 6 months postoperatively. In both cases, the graft appeared grossly intact with only minor grade I Outerbridge changes present at the defect site. Although the patients seemed to demonstrate improvement in symptoms following the procedure, routine follow-up X-rays of the shoulder showed progression of osteoarthritic changes in all cases. Furthermore, 1 of the 8 patients had a poor result with respect to function of the knee as measured by the Lysholm score. The patient subsequently underwent two additional surgical debridement procedures for the knee due to recurrent pain and effusion.

The senior authors of the previous paper later published a long-term follow study on the same group of patients with a mean follow-up time of 8.75 years [66]. One of the eight patients was lost to follow-up. They found that the Constant scores continued to improve until the final follow-up and the Lysholm scores remained steady from the previously reported values. Although there was a significant progression of osteoarthritis of the glenohumeral joint from preoperative level to final follow-up, it did not appear to be directly related to the defect size, number of plugs used, or the Constant score. Follow-up MRI scans showed successful osseous integration of all the plugs and congruent joint surfaces in all but one patient.

Although the procedure is most commonly described using an open approach for the shoulder, Park et al. reported their experience utilizing an all arthroscopic approach for both the knee and shoulder [67]. In the case report, a 10 mm osteochondral plug was harvested from the non-weight-bearing portion of the lateral femoral condyle of the ipsilateral knee and transferred to the 9 mm defect in the humeral head. A second look arthroscopy was performed for both the knee and the shoulder and the donor and defect site were both healed and covered with congruent articular cartilage. At the time of final follow-up (2.5 years), the patient was asymptomatic with good functional use of the shoulder.

23.11 Fresh Osteochondral Allograft Transfer

Fresh osteochondral allograft transplantation is a technique that closely resembles that of the osteochondral autograft transfer. Instead of using plugs harvested from another site in the body, fresh allograft tissue is used. This prevents the complications associated with donor site morbidity and results in a shorter surgical time. The use of allograft tissue also allows for the treatment of larger defects since there is no restriction on the size or quantity of osteochondral plugs harvested [4]. As such, osteochondral allograft transplantation is often used for defects larger than 2 cm² and uncontained defects at the periphery of the humeral head or glenoid (Fig. 23.1).

In order to match the geometry of the defect site as closely as possible and prevent incongruence of the articular surface, the allografts should be side- and size-matched to the patient based on preoperative imaging. For the glenohumeral joint this often requires obtaining a CT scan for maximal accuracy however standard radiographs may suffice in other joints such as the knee [68]. While allograft humeral heads are readily available, allograft glenoid is not commercially available at this point. Since the contour and geometry of the distal tibial closely match that of the glenoid, distal tibia allografts may be used for glenoid defects. In order to ensure maximal donor chondrocyte viability, the graft must be implanted no longer than 28 days after procurement. Given that the testing process for infectious organisms can take up to 14 days, this only leaves the surgeon with a 2-week window to use the graft; therefore, all operative arrangements and insurance authorization must be finalized before accepting the graft [68].

While the outcomes of osteochondral allograft transplantation for treatment of shoulder instability are well-documented, there is a relative paucity of literature in its use for the treatment of symptomatic chondral defects of the glenohumeral joint [69–71]. Specifically, a systematic review analyzing the outcomes of osteochondral allograft transplantation for treatment of large Hill-Sachs lesions in the setting of instability revealed improvement in range of motion, ASES scores, patient satisfaction, and return to work rates [72]. On a technical note, Wang et al. describe their technique for the treatment of a posterior chondral defect of the humeral head caused by anchor arthropathy after failed SLAP repair [73]. They utilized an open deltopectoral approach to the shoulder and used a cannulated coring drill system to harvest the graft and prepare the defect site. The postoperative rehab protocol included a 4 week period of sling immobilization in order to allow healing of the subscapularis repair. Aggressive strengthening involving internal rotation was started at 12 weeks. A virtually identical technique was used by Johnson et al. to treat a 2 cm diameter OCD lesion of the humeral head in a 19-year-old patient. At the time of final follow-up 3 years postoperatively, the patient had a full pain-free arc of motion and had returned to all preinjury activities without any difficulty [22].

In the only case series related to the treatment of symptomatic glenohumeral chondral defects with fresh osteochondral allograft, Riff et al. reported their midterm outcomes (mean follow-up time of 67 months) for 20 patients [74]. The average age of the patients was 24.8 years and 11 patients underwent concomi-tant glenoid surgery (microfracture or soft tissue resurfacing) in addition to humeral head osteochondral allograft transplantation. Eighteen out of 20 patients had successful graft incorporation and four patients went on to shoulder arthroplasty at a mean of 25 months postoperatively. Eleven out of the 20 patients reported that they were satisfied with their outcome. The authors also reported significant improvement in several patient-reported outcome scores such as the VAS, SST, ASES, and SF-12. Of note, patients who had a history of intra-articular pain pump use experienced lower satisfaction and trended towards worse outcomes compared to the rest of the cohort.

23.12 Autologous Chondrocyte Implantation

Autologous Chondrocyte Implantation (ACI) is a technique that was first described by Brittberg et al. in 1994 in which cartilage defects are covered by autologous chondrocytes that have been cultured in vitro [75]. It is a two-stage procedure. The first stage involves harvesting a cartilage biopsy from a non-weight-bearing portion of a joint, most commonly from the intercondylar notch in the knee. The biopsy is transferred to a laboratory where the chondrocytes are isolated via enzymatic digestion and expanded to achieve a final concentration of ideally $2–3 \times 10^7$ chondrocytes/ml. There have been several iterations of the ACI technique however in the current (third-generation) version, the chondrocytes are then directly seeded in a monolayer onto a biodegradable porcine type I/III collagen scaffold which can be placed directly on the chondral defect and secured with fibrin glue or sutures. In previous generations of the technique, a periosteal flap had to be harvested and placed over the defect to contain the chondrocytes since they were in a liquid suspension. This technology produces a 3D collagen scaffold with the chondrocytes cultured directly into the membrane. This results in equal distribution of the chondrocytes throughout the membrane and the 3D nature of this construct prevents chondrocyte dedifferentiation and loss of phenotype [76, 77]. This technique is most applicable to larger chondral lesions greater than 2 cm² which do not have significant involvement of the underlying subchondral bone. The advantage of this technique over others such as microfracture is that it results in the formation of hyaline-like cartilage as opposed to fibrocartilage [78].

Although ACI has been proven to be an effective surgical option for the treatment of chondral lesions in the knee, there is much less data on outcomes when treating these lesions in the glenohumeral joint [79–81]. One of the first descriptions for the use of ACI in the shoulder was a case report published by Romeo et al. [82] The authors performed first-generation ACI for treatment of 3.3 × 1.5 cm defect of the humeral head. Twelve months postoperatively, the patient had a full, pain-free range of motion and no complaints of pain at rest or during activities.

In a case series of four patients, Buchmann et al. utilized third-generation ACI for treatment of full-thickness chondral defects of the humeral head and glenoid [83]. Three of the patients had humeral lesions measuring 6 cm² and the other patient had a glenoid lesion measuring 2 cm². Cartilage biopsies were obtained from the ipsilateral shoulder during an initial diagnostic arthroscopy procedure and the subsequent cartilage procedure was performed by the senior author using an open deltopectoral approach to the shoulder. At a mean follow-up of 41.3 months, all the patients had satisfactory shoulder function with mean VAS, Constant, and ASES scores of 0.3, 83.3, and 95.3, respectively. Of note, preoperative scores were not reported which makes it difficult to appreciate how much benefit the patients experienced as a result of the surgery. All of the patients underwent follow-up MRI which showed satisfactory defect coverage with fibrocartilaginous repair tissue.

In a slightly larger case series of seven patients, Boehm et al. performed ACI using 3D spheroids to treat full-thickness chondral defects of the humeral head. The median defect size was 3 cm² and the procedure was performed arthroscopically in some cases and open in others, depending on the ability to easily access the lesions arthroscopically. At a median follow-up time of 32 months, the median Subjective Shoulder Value was 95% as compared to a median preoperative value of 60%. Additionally, the median postoperative VAS, Constant, and ASES scores were 0, 95, and 97, respectively. A second look arthroscopy was performed in five patients at a median of 6 months postoperatively. On arthroscopic examination, compete coverage of the defect was observed in four patients and the other patient has a persistent 0.25 cm² full-thickness defect. All of the patients expressed that they would repeat and recommend the procedure.

23.13 Resurfacing

While resurfacing the humeral head or glenoid is typically reserved for young patients with advanced glenohumeral arthritis, it can also be considered for patients with focal chondral defects. Some situations where it may be a reasonable option include large osteochondral defects, bipolar kissing lesions, and in the revision setting when other cartilage procedures have failed. Resurfacing on the humeral side involves placement of a stemless, low-profile metal prosthesis, and on the glenoid side might involve a biologic patch such as a tendon, dermal graft, meniscus, or an inlay glenoid.

Although there are multiple implant choices available for resurfacing of the humeral head such as inlay, stemless onlay, and stemmed onlay, the inlay prosthesis is perhaps most suited for the management of focal chondral defects since it allows for precise patient-specific surface reconstruction [84]. The HemiCAP inlay prosthesis, which was FDA approved in 2014, has implants with diameters ranging in size from 25–40 and multiple offset options for each diameter implant. The inlay design of the implant allows the prosthesis to match the contour of the patients remaining native cartilage (Fig. 23.2). Sweet et al. reported on the outcomes of 19 patients after per-

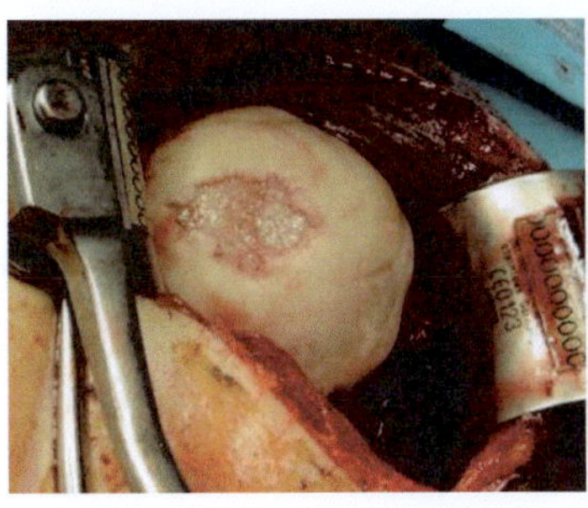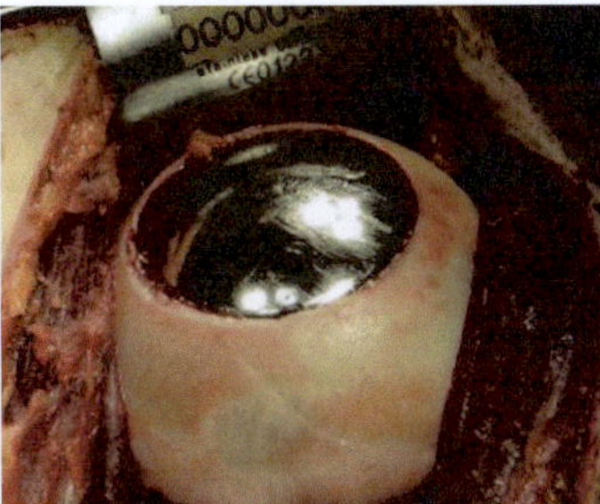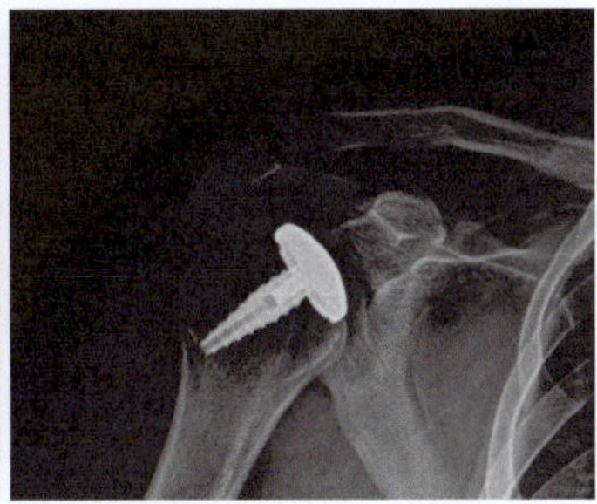

Fig. 23.2 Arthrosurface Humeral Inlay Prosthesis. *Left*: Full-thickness chondral defect seen in the humeral head. *Middle*: Final appearance of the implanted prosthesis. *Right*: Postoperative radiographs showing congruence of prosthesis with surrounding humeral head

forming humeral head resurfacing using the HemiCAP implant for management of humeral head defects [84]. At a mean follow-up of 32.7 months, there was a statistically significant improvement in mean ASES, SST, and VAS scores as well as in degrees of forward flexion and eternal rotation. Follow-up radiographic evaluation showed no evidence of component loosening or failure and 90% of patients were satisfied with their choice of procedure. There were three reported complications including subsequent rotator cuff tear, progression of glenoid wear, and deep infection resulting in rupture of the subscapularis. Multiple other case reports of inlay partial resurfacing have been reported for the management of humeral head pathology with generally favorable outcomes [85–87].

Biologic resurfacing of the glenoid was first described by Burkhead and Hutton in 1995 with the use of an autologous facia lata graft for the glenoid in combination with humeral head replacement [39, 88]. Other commonly used biologic resurfacing options include lateral meniscus allograft, Achilles tendon allograft, acellular dermal allograft, and shoulder capsular tissue. Krishnan et al. have published their results and technique for biologic resurfacing of the glenoid in combination with humeral head replacement [89, 90]. In their case series of 36 shoulders, biologic resurfacing was performed using anterior capsule in 7 patients, autogenous facia lata in 11, and Achilles allograft in 18. They reported excellent clinical outcomes in 18 out of 36 shoulders and found a trend towards better outcomes with the use of Achilles tendon allograft as the resurfacing material. In a systematic review, Meaike et al. examined the clinical outcomes after soft tissue resurfacing for glenohumeral arthritis [91]. Eleven studies were included and the minimum follow-up was 24 months. The glenoid grafts included all of the previously mentioned sources; however, the two most commonly used were lateral meniscus allograft (44.3%) and human acellular dermal allograft (25.4%). Although the studies reported significantly improved ASES, SST, and VAS scores postoperatively, there was a complication rate of over 36% and a revision rate of 34%. The results indicate that while biologic resurfacing of the glenoid can result in significant functional improvement in certain patients, patients should be counseled about the potential for complications and the need for revision surgery.

23.14 Conclusion

While articular cartilage defects in the glenohumeral joint tend to be better-tolerated and less symptomatic than those in weight-bearing joints, such as the knee, they can occasionally be the source of significant discomfort, mechanical symptoms, and disability for patients. A comprehensive workup including history, physical examination, and advanced imaging is crucial in the diagnosis of a symptomatic chondral defect and ruling out concomitant pathology as the etiology of the patient's symptoms. When the diagnosis of a symptomatic lesion is established, the treating surgeon has many options available for the management of these lesions, both nonoperative and surgical. Thus, treatment should be tailored to the individual needs of each patient. Factors such as lesion size, location, depth as well as patient-specific factors such as expectations, comorbidities, and overall demand should all be taken into consideration. Although many of the procedures discussed in this chapter have been validated through high-quality studies performed in other joints, such as the knee, the data as it pertains to the glenohumeral joint is scarce and consists mainly of level IV case series. Although the clinical results from these studies are generally favorable, additional higher quality studies are needed to more accurately determine the efficacy of the procedures and to compare the various treatment options against each other.

References

1. Cole BJ, Yanke A, Provencher MT. Nonarthroplasty alternatives for the treatment of glenohumeral arthritis. J Shoulder Elb Surg. 2007; https://doi.org/10.1016/j.jse.2007.03.011.
2. Gross CE, Chalmers PN, Chahal J, Van Thiel G, Bach BR, Cole BJ, Romeo AA. Operative treatment of chondral defects in the glenohumeral joint. Arthroscopy. 2012; https://doi.org/10.1016/j.arthro.2012.03.026.
3. McCarty LP, Cole BJ. Nonarthroplasty treatment of glenohumeral cartilage lesions. Arthroscopy. 2005; https://doi.org/10.1016/j.arthro.2005.06.023.
4. Seidl AJ, Kraeutler MJ. Management of articular cartilage defects in the glenohumeral joint. J Am Acad Orthop Surg. 2018; https://doi.org/10.5435/JAAOS-D-17-00057.
5. Elser F, Braun S, Dewing CB, Millett PJ. Glenohumeral joint preservation: current options for managing articular cartilage lesions in young, active patients. Arthroscopy. 2010; https://doi.org/10.1016/j.arthro.2009.10.017.
6. Saltzman BM, Leroux T, Cole BJ. Management and surgical options for articular defects in the shoulder. Clin Sports Med. 2017; https://doi.org/10.1016/j.csm.2017.02.009.
7. Halder AM, Itoi E, An KN. Anatomy and biomechanics of the shoulder. Orthop Clin North Am. 2000; https://doi.org/10.1016/S0030-5898(05)70138-3.
8. Soslowsky LJ, Flatow EL, Bigliani LU, Mow VC. Articular geometry of the glenohumeral joint. Clin Orthop Relat Res. 1992; https://doi.org/10.1097/00003086-199212000-00023.
9. Sophia Fox AJ, Bedi A, Rodeo SA. The basic science of articular cartilage: structure, composition, and function. Sports Health. 2009; https://doi.org/10.1177/1941738109350438.
10. Taylor DC, Arciero RA. Pathologic changes associated with shoulder dislocations. Arthroscopic and physical examination findings in first-time, traumatic anterior dislocations. Am J Sports Med. 1997; https://doi.org/10.1177/036354659702500306.
11. Calandra JJ, Baker CL, Uribe J. The incidence of Hill-Sachs lesions in initial anterior shoulder dislocations. Arthroscopy. 1989; https://doi.org/10.1016/0749-8063(89)90138-2.
12. Hintermann B, Gächter A. Arthroscopic findings after shoulder dislocation. Am J Sports Med. 1995; https://doi.org/10.1177/036354659502300505.
13. Werner AW, Lichtenberg S, Schmitz H, Nikolic A, Habermeyer P. Arthroscopic findings in atraumatic shoulder instability. Arthroscopy. 2004; https://doi.org/10.1016/j.arthro.2003.11.037.
14. Hill HA, Sachs MD. The grooved defect of the humeral head. Radiology. 1940; https://doi.org/10.1148/35.6.690.
15. Provencher MT, Frank RM, LeClere LE, Metzger PD, Ryu JJ, Bernhardson LTA, Romeo AA. The Hill-Sachs lesion: diagnosis, classification, and management. J Am Acad Orthop Surg. 2012; https://doi.org/10.5435/JAAOS-20-04-242.
16. Miller C, Savoie FH. Glenohumeral abnormalities associated with full-thickness tears of the rotator cuff. Orthop Rev. 1994;23:159–62.
17. Gartsman GM, Taverna E. The incidence of glenohumeral joint abnormalities associated with full-thickness, reparable rotator cuff tears. Arthroscopy. 1997; https://doi.org/10.1016/S0749-8063(97)90123-7.
18. Paley KJ, Jobe FW, Pink MM, Kvitne RS, ElAttrache NS. Arthroscopic findings in the overhand throwing athlete: evidence for posterior internal impingement of the rotator cuff. Arthroscopy. 2000; https://doi.org/10.1016/S0749-8063(00)90125-7.
19. Patzer T, Lichtenberg S, Kircher J, Magosch P, Habermeyer P. Influence of SLAP lesions on chondral lesions of the glenohumeral joint. Knee Surg Sport Traumatol Arthrosc. 2010; https://doi.org/10.1007/s00167-009-0938-2.
20. Mahirogullari M, Chloros GD, Wiesler ER, Ferguson C, Poehling GG. Osteochondritis dissecans of the humeral head. Joint Bone Spine. 2008; https://doi.org/10.1016/j.jbspin.2007.04.025.
21. Hamada S, Hamada M, Nishiue S, Doi T. Osteochondritis dissecans of the humeral head. Arthroscopy. 1992; https://doi.org/10.1016/0749-8063(92)90148-5.
22. Johnson DL, Warner JJP. Osteochondritis dissecans of the humeral head: treatment with a matched osteochondral allograft. J Shoulder Elb Surg. 1997; https://doi.org/10.1016/S1058-2746(97)90038-0.
23. Jafari D, Shariatzadeh H, Mazhar FN, Okhovatpour MA, Razavipour M. Osteochondritis dissecans of the humeral head: a case report and review of the

literature. Arch Bone Joint Surg. 2017; https://doi.org/10.22038/abjs.2016.7862.

24. Solomon DJ, Navaie M, Stedje-Larsen ET, Smith JC, Provencher MT. Glenohumeral chondrolysis after arthroscopy: a systematic review of potential contributors and causal pathways. Arthroscopy. 2009; https://doi.org/10.1016/j.arthro.2009.06.001.

25. Petty DH, Jazrawi LM, Estrada LS, Andrews JR. Glenohumeralysis Chondrolysis after shoulder arthroscopy: case reports and review of the literature. Am J Sports Med. 2004; https://doi.org/10.1177/0363546503262176.

26. Good CR, Shindle MK, Kelly BT, Wanich T, Warren RF. Glenohumeral Chondrolysis after shoulder arthroscopy with thermal capsulorrhaphy. Arthroscopy. 2007; https://doi.org/10.1016/j.arthro.2007.03.092.

27. Busfield BT, Romero DM. Pain pump use after shoulder arthroscopy as a cause of glenohumeral chondrolysis. Arthroscopy. 2009; https://doi.org/10.1016/j.arthro.2009.01.019.

28. Outerbridge RE. The etiology of chondromalacia patellae. J Bone Joint Surg Br. 1961; https://doi.org/10.1302/0301-620X.43B4.752.

29. Brittberg M, Winalski CS. Evaluation of cartilage injuries and repair. J Bone Joint Surg Am. 2003; https://doi.org/10.2106/00004623-200300002-00008.

30. Ellman H, Harris E, Kay SP. Early degenerative joint disease simulating impingement syndrome: arthroscopic findings. Arthroscopy. 1992; https://doi.org/10.1016/0749-8063(92)90012-Z.

31. Lecouvet FE, Dorzée B, Dubuc JE, Berg BCV, Jamart J, Malghem J. Cartilage lesions of the glenohumeral joint: diagnostic effectiveness of multidetector spiral CT arthrography and comparison with arthroscopy. Eur Radiol. 2007; https://doi.org/10.1007/s00330-006-0523-8.

32. Spencer BA, Dolinskas CA, Seymour PA, Thomas SJ, Abboud JA. Glenohumeral articular cartilage lesions: prospective comparison of non-contrast magnetic resonance imaging and findings at arthroscopy. Arthroscopy. 2013; https://doi.org/10.1016/j.arthro.2013.05.023.

33. Carroll KW, Helms CA, Speer KP. Focal articular cartilage lesions of the superior humeral head: MR imaging findings in seven patients. Am J Roentgenol. 2001; https://doi.org/10.2214/ajr.176.2.1760393.

34. Denti M, Monteleone M, Trevisan C, De Romedis B, Barmettler F. Magnetic resonance imaging versus arthroscopy for the investigation of the osteochondral humeral defect in anterior shoulder instability - a double-blind prospective study. Knee Surg Sport Traumatol Arthrosc. 1995; https://doi.org/10.1007/BF01565481.

35. Guntern DV, Pfirrmann CWA, Schmid MR, Zanetti M, Binkert CA, Schneeberger AG, Hodler J. Articular cartilage lesions of the glenohumeral joint: diagnostic effectiveness of MR arthrography and prevalence in patients with subacromial impingement syndrome. Radiology. 2003; https://doi.org/10.1148/radiol.2261012090.

36. Gold GE, Reeder SB, Beaulieu CF. Advanced MR imaging of the shoulder: dedicated cartilage techniques. Magn Reson Imaging Clin N Am. 2004; https://doi.org/10.1016/j.mric.2004.01.007.

37. Jazrawi LM, Alaia MJ, Chang G, FitzGerald EF, Recht MP. Advances in magnetic resonance imaging of articular cartilage. J Am Acad Orthop Surg. 2011; https://doi.org/10.5435/00124635-201107000-00005.

38. Gold GE, Chen CA, Koo S, Hargreaves BA, Bangerter NK. Recent advances in MRI of articular cartilage. Am J Roentgenol. 2009; https://doi.org/10.2214/AJR.09.3042.

39. Ansok CB, Muh SJ. Optimal management of glenohumeral osteoarthritis. Orthop Res Rev. 2018; https://doi.org/10.2147/ORR.S134732.

40. Sinha I, Lee M, Cobiella C. Management of osteoarthritis of the glenohumeral joint. Br J Hosp Med. 2008; https://doi.org/10.12968/hmed.2008.69.5.29358.

41. Jong BY, Goel DP. Biologic options for Glenohumeral arthritis. Clin Sports Med. 2018; https://doi.org/10.1016/j.csm.2018.06.001.

42. Carr JB, Rodeo SA. The role of biologic agents in the management of common shoulder pathologies: current state and future directions. J Shoulder Elb Surg. 2019; https://doi.org/10.1016/j.jse.2019.07.025.

43. Blaine T, Moskowitz R, Udell J, Skyhar M, Levin R, Friedlander J, Daley M, Altman R. Treatment of persistent shoulder pain with sodium hyaluronate: a randomized, controlled trial. A multicenter study. J Bone Joint Surg Am. 2008; https://doi.org/10.2106/JBJS.F.01116.

44. Zhang B, Thayaparan A, Horner N, Bedi A, Alolabi B, Khan M. Outcomes of hyaluronic acid injections for glenohumeral osteoarthritis: a systematic review and meta-analysis. J Shoulder Elb Surg. 2019; https://doi.org/10.1016/j.jse.2018.09.011.

45. Merolla G, Sperling JW, Paladini P, Porcellini G. Efficacy of Hylan G-F 20 versus 6-methylprednisolone acetate in painful shoulder osteoarthritis: a retrospective controlled trial. Musculoskelet Surg. 2011; https://doi.org/10.1007/s12306-011-0138-3.

46. Ben FJ, Barnard A. To evaluate the effect of combining photo-activation therapy with platelet-rich plasma injections for the novel treatment of osteoarthritis. BMJ Case Rep. 2013; https://doi.org/10.1136/bcr-2012-007463.

47. Rossi LA, Piuzzi NS, Shapiro SA. Glenohumeral osteoarthritis: the role for orthobiologic therapies: platelet-rich plasma and cell therapies. JBJS Rev. 2020; https://doi.org/10.2106/JBJS.RVW.19.00075.

48. Freitag J, Ford J, Bates D, Boyd R, Hahne A, Wang Y, Cicuttini F, Huguenin L, Norsworthy C, Shah K. Adipose derived mesenchymal stem cell therapy in the treatment of isolated knee chondral lesions: design of a randomised controlled pilot study comparing arthroscopic microfracture versus arthroscopic microfracture combined with postoperative mesenchymal.

BMJ Open. 2015; https://doi.org/10.1136/bmjopen-2015-009332.

49. Shapiro SA, Kazmerchak SE, Heckman MG, Zubair AC, O'Connor MI. A prospective, single-blind, placebo-controlled trial of bone marrow aspirate concentrate for knee osteoarthritis. Am J Sports Med. 2017; https://doi.org/10.1177/0363546516662455.

50. Chahla J, Dean CS, Moatshe G, Pascual-Garrido C, Serra Cruz R, LaPrade RF. Concentrated bone marrow aspirate for the treatment of chondral injuries and osteoarthritis of the knee: a systematic review of outcomes. Orthop J Sport Med. 2016; https://doi.org/10.1177/2325967115625481.

51. Cameron BD, Galatz LM, Ramsey ML, Williams GR, Iannotti JP. Non-prosthetic management of grade IV osteochondral lesions of the glenohumeral joint. J Shoulder Elb Surg. 2002; https://doi.org/10.1067/mse.2002.120143.

52. Skelley NW, Namdari S, Chamberlain AM, Keener JD, Galatz LM, Yamaguchi K. Arthroscopic debridement and capsular release for the treatment of shoulder osteoarthritis. Arthroscopy. 2015; https://doi.org/10.1016/j.arthro.2014.08.025.

53. Kerr BJ, McCarty EC. Outcome of arthroscopic débridement is worse for patients with glenohumeral arthritis of both sides of the joint. Clin Orthop Relat Res. 2008; https://doi.org/10.1007/s11999-007-0088-0.

54. Weinstein DM, Bucchieri JS, Pollock RG, Flatow EL, Bigliani LU. Arthroscopic debridement of the shoulder for osteoarthritis. Arthroscopy. 2000; https://doi.org/10.1053/jars.2000.5042.

55. Namdari S, Skelley N, Keener JD, Galatz LM, Yamaguchi K. What is the role of arthroscopic debridement for glenohumeral arthritis? A critical examination of the literature. Arthroscopy. 2013; https://doi.org/10.1016/j.arthro.2013.02.022.

56. Van Thiel GS, Sheehan S, Frank RM, Slabaugh M, Cole BJ, Nicholson GP, Romeo AA, Verma NN. Retrospective analysis of arthroscopic management of glenohumeral degenerative disease. Arthroscopy. 2010; https://doi.org/10.1016/j.arthro.2010.02.026.

57. Williams BT, Beletsky A, Kunze KN, Polce EM, Cole BJ, Verma NN, Chahla J. Outcomes and survivorship after arthroscopic treatment of glenohumeral arthritis: a systematic review. Arthroscopy. 2020; https://doi.org/10.1016/j.arthro.2020.02.036.

58. Steadman JR, Rodkey WG, Rodrigo JJ. Microfracture: surgical technique and rehabilitation to treat chondral defects. Clin Orthop Relat Res. 2001; https://doi.org/10.1097/00003086-200110001-00033.

59. Steadman JR, Rodkey WG, Briggs KK. Microfracture to treat full-thickness chondral defects: surgical technique, rehabilitation, and outcomes. J Knee Surg. 2002;15:170–6.

60. Hensley CP, Sum J. Physical therapy intervention for a former power lifter after arthroscopic microfracture procedure for grade iv glenohumeral chondral defects. Int J Sports Phys Ther. 2011;6:10–26.

61. Millett PJ, Huffard BH, Horan MP, Hawkins RJ, Steadman JR. Outcomes of full-thickness articular cartilage injuries of the shoulder treated with microfracture. Arthroscopy. 2009; https://doi.org/10.1016/j.arthro.2009.02.009.

62. Frank RM, Van Thiel GS, Slabaugh MA, Romeo AA, Cole BJ, Verma NN. Clinical outcomes after microfracture of the Glenohumeral joint. Am J Sports Med. 2010; https://doi.org/10.1177/0363546509350304.

63. Wang KC, Frank RM, Cotter EJ, Davey A, Meyer MA, Hannon CP, Leroux T, Romeo AA, Cole BJ. Long-term clinical outcomes after microfracture of the Glenohumeral joint: average 10-year follow-up. Am J Sports Med. 2018; https://doi.org/10.1177/0363546517750627.

64. Snow M, Funk L. Microfracture of chondral lesions of the glenohumeral joint. Int J Shoulder Surg. 2008; https://doi.org/10.4103/0973-6042.44142.

65. Scheibel M, Bartl C, Magosch P, Lichtenberg S, Habermeyer P. Osteochondral autologous transplantation for the treatment of full-thickness articular cartilage defects of the shoulder. J Bone Joint Surg Br. 2004; https://doi.org/10.1302/0301-620X.86B7.14941.

66. Kircher J, Patzer T, Magosch P, Lichtenberg S, Habermeyer P. Osteochondral autologous transplantation for the treatment of full-thickness cartilage defects of the shoulder: results at nine years. J Bone Joint Surg Br. 2009; https://doi.org/10.1302/0301-620X.91B4.21838.

67. Park TS, Kim TS, Cho JH. Arthroscopic osteochondral autograft transfer in the treatment of an osteochondral defect of the humeral head: report of one case. J Shoulder Elb Surg. 2006; https://doi.org/10.1016/j.jse.2005.10.008.

68. Beer AJ, Tauro TM, Redondo ML, Christian DR, Cole BJ, Frank RM. Use of allografts in Orthopaedic surgery: safety, procurement, storage, and outcomes. Orthop J Sport Med. 2019; https://doi.org/10.1177/2325967119891435.

69. Chapovsky F, Kelly JD IV. Osteochondral allograft transplantation for treatment of glenohumeral instability. Arthroscopy. 2005; https://doi.org/10.1016/j.arthro.2005.04.005.

70. Provencher MT, Sanchez G, Schantz K, Ferrari M, Sanchez A, Frangiamore S, Mannava S. Anatomic humeral head reconstruction with fresh osteochondral talus allograft for recurrent Glenohumeral instability with reverse Hill-Sachs lesion. Arthrosc Tech. 2017; https://doi.org/10.1016/j.eats.2016.10.017.

71. Petrera M, Veillette CJ, Taylor DW, Park SS, Theodoropoulos JS. Use of fresh osteochondral glenoid allograft to treat posteroinferior bone loss in chronic posterior shoulder instability. Am J Orthop (Belle Mead NJ). 2013;42:78–82.

72. Saltzman BM, Riboh JC, Cole BJ, Yanke AB. Humeral head reconstruction with osteochondral allograft transplantation. Arthroscopy. 2015; https://doi.org/10.1016/j.arthro.2015.03.021.

73. Wang KC, Waterman BR, Cotter EJ, Frank RM, Cole BJ. Fresh osteochondral allograft transplantation for focal chondral defect of the humerus associated with anchor arthropathy and failed SLAP

repair. Arthrosc Tech. 2017; https://doi.org/10.1016/j.eats.2017.06.008.

74. Riff AJ, Yanke AB, Shin JJ, Romeo AA, Cole BJ. Midterm results of osteochondral allograft transplantation to the humeral head. J Shoulder Elb Surg. 2017; https://doi.org/10.1016/j.jse.2016.11.053.

75. Brittberg M, Lindahl A, Nilsson A, Ohlsson C, Isaksson O, Peterson L. Treatment of deep cartilage defects in the knee with autologous chondrocyte transplantation. N Engl J Med. 1994; https://doi.org/10.1056/NEJM199410063311401.

76. Miller MD, Thompson SR. DeLee, Drez and Miller's orthopaedic sports medicine: principles and practice. Philadelphia: PA Saunders; 2003.

77. Bartlett W, Skinner JA, Gooding CR, Carrington RWJ, Flanagan AM, Briggs TWR, Bentley G. Autologous chondrocyte implantation versus matrix-induced autologous chondrocyte implantation for osteochondral defects of the knee. A prospective, randomised study. J Bone Joint Surg Br. 2005; https://doi.org/10.1302/0301-620X.87B5.15905.

78. Zheng MH, Willers C, Kirilak L, Yates P, Xu J, Wood D, Shimmin A. Matrix-induced autologous chondrocyte implantation (MACI®): biological and histological assessment. Tissue Eng. 2007; https://doi.org/10.1089/ten.2006.0246.

79. Knutsen G, Drogset JO, Engebretsen L, Grøntvedt T, Isaksen V, Ludvigsen TC, Roberts S, Solheim E, Strand T, Johansen O. A randomized trial comparing autologous chondrocyte implantation with microfracture: findings at five years. J Bone Joint Surg Am. 2007; https://doi.org/10.2106/JBJS.G.00003.

80. Saris DBF, Vanlauwe J, Victor J, et al. Characterized chondrocyte implantation results in better structural repair when treating symptomatic cartilage defects of the knee in a randomized controlled trial versus microfracture. Am J Sports Med. 2008; https://doi.org/10.1177/0363546507311095.

81. Peterson L, Minas T, Brittberg M, Nilsson A, Sjögren-Jansson E, Lindahl A. Two-to 9-year outcome after autologous chondrocyte transplantation of the knee. Clin Orthop Relat Res. 2000; https://doi.org/10.1097/00003086-200005000-00020.

82. Romeo AA, Cole BJ, Mazzocca AD, Fox JA, Freeman KB, Joy E. Autologous chondrocyte repair of an articular defect in the humeral head. Arthroscopy. 2002; https://doi.org/10.1053/jars.2002.36144.

83. Buchmann S, Salzmann GM, Glanzmann MC, Wörtler K, Vogt S, Imhoff AB. Early clinical and structural results after autologous chondrocyte transplantation at the glenohumeral joint. J Shoulder Elb Surg. 2012; https://doi.org/10.1016/j.jse.2011.07.030.

84. Sweet SJ, Takara T, Ho L, Tibone JE. Primary partial humeral head resurfacing: outcomes with the Hemi-CAP implant. Am J Sports Med. 2015; https://doi.org/10.1177/0363546514562547.

85. Bixby EC, Sonnenfeld JJ, Alrabaa RG, Trofa DP, Jobin CM. Partial humeral head resurfacing for avascular necrosis. Arthrosc Tech. 2020; https://doi.org/10.1016/j.eats.2019.09.025.

86. Uribe JW, Botto-van Bemden A. Partial humeral head resurfacing for osteonecrosis. J Shoulder Elb Surg. 2009; https://doi.org/10.1016/j.jse.2008.10.016.

87. Bessette MC, Frisch NC, Kodali P, Jones MH, Miniaci A. Partial resurfacing for humeral head defects associated with recurrent shoulder instability. Orthopedics. 2017; https://doi.org/10.3928/01477447-20171012-01.

88. Burkhead WZ, Hutton KS. Biologic resurfacing of the glenoid with hemiarthroplasty of the shoulder. J Shoulder Elb Surg. 1995; https://doi.org/10.1016/S1058-2746(05)80019-9.

89. Krishnan SG, Reineck JR, Nowinski RJ, Harrison D, Burkhead WZ. Humeral hemiarthroplasty with biologic resurfacing of the glenoid for glenohumeral arthritis. J Bone Joint Surg Am. 2008; https://doi.org/10.2106/JBJS.G.01220.

90. Krishnan SG, Nowinski RJ, Harrison D, Burkhead WZ. Humeral hemiarthroplasty with biologic resurfacing of the glenoid for glenohumeral arthritis: two to fifteen-year outcomes. J Bone Joint Surg Am. 2007; https://doi.org/10.2106/JBJS.E.01291.

91. Meaike JJ, Patterson DC, Anthony SG, Parsons BO, Cagle PJ. Soft tissue resurfacing for glenohumeral arthritis: a systematic review. Shoulder Elbow. 2020; https://doi.org/10.1177/1758573219849606.

Nontraumatic Shoulder Osteochondral Defects

Aleksandra Sibilska, Katarzyna Herman, and Adam Kwapisz

24.1 Introduction

Chondral lesions of glenohumeral joint are a substantial issue, though not as frequent as in knee joint and so far, the literature on this matter is rather limited. Unlike the knee joint, in non-weight-bearing glenohumeral joint chondral lesions are usually observed with other shoulder pathologies or correlates with a traumatic event. This chapter focuses on the nontraumatic osteochondral defects that are less commonly observed. The etiology of such may vary from focal osteonecrosis, osteoarthritis, genetic disorders, nontraumatic shoulder instability or microinstability, septic arthritis, rotator cuff arthropathy, loose bodies to degenerative changes, or osteochondrosis dissecans [1–4].

24.2 Etiology

The overall incidence of focal chondral lesions in publications varies from 5% to 29%, but much depends on the pathologies encountered with the lesions [5]. Available data does not specify the incidence of nontraumatic focal chondral lesions, but in some cases, those may be coexisting with an underlying pathology. For instance, frequency of symptomatic glenohumeral cartilage defects among patients with rotator cuff tears and overhead athletes was reported as 13–17% whereas other papers reported up to 29% prevalence of humeral cartilage lesions and 15% of glenoid cartilage lesions in patients undergoing shoulder arthroscopy due to subacromial impingement [6, 7]. Cases of chondral lesions in glenohumeral joint due to osteochondrosis dissecans are also described however they are not frequent as in knee or ankle [8]. If reported, they typically occur in glenoid fossa, but few authors described also cases with humeral head localization [9, 10].

Osteonecrosis (ON) of the humeral head is one of the causes of osteochondral defects and it is the second most common site of osteonecrosis after the femoral head. In an epidemiological study, Cooper et al. reported overall humeral head ON to be 2.3% [11]. It is also more frequent in patients with multifocal osteonecrosis, rather than as an isolated case. Most commonly ON lesions are located in the superior aspect of the humeral head [12]. These lesions form due to an ischemic event, a subsequent necrosis that leads to subchondral bone destruction and permanent changes prompted by normal forces acting upon an impaired bone [13, 14].

Many possible underlying causes of this disease are described, most common being corticosteroid therapy administered for various reasons

A. Sibilska · A. Kwapisz (✉)
Clinic of Orthopedics and Pediatric Orthopedics,
Medical University of Lodz, Lodz, Poland

K. Herman
Department of Orthopedics and Traumatology,
Brothers Hospitallers Hospital, Katowice, Poland

© ISAKOS 2022
A. Gobbi et al. (eds.), *Joint Function Preservation*, https://doi.org/10.1007/978-3-030-82958-2_24

[15]., Mont et al. found it to be associated with 82% of all humeral head ON cases in their study group [16]. The time between the onset of steroid therapy and diagnosis is estimated to be an average 15 months [17]. Less commonly the ON can be associated with alcohol abuse, smoking, hemoglobinopathies (Sickle-cell disease), pancreatitis, hyperlipidemia, Gaucher disease, dysbarism, connective tissue disorders, radiation therapy, pancreatitis, and pregnancy [15].

24.3 Diagnosis

Knowledge of the anatomy of glenohumeral joint surfaces will ease correct diagnosis. Remembering that humeral head chondral surface is the thickest in the center whereas in the glenoid cartilage is the thickest cartilage peripherally [18]. These data should be taken into account in the diagnosis and decision-making, together with knowledge of glenoid version (1.5° retroversion) and inclination (4.2° superiorly) are important to consider for as well for approaching the joint surgically [4, 19].

During the diagnosing process, thorough medical history and chief complaints should be obtained. Considering that chondral defects of glenohumeral joint are generally well tolerated and asymptomatic, diagnosis may be challenging. Clinical diagnosis might be challenging, especially when the pain is ambiguous and not clearly located, similarly to other shoulder pathologies [20]. It is worth remembering that sometimes it is the coexisting pathology that can produce symptoms.

In some patients, chondral lesions may result in unremitting, activity-related pain. Over time, it may cause reduction of shoulder range of motion. Occasionally patients describe some grinding or catching in the joint while doing some specific activities [8, 21–23]. Ellman et al. described compression-rotation test helping to distinguish between symptomatic subacromial impingement syndrome and one from glenohumeral cartilage lesion. To perform the test, the patient is placed in the lateral recumbent position on the unaffected side and the examiner compresses the humeral head into the glenoid while the patient internally and externally rotates the arm. Provocation of pain with this maneuver is suggestive of chondral pathology. The test can be made more specific if Neer's impingement test is conducted first, particularly if pain with forward flexion is eliminated with subacromial injection of local anesthetic and the compression-rotation test continues to elicit discomfort [24, 25]. Moreover, sleep may be affected. In patients reporting a nontraumatic onset of symptoms, it is important to analyze patients' comorbidities and medications. All information obtained during the medical interview and examinations along with patients' age, activity level, and expectations should be taken into consideration while choosing the proper treatment method.

24.4 Classification and Imaging

The primary imaging method is plain radiographs including true anteroposterior view (Grashey), scapular Y (Neer), and axillary view (Fig. 24.1). The Stryker Notch and West Point may also be helpful in patients with instability history [4, 26]. They are helpful to visualize posterosuperior humeral head and anteroinferior glenoid rim. Yet magnetic resonance (MRI) is the best imaging option for chondral injury however not ideal. One study reported 60% accuracy and 87% sensitivity of MRI for diagnosing osteochondral lesions of the humeral head in patients with anterior shoulder instability [27]. Seen on MRI, chondral defects are present as a contour deformity with areas of abnormal signal intensity [7] (Fig. 24.2). When there was some trauma incidence, subchondral bone edema may be observed. Moreover, delayed gadolinium-enhanced MRI of cartilage, T1rho, T2, and T2 mapping are being described in the literature to better evaluating the formation of articular cartilage [4, 28].

Humeral head ON may be described by Cruess classification [29]. First stage can only be diagnosed on MRI scans, second may be visible on plain radiographs as sclerotic changes of the middle superior portion of the head. The crescent sign suggesting a subchondral fracture is first

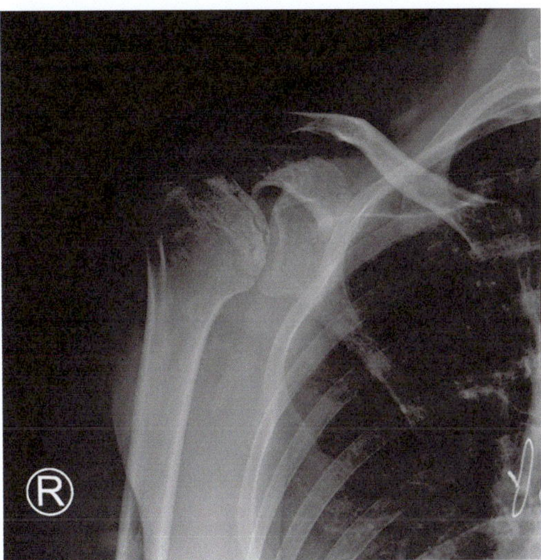

Fig. 24.1 An AP view of the right shoulder of the 38-years-old female. The gross focal chondral defect is visible on the humeral head surface. As far as the lesion size is concerned, this particular case may require either osteochondral autograft or partial resurfacing

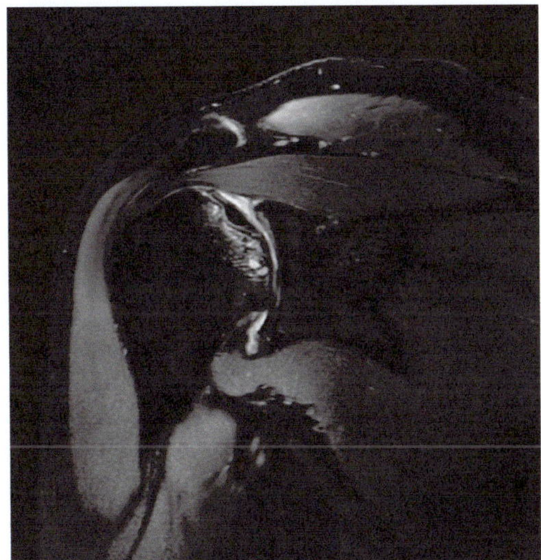

Fig. 24.2 A coronal cut of the right shoulder MRI of the 36-years-old male. Signs of the humeral head are visible with possible secondary degeneration of the glenoid cartilage. In that case, either hemiarthroplasty or TSA may be required

sign of progressing collapse of the head, which becomes obvious in fourth stage visible as deformation and flattening. Fifth stage is characterized by collapsed head and progressing degenerative changes of the glenoid.

As for focal chondral defects in the shoulder, there is no dedicated classification or scoring system however generally Outerbridge system, primarily designed for the patellofemoral joint, can be used to describe the lesions. However, International Cartilage Regeneration and Joint Preservation Society (ICRS) published the ICRS Hyaline Cartilage Lesion Classification System, which is modification of the Outerbridge classification and is now commonly used [30]. Stage 1a is macroscopically normal cartilage with fibrillation or softening and 1b if there are lacerations present. Defects that are deeper but not extending 50% of cartilage thickness can be classified as stage 2, if the damage exceeds 50% should be classified as stage 3. Severe cartilage defects extending into the subchondral bone are classified as ICRS 4 [31].

ICRS also provided a classification for OCD, where a stable lesion covered with undamaged cartilage is classified as ICRS OCD I. A partial discontinuity in the lesion yet stable when probed is ICRS OCD II, lesions with a complete discontinuity that are unstable should be classified as ICRS OCD III. Empty lesion or a lesion with a dislocated fragment or a completely loose one within the lesion is ICRS OCD IV [31].

It is worth remembering that some findings in the joint may be taken as pathology, such as a bare spot on glenoid or a bare area on the humeral head. These are anatomical and normal findings and should not be mistaken for cartilage defects [4].

24.5 Nonoperative Treatment

There are no precise treatment guidelines dedicated to chondral defects. The first line of treatment is always nonoperative treatment, but the further choice of the surgical procedure is made without one unified protocol.

It is believed that nonsurgical treatment will help to relieve symptoms and should be first-line treatment. Nonsurgical treatment includes nonsteroidal anti-inflammatory medications, corticosteroid injections (may also be diagnostic, if local anesthetic is administered), and physical therapy [32]. According to Dacre's et al. conclusion, corticosteroid injections are notably effective in patients with inflammatory arthropathies and might be even more cost-effective treatment option than physical therapy [33]. However, corticosteroids are also a major cause of nontraumatic ON of the humeral head and they should be administered only in selected group of patients. For these cases, tri-amcinolone acetonide seems to be a viable option due to prolonged presence in the joint, reduced systemic exposure, and no deleterious effect on cartilage compared with other available corticosteroids [34, 35]. Administration of local anesthetics should be reserved for selected diagnostic purposes only, as they were reported to have dose- and time-dependent destructive effects on chondrocytes. If needed, ropivacaine at concentrations of 0.5% or less is least chondrotoxic of anesthetics, while lidocaine should be avoided [36, 37].

Nonoperative care in ON can be considered in the early stages (I and II), it starts with patient education focused on known risk factors, patients should discontinue alcohol abuse and smoking [15]. However, Herningou et al. reported that regression or decrease in size of the lesion was never observed in their study group when the lesion was symptomatic at the initial visit [17].

As for physical therapy, according to literature, the main emphasis should be on scapulothoracic and glenohumeral strengthening, range of motion, glenohumeral capsular stretching, and rotator cuff strengthening [24]. Such an intervention has particularly good effects in patients who report restricted range of motions, but without a large pain component.

24.6 Operative Treatment

The identification of chondral defect alone should not be an indication for a surgical procedure. If the patient still experiences pain after attempting conservative treatment, surgical treatment should be considered. Chief complaints, the size of the lesion, its location, and the patient's expectations should be taken into account while decision-making. Young patients, with high expectations of treatment results diagnosed with glenohumeral focal chondral defects are worth considering the definitive surgical treatment. However, patients diagnosed with ON of the humeral heal should be managed differently, according to the diagnosed stage of the disease.

Among the methods of surgical treatment of focal defects arthroscopic debridement, microfracture, osteochondral autograft transplantation, osteochondral allograft, and autologous chondrocyte implantation are described and practiced in focal lesions. Few authors have also made a different classification, categorizing these methods as palliative, primary repair, reparative, restorative, or reconstructive [4, 24]. In cases of ON depending on the stage, the literature identifies core decompression, humeral head resurfacing, hemiarthroplasty, and total shoulder arthroplasty (TSA).

24.6.1 Focal Chondral Defects Treatment

24.6.1.1 Arthroscopic Debridement

Arthroscopic lavage and debridement are considered as a palliative method, mainly to alleviate symptoms and not to block the chances of further surgical interventions if needed. Its role is rather limited in case of chondral defects management however few authors reported quite successful outcomes usually when applying this procedure in generalized osteoarthritis. What is significant in choosing this method, in one of the papers, written by van Thiel et al., it is concluded that the presence of grade 4 bipolar arthritis (according to Outerbridge classification), joint space of less than 2 mm and large osteophytes are significant risk factors for progressing to shoulder arthroplasty [38].

Also, worth considering is the Comprehensive Arthroscopic Management (CAM) Procedure described by Millet et al. The CAM procedure involves the combination of glenohumeral chondroplasty; removal of loose bodies if present; humeral osteoplasty and osteophyte resection (goat's beard deformity); anterior, posterior, and inferior capsular release; subacromial decompression; axillary nerve neurolysis; and biceps tenodesis [39, 40]. The CAM procedure results were published a few years later by Mitchell et al. who performed this procedure on a consecutive series of 46 patients (49 shoulders) with advanced glenohumeral osteoarthritis (GHOA) who met the criteria for shoulder arthroplasty however opted for arthroscopic treatment. Eventually, they noted significant improvements in midterm clinical outcomes and high patient satisfaction after the arthroscopic CAM procedure for GHOA, with 76.9% survivorship at a minimum of 5 years postoperatively. Analyzing these results and the data of other authors, Mitchell et al. summarized in another study that extremely narrowed joint space (1.3 vs 2.6 mm), Kellgren-Lawrence grade of 4, and Walch glenoid type B2 or C were found to be predictive of early progression to TSA what is crucial in predicting possible failure factors [41, 42].

holes are spaced 3–4 mm apart and they are around 2–4 mm depth [5]. Proper placement technique is crucial to prevent from fractures between holes [8]. The idea of this technique is to create a canal for the bone marrow and to stimulate healing by the inflow of mesenchymal stem cells, growth factors, fibrin, and platelets that should form a clot covering the lesion [4, 46]. The authors emphasize excellent blood supply and the likely good results when it comes to glenohumeral joint [20]. In the treatment of knee cartilage better clinical results were described in younger patients (<30 y.o.) with small (<4 cm^2) and solitary lesions and though outcomes are promising in short-term follow-up, deterioration of the results may be expected after 2 and 5 years after surgery [47]. Many authors suggest small, focal, symptomatic lesions that failed conserva-

Author	Year	Method	Patients	Mean follow-up	Mean age	Results
Kerr et al. [43]	2008	Arthroscopic debridement with or without capsular release	20 shoulders (19 patients) all with lesions with Outerbridge grade 2–4	20 months	38	Three patients required shoulder arthroplasty; 88% of patients had significant pain relief
Skelley et al. [44]	2015	Arthroscopic debridement and capsular release	33 patients (Outerbridge grade 2–4)	8.8 months	55	14 patients required total shoulder arthroscopy
Cameron et al. [45]	2002	Arthroscopic debridement with/ without capsular release	61 patients	2 years (45 patients)	49.5	Significant improvement in pain and function noted in 88% patients
Van Thiel et al. [38]	2010	Arthroscopic debridement	81 patients	27 months	47	16 patients (22%) underwent arthroplasty at a mean of 10.1 months after debridement
Millett et al. [39]	2013	CAM procedure	29 (30 shoulders)	2.6 years	52	The ASES scores significantly improved and pain levels decreased. Six shoulders progressed to arthroplasty at a mean of 1.9 year.

24.6.1.2 Microfracture

Microfracture technique is considered as first-line treatment for isolated full-thickness small cartilage defect. Published results are most commonly described in the treatment of chondral defects in the knee but rarely in the shoulder. Same as with debridement, it may be done arthroscopically. Commonly microfracture's

tive management in young patients are good indications for this technique [48, 49]. Contraindications include ligamentous laxity, degenerative joint osteoarthritis, partial thickness lesions, or lesions associated with large bony defects [48, 50]. Moreover, this technique is usually performed with concomitant interventions that address particular pathology.

Authors	Year	Method	Patients	Mean follow-up	Mean age	Results
Siebold et al. [51]	2003	Microfracture with concurrent procedures if needed (posterior capsule shift, anchor removal, labral augmentation)	5	25.8 months	32 years	Constant score significantly improved (from 43.4% to 81.8%), RTG and MRI showed progression of arthritis in two patients
Snow and Funk [20]	2008	Microfracture with concurrent procedures if needed (capsular plication, anterior stabilization)	8	15.4 months	37 years	Mean constant score improved from 4388 to 9025
Millet et al. [48]	2009	Microfracture with concurrent procedures if needed (instability procedures, SAD, capsular release, SLAP repair, bicep release)	31 shoulders (30 patients)	47 months	43 years	Mean ASES score improved by 20 points; failure in 6 of 31 shoulders
Frank et al. [5]	2010	Microfracture	17 shoulders (16 patients)	27.8 months	37 year	Significant mean ASES improvement (from 44.3 to 86.3); 3 failures

24.6.1.3 Osteochondral Autograft Transplantation/Transfer

Osteochondral autograft transfer is based on harvesting one to three osteochondral autograft plugs and transfer them into chondral defects donor site morbidity, possible dead space between circular grafts, graft integration, and mechanical and geometrical differences between the recipient and donor cartilages. This technique is usually considered as a second line of treatment.

Authors	Year	Method	Patients	Mean follow-up	Mean age	Results
Park et al. [54]	2006	Arthroscopic osteochondral autograft transfer	1	31 months	13 years	Final follow-up at 31 months—no symptoms and good functional results with radiographic resolution
Scheibel et al. [52]	2004	Osteochondral autologous transplantation; 4 patients have concurrent procedures	8	32.6 months	43.1 years	Significant improvements in the mean constant score with MRI showing good osseointegration and congruent cartilage in all but one patient

[52]. This single-stage procedure, unlike debridement and microfracture, requires open surgery to have full access to the joint to transfer the osteochondral plugs [8]. According to literature, the ideal osteochondral defect size for osteochondral autologous transplantation to the shoulder is between 10 and 20 mm in diameter or an area of 1.0–1.5 cm [1, 52]. Usually, it is used in cases with anterior shoulder instability with Hill Sachs lesions [53]. Among the disadvantages of this method are

24.6.1.4 Osteochondral Allograft

The approach of procedure and the management of postoperative rehabilitation is similar to described above osteochondral autografting [4]. Typically, it is performed from deltopectoral approach however arthroscopic techniques for osteochondral allograft transplantation using tibial plateau allograft have been described [55]. The goal of osteochondral allografting is to reconstruct the congruency of the articular surfaces [56]. Clearly allografting requires suitable cadaver donor with similar articular geometry for a proper fit. The advantages of allografting are shorter surgical time and decreased morbidity [8]. As mentioned by Saltzman et al. there are doubts whether reaming may cause a cortical blowout hence whether proper depth of reaming may occur to provide a stable press-fit of an osteochondral graft [4]. Saltzman et al. evaluated osteochondral allograft transplantation in terms of large Hill Sachs lesions due to instability. They reported significant improvement in shoulder motion with high rates of return to work and satisfaction [57]. Although osteochondral allograft appears to be highly cost-effective method of chondral lesions treatment [58], this procedure has limitations due to the chondrocyte vitality. OA can be preserved for 28 days at 37°, after that structural changes can be identified, and viability of the chondrocytes starts to decrease [59]. Moreover, tissue banks currently need at least 14 days for completion of microbiologic and serologic testing of the OCA tissue [60].

in vitro. In the next step, the chondrocytes are administered into the previously prepared defect. Loose cartilage and calcified bone from the bottom of the lesion should be removed carefully, without damaging the subchondral bone plate. It is crucial that the borders of the lesion are perpendicular to the bone. Depending on the lesion size and location arthroscopy or arthrotomy may be used however arthroscopy should only be used when the lesion is well visualized and accessible. Depending on the technique, chondrocytes grown from biopsy sample may be injected under a periosteal patch or seeded onto a collagen membrane which is properly adjusted to the size and shape of the damage [8]. The literature on autologous chondrocyte implantation applied in the shoulder joint is rather limited, but good results in the knee joint encourage further research. Due to chondrocytes' potential to produce anabolic growth factors to promote cell survival and to induce chondrocytes' further proliferation

Authors	Year	Method	Patients	Mean follow-up	Mean age	Results
Camp et al. [61]	25	Tibial osteochondral allograft	1	1 years	25 years	Significant improvement in QuickDash score, subjective shoulder value, and ASES score

24.6.1.5 Autologous Chondrocyte Implantation

Autologous chondrocyte implantation is a two-step method. In the first stage, a healthy biopsy sample of cartilage is taken (for instance from the edge of the defect) [62]. Then the sample containing chondrocytes is properly prepared

some researchers believe that it is a promising method [4]. Given the cases reported until now, it appears that a good indication is restricted, unipolar, superficial defect without subchondral bone involvement. This method is usually performed in relatively young patients as a second line of treatment [4, 53, 61].

Authors	Year	Method	Patients	Mean follow-up	Mean age	Results
Romeo et al. [23]	2002	Autologous chondrocyte implantation	1	12 months	16 years	12 months postoperative full and painless ROM
Buchmann et al. [62]	2012	Autologous chondrocyte implantation	4	41.3 months	29.3 years	Mean VAS score postoperative 0.3; mean ASES 95.3; satisfactory coverage of the defect on MRI

24.6.2 Osteonecrosis of the Humeral Head Treatment

24.6.2.1 Core Decompression

Core decompression is known for its use in the treatment of osteonecrosis of the femoral head and the best clinical effect is achieved when it is performed in the earliest stages of this disease. In theory, the goal of the procedure is to reduce the intra-osseous pressure to restore the normal vascular flow. A percutaneous small-diameter perforation decompression technique in stage I or II was described by Harrald et al. showing a good functional outcome in 25 of 26 shoulders [63] La Porte et al. used an open technique of core decompression via 2–3 cm incision in the anterior axillary fold. Authors have reported good and excellent UCLA scores in 75% of the patients at final follow-up of mean 10 years for a group of 43 patients. The study group was divided by stages of the ON (Ficat and Arlat scale), authors observed the earlier stage the better the result: I—94%, II—88%, III—70%, IV—14%, also slightly better results were reported in patients who did not take corticosteroid medication [64]. Mont et al. reported the results for 30 shoulders in early stages of ON (I and II). In 20 shoulders an excellent improvement was seen however eight shoulders improvement was poor leading to hemiarthroplasty or TSA [65]. However, L'Insalata reported that core decompression was not effective in preventing the progression of stage III ON [66]. What is more, Kennon et al. described progression of 7 out of 8 cases from stage I/II to stage III/IV that underwent core decompression over a course of 1 year follow-up [67].

24.6.2.2 Humeral Head Resurfacing

Primary indication for humeral head resurfacing arthroplasty is a severe chondral or osteochondral lesion of the humeral head with significant shoulder pain that is irresponsive to nonoperative treatment. It is a less invasive procedure that maintains native joint biomechanics and leaves an opportunity for TSA as a salvage procedure in case of failure. Sufficient quantity and quality of the bone at the epiphyseal portion of the humerus are needed to ensure a stable fixation of the implant [68]. Depending on the lesion size, the resurfacing may be total or partial also known as "inlay arthroplasty." If an inlay method is chosen an implant is placed into the joint contour to recreate the joint surface in the specific location of the lesion. To attain a patient-specific surface reconstruction, every implant diameter has a component group with various offsets [69]. However, the available data on this method of treatment for ON is limited and reported only in a small group of patients or case reports. In a prospective series of 12 cases of advanced stage ON, Uribe et al. have reported partial resurfacing of the humeral head to be effective in relieving pain and restoring function. Patients achieved good to excellent results for the Western Ontario Osteoarthritis of the Shoulder index, Shoulder Score Index, and Constant score and on examination Forward elevation improved from mean of 94 degrees preoperatively to 142° postoperatively [70]. Ranalletta et al. reported significant improvement in functional scores and mobility in a group of nine patients treated with partial resurfacing of the humeral head with an average follow-up of 44 months. However, one patient presented symptomatic glenoid wear throughout follow-up and required revision surgery [71].

Surface replacement arthroplasty for glenohumeral arthropathy may be considered in coexisting humeral head and glenoid destruction. The inlay method allows to implant both components, humeral and glenoid, flush with the adjacent bone surface [72]. Authors reported good clinical outcomes, their nonradiographic loosening, and a high rate of return to sports in OA patients [72, 73] however data on this type of treatment in ON is limited.

The senior author uses both methods depending on the size of the lesion. In each case, the detailed planning based on CT and MRI is necessary to properly address cartilage degeneration. We aim to use as cartilage preserving method as possible however one should be aware of the size limits of the implants designed to perform partial inlay resurfacing. Thus, once the implant cannot fully cover the lesion or/and the lesion is larger than 50% of the native cartilage area, the total resurfacing should be taken into consideration. We also recommend considering stemless TSA in older and less active/demanding patients. However, more data in the literature need to be available to make a stronger recommendation.

24.6.2.3 Hemiarthroplasty and Total Shoulder Arthroplasty

Only 5% of all shoulder arthroplasties are performed due to ON of the humeral head [15]. That is why the literature on that matter is limited and many of the studies include only a few cases and/or a short-term follow-up. The choice whether to use hemiarthroplasty or TSA is usually based on many factors, but always overall clinical conditions of the patient should be taken into account. Obviously, the arthroplasty is absolutely contraindicated in cases of active infection and relatively contraindicated in brachial plexopathy and in concomitant deltoid and rotator cuff insufficiency. In a systematic review analyzing treatment for atraumatic ON Franchesi et al. reported that functional results were independent of implant type, but a higher complication rate seems to be associated with TSA. That is why the authors concluded that it should be recommended only in stage V ON and in cases where hemiarthroplasty cannot be done [74].

Hattrup and Cofield evaluated 127 shoulders, including 71 humeral head replacements and 56 TSA. In 88 patients available for final follow-up average 8.9 years, they found that nearly 80% of treated shoulders had subjective improvement and 77% reported little or no pain. Both groups had comparable postoperative ASES scores: 63 in hemiarthoplasty group and 62 points in TSA group, and in terms of ROM, no significant differences were reported. The authors also divided the outcomes by etiology, comparing traumatic ON and steroid ON demonstrating a better improvement in the ASES score in steroid-induced ON [75].

Feeley et al. compared hemiarthroplasty and TSA in treatment of ON in a retrospective study: 37 patients were treated with a hemiarthroplasty and 27 with a TSA. The average follow-up for each group was: hemiarthroplasty—53 months and TSA—60 months. Similar range of motion

and outcomes scores were reported in both groups. However, TSA was associated with a higher complication rate, 1 reoperation was for a superficial infection, 1 for a loose humeral component, and 4 for a loose glenoid component. In hemiarthroplasty two patients needed reoperation due to excessive glenoid wear [76]. Ristow et al. evaluated 29 cases of ON treated both with hemiarthroplasty (19 shoulders) and TSA (10 shoulders). There were no reoperations and only minor complications occurred in two cases. It was concluded that patients who underwent total shoulder arthroplasty (TSA) had higher median outcome scores and greater improvement in all scoring methods compared with hemiarthroplasty patients [77].

In a short-term observation, Navarro et al. found no statistical differences in complication rate between patients undergoing TSA for ON and osteoarthritis (OA) [78]. On the contrary, Tyrrell Burrus et al. shown that patients with humeral head ON undergoing TSA have significantly higher rates of various postoperative complications compared to patients without a diagnosis of ON. Based on the underlying cause of ON, the rates of complication varied but were highest in patients with steroid-associated and post-traumatic AVN. The authors found that the infection rate was seven times higher (10%) in patients with-steroid related ON. What is more, periprosthetic fractures were four times more frequent in ON group, and the revision rate was two times higher [79].

As mentioned in the previous paragraph the senior author aims to perform as cartilage preserving surgery as possible. However, in older and less active/demanding patients the shoulder replacement might be the only possible option. With the variety of available systems and designs, we prefer to use bone preserving implants when possible. This can be provided by total resurfacing with the inlay glenoid implant however not much data is still available in the literature. Once this method is technically inappropriate, we aim to use stemless TSA. However, in older patients with questionable bone quality, the stemmed humeral implant might be a better option. As in our practice, most of the patients qualified for different than resurfacing methods have some degree of glenoid wear, we very rarely see the indications to perform hemiarthroplasty.

24.7 Conclusion

To the authors best knowledge, there is no shoulder osteochondral defects treatment strategy published in the literature. The diversity of recently available techniques options makes it possible to customize the treatment approach due to the type of lesion as well as the patient's characteristic. Nevertheless, in more severe cases, the arthroplasty may be the only option. Therefore, the early identification of the pathology remains crucial to allow for less invasive sparing techniques. Our algorithm is a unique background for further discussion over the proper management of the non-traumatic shoulder lesions.

References

1. Patzer T, Lichtenberg S, Kircher J, Magosch P, Habermeyer P. Influence of SLAP lesions on chondral lesions of the glenohumeral joint. Knee Surg Sports Traumatol Arthrosc. 2010;18(7):982–7.
2. Provencher MT, Barker JU, Strauss EJ, Frank RM, Romeo AA, Matsen FA, et al. Glenohumeral arthritis in the young adult. Instr Course Lect. 2011;60:137–53.
3. Shin JJ, Mellano C, Cvetanovich GL, Frank RM, Cole BJ. Treatment of glenoid chondral defect using micronized allogeneic cartilage matrix implantation. Arthrosc Tech. 2014;3(4):e519–22.
4. Saltzman BM, Leroux T, Cole BJ. Management and surgical options for articular defects in the shoulder. Clin Sports Med. 2017;36(3):549–72.
5. Frank RM, Van Thiel GS, Slabaugh MA, Romeo AA, Cole BJ, Verma NN. Clinical outcomes after microfracture of the glenohumeral joint. Am J Sports Med. 2010;38(4):772–81.
6. Paley KJ, Jobe FW, Pink MM, Kvitne RS, ElAttrache NS. Arthroscopic findings in the overhand throwing athlete: evidence for posterior internal impingement of the rotator cuff. Arthroscopy. 2000;16(1):35–40.
7. Guntern DV, Pfirrmann CWA, Schmid MR, Zanetti M, Binkert CA, Schneeberger AG, et al. Articular cartilage lesions of the glenohumeral joint: diagnostic effectiveness of MR arthrography and prevalence in patients with subacromial impingement syndrome. Radiology. 2003;226(1):165–70.
8. Seidl AJ, Kraeutler MJ. Management of articular cartilage defects in the glenohumeral joint. J Am Acad Orthop Surg. 2018;26(11):e230–7.
9. Lunden JB, Legrand AB. Osteochondritis dissecans of the humeral head. J Orthop Sports Phys Ther. 2012;42(10):886.
10. Mima Y, Matsumura N, Ogawa K, Iwamoto T, Ochi K, Sato K, et al. Osteochondritis dissecans on the medial aspect of the humeral head. Int J Shoulder Surg. 2016;10(2):89–91.
11. Cooper C, Steinbuch M, Stevenson R, Miday R, Watts NB. The epidemiology of osteonecrosis: findings from the GPRD and THIN databases in the UK. Osteoporos Int. 2010;21(4):569–77.
12. Lee JA, Farooki S, Ashman CJ, Yu JS. MR patterns of involvement of humeral head osteonecrosis. J Comput Assist Tomogr. 2002;26(5):839–42.
13. Guerado E, Caso E. The physiopathology of avascular necrosis of the femoral head: an update. Injury. 2016;47:S16–26.
14. Shah KN, Racine J, Jones LC, Aaron RK. Pathophysiology and risk factors for osteonecrosis. Curr Rev Musculoskelet Med. 2015;8(3):201–9.
15. Hasan SS, Romeo AA. Nontraumatic osteonecrosis of the humeral head. J Shoulder Elb Surg. 2002;11(3):281–98.
16. Mont MA, Payman RK, Laporte DM, Petri M, Jones LC, Hungerford DS. Atraumatic osteonecrosis of the humeral head. J Rheumatol. 2000;27(7):1766–73.
17. Hernigou P, Flouzat-Lachaniette C-H, Roussignol X, Poignard A. The natural progression of shoulder osteonecrosis related to corticosteroid treatment. Clin Orthop Relat Res. 2010;468(7):1809–16.
18. Fox JA, Cole BJ, Romeo AA, Meininger AK, Williams JM, Glenn RE, et al. Articular cartilage thickness of the humeral head: an anatomic study. Orthopedics. 2008;31(3):216.
19. Bicos J, Mazzocca A, Romeo AA. The glenoid center line. Orthopedics. 2005;28(6):581–5.
20. Snow M, Funk L. Microfracture of chondral lesions of the glenohumeral joint. Int J Shoulder Surg. 2008;2(4):72–6.
21. Gross CE, Chalmers PN, Chahal J, Van Thiel G, Bach BR, Cole BJ, et al. Operative treatment of chondral defects in the glenohumeral joint. Arthroscopy. 2012;28(12):1889–901.
22. Johnson DL, Warner JJ. Osteochondritis dissecans of the humeral head: treatment with a matched osteochondral allograft. J Shoulder Elb Surg. 1997;6(2):160–3.
23. Romeo AA, Cole BJ, Mazzocca AD, Fox JA, Freeman KB, Joy E. Autologous chondrocyte repair of an articular defect in the humeral head. Arthroscopy. 2002;18(8):925–9.
24. McCarty LP, Cole BJ. Nonarthroplasty treatment of glenohumeral cartilage lesions. Arthroscopy. 2005;21(9):1131–42.
25. Ellman H, Harris E, Kay SP. Early degenerative joint disease simulating impingement syndrome: arthroscopic findings. Arthroscopy. 1992;8(4):482–7.
26. Kang RW, Frank RM, Nho SJ, Ghodadra NS, Verma NN, Romeo AA, et al. Complications associated with anterior shoulder instability repair. Arthroscopy. 2009;25(8):909–20.

27. Denti M, Monteleone M, Trevisan C, De Romedis B, Barmettler F. Magnetic resonance imaging versus arthroscopy for the investigation of the osteochondral humeral defect in anterior shoulder instability. A double-blind prospective study. Knee Surg Sports Traumatol Arthrosc. 1995;3(3):184–6.

28. Potter HG, Chong LR. Magnetic resonance imaging assessment of chondral lesions and repair. J Bone Joint Surg Am. 2009;91(Suppl 1):126–31.

29. Cruess RL. Experience with steroid-induced avascular necrosis of the shoulder and etiologic considerations regarding osteonecrosis of the hip. Clin Orthop Relat Res. 1978;130:86–93.

30. Hünnebeck SM, Magosch P, Habermeyer P, Loew M, Lichtenberg S. Chondral defects of the glenohumeral joint: long-term outcome after microfracturing of the shoulder. Obere Extrem. 2017;12(3):165–70.

31. Brittberg M, Winalski CS. Evaluation of cartilage injuries and repair. J Bone Joint Surg Am. 2003;85-A(Suppl 2):58–69.

32. Bhatia DN, van Rooyen KS, du Toit DF, et al. Arthroscopic technique of interposition arthroplasty of the glenohumeral joint. Arthroscopy. 2006;22:570. e1–5.

33. Dacre JE, Beeney N, Scott DL. Injections and physiotherapy for the painful stiff shoulder. Ann Rheum Dis. 1989;48(4):322–5.

34. Paik J, Duggan ST, Keam SJ. Triamcinolone acetonide extended-release: a review in osteoarthritis pain of the knee. Drugs. 2019;79(4):455–62.

35. Kraus V, Conaghan P, Aazami H, Mehra P, Kivitz A, Lufkin J, Hauben J, Johnson J, Bodick N. Synovial and systemic pharmacokinetics (PK) of triamcinolone acetonide (TA) following intra-articular (IA) injection of an extended-release microsphere-based formulation (FX006) or standard crystalline suspension in patients with knee osteoarthritis (OA). Osteoarthr Cartil. 2018;26:34–42.

36. Kreuz PC, Steinwachs M, Angele P. Single-dose local anesthetics exhibit a type-, dose-, and time-dependent chondrotoxic effect on chondrocytes and cartilage: a systematic review of the current literature. Knee Surg Sports Traumatol Arthrosc. 2017;26:819–30.

37. Jayaram P, Kennedy DJ, Yeh P, Dragoo J. Chondrotoxic effects of local anesthetics on human knee articular cartilage: a systematic review. PM R. 2019;11(4):379–400.

38. Van Thiel GS, Sheehan S, Frank RM, Slabaugh M, Cole BJ, Nicholson GP, et al. Retrospective analysis of arthroscopic management of glenohumeral degenerative disease. Arthroscopy. 2010;26(11):1451–5.

39. Millett PJ, Horan MP, Pennock AT, Rios D. Comprehensive arthroscopic management (CAM) procedure: clinical results of a joint-preserving arthroscopic treatment for young, active patients with advanced shoulder osteoarthritis. Arthroscopy. 2013;29(3):440–8.

40. Mook WR, Petri M, Greenspoon JA, Millett PJ. The comprehensive arthroscopic management procedure for treatment of Glenohumeral osteoarthritis. Arthrosc Tech. 2015;4(5):e435–41.

41. Mitchell JJ, Warner BT, Horan MP, Raynor MB, Menge TJ, Greenspoon JA, et al. Comprehensive arthroscopic management of glenohumeral osteoarthritis: preoperative factors predictive of treatment failure. Am J Sports Med. 2017;45(4):794–802.

42. Mitchell JJ, Horan MP, Greenspoon JA, Menge TJ, Tahal DS, Millett PJ. Survivorship and patient-reported outcomes after comprehensive arthroscopic management of glenohumeral osteoarthritis: minimum 5-year follow-up. Am J Sports Med. 2016;44(12):3206–13.

43. Kerr BJ, McCarty EC. Outcome of arthroscopic débridement is worse for patients with glenohumeral arthritis of both sides of the joint. Clin Orthop Relat Res. 2008;466(3):634–8.

44. Skelley NW, Namdari S, Chamberlain AM, Keener JD, Galatz LM, Yamaguchi K. Arthroscopic debridement and capsular release for the treatment of shoulder osteoarthritis. Arthroscopy. 2015;31(3):494–500.

45. Cameron BD, Galatz LM, Ramsey ML, et al. Nonprosthetic management of grade IV osteochondral lesions of the glenohumeral joint. J Shoulder Elb Surg. 2002;11(1):25–32.

46. Steadman JR, Rodkey WG, Briggs KK, Rodrigo JJ. [The microfracture technic in the management of complete cartilage defects in the knee joint]. Orthopade. 1999;28(1):26–32.

47. Gobbi A, Karnatzikos G, Kumar A. Long-term results after microfracture treatment for full-thickness knee chondral lesions in athletes. Knee Surg Sports Traumatol Arthrosc. 2014;22(9):1986–96.

48. Millett PJ, Huffard BH, Horan MP, Hawkins RJ, Steadman JR. Outcomes of full-thickness articular cartilage injuries of the shoulder treated with microfracture. Arthroscopy. 2009;25:856–63.

49. Mithoefer K, Williams RJ 3rd, Warren RF, et al. Chondral resurfacing of articular cartilage defects in the knee with microfracture technique. Surgical technique. J Bone Joint Surg Am. 2006;68(Suppl 1, Pt):294–304.

50. Slabaugh MA, Frank RM, Cole BJ. Resurfacing of isolated articular cartilage defects in the glenohumeral joint with microfracture: a surgical technique & case report. Am J Orthop. 2010;39(7):326–32.

51. Siebold R, Lichtenberg S, Habermeyer P. Combination of microfracture and periostal-flap for the treatment of focal full thickness articular cartilage lesions of the shoulder: a prospective study. Knee Surg Sports Traumatol Arthrosc. 2003;11(3):183–9.

52. Scheibel M, Bartl C, Magosch P, Lichtenberg S, Habermeyer P. Osteochondral autologous transplantation for the treatment of full-thickness articular cartilage defects of the shoulder. J Bone Joint Surg Br. 2004;86(7):991–7.

53. Fox JA, Cole BJ, Pylawka TK, Romeo AA. Chapter 20: cartilage injuries in the shoulder. In: Cartilage injury in the athlete. New York, NY: Thieme Medical Publishers; 2006. p. 217–31.

54. Park T-S, Kim T-S, Cho J-H. Arthroscopic osteochondral autograft transfer in the treatment of an osteochondral defect of the humeral head: report of one case. J Shoulder Elb Surg. 2006;15(6):e31–6.

55. Gobezie R, Lenarz CJ, Wanner JP, Streit JJ. All-arthroscopic biologic total shoulder resurfacing. Arthroscopy. 2011;27(11):1588–93.

56. Provencher MT, Barker JU, Strauss EJ, Frank RM, Romeo AA, Matsen FA, et al. Glenohumeral arthritis in the young adult. Instr Course Lect. 2011;60:137–53.

57. Saltzman BM, Riboh JC, Cole BJ, Yanke AB. Humeral head reconstruction with osteochondral allograft transplantation. Arthroscopy. 2015;31(9):1827–34.

58. Mistry H, Metcalfe A, Smith N, Loveman E, Colquitt J, Royle P, et al. The cost-effectiveness of osteochondral allograft transplantation in the knee. Knee Surg Sports Traumatol Arthrosc. 2019;27(6):1739–53.

59. Pallante AL, Bae WC, Chen AC, Görtz S, Bugbee WD, Sah RL. Chondrocyte viability is higher after prolonged storage at 37 degrees C than at 4 degrees C for osteochondral grafts. Am J Sports Med. 2009;37(Suppl 1):24S–32S.

60. Görtz S, Bugbee WD. Fresh osteochondral allografts: graft processing and clinical applications. J Knee Surg. 2006;19(3):231–40.

61. Camp CL, Barlow JD, Krych AJ. Transplantation of a tibial osteochondral allograft to restore a large glenoid osteochondral defect. Orthopedics. 2015;38(2):e147–52.

62. Buchmann S, Salzmann GM, Glanzmann MC, Wörtler K, Vogt S, Imhoff AB. Early clinical and structural results after autologous chondrocyte transplantation at the glenohumeral joint. J Shoulder Elb Surg. 2012;21(9):1213–21.

63. Harreld KL, Marulanda GA, Ulrich SD, Marker DR, Seyler TM, Mont MA. Small-diameter percutaneous decompression for osteonecrosis of the shoulder. Am J Orthop. 2009;38(7):348–54.

64. LaPorte DM, Mont MA, Mohan V, Pierre-Jacques H, Jones LC, Hungerford DS. Osteonecrosis of the humeral head treated by core decompression. Clin Orthop Relat Res. 1998;355:254–60.

65. Mont MA, Maar DC, Urquhart MW, Lennox D, Hungerford DS. Avascular necrosis of the humeral head treated by core decompression. A retrospective review. J Bone Joint Surg Br. 1993;75(5):785–8.

66. L'Insalata JC, Pagnani MJ, Warren RF, Dines DM. Humeral head osteonecrosis: clinical course and radiographic predictors of outcome. J Shoulder Elb Surg. 1996;5(5):355–61.

67. Kennon JC, Smith JP, Crosby LA. Core decompression and arthroplasty outcomes for atraumatic osteonecrosis of the humeral head. J Shoulder Elb Surg. 2016;25(9):1442–8.

68. Scalise J, Miniaci A, Iannotti J. Resurfacing arthroplasty of the humerus: indications, surgical technique, and clinical results. Tech Shoulder Elb Surg. 2007;8(3):152–60.

69. Sweet SJ, Takara T, Ho L, Tibone JE. Primary partial humeral head resurfacing: outcomes with the HemiCAP implant. Am J Sports Med. 2015;43(3):579–87.

70. Uribe JW, Botto-van BA. Partial humeral head resurfacing for osteonecrosis. J Shoulder Elb Surg. 2009;18(5):711–6.

71. Ranalletta M, Bertona A, Tanoira I, Rossi LA, Bongiovanni S, Maignón GD. Results of partial resurfacing of humeral head in patients with avascular necrosis. Rev Esp Cir Ortop Traumatol. 2019;63(1):29–34.

72. Cvetanovich GL, Naylor AJ, O'Brien MC, Waterman BR, Garcia GH, Nicholson GP. Anatomic total shoulder arthroplasty with an inlay glenoid component: clinical outcomes and return to activity. J Shoulder Elb Surg. 2020;29(6):1188–96.

73. Egger AC, Peterson J, Jones MH, Miniaci A. Total shoulder arthroplasty with nonspherical humeral head and inlay glenoid replacement: clinical results comparing concentric and nonconcentric glenoid stages in primary shoulder arthritis. JSES Open Access. 2019;3(3):145–53.

74. Franceschi F, Franceschetti E, Paciotti M, Torre G, Samuelsson K, Papalia R, et al. Surgical management of osteonecrosis of the humeral head: a systematic review. Knee Surg Sports Traumatol Arthrosc. 2017;25(10):3270–8.

75. Hattrup SJ, Cofield RH. Osteonecrosis of the humeral head: results of replacement. J Shoulder Elb Surg. 2000;9(3):177–82.

76. Feeley BT, Fealy S, Dines DM, Warren RF, Craig EV. Hemiarthroplasty and total shoulder arthroplasty for avascular necrosis of the humeral head. J Shoulder Elb Surg. 2008;17(5):689–94.

77. Ristow JJ, Ellison CM, Mickschl DJ, Berg KC, Haidet KC, Gray JR, et al. Outcomes of shoulder replacement in humeral head avascular necrosis. J Shoulder Elb Surg. 2019;28(1):9–14.

78. Navarro SM, Haeberle HS, Khlopas A, Newman JM, Karnuta JM, Mont MA, Ramkumar PN. Short-term outcomes after anatomic total shoulder arthroplasty in patients with osteoarthritis versus osteonecrosis. Ann Transl Med. 2019;7:48.

79. Tyrrell Burrus M, Cancienne JM, Boatright JD, Yang S, Brockmeier SF, Werner BC. Shoulder arthroplasty for humeral head avascular necrosis is associated with increased postoperative complications. HSS J. 2018;14(1):2–8.

OCD of the Elbow: Treatment by Autograft

L. A. Pederzini, M. Bartoli, A. Cheli, and A. Celli

25.1 Introduction

Osteochondritis dissecans (OCD) is an inflammatory disease involving bones and cartilage, leading to the deterioration of the joints and to the detachment of one or several fragments due to necrosis [1, 2].

Etiopathogenesis is currently not clearly defined; the most acknowledged theory is that repetitive microtraumas and overuse lead to localized venous hypertension, ischemia, edema of the fat cells, and subsequent fat and osseous necrosis [3].

It is commonly accepted that OCD causes elbow pain and impairment, normally affecting the humeral capitulum although the process has also been detected in the trochlea, radial head, and olecranon [4].

Overloading areas on the articular surface change with sport-specific gestures thereby varying the location of OCD lesions [5]. In throwing sports, the elbow is susceptible to a valgus overload (ex. Baseball players); gymnasts often weight-bear in maximum elbow extension with axial force applied; as a consequence, gymnasts generally have lesions approximately 30° more posterior on the capitulum compared to launch sportsmen [6].

OCD commonly strikes young adolescents and athletes aged 13–16.

Persistent pain, gradual decrease in function, and restricted range of motion are the typical symptoms of the advanced stages, reaching up to 30° limitation in extension [7]. Further symptoms include posterolateral crepitation or popping and restricted activities, impacting on physical activities. Physical examination also detects secondary joint contractures and lateral joint edema.

Differential diagnosis with Panner's disease is mandatory, considering the typical age of onset (boys aged younger than 10 years old with no history of trauma [8, 9]) and imaging pattern: entire ossification center should be involved.

25.2 Imaging

Routine radiographs may appear normal or show minimal changes in opacity in the early stages.

In order to detect the early signs, an AP X-Ray of a 45° elbow flexion could be useful [10].

Loose bodies and joint irregularities could be present in the progressive stages; frequent reduction in ROM is also reported and should be significant for an unstable lesion.

The best option for an accurate diagnosis is MRI in order to detect lesions, although not frequently qualitatively adequate to define the

L. A. Pederzini · M. Bartoli (✉) · A. Cheli
Orthopedics and Trauma Unit, Nuovo Ospedale
Civile di Sassuolo, Modena, Italy

A. Celli
Orthopedics and Trauma Unit, Hesperia Hospital,
Modena, Italy

© ISAKOS 2022
A. Gobbi et al. (eds.), *Joint Function Preservation*, https://doi.org/10.1007/978-3-030-82958-2_25

cartilage cap firmness, which is essential in choosing surgical and nonsurgical treatment.

Preoperative MRI showed a sensitivity of 84%, with particular mention of T2 sequences findings: high signal intensity interface and line through the articular cartilage; CT scans could be useful to characterize a minimally displaced fragment with bony parts [11]. A complete imaging study is recommended, supported by accurate clinical evaluation and following progression criteria: Rx, MRI, and finally CT scan if strictly necessary.

25.3 Treatment

Surgeons may be faced with difficulties regarding OCD lesions, as any other pathology involving the cartilage, considering the right choice of therapy prior to a difficult choice concerning the adequate selection of treatment.

Recent opinions about conservative treatment are rather consistent concerning a solid option when the diagnosis is made on a recent and stable lesion. The clinical outcome exacerbation has been reported for bigger and deeper lesions; a conservative approach seems to be contraindicated in lateral column involvement [12].

Many authors documented that unstable lesions had a non-acceptable outcome if treated nonsurgically; worst outcomes were seen in closed physes patients, in old lesions, and in case of bad compliance with physician indications [13].

The literature reports many arthroscopic and open techniques regarding the surgical treatment of advanced, unstable lesions, or failure of prolonged nonsurgical treatment.

An advantageous option consists of open or arthroscopic fixation of the fragment and literature reports many techniques with considerable results [14–16]; however, a second operation is necessary to remove the fixation device. Furthermore, in 2008, Nobuta et al. demonstrated the scarcity of the healing procedure of lesions deeper than 9 mm compared to superficial ones if fixed [17].

The removal of loose bodies, debridement, micro-fractures still remain good arthroscopic options with long-term follow-up disposable data, but, some reviews report a higher rate of reoperation for these easier procedures and poor recovery of sports activities at the previous level [18, 19]. Recent literature [5] suggests to use these techniques solely for lesions smaller than 1 cm^2 and with an intact lateral buttress.

Major lesions did not heal appropriately and the surface was not so similar to the original one [9, 11] still entailing a high risk of consequent arthritis.

Other open surgery options are osteochondral transplant (OAT), mosaicplasty, autologous chondrocyte transplant associated with or without wedge osteotomies of the lateral condyle.

OAT technique is the sole method to provide the damaged articular surface with a hyaline cartilage top layer consistent with the original cover when the fragment cannot be repaired. Indeed, healing time is reasonable and the operation costs are undoubtedly low.

In 2006, Yamamoto et al. reported the first and encouraging clinical results; he achieved outstanding functional outcomes in an open procedure performed in young throwing athletes (89%) [20]. Simultaneously, Tsuda et al. published a small case series: three non-throwing athletes had considerable results at a short follow-up after being treated arthroscopically [21]. Earlier Japanese reports refer minor experiences. Over the past 10 years, the literature has demonstrated the midterm efficacy and consistency of this technique if carried out on the elbow, as previously performed on the knee [22].

The surgeon can use one cylinder or further plugs (mosaicplasty) to assess the size of the lesion.

Our choice to use a bigger single cylinder is based on partial scientific evidence of a higher degree of stability and greater complete healing rate of the transplant [23]. No reliable data exist to definitely demonstrate that a single or a lower number of big plugs leads to better long-term results.

The selection of the donor site is still debated.

The plug can be shaped to match the accurate features of the recipient area, but cartilage thickness of the plug is difficult to tailor. In light of this, several MRI findings have acknowledged the area of the inferior medial trochlear ridge of the knee as being more consistent with the cartilage thickness. In addition, an osteochondral plug from the fifth or sixth rib, has more recently been applied to the elbow. Reported advantages include a broader donor surface in order to allow larger plug harvesting (>15 mm), less donor site morbidity, and a similar structure to the subchondral bone and cartilage present in synovial joints, such as the knee [5].

It has been demonstrated that autograft rib plugs are a valid alternative for surgeons who are familiar with the costal anatomy and are aware of the possible risks of hurting the thoracic cavity and underlying pleura.

Reassuring reports have been found about local morbidity about the lateral aspect of the trochlea of the ipsilateral knee which we have chosen. Further hypotheses have recently been taken into account, although the findings are not as favorable [24].

OAT seems to guarantee the best recovery in sports and manual work, restoring the cartilaginous surface of the joint.

In 2015, Gancarczyk et al. published an in vitro study concerning an elbow arthroscopy on 21 specimens. By carrying out a proper OAT procedure, he showed was the possibility to achieve the right positioning of the recipient tunnels at the capitulum humeri [25].

Our team has been effectively performing this mini-invasive operation since 2010, promoting early post-op recovery and decreasing the risks of stiffness, mainly in young people.

In 2016, in a systematic review, Westermann et al. demonstrated more effective results for OAT rather than other techniques, regarding the percentage of athletes resuming sports. He also emphasized the lack of study populations treated consistently as well as outcome evaluations achieved by using validated scales [18].

Recently, our group has published a paper describing the arthroscopic procedure for posterior and central OCD, treated with single plug OAT harvested from the superolateral aspect of the trochlea in the omolateral knee [26].

Nine patients were included in the casuistry reaching a minimum FU of 2.5 years.

The average range of motion improved by 17.9° in extension and 10.6° in flexion. All patients achieved full flexion in comparison with the contralateral side and full extension was 7 over 9. Prono-supination remained complete after the operation.

The average VAS improved by 7.11 and the mean post-op value was 0.67.

Total recovery of ADL performance in 9 out of 9 patients was demonstrated by the MEPS score, with a full score in 8 out of 9 patients.

Five over nine patients reached the best score at the Quick-DASH.

All patients resumed sports 6 months after surgery: one of them at a higher level than before, two of them to a noncompetitive level by choice.

25.4 Surgical Arthroscopic Technique [27]

The patient is arranged in a lateral decubitus position with a 90° shoulder abduction and a 90° flexion of the elbow on an arm holder. The ipsilateral hip is free to be abducted when necessary during the surgical procedure (Fig. 25.1).

General anesthesia is largely preferred after the insertion of a peri-neural catheter in the involved arm. After surgery, a peripheral block is advisable in order to manage pain.

The surgeon examines the patient under anesthesia to evaluate the ROM and stability and applies a tourniquet which is inflated to 250 mmHg.

It is useful to outline the medial and lateral epicondyles, the ulnar nerve, the radial head, and the posterior soft spots with a skin marker.

An 18-needle is inserted in the elbow through the soft spot in the center (UK) of the triangle created by the lateral epicondyle, the radial head, and the olecranon. Joint distension is obtained, by injecting 20 ml of normal saline to divide the anterior neurovascular structures and a skin

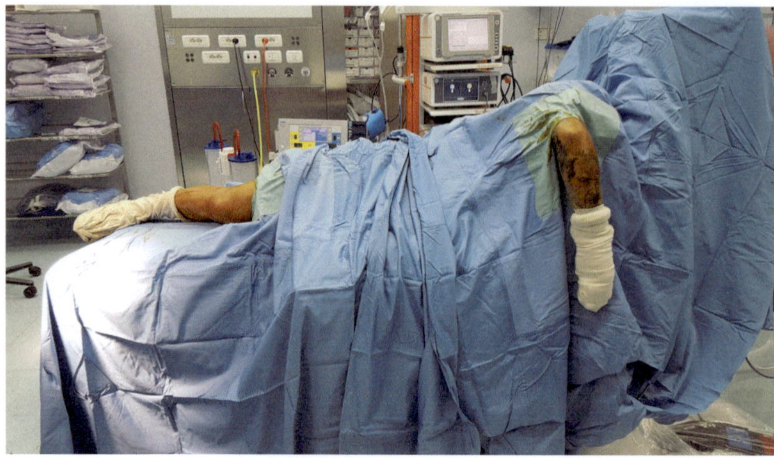

Fig. 25.1 Patient setting: lateral decubitus

incision is performed followed by blunt dissection of the soft tissues using a fine hemostat.

Five portals are made in the same way: three posterior and two anterior.

The joint is distended by using a pump set on 35–50 mmHg.

The second portal is 1.5 cm proximal to the first one and provides a complete view of the proximal radio-ulnar joint and the use of a shaver on the posterior aspect of the radial head.

The third posterior portal is placed in the olecranon fossa, 2–3 cm proximal to the olecranon tip, just medial to the triceps tendon.

Firstly, the surgeon should identify the OCD lesion using the first portal; switching the scope to the second portal, the surgeon must assess if the lesion is accessible perpendicularly with a spinal needle. If the area is excessively frontal on the condyle (with the elbow flexed at 90°), the surgeon cannot perform the technique arthroscopically and the procedure must be completed by opening the joint anteriorly.

If the lesion is on the posterior side of the condyle or it is possible to flex it to the back with the elbow, the surgeon may perform an arthroscopy.

Because of the detection of a lesion lower than 1 cm in diameter and with an unstable cap, after possible loose body removal, the surgeon performs mapping and drilling to correctly prepare the lesion in order to carry out the osteochondral transplant.

A 6.5–8.5 mm cylindrical graft is harvested arthroscopically from the side trochlea of the ipsilateral knee using Mosaicplasty instruments

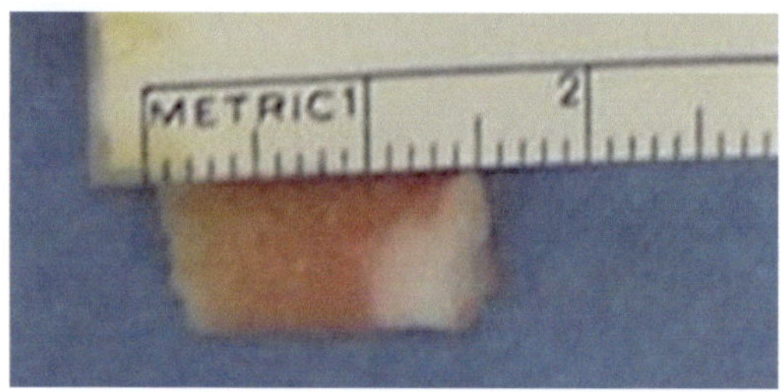

Fig. 25.2 Osteochondral cylinder: ready to be implanted

by Smith and Nephew, and is placed into the previously prepared area that is entirely covered (Fig. 25.2).

The graft is stabilized and press-fitted (Fig. 25.3).

Washout of the joint and the positioning of an intra-articular drainage for 2 or 3 days are recommended.

A long arm splint is positioned posteriorly to immobilize the arm at 90° flexion and neutral rotation.

Difficulties of this type of elbow arthroscopy are similar to other procedures; neurological difficulties are the most commonly reported, although often transient.

25.4.1 Postoperative Protocol

The involved arm is immobilized in a cast or in a brace for 3 weeks, whereas mobilization of the wrist is permitted. The donor knee should be pro-

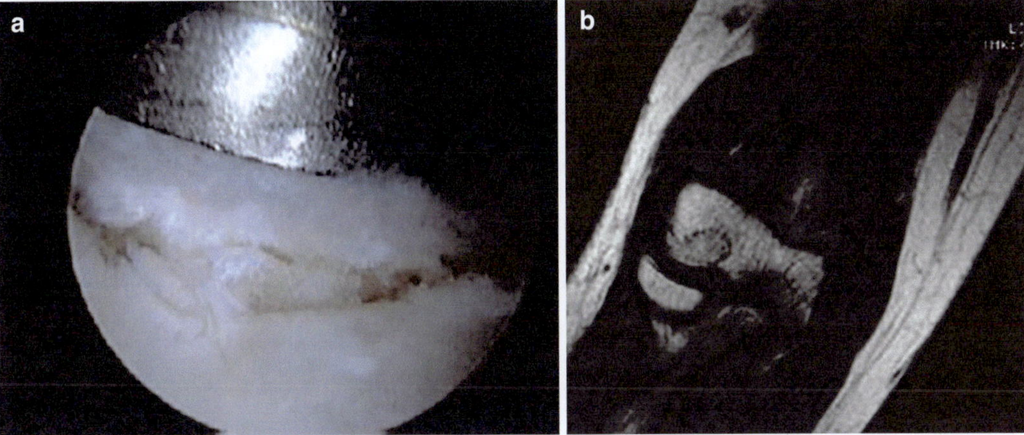

Fig. 25.3 (**a**) intra-operative view of the transplant. (**b**) MRI follow-up

tected from a full load with one crutch on the opposite hand for 3/4 weeks. Leg extension in open chain, squatting, or kneeling movements cannot be carried out for at least 1 month.

A week later, manual therapy is administered with radio-humeral distraction and a careful passive mobilization in order to attain full ROM, restraining from valgus stress, and closed chain exercises. Progressive recovery of proprioception and kinesthesia are further aims. Laser therapy and PEMF could be helpful. Furthermore, ice therapy is recommended.

When full ROM is achieved and residual pain disappears, active exercises are added to strengthen muscles.

No complications following surgery were referred, not even in the donor site.

Currently, autologous transplant of an osteochondral plug appears to be a safe and suitable technique in active patients with capitulum humeri OCD.

The lack of patient complaints regarding the healthy donor site is another aspect to consider in preoperative decision-making.

Currently, in accordance with the existing literature, this surgical technique seems to be the best option for a successful, single phase and stable treatment in such lesions types. Although more challenging, arthroscopy is an advantageous tool in order to decrease the invasiveness of the operation, with all the resulting benefits.

References

1. Schenk RC Jr, Goodnight JM. Ostheocondritis dissecans. J Bone and Joint Surg Am. 1996;78:439–56.
2. Baumgarten TE, Andrews JR, Satterwhite YE. The arthroscopic classification and treatment of osteochondritis dissecans of the capitellum. Am J Sports Med. 1998;26(4):520–3. https://doi.org/10.1177/0363 5465980260040801.
3. Eygendaal D, Bain G, Pederzini L, et al. Osteochondritis dissecans of the elbow: state of the art. J ISAKOS Joint Disord Orthop Sports Med. 2017;2:47–57.
4. Bradley JP, Petrie RS. Osteochondritis dissecans of the humeral capitellum. Diagnosis and treatment. Clin Sports Med. 2001;20(3):565–90.
5. Logli AL, Bernard CD, O'Driscoll SW, Sanchez-Sotelo J, Morrey ME, Krych AJ, Camp CL. Osteochondritis dissecans lesions of the capitellum in overhead athletes: a review of current evidence and proposed treatment algorithm. Curr Rev Musculoskelet Med. 2019;12(1):1–126.
6. Kajiyama S, Muroi S, Sugaya H, Takahashi N, Matsuki K, Kawai N, et al. Osteochondritis dissecans of the humeral capitellum in young athletes: comparison between baseball players and gymnasts. Orthop J Sports Med. 2017;5(3):2325967117692513.
7. Baumgarten TE, Andrews JR, Satterwhite YE. The arthroscopic classification of osteochondritis dissecans of the capitellum. Am J Sports Med. 1998;26:520–3.
8. Panner HJ. A peculiar affection of the capitellum humeri, resembling calve-Perthes' disease of the hip. Acta Radiol. 1927;8:617–8.
9. Smith MGH. Osteochondritis of the humeral capitellum. J Bone Joint Surg Br. 1964;46:50–4.
10. Takahara M, Ogino T, Takagi M, Tsuchida H, Orui H, Nambu T. Natural progression of osteochondritis

dissecans of the humeral capitellum: initial observations. Radiology. 2000;216:207–12.

11. Satake H, Takahara M, Harada M, Maruyama M. Preoperative imaging criteria for unstable osteochondritis dissecans of the capitellum. Clin Orthop Relat Res. 2013;471(4):1137–43.

12. Brownlow HC, O'Connor-Read LM, Perko M. Arthroscopic treatment of osteochondritis dissecans of the capitellum. Knee Surg Sports Traumatol Arthrosc. 2006;14:198–202.

13. Mihara K, Tsutsui H, Nishinaka N, Yamaguchi K. Nonoperative treatment for osteochondritis dissecans of the capitellum. Am J Sports Med. 2009;37(2):298–304.

14. Graaff F, Krijnen MR, Poolman RW, Willems WJ. Arthroscopic surgery in athletes with osteochondritis dissecans of the elbow. Arthroscopy. 2011;27:986–93.

15. Pill SG, Ganley TJ, Flynn JM, Gregg JR. Osteochondritis dissecans of the capitellum: arthroscopic-assisted treatment of large, full-thickness defects in young patients. Arthroscopy. 2003;19:222–5.

16. Koehler SM, Walsh A, Lovy AJ, Pruzansky JS, Shukla DR, Hausman MR. Outcomes of arthroscopic treatment of osteochondritis dissecans of the capitellum and description of the technique. J Shoulder Elb Surg. 2015 Oct;24(10):1607–12.

17. Nobuta S, Ogawa K, Sato K, Nakagawa T, Hatori M, Itoi E. Clinical outcome of fragment fixation for osteochondritis dissecans of the elbow. Ups J Med Sci. 2008;113:201–8.

18. Westermann RW, Hancock KJ, Buckwalter JA, Kopp B, Glass N, Wolf BR. Return to sport after operative management of osteochondritis dissecans of the capitellum. A systematic review and meta-analysis. Orthop J Sports Med. 2016;4:1–8.

19. Kirsch JM, Thomas J, Bedi A, Lawton JN. Current concepts: osteochondritis Dissecans of the Capitellum and the role of osteochondral autograft transplantation. Hand (N Y). 2016;11(4):396–402.

20. Yamamoto Y, Ishibashi Y, Tsuda E, Sato H, Toh S. Osteochondral autograft transplantation for osteochondritis dissecans of the elbow in juvenile baseball players: minimum 2-year follow-up. Am J Sports Med. 2006;34(5):714–20.

21. Tsuda E, Ishibashi Y, Sato H, Yamamoto Y, Toh S. Osteochondral autograft transplantation for osteochondritis dissecans of the capitellum in nonthrowing athletes. Arthroscopy. 2005;21(10):1270.

22. Kosaka M, Nakase J, Takahashi R, Toratani T, Ohashi Y, Kitaoka K, Tsuchiya H. Outcomes and failure factors in surgical treatment for osteochondritis dissecans of the capitellum. J Pediatr Orthop. 2013;33(7):719–24.

23. Hangody L, Dobos J, Balo E, Panics G, Hangody LR, Berkes I. Clinical experiences with autologous osteochondral mosaicplasty in an athletic population: a 17-year prospective multicenter study. Am J Sports Med. 2010;38:1125–33.

24. Maruyama M, Harada M, Satake H, Uno T, Takagi M, Takahara M. Bone-peg grafting for osteochondritis dissecans of the humeral capitellum. J Orthop Surg. 2016;24:51–6.

25. Gancarczyk SM, Makhni EC, Lombardi JM, Popkin CA, Ahmad CS. Arthroscopic articular reconstruction of capitellar osteochondral defects. Am J Sports Med. 2015;43:2452–8.

26. Pederzini LA, Bartoli M, Cheli A, Nicoletta F, Severini G. Encouraging mid-term outcomes for arthroscopic autologous osteochondral transplant (OAT) in capitellum osteochondritis dissecans (OCD). Knee Surg Sports Traumatol Arthrosc. 2019;27(10):3291–6.

27. Pederzini LA, Tripoli E, DI Palma F, Nicoletta F. Chpt. 2 Arthroscopic treatment of osteochondritis dissecans using mosaicplasty. In: Kelberine FM, Khan W, editors. Elbow surgery. London: JP Medical Ltd.; 2015.

Elbow Osteochondral Unit Function

26

Carina Cohen, Gyoguevara Patriota,
Guilherme Augusto Stirma, and Benno Ejnisman

26.1 The Elbow Anatomy

The elbow is a trocho-ginglymus joint that has three joints: the radiocapitellar, proximal radioulnar, and ulnohumeral joints located within a synovial-lined joint capsule. The ulnohumeral joint, a hinge joint, is formed by the articulation of the ulnar trochlear notch with the central waist of the humeral trochlea. The proximal radioulnar joint consists of the radial head, whose outer circumference articulates with radial notch, a small depression along the lateral surface of the coronoid process of the ulna. The radiocapitellar joint is formed by the articulation of the concave surface of the radial head with the convex cartilage-covered capitellum [1]. Magnetic Resonance Image (MRI) is considered the best noninvasive method for evaluating articular cartilage due to its high contrast of soft tissues [1] (Fig. 26.1).

26.2 The Elbow Articular Surface

The articular surfaces are covered with hyaline cartilage, which is firm, aneural, alymphatic, and avascular with important biomechanical properties of resilience, load-bearing, and durability that depend primarily on the chemical nature and complex spatial arrangement of the cartilaginous extracellular matrix [2]. On microscopic examination articular surface is characterized by abundant shiny extracellular matrix, with sparse cells isolated in well-defined spaces [3]. The constituents of this matrix are water (60–80%) and macromolecules, which include proteoglycans, collagens (forms the fibrillar meshwork), and other proteins (20–40%) [2].

The cartilage is complex and can be divided into three zones: superficial zone, middle zone, and deep zone. The superficial is in contact with the synovial fluid and has a higher density, lower rates of glycosaminoglycans (GAGs). The deep zone is in contact with the subchondral bone; the characteristic is less cell density and an increase in GAGs. Regarding cell distribution and morphology, chondrocytes in the superficial zone are smaller and flattened, while in the deep zone they are larger and rounded [4].

26.3 Function

The most important functions of elbow cartilage are to provide a smooth surface for low-friction articulation and to facilitate the load transmission to the underlying subchondral bone. The specific characteristics of cartilage allow to carry high contact forces and to disperse the resulting compressive stresses to the underlying subchondral

C. Cohen (✉) · G. Patriota · G. A. Stirma
B. Ejnisman
Department of Orthopedics and Traumatology, Sports Traumatology Center (CETE), Federal University of São Paulo (UNIFESP), São Paulo, Brazil

© ISAKOS 2022
A. Gobbi et al. (eds.), *Joint Function Preservation*, https://doi.org/10.1007/978-3-030-82958-2_26

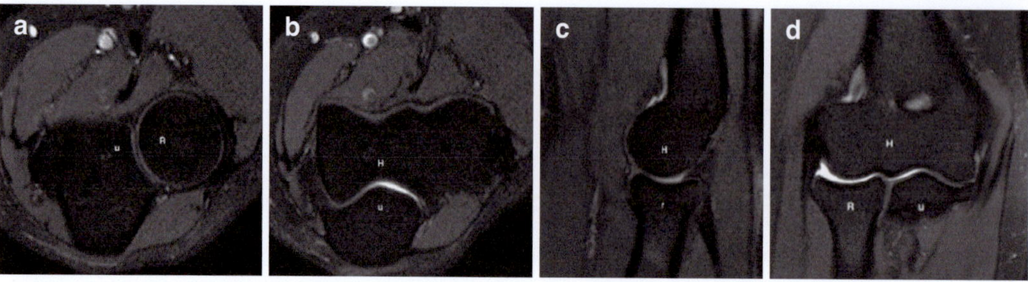

Fig. 26.1 Elbow joint surface in the Magnetic Resonance Image (MRI). T2-weighted sequences better assess subchondral bone and the interface between cartilage and synovial fluid [1]. (**a**) Radioulnar joint (Axial T2); (**b**) Ulnotrochlear joint (Axial T2); (**c**) Radiocapitllar joint (Sagittal T2); (**d**) All joints of elbow (Coronal T2). Ulna (u); Radio (R); Humerus (H)

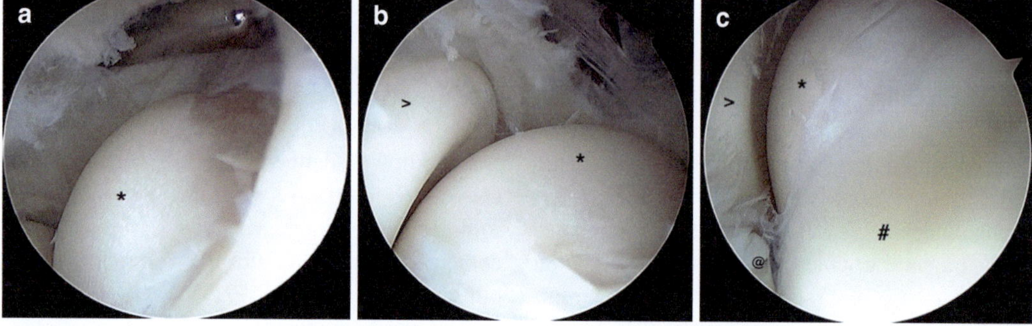

Fig. 26.2 Images of the elbow joint surface under arthroscopic view. (**a**) Cartilage surface of capitellum (*); (**b**) Radiocapitellar joint; (**c**) Ulnohumeral; and radiocapitellar joint. Capitellum (*); Radial head (>); Coronoid process (@); Trochlea (#)

bone [5]. Synovial fluid (SF) also plays an important role in the biomechanical behavior, nutrition, and lubrication of the articular cartilage, being the major source of nutrients. SF is a dynamic reservoir of proteins derived from cartilage and synovial tissue [5].

The subchondral region has high number of arterial and venous vessels, as well as nerves with branches into the calcified cartilage. The subchondral bone protects the hyaline cartilage against damage caused by excessive loads, and attenuate about 30% of the loads, while cartilage only 1–3% [6]. The load transfer from the articular cartilage to the diaphyseal cortex creates largest shear stresses in the subchondral region (compressive forces into tensile stresses) [7].

There are few studies on the distribution of cartilage in the elbow. The radiocapitellar joint presents thicker cartilage on the outer edge of the radial head [8, 9]. The thicker cartilage in the ulnotrochlear joint is distributed from the anterolateral edge of the coronoid to the anteroproximal edge of the proximal sigmoid notch and faces the intermediate region of the distal part of the humerus. In the proximal radioulnar joint, there is a thicker cartilage area in the anteroproximal edge of the proximal sigmoid notch, which is faced by the articular zone cartilage on the proximal circumferences [9] (Fig. 26.2).

26.4 Regeneration of Cartilage

The lesions in the articular cartilage are devoid of spontaneous healing and so the tissue will repair with fibrocartilage. This tissue does not have the same biomechanical, anatomical properties, and resistance of hyaline cartilage; therefore, can cause early degenerative changes [4].

The chondrocytes are specialized in producing collagen and proteoglycans, besides having an intimate relationship with the synovial cells. In some pathologies such as osteoarthritis, synovial cells stimulate chondrocytes to produce atypical tissue (fibrocartilage) [10]. The collagen fibers are perpendicular in the deep zone and parallel in the superficial zone. The distribution and organization of proteoglycan among the fibers establish biomechanical characteristics which give rigidity to compression, resilience, and resistance [4].

Due to local overload and genetic factors, damage to fibrils and chondrocytes reaches a critical point at which they cannot withstand pressure. The superficial cartilage is damaged and removed thus causing even more focal overload. The surface becomes uneven and cracked, known as fibrillation. Fibrillation causes a decrease in thickness and the consequent overload will lead to osteoarthritis [10].

26.5 Osteochondral Injury

An osteochondral injury involves the separation of an articular cartilage segment along with its underlying bone [11]. Such injuries mostly occur when there is prolonged and repetitive joint overload or a sudden high impact produces compressive stress to the tissue and shear stress at the subchondral bone junction [12]. Many causes have been proposed, including vascular insult, trauma, endocrinopathies, genetics, ossification abnormalities, hormonal abnormalities, and multifactorial causes [11].

Repetitive microinjuries to the subchondral bone and calcified cartilage initiate a repair mechanism, with activation of osteoclasts, osteoblasts, and fibrovascular tissue ultimately resulting in the formation of new bone (subchondral sclerosis) and establishment of a new, cartilaginous mineralization zone (duplication or triplication). This process also leaves behind deep zones of uncalcified cartilage [13]. If overloading of cartilage and subchondral bone continues, super-

ficial parts of the uncalcified cartilage are stripped and their surfaces become irregular. These areas progressively increase in size with time and continuous overloading, resulting in complete cartilage loss [13].

The elbow is used in many different activities such as throwing, tennis, golf swings, and volleyball therefore most elbow complaints are related to sports. Most elbow injuries occur as a result of repetitive use, although traumatic injuries are also common [14].

The surfaces of diarthrodial joints are more susceptible to an osteochondral injury and the most involved areas include the knee, elbow, and ankle [12]. They can occur in all articular surfaces of the elbow, including the radial head, trochlea, and olecranon, but most reported injuries involve the capitellum [12].

26.6 Classifications of Osteochondral Injury

The review of literature showed that there are different classifications for osteochondral injuries however there is no specific osteochondral classification for the elbow.

Outerbridge, in 1961, published the most popular arthroscopic classifications for cartilage injury (created for knee, but also used for elbow, shoulder, and hip). This classification is graduated in five degrees according to the table(Table 26.1) [15].

In 2000, the International Cartilage Repair Society (ICRS) developed a new arthroscopic classification during the ICRS Standards Workshop, in Switzerland (Table 26.2) [16].

Table 26.1 Outerbridge classification [15]

Grade I	Chondral lesions characterized by softening and swelling
Grade II	Lesion describes a partial-thickness defect (fissures <0.5 in. in diameter or that do not reach subchondral bone)
Grade III	Fissures of the cartilage >0.5 in. (reaching subchondral bone)
Grade IV	Erosion of the articular cartilage that exposes subchondral bone

Table 26.2 ICRS Classification [16]

Grade 0	Normal
Grade I	Nearly Normal (soft indentation and/or superficial fissures and cracks)
Grade II	Abnormal (lesions extending less than 50% of cartilage depth)
Grade III	Severely abnormal (cartilage defects >50% of cartilage depth)
Grade IV	Severely abnormal (cartilage defects extending through the subchondral bone)

26.7 Specific Elbow Condral Lesions

26.7.1 Osteochondritis Dissecans/ Panner's Disease

The osteochondritis dissecans of the elbow (ODE) is more prevalent in the capitellum. This chondral alteration mostly happens in young athletes, baseball players, and gymnasts (Fig. 26.3).

The evolution can lead to osteoarthritis of the elbow. There are many theories for its origin, such as ischemia, genetics, and recurrent microtrauma, but the etiology of ODE remains uncertain [17]. The most accepted theory is fatigue of the medial collateral ligament in patients who use the pitch repeatedly, creating an increase in valgus compression forces in the radiocapitellar joint [18]. The stress evolves to a focal lesion in subchondral bone characterized by avascular necrosis and subchondral bone changes. As a result, it will progress to loss of the overlying articular cartilage and formation of free bodies [19] (Fig. 26.4).

It is important to differentiate ODE from Panner's disease. Usually, trauma is not present in Panner's history. Also, patients are younger compared to ODE. Besides fragmentation, there is rarefaction involving the entire ossification nucleus of the capitellum; however, it is a self-limiting process, being solved by reossification [20].

The patient with ODE presents progressive worsening in pain and stiffness. Occasionally can present symptoms and mechanical changes, crackles, pops, and loose bodies [21]. Large lesions and old patients are associated with worst symptoms and radiographic changes [22].

Some lesions may heal spontaneously, especially in skeletally immature patients. The treatment consists of prevention of pitches, bending, excessive force at the elbow, arm, and drop weightlifting [23]. Stable lesions should receive conservative treatment and healing process must be accompanied by radiographs. Surgical treatment should be considered when there is no radiographic improvement at 3 months [17]. Matsuura et al. analyzed 176 patients with a mean age of 12.8 years who played baseball and had ODE. Conservative treatment was carried out for 6 months and proposed to 101 patients, 84 were diagnosed as stage I and 17 were in stage II. About 90% (76 cases) with radiolucent lesions, in stage I, healed on average of 14.9 months. In stage II, 52.9% of the 17 cases healed in approximately 12.3 months [18].

There are several surgical options for ODE. Indications for operative treatment are failure in conservative treatment, unstable injuries, mechanical symptoms such as crackles, stiffness, loose bodies, and pain for daily activities [23]. Arthroscopic loose bodies removal associated with abrasive chondroplasty and microfractures is a surgical option, but the results are variable. Takahara et al. found that patients undergoing the removal of the loose bodies, when greater than 50% of the surface (about 12 mm) of the capitellum, were associated with worst outcomes [24]. Tis et al. studied the arthroscopic treatment of ODE for debridement and microfractures, without reinsertion of the fragment to the subchondral bone. By this technique, pain persisted in 33% of patients and, although some returned to physical activity, many patients did not reach pre-injury activity levels [25]. Microfractures are still a very common procedure for focal lesions in arthroscopy procedure. New technologies and techniques, as autologous chondrocyte implantation, incorporate synthetic and biological grafts (with or without the addition of cells) aiming better adhesion, organization, migration, and differentiation of mesenchymal cells into chondrocytes, enabling better cartilage regeneration [18].

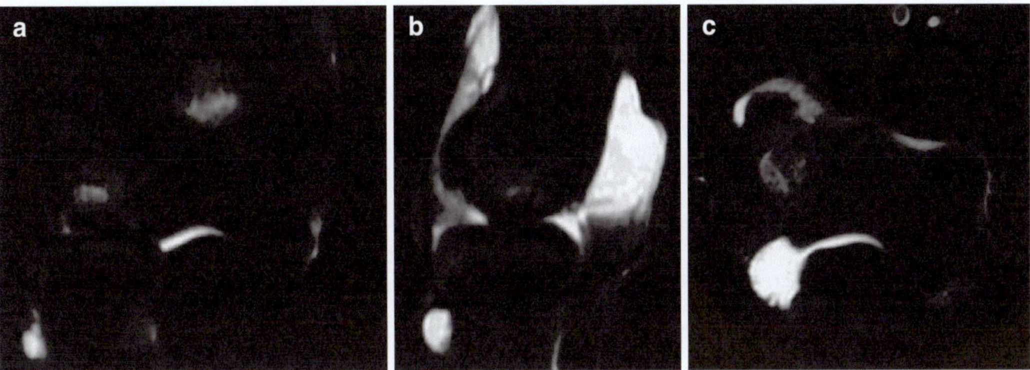

Fig. 26.3 A 16-year-old female, tennis player, with a grade II osteochondritis dissecans lesion of the right elbow with a closed capitellar growth plate. She complained of 1–2 years of pain and mechanical symptoms. (**a**) Coronal view on MRI; (**b**) Sagital view; and (**c**) Axial view

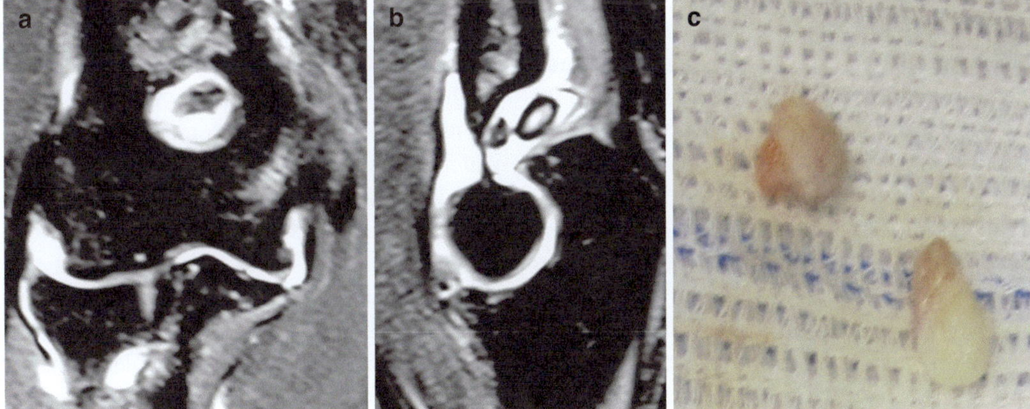

Fig. 26.4 A female patient who complained of pain, blockage, and stiffness symptoms. (**a**) Coronal view on MRI with loose bodies; (**b**) Sagital view; and (**c**) Loose Bodies after arthroscopy procedure

Bone fixation can be indicated when osteochondral fragments are stable, and fixation is possible with bone plugs, metal screws, and bioabsorbable screws. Success rate is approximately 80% and reossification is observed in 44–100% [21].

Joint reconstruction osteochondral plugs from the knee indication include large lesions (>12 mm in diameter or more than 50% of articular surface) and unstable lesions. Takara et al. reviewed 33 patients who underwent joint reconstruction. At follow-up period of 28.4 months, approximately 94% of the cases had returned to pre-injury game level with an average of 6.9 months [17], and a systematic review showed that 94% of the athletes at 30.2 months follow-up, returned to sports with an average of 5.6 months [26]. Despite encouraging results, the procedures using grafts still show difficulties and morbidity in donor site, death of chondrocytes; mechanical and biological integrity, cell viability, risk of disease transmission; loss of chondrogenic phenotype. As an alternative, osteochondral allograft transplantation (OCA) may be obtained from cadaveric capitellum or femoral condyle. OCA keeps most benefits of osteochondral autograft transfer while cutting out the morbidity of donor site. In a study 9 baseball players treated with OCA, follow-up of 48.3 months, describe an improvement in pain and all clinical outcome scores [27]. Also, in the last few decades, 3D tissue engineering has emerged as a strategy for cartilage regeneration. The bioprinting 3D cutting is a deposition process of blends of cells and biomaterials in layers to form ordered and predetermined tissue, recreating the physical environment, matrix, and elbow anatomy [28]. Treatment of ODE is summarized in Fig. 26.5.

26.7.2 Degenerative Joint Disease

The primary osteoarthritis (OA) of the elbow is rare, corresponding to less than 2% of all degenerative joints cases [29]. OA is characterized by loss of range of motion, chronic pain, stiffness, and impact on patient quality of life due to difficulty in performing daily activities and sports practices [30]. Although this condition in the elbow is uncommon compared to the knee and hip, can be common in young athletes, middle-aged patients, and heavy workers. Patients with initial OA claim pain in the extremes of flexion and extension movement, and later to blockage, stiffness, and cubital tunnel syndrome [31].

The pathological processes related to OA of the elbow is a fragile and fragmented cartilage that releases free bodies, forming reactive bone and osteophytes, sclerosis, subarticular cyst formation, fibrosis, and the consequent progressive joint stiffness [32]. Its etiology has not yet been fully elucidated. The studies proposed environmental factors as possible primary etiology since no morphological characteristics were recognized as predisposition to the development of this pathology [33].

26.7.3 Joint Disease after High-Energy Trauma

High-energy elbow fractures, especially intra-articular fractures are associated with post-traumatic osteoarthritis (Fig. 26.6). The articular degeneration generally starts at radiocapitellar joint then goes to the ulnohumeral joint [21]. The mechanisms to develop osteoarthritis are complex, there is an association between the injury pattern and the energy absolved. Chondrocyte death begins after joint damage. Apoptosis markers increase, changing the composition of synovial fluid, resulting in necrosis and degradation of chondrocytes [34]. It is important to remember that associated with chondral injury, there is a large release of cytokines and pro-inflammatory metalloproteinases, that are involved in the direct or indirect activation of fibroblasts [35]. In addition to the response of the articular joint for unknown reasons, the injury can cause contracture of the joint capsule, bone formation in the capsule or musculature, so-called heterotopic calcification.

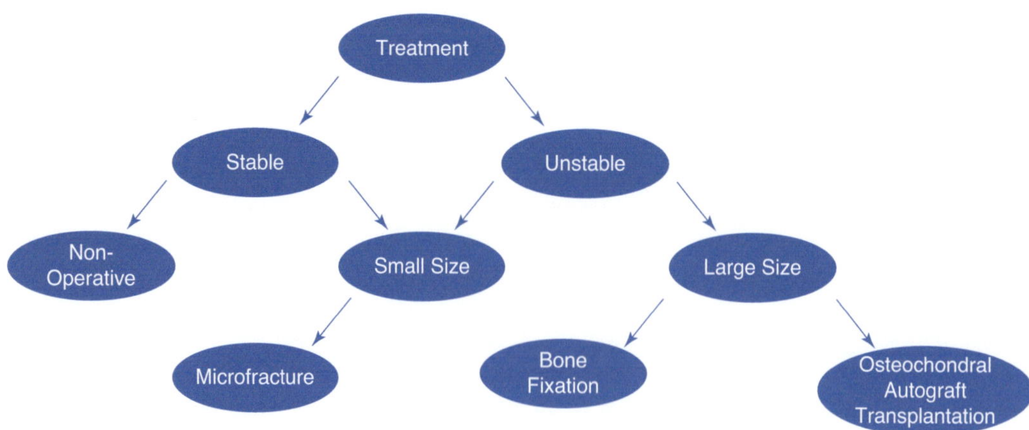

Fig. 26.5 Flowchart—Treatment of ODE

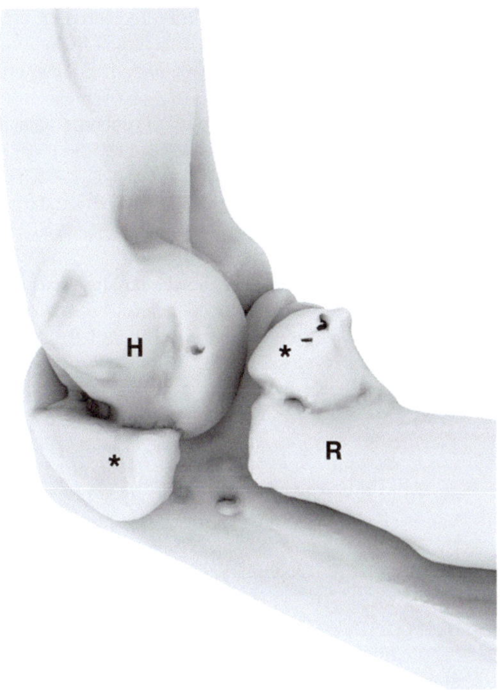

Fig. 26.6 High energy trauma causing intra-articular fractures that results in post-traumatic osteoarthritis. Radio (R); Humerus (H); Fragments of radio (*)

26.7.4 Treatment Options for OA

Patients with OA are initially treated with medication, anti-inflammatory drugs, and physiotherapy then, after 3–6 months, if there is no improvement in pain, surgery may be considered. Intra-articular corticosteroid and hyaluronic acid injections can be used; however, Van Brakel et al. reported pain relief with poor functional improvement at 3 months and no benefits at 6 months with hyaluronic acid injections [36].

The possible procedures are arthroscopic release, open debridement with ulnohumeral arthroplasty, interposition arthroplasty, and total elbow arthroplasty (TEA). The literature recommends treating with arthroscopy the milder cases of elbow osteoarthritis. Open debridement, elbow interposition arthroplasty (EIA), or total elbow arthroplasty for advanced OA or severe elbow deformity cases. In a systematic review, Cohen et al. showed that both procedures are powerful in improving pain, range of motion, and patient satisfaction [37].

Open debridement has historically been used in heavy workers and younger patients with pain and stiffness of the elbow, but with the advancement in elbow arthroscopy, the scope allows a better joint evaluation minimizing soft-tissue damage, lower risk of infection, and pain after procedure. Sochacki et al. in a systematic review showed elbow arthroscopic debridement for primary degenerative osteoarthritis results in improvement in range of motion, function, clinical outcomes of the elbow, while having low complication and reoperation rates [38].

The TEA is the gold standard treatment for severe osteoarthritis in elderly and low-demand patients demonstrating an increase in functional scores [39]. However, outcome reports of TEA for Rheumatoid arthritis (RA) and post-traumatic-related conditions reveal quite distinct results. The patients with post-traumatic arthritis have an increased risk of axle failure, component disassembly, component fracture, subsequent bushing wear, and lower risk of septic loosening [40].

Elbow interposition arthroplasty is indicated for patients with severe arthritis too young or active for TEA. The EIA could be more durable than TEA in these patients with less bone resection and is considered as salvage treatment for post-traumatic arthritis in patients with active lifestyles and heavy workers. In a series, 85% of EIAs performed for post-traumatic arthritis had good or excellent results. However, 37% demanded revision within a mean of 7 years [41].

26.8　Conclusion

There are few studies of elbow osteochondral injuries and literature on prevalence is scarce. Since it is a joint with low mechanical load, they are usually mild chondral lesions that may present with few symptoms or be asymptomatic therefore being underdiagnosed. Once diagnosed, they are mostly treated conservative and indications for surgery are restricted for the treatment of severe joint sequelae and joint incongruity in young people.

References

1. Daniels DL, Mallisee TA, Erickson SJ, Boynton MD, Carrera GF. The elbow joint: osseous and ligamentous structures. Radiographics. 1998;18(1):229–36. https://doi.org/10.1148/radiographics.18.1.9460127.

2. Buckwalter JA, Rosenberg LC, Hunziker EB. Articular cartilage: composition, structure, response to injury and methods of facilitating repairs. In: Ewing JF, editor. Articular cartilage and knee joint function: basic science and arthroscopy. New York: Raven Press; 1990.

3. Bullough PG. Atlas of orthopaedic pathology. 2nd ed. New York: Gower; 1992.

4. Di Bella C, Fosang A, Donati DM, Wallace GG, Choong PFM. 3D bioprinting of cartilage for orthopedic surgeons: reading between the lines. Front Surg. 2015;2(8):1–7. https://doi.org/10.3389/fsurg.2015.00039.

5. Carballo CB, Nakagawa Y, Sekiya I, Rodeo SA. Basic science of articular cartilage. Clin Sports Med. 2017;36(3):413–25. https://doi.org/10.1016/j.csm.2017.02.001.

6. Brown TD, Radin EL, Martin RB, Burr DB. Finite element studies of some juxtarticular stress changes due to localized subchondral stiffening. J Biomech. 1984;17:11–24.

7. Lane LB, Villacin A, Bullough PG. The vascularity and remodeling of subchondral bone and calcifield cartilage in adult femoral and humeral heads. J Bone Joint Surg. 1977;59B:272–8.

8. Yeung C, Deluce S, Willing R, Johnson M, King GJ, Athwal GS. Regional variations in cartilage thickness of the radial head: implications for prosthesis design. J Hand Surg Am. 2015;40(12):2364–71.e1. https://doi.org/10.1016/j.jhsa.2015.09.005.

9. Miyamura S, Sakai T, Oka K, Abe S, Shigi A, Tanaka H, Shimada S, Mae T, Sugamoto K, Yoshikawa H, Murase T. Regional distribution of articular cartilage thickness in the elbow joint: a 3 dimensional study in elderly humans. JBJS Open Access. 2019;4(3) https://doi.org/10.2106/JBJS.OA.19.00011. pii: e0011.1–11

10. Imhof H, Breitenseher M, Kainberger F, Rand T, Trattnig S. Importance of subchondral bone to articular cartilage in health and disease. Top Magn Reson Imaging. 1999;10(3):180–92. https://doi.org/10.1097/00002142-199906000-00002.

11. Birk GT, DeLee JC. Osteochondral injuries: clinical findings. In: osteochondral injuries of the knee. Clin Sports Med. 2001;20(2):279–86. https://doi.org/10.1016/s0278-5919(05)70306-9.

12. GAM F, Noyes FR, editors. Biology and biomechanics of the traumatized synovial joint the knee as a model. Rosemont, IL: American Academy of Orthopaedic Surgeons; 1992. p. 597–9.

13. Lane LB, Bullough PG. Age related changes in the thickness of the calcified cartilage zone and the number of tidemarks in adult human cartilage. J Bone Joint Surg. 1980;62B:372–5.

14. Pope TL Jr, Bloem HL, Beltran J, Morrison WB, Wilson DJ. Normal elbow. In: Imaging of the musculoskeletal system, vol. 1. Philadelphia, PA: Saunders; 2008. p. 221.

15. Outerbridge RE. The etiology of chondromalacia patellae. J Bone Joint Surg Br. 1961;43:752–7.

16. ICRS Cartilage Injury Evaluation Package; 2000. The International Cartilage Repair Society website: http://www.cartilage.org/_files/contentmanagement/ICRS_evaluation.pdf.

17. Maruyama M, Takahara M, Satake H. Diagnosis and treatment of osteochondritis dissecans of the humeral capitellum. J Orthop Sci. 2018;23(2):213–9. https://doi.org/10.1016/j.jos.2017.11.013.

18. Matsuura T, Kashiwaguchi S, Iwase T, Takeda Y, Yasui N. Conservative treatment for osteochondrosis of the humeral capitellum. Am J Sports Med. 2008;36(5):868–72. https://doi.org/10.1177/0363546507312168.

19. Van Bergen CJA, Van Den Ende KIM, Ten Brinke B, Eygendaal D. Osteochondritis dissecans of the capitellum in adolescents. World J Orthop. 2016;7(2):102–8. https://doi.org/10.5312/wjo.v7.i2.102.

20. Birk GT, Delee JC. Osteochondral injuries - clinical findings. Clin Sports Med. 2001;20(2):279–86. https://doi.org/10.1016/s0278-5919(05)70306-9.

21. Churchill RW, Munoz J, Ahmad CS. Osteochondritis dissecans of the elbow. Curr Rev Musculoskelet Med. 2016;9(2):232–9. https://doi.org/10.1007/s12178-016-9342-y.

22. Heijink A, Vanhees M, van den Ende K, van den Bekerom MP, van Riet RP, Van Dijk CN, Eygendaal D. Biomechanical considerations in the pathogenesis of osteoarthritis of the elbow. Knee Surg Sports Traumatol Arthrosc. 2016;24(7):2313–8. https://doi.org/10.1007/s00167-015-3518-7.

23. Logli AL, Bernard CD, O'Driscoll SW, Sanchez-Sotelo J, Morrey ME, Krych AJ, Camp CL. Osteochondritis dissecans lesions of the capitellum in overhead athletes: a review of current evidence and proposed treatment algorithm. Curr Rev Musculoskelet Med. 2019;12(1):1–12. https://doi.org/10.1007/s12178-019-09528-8.

24. Takahara M, Mura N, Sasaki J, Harada M, Ogino T. Classification, treatment, and outcome of osteochondritis dissecans of the humeral capitellum. J Bone Joint Surg. 2008;90(2):47–62. https://doi.org/10.2106/JBJS.G.01135.

25. Tis JE, Edmonds EW, Bastrom T, Chambers HG. Short-term results of arthroscopic treatment of osteochondritis dissecans in skeletally immature patients. J Pediatr Orthop. 2012;32(3):226–31. https://doi.org/10.1097/BPO.0b013e31824afeb8.

26. Kirsch JM, Thomas JR, Khan M, Townsend WA, Lawton JN, Bedi A. Return to play after osteochondral autograft transplantation of the capitellum: a systematic review. Arthroscopy. 2017;33(7):1412–1420.e1. https://doi.org/10.1016/j.arthro.2017.01.046.

27. Logli AL, Bernard CD, O'Driscoll SW, Sanchez-Sotelo J, Morrey ME, Krych AJ, et al. Osteochondritis

dissecans lesions of the capitellum in overhead athletes: a review of current evidence and proposed treatment algorithm. Curr Rev Musculoskelet Med. 2019;12(1):1–12.

28. Roseti L, Cavallo C, Desando G, Parisi V, Petretta M, Bartolotti I, Grigolo B. Three-dimensional bioprinting of cartilage by the use of stem cells: a strategy to improve regeneration. Materials. 2018;11(9):1–20. https://doi.org/10.3390/ma11091749.

29. Merolla G, Buononato C, Chillemi C, Paladini P, Porcellini G. Arthroscopic joint debridement and capsular release in primary and post-traumatic elbow osteoarthritis: a retrospective blinded cohort study with minimum 24-month follow-up. Musculoskelet Surg. 2015;99:83–90. https://doi.org/10.1007/s12306-015-0365-0.

30. Poonit K, Zhou X, Zhao B, Sun C, Yao C, Zhang F, Yan H. Treatment of osteoarthritis of the elbow with open or arthroscopic debridement: a narrative review. BMC Musculoskelet Disord. 2018;19(1):1–10. https://doi.org/10.1186/s12891-018-2318-x.

31. Kim DM, Han M, Jeon IH, Shin MJ, Koh KH. Range-of-motion improvement and complication rate in open and arthroscopic osteocapsular arthroplasty for primary osteoarthritis of the elbow: a systematic review. Int Orthop. 2020;44(2):329–39. https://doi.org/10.1007/s00264-019-04458-z.

32. Adams JE, Wolff LH, Merten SM, Steinmann SP. Osteoarthritis of the elbow: results of arthroscopic osteophyte resection and capsulectomy. J Shoulder Elb Surg. 2008;17(1):126–31. https://doi.org/10.1016/j.jse.2007.04.005.

33. Gallo RA, Payatakes A, Sotereanos DG. Surgical options for the arthritic elbow. J Hand Surg. 2008;33(5):746–59. https://doi.org/10.1016/j.jhsa.2007.12.022.

34. Phen HM, Schenker ML. Minimizing posttraumatic osteoarthritis after high-energy intra-articular fracture. Orthop Clin N Am. 2019;50(4).433–43. https://doi.org/10.1016/j.ocl.2019.05.002.

35. Wahl EP, Lampley AJ, Chen A, Adams SB, Nettles DL, Richard MJ. Inflammatory cytokines and matrix metalloproteinases in the synovial fluid after intra-articular elbow fracture. J Shoulder Elb Surg. 2020;29(4):736–42. https://doi.org/10.1016/j.jse.2019.09.024.

36. Van Brakel RWED. Intra-articular injection of hyaluronic acid is not effective for the treatment of post-traumatic osteoarthritis of the elbow. Arthroscopy. 2006;22(11):1199–203. https://doi.org/10.1016/j.arthro.2006.07.023.

37. Cohen AP, Redden JF, Stanley D. Treatment of osteoarthritis of the elbow: a comparison of open and arthroscopic debridement. Arthroscopy. 2000;16(7):701–6. https://doi.org/10.1053/jars.2000.8952.

38. Sochacki KR, Jack RA, Hirase T, McCulloch PC, Lintner DM, Liberman SR, Harris JD. Arthroscopic debridement for primary degenerative osteoarthritis of the elbow leads to significant improvement in range of motion and clinical outcomes: a systematic review. Arthroscopy. 2017;33(12):2255–62. https://doi.org/10.1016/j.arthro.2017.08.247.

39. Kodama A, Mizuseki T, Adachi N. Macroscopic investigation of failed kudo type 5 total elbow arthroplasty. J Shoulder Elb Surg. 2018;27(8):1380–5. https://doi.org/10.1016/j.jse.2018.05.002.

40. Wang JH, Ma HH, Chou TFA, Tsai SW, Chen CF, Wu PK, et al. Outcomes following total elbow arthroplasty for rheumatoid arthritis versus post-traumatic conditions: a systematic review and meta-analysis. Bone Joint J. 2019;101-B(12):1489–97. https://doi.org/10.1302/0301-620X.101B12.BJJ-2019-0799.R1.

41. Larson AN, Adams RA, Morrey BF. Revision interposition arthroplasty of the elbow. J Bone Joint Surg Br. 2010 Sep;92(9):1273–7. https://doi.org/10.1302/0301-620X.92B9.24039.

Wrist Osteochondral Unit Function and Treatment

Riccardo D'Ambrosi

27.1 Etiology

According to the American Academy of Orthopedic Surgeons (AAOS), wrist arthritis (WA) can be divided into three subcategories [1]:

1. Primary osteoarthritis (OA): This is the most common cause of wrist pain in the elderly population, although it can affect any age group. Aging, hereditary factors, high body mass index (BMI), joint anatomy, and gender are the risk factors linked to the development of osteoarthritis.
2. Rheumatoid arthritis (RA) is an inflammatory condition that affects the peripheral joints symmetrically. The exact etiology of rheumatoid arthritis remains unknown but is due to multifactorial factors [2, 3].
3. Post-traumatic arthritis due to traumatic events such as injuries to the ligaments or fracture. Despite adequate treatment, damage to the cartilage increases the risk of developing arthritis over time [4].

Other causes of WA can be infection, crystal-induced arthritis, reactive arthritis, and systemic diseases like sarcoid arthropathy, myelodysplastic, and leukemic disorders.

27.2 Epidemiology

Although the wrist is not a weight-bearing joint, it has a significant function in daily activities, and that predisposes it to trauma and fractures. On an estimate, one in seven persons in the United States reports WA (13.6%). The prevalence of RA affecting the wrist is 2.5 million people in the United States and approximately 75% in the general population. Gout affects the wrist in 0.28% of the population [5].

27.3 Pathophysiology

The pathophysiology of wrist arthritis depends on the type of arthritis affecting the wrist.

- OA, a disease previously thought to be due to wear and tear, has more complex pathogenesis related to biomechanical and chemical factors such as matrices metalloproteinases (MMPs), cysteine proteinases, serine proteinases, and proinflammatory cytokines.
- RA results from a complex interaction between genetic and environmental factors that leads to a breakdown of immune tolerance and synovial inflammation.
- In post-traumatic OA, the mechanics of the wrist and ligaments will change, and loading factors are redirected or misdirected, resulting in damage to cartilage,

R. D'Ambrosi (✉)
IRCCS Istituto Ortopedico Galeazzi, Milan, Italy

© ISAKOS 2022
A. Gobbi et al. (eds.), *Joint Function Preservation*, https://doi.org/10.1007/978-3-030-82958-2_27

## 27.4	Medical History

Patients with WA mainly refer pain that is diffuse in the whole joint. The character of the pain depends on the type of arthritis. For example, pain due to OA is worse with joint use and is relieved by rest. Pain due to RA is often associated with stiffness in the morning and gets better as the day goes on. Symmetric involvement of the joints is also more common in RA. Other than pain, swelling is an important feature of arthritis. Swelling can be due to effusion or synovial hypertrophy. Redness and warmth of the joints, along with swelling in a nontraumatic wrist joint suggest an inflammatory disease or an infection.

## 27.5	Physical Examination

Physical examination includes inspection, palpation, range of motion, and special tests.

- Inspection: Swelling and deformities are the two important findings associated with WA. Regarding swelling, it is important to distinguish between a joint effusion from tenosynovitis or a localized mass. Arthritis usually produces a diffuse circumferential swelling. Chronic inflammation in diseases like rheumatoid arthritis can cause deformities like volar subluxation of the carpus, carpal collapse, and radial deviation of the carpus. It can also result in instability with dorsal subluxation of the ulnar head, which causes "piano key" like movement with downward pressure.
- Palpation: Palpation helps in identifying the specific painful area. The wrist is best palpated in slight flexion and feeling the dorsal surface of the wrist with the thumb while supporting the wrist with the fingers of both hands. Dorsal instability is a sign of joint effusion. Instability can be tested by looking for transmission of pressure from one hand placed at one side of the joint to the second hand placed on the opposite side.
- Range of Motion (ROM): Clinicians should test the active range of motion first. The range of motion tested at the wrists is flexion, exten-

sion, radial, and ulnar deviation. The normal range of flexion is 65–80° of flexion, 55–75° of extension, 30–45° of ulnar deviation, and 15–25° of radial deviation.
- Special tests: Tinel sign, Carpal compression test, Phalen test, Finkelstein test, etc., excluding causes other than arthritis in a patient with wrist pain.

## 27.6	Evaluation

Evaluation of wrist arthritis includes a complete medical history, starting from the onset of symptoms, location, nature, duration, aggravating, and easing factors. If the pain is chronic, triggers causing recent exacerbations should be enquired.

Radiographs of the joint: Conventional radiography is the most widely used imaging modality and allows for the detection of bone pathologies like fracture, erosions, osteonecrosis, osteoarthritis, or a juxta-articular bone tumor. Characteristic features of OA include marginal osteophytes, joint space narrowing, subchondral sclerosis, and cysts.

Ultrasonography is unhelpful in checking the bones or deep parts of the joints and is operator-dependent, but it might show OA-associated structural changes, osteophytes, crystal deposition and is also useful for detecting synovial inflammation, joint effusion, and erosions [6, 7].

Laboratory tests: Leukocytosis supports the possibility of infection. Cultures of blood, urine, or other possible primary sites of infection are mandatory when a septic joint is being considered. Rheumatoid factor and anti-CCP (cyclic citrullinated peptide) antibodies should be ordered if there is clinical suspicion for rheumatoid arthritis. A serum uric acid level is often ordered by clinicians when gout is suspected, but it is not reliable as it may be spuriously elevated in acute inflammatory conditions or acutely diminished during a true gout attack.

Synovial fluid analysis: A joint arthrocentesis and synovial fluid analysis are mandatory if an infection is suspected. Such patients should also be started on empiric antibiotic therapy as soon as possible after the synovial fluid sample is obtained.

27.7 Treatment/Management

Like the pathogenesis, treatment of WA greatly depends upon the type of arthritis.

OA: Nonsurgical management includes, NSAIDs, and other analgesic medications, avoiding activities causing exacerbation of the pain, immobilization of the joint with wrist splints, especially during daytime and during activities, physical therapy, and local corticosteroid injections. Systemic steroids have no role and should be avoided. Pills containing hyaluronic acid and glucosamine are ineffective and have a placebo effect [8].

RA: Disease-modifying antirheumatic drug (DMARD) therapy is the cornerstone in the management of RA. Antiinflammatory therapies, including systemic and intra-articular glucocorticoids and NSAIDs, are used primarily as adjuncts for temporary control of disease activity in patients in whom treatment is being started with DMARDs or during disease flares and modification of the DMARD regimen. Methotrexate, hydroxychloroquine, sulfasalazine, and leflunomide are the major traditional DMARDs. Biologic agents like anti-TNF-alpha agents, including etanercept, infliximab, and adalimumab, tocilizumab(IL-6 inhibitor), tofacitinib (JAK inhibitor), and rituximab(anti-CD-20 monoclonal antibody) are all used for the treatment of RA.

Surgical treatment is indicated when disabling pain emerges despite conservative and nonsurgical treatments. There are many surgical approaches available, like wrist denervation, ulnar resection (removes the pressure from wrist), or synovectomy, but the ones used most often include proximal row carpectomy, wrist fusion, and wrist replacement [1].

Carpectomy involves the removal of the proximal carpal bones close to the forearm to ease pain and sustain wrist motion. Fusion or arthrodesis is a welding process that removes the damaged cartilage and attaches wrist bones to make sure they heal as a single and solid bone that does not cause pain. Fusion will reduce the range of motion but eliminate the pain. In wrist replacement, the surgical procedure involves the removal of the damaged wrist cartilages and bones and replacement with plastic or metal joint. The goal is to restore function, regain range of motion, and reduce the pain. The implants have not resulted in gratifying results such as those with knee or hip replacement [9–12].

27.8 Arthroscopy in Arthritis

27.8.1 Arthrosis of the Proximal Pole of the Hamate (Uncinate Bone)

Cartilage erosion of the proximal pole of the hamate is a common site of arthrosis within the wrist. Viegas et al. [13] described two types of lunate morphology based on the presence or absence of a separate hamate facet on the distal lunate articular surface. A type I variant (no hamate facet) was evident in 34.5% of the dissected specimens, and a type II variant (distinct hamate facet) was evident in 65.5% of the dissected specimens. Significant cartilage erosion at the proximal pole of the hamate was identified at dissection in 44.4% of the type II lunates but in none of the type I lunates [14]. Chondral lesions of the midcarpal joints in type I lunates were always associated with other ligamentous and/or osteochondral lesions, whereas the same lesions could be found isolated in type II lunates. Harley et al. noted a strong association between hamate arthrosis and lunotriquetral interosseous ligament (LTIL) tears and coined the acronym "HALT" for hamate arthrosis lunate ligament tear. In a biomechanical study, they found that resection of 2.4 mm of the proximal pole of the hamate completely unloaded the hamato lunate articulation [15].

27.8.1.1 Indications

An arthroscopic resection of the proximal pole of the hamate is indicated in patients with persistent ulnar-sided wrist pain who have failed an adequate trial of conservative treatment. The patients may have tenderness and swelling distal to the triangular fibrocartilage complex (TFCC) and pain with wrist extension and ulnar deviation.

Plain radiographs have a low sensitivity for making the diagnosis. An MRI is recommended, as the history and physical findings are not diagnostic of the condition. Cartilage-sensitive sequencing will identify cartilage loss on the proximal hamate pole, and more severe cases will show edema within the hamate. Often the dual facet is diagnosed as an incidental finding at the time of arthroscopy and varying degrees of hamate chondromalacia may be found in association with other ulnar-sided wrist pathology. In these cases, an arthroscopic resection is not indicated unless the patient has ulnar-sided pain and tenderness because it is often an asymptomatic finding.

27.8.1.2 Contraindications

Inflammatory arthritis, or autoimmune disease, that involves the wrist is a contraindication. There is no data to compare resection of the proximal pole of the hamate with subchondral drilling of the chondral lesion or nonoperative treatment in a patient with focal chondromalacia. Patients with midcarpal degenerative arthritis and/or interosseous ligament tears will have compromised outcomes and may be more suitable for partial fusions or a proximal row carpectomy.

27.8.2 Chondral Defects

27.8.2.1 Indications

Articular cartilage damage is a common cause of wrist pain and may result from post-traumatic osteochondral fractures, chronic carpal instability, or attrition. Loose bodies commonly result from osteoarthritis, but may also be associated with AVN, primary synovial chondromatosis, or trauma. Loose bodies give rise to pain and locking, which is relieved following an arthroscopic removal. Articular defects are often undetected by preoperative imaging studies and are best seen at the time of arthroscopy. Culp et al. have provided a modified Outerbridge classification for chondral lesions in the wrist where grade I represents softening of the hyaline surface, grade II consists of fibrillation and fissuring, grade III represents a fibrillated lesion of varying depth in the articular surface, and grade IV has a full-

thickness defect down to bone [16]. Grade I through III lesions are treated with debridement and localized synovectomy. Localized grade IV lesions are treated with abrasion chondroplasty and subchondral drilling.

27.8.2.2 Contraindications

These treatments are contraindicated if there is widespread cartilage loss.

27.9 Complications

The complications of wrist arthritis are mostly due to the various surgical treatments of the disease. Continued wrist pain due to a nonunion or fibrous union is a potential complication of fusion surgeries. As with all orthopedic surgical procedures, there is a risk of prosthetic infection, neurovascular injury as a complication of the procedure itself, and implant failure or loosening.

27.10 Conclusion

Wrist arthritis can be challenging in terms of both diagnosis and management. Since hand motions are essential for many higher functions, clinicians must pay close attention to details while addressing wrist arthritis in a time-sensitive manner to prevent disability. One should consider an interprofessional approach involving rheumatologists, orthopedic surgeons, physical therapists, and occupational therapists whenever appropriate.

References

1. Kurdi AJ, Voss BA, Tzeng TH, Scaife SL, El-Othmani MM, Saleh KJ. Rheumatoid arthritis vs osteoarthritis: comparison of demographics and trends of joint replacement data from the nationwide inpatient sample. Am J Orthop. 2018;47(7). https://doi.org/10.12788/ajo.2018.0050.
2. Ben Achour W, Bouaziz M, Mechri M, Zouari B, Bahlous A, Abdelmoula L, Laadhar L, Sellami M, Sahli H, Cheour E. A cross sectional study of bone and cartilage biomarkers: correlation with struc-

tural damage in rheumatoid arthritis. Libyan J Med. 2018;13(1):1512330.

3. Guo Q, Wang Y, Xu D, Nossent J, Pavlos NJ, Xu J. Rheumatoid arthritis: pathological mechanisms and modern pharmacologic therapies. Bone Res. 2018;6:15.

4. Lotz MK, Kraus VB. New developments in osteoarthritis. Posttraumatic osteoarthritis: pathogenesis and pharmacological treatment options. Arthritis Res Ther. 2010;12(3):211.

5. Chen JH, Huang KC, Huang CC, Lai HM, Chou WY, Chen YC. High power Doppler ultrasound score is associated with the risk of triangular fibrocartilage complex (TFCC) tears in severe rheumatoid arthritis. J Investig Med. 2019;67(2):327–30.

6. Samanta M, Mitra S, Samui PP, Mondal RK, Hazra A, Sabui TK. Evaluation of joint cartilage thickness in healthy children by ultrasound: an experience from a developing nation. Int J Rheum Dis. 2018;21(12):2089–94.

7. Rosendahl K, Bruserud IS, Oehme N, Júlíusson PB, de Horatio LT, Müller LO, Magni-Manzoni S. Normative ultrasound references for the paediatric wrist; dorsal soft tissues. RMD Open. 2018;4(1):e000642.

8. Romanowski MW, Majchrzak M, Kostiukow A, Malak R, Samborski W. Effect of manual therapy on pain sensation and hand dexterity and grip strength in a patient with RA. Didactic case study. Ortop Traumatol Rehabil. 2018;20(2):157–62.

9. Gaspar MP, Pham PP, Pankiw CD, Jacoby SM, Shin EK, Osterman AL, Kane PM. Mid-term outcomes of routine proximal row carpectomy compared with proximal row carpectomy with dorsal capsular interposition arthroplasty for the treatment of late-stage arthropathy of the wrist. Bone Joint J. 2018;100-B(2):197–204.

10. Adams BD. Surgical management of the arthritic wrist. Instr Course Lect. 2004;53:41–5.

11. Kistler U, Weiss AP, Simmen BR, Herren DB. Long-term results of silicone wrist arthroplasty in patients with rheumatoid arthritis. J Hand Surg Am. 2005;30(6):1282–7.

12. Field J, Herbert TJ, Prosser R. Total wrist fusion. A functional assessment. J Hand Surg Br. 1996;21(4):429–33.

13. Viegas SF, Patterson RM, Hokanson JA, et al. Wrist anatomy: incidence, distribution, and correlation of anatomic variations, tears, and arthrosis. J Hand Surg Am. 1993;18:463–75.

14. Viegas SF, Wagner K, Patterson R, et al. Medial (hamate) facet of the lunate. J Hand Surg [Am]. 1990;15:564–71.

15. Harley BJ, Werner FW, Boles SD, et al. Arthroscopic resection of arthrosis of the proximal hamate: a clinical and biomechanical study. J Hand Surg Am. 2004;29:661–7.

16. Culp RW, Osterman AL, Kaufmann RA. Wrist arthroscopy: operative procedures. In: Green DP, Hotchkiss RN, Pederson WC, Wolfe SW, editors. Green's operative hand surgery. Philadelphia: Elsevier; 2005. p. 781–803.

D. Camacho and R. Mardones

28.1 Introduction

28.1.1 The Osteochondral Unit

Classically, there was the belief that joint disease and osteoarthritis (OA) was limited to articular cartilage injury, and treatment strategies were focused on the repair of the articular cartilage only. Currently, OA and other similar disorders are considered an "organ disease of the whole joint." This has lead to an understanding that damage to the articular surface can lead to, be caused by, or occur in parallel with, damage to other tissues in the joint, such as underlying subchondral bone [1].

The osteochondral unit (OCU), which is a functional unit represented by articular cartilage and underlying subchondral one, is essential to maintaining the integrity and health of the joint and is composed of (1) hyaline cartilage, (2) the boundary between hyaline and calcified cartilage (tidemark), (3) calcified cartilage (connected to the subchondral cortical bone), and (4) the subchondral plate (subchondral cortical bone) which continues with the metaphyseal trabecular bone [1]. These tissue layers are interdependent mechanically, physiologically, and biochemically. Damage to one component of the OCU can impact the function of the other components, initiating a cascade of repair and remodeling processes that often have detrimental effects on the long-term health and function of the joint, potentially leading to OA.

It is well-known that there is a continuous exchange of nutrients, cytokines, prostaglandins, and other bioactive factors between bone and overlying cartilage. More than a decade ago, Pan [2] and other authors demonstrated the transportation of fluorescent dyes from the subchondral circulation to cartilage, and this diffusion is elevated in OA due to the increasing subchondral bone porosity [1, 2].

Pain is typically the primary symptom of osteochondral lesions, including OA, and is closely related to the OCU. Healthy hyaline cartilage has no nociceptors therefore pain originates from the subchondral bone or associated joint soft tissues (joint capsule, synovium) [1]. It has been shown that subchondral bone lesions correlate better with joint pain compared to synovitis [3].

D. Camacho
Centro de Cadera Clinica Las Condes, Instituto Traumatologico, Universidad de Chile,
Las Condes, Santiago, Chile
e-mail: dcamacho@cdma.cl

R. Mardones (✉)
Centro de Cadera Clinica Las Condes, Las Condes,
Santiago, Chile
e-mail: rmardones@cdma.cl;
rmardones@clinicalascondes.cl

© ISAKOS 2022
A. Gobbi et al. (eds.), *Joint Function Preservation*, https://doi.org/10.1007/978-3-030-82958-2_28

Chondral lesions do not have the capacity to fully self-heal. Full-thickness lesions with subchondral bone involvement allow for some degree of healing, with migration of bone marrow mesenchymal cells (BM-MSC) and formation of an inflammatory "super clot." Newly formed fibrocartilage tissue differs in structure from the original hyaline articular cartilage, containing predominantly type I collagen as opposed to type II collagen which is ubiquitous in hyaline cartilage [4].

28.2 Hip Particularities

28.2.1 Anatomy

The human pelvis is adapted to the unique demands of bipedal locomotion. The characteristic long and narrow pelvis of quadrupeds became short and wide to facilitate mediolateral balancing of the trunk in the single-standing phase of walking [5].

The hip is a ball-in-socket, weight-bearing joint; the femoral head is almost completely covered by cartilage, while the cartilage of the acetabulum is horseshoe-shaped, covering approximately 3/4 of its surface [6]. Normal hyaline cartilage thickness of the hip varies across the articular surface [7, 8]. Femoral head cartilage is thickest at the center, with a mean thickness of 2.8 mm (range 1.5–5 mm), while acetabular cartilage is thickest near the labrum, with a mean depth of 3 mm (range 1.4–4.8 mm). The cartilage is thinner about the periphery of the femoral head (mean thickness of 1 mm) and at the medial-anterior-superior area of the acetabulum (mean thickness 1.3 mm) [7, 8].

28.2.2 Hip Osteochondral Injuries

Osteochondral injuries in the hip can arise from multiple pathologic conditions such as femoroacetabular impingement (FAI), femoral head osteonecrosis (ON), developmental dysplasia (DDH), acetabular osteochondritis dissecans (OCD), traumatic injury, and osteoarthritis (OA) [9]. Each type of lesion can present with a characteristic injury pattern.

Osteochondral lesions of the hip are not distributed evenly between the acetabulum and femoral head. The acetabulum is the most common location of chondral damage in the hip, with the anterosuperior acetabulum reportedly involved in up to 88% of chondral defects [9]. In a cohort of FAI cases previously examined by our group [10], 100% of cases had chondral delamination in zones 2, 3, and 4 according to the geographic description [11] (Fig. 28.1). Furthermore, 88.2% suffered from labral tears at the adjacent labrum-cartilage junction.

Hip morphology in the presence of hip pathology directly influences the location of osteochondral damage. Kaya et al. [12] described a hip morphology-specific pattern of cartilage damage. In FAI, they reported that the most affected areas were anterosuperior in the acetabulum and anterolateral in the femoral head. In cases of DDH, the most affected area was the acetabular margin, and in patients with borderline dysplasia, the pericotyloid fossa of the acetabulum and apex of the femoral head showed greater cartilage damage. Regarding the degree of chondral damage, full-thickness

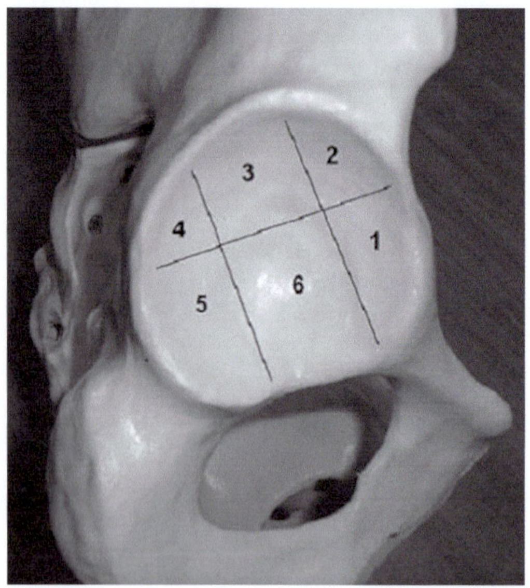

Fig. 28.1 Geographic acetabular zone description [11]

defects are more frequent in FAI and DDH, while partial-thickness defects are more dominant in cases of hip joint laxity [12].

Besides the classic chondral lesion descriptions used for joints in general and classified by traditional systems (Outerbridge, ICRS, etc.), there is a chondral delamination type of lesion in the hip, with detachment of "healthy cartilage" from the underlying subchondral bone. This injury is practically exclusive to the hip.

Many advances in the imaging and treatment of knee cartilage injury have been extrapolated to the hip joint. Importantly however cartilage anatomy and biomechanics are disparate between the hip and knee joints, making it difficult to predict the outcomes of some cartilage repair techniques used within the hip that have been extensively studied in the knee.

28.3 Hip Osteochondral Unit Treatment

Biologic injection therapies and surgical procedures have been used to treat hip OCU pathologies. Biological injection treatments include the use of orthobiologics, which are tissue materials derived from biological substances that are naturally present in blood. These biologic therapies aim to improve the regenerative capacity of tissues associated with musculoskeletal injury. The most widely used and studied orthobiologics for osteochondral damage include hyaluronic acid (HA), platelet-rich plasma (PRP), stem cells isolated from bone marrow as bone marrow aspirate concentrate (BMAC), and intra-articular injected expanded mesenchymal stem cells (MSC) [13]. Surgical treatments include microfracture alone or microfracture with augmentation, autologous chondrocyte implantation (ACI), matrix-induced autologous chondrocyte implantation (MACI), autologous matrix-induced chondrogenesis (AMIC), matrix-associated stem cell transplantation (MAST), mosaicplasty, osteochondral autograft transfer (OATS) and allograft transplantation [14], and the development of prosthetic biocomposites.

28.4 Biological Treatments

28.4.1 Hyaluronic Acid (HA)

High molecular weight glycosaminoglycans play an essential role in joint lubrication and shock absorption. Intra-articular HA injections work through several mechanisms: inducing the production of HA from chondrocytes and synoviocytes (viscoinduction); preventing cartilage fragmentation (chondroprotection) by interacting with CD44 receptors, inhibiting interleukin-1ß, and decreasing the production of matrix metalloproteinase and cartilage catabolism [13, 15]; and providing protection from mechanical stress (viscosupplementation) [13, 15, 16].

Migliore et al. [17] evaluated 120 patients and reported a significant reduction in Lequesne index scores and pain visual analog scale (VAS) at 3 months after HA injection. These results were maintained over time through cyclical repetition of intra-articular injections (one injection every 6 months). It is important to note that the wide variations in preparations of injectable HA, the heterogeneity in their characteristics such as mean molecular weight, the paucity of randomized studies (particularly in the hip), and the lack of outcome consistency in the literature make clinical recommendations regarding this treatment difficult [18]. The American Academy of Orthopaedic Surgeons (AAOS) clinical practice guidelines for osteoarthritis (OA) state that the use of HA in OA is not supported [19].

28.4.2 Platelet-Rich Plasma (PRP)

PRP contains high concentrations of platelets, which release more than 300 proteins, including cytokines and growth factors involved in tissue healing and regeneration. These biologic factors impart a strong chemotactic effect on chondrocytes and mesenchymal stem cells/signaling cells, and inhibit the production of matrix metalloproteinase 13 and nuclear factor-kappa B, decreasing the inflammatory environment associated with OA [13, 15].

The published results of clinical PRP use remain inconsistent, with conflicting results described in the treatment of the same pathologies. There are several considerations related to clinical evaluation of PRP efficacy that likely influence results in the literature, such as patient demographic differences between trials, variations in commercial kit preparation, and variability in the applied concentrations of PRP constituents [20].

The use of PRP in the treatment of chondral lesions has not been thoroughly examined in the literature, and evidence for use specifically in the hip joint is particularly lacking; however, the relatively low-cost and minimal risk associated with the procedure in association with some promising published results support its use in clinical practice [13, 20].

An experimental study conducted in sheep demonstrated that the use of a PRP clot in association with microfracture achieved complete filling of chondral defects with macroscopic, biomechanic, and microscopic characteristics of repair tissue similar to normal hyaline cartilage [21]. Clinical results and the quality of repair tissue achieved by combining microfracture with PRP injection have demonstrated better results when compared to microfracture-only treatment [13]. Furthermore, Sanchez et al. [22] evaluated the effect of hip intra-articular injection of PRP in 40 patients with severe OA and reported a clinically significant reduction in pain and improved function after short-term follow-up of 6 months.

28.4.3 Bone Marrow Aspirate Concentrate (BMAC) and Mesenchymal Stem Cells (MSCs)

Chahla et al. [14] described the role of stem cells in cartilage regeneration to be similar to the director of an orchestra, providing messages to the rest of the tissues in order to repair the damage. BMAC is a source of MSCs, with stem cell concentrations estimated to be in the range of 0.001–0.01%. Additionally, BMAC is a rich source of growth factors that contribute to chondrogenesis and repair processes [14]. To achieve increased concentrations of stem cells for clinical use, stem cells are isolated from bone marrow aspirate and then seeded and expanded for 2–6 weeks. This procedure generates an isolate containing 20–200 million cells per milliliter. The ideal dose, frequency, and number of injections remain unclear, but some studies suggest that a higher concentration of stem cells results in improved clinical outcomes [14, 23]. As in the case of other treatments, most studies examining the use of stem cells to treat chondral injury have been performed in other joints, such as the knee, with very few hip-related studies.

Our group has examined the use of stem cell therapy to treat hip pathology. In a previous study [24], our group evaluated the use of expanded MSCs in focal chondral defects in 20 patients (29 hips) with FAI and focal Outerbridge grade III–IV chondral injury, and with or without mild to moderate OA (Tönnis scale I–III). Mean follow-up was 24 months. All patients underwent arthroscopic FAI treatment, and in the same surgical setting, 80 cc of BMA was aspirated from the anterior iliac crest. Mononuclear cells were isolated and seeded. When the differentiated cultures reached the required cell number (20×10^6 cells) (within 2–3 weeks), each patient received three intra-articular injections of 20×10^6 cells in 1.5 cc, once per week, 4–6 weeks postoperatively under radioscopic guidance. The modified Harris Hips Score (mHHS) and Western Ontario and McMaster Universities Osteoarthritis Index (WOMAC) score improved from preoperative medians of 64.3 and 73 to postoperative medians of 91 and 97, respectively, at final follow-up ($p < 0.05$). The median VAS score improved from 6 to 2. Thirteen percent of the hips received a total hip arthroplasty (THA) at a median of 9 months (range 6–36 months) postoperatively. There were no major complications.

Regarding hip OA, we recently described our results using the technique of intra-articular infusion of ex vivo expanded autologous bone marrow-derived MSCs for diffuse hip chondral damage and mild to moderate osteoarthritis in active patients seeking a non-arthroplasty treatment [20, 25]. Thirteen hips in 10 patients with a mean age of 49-years old (range 24–60 years), with radiological and symptomatic mild to moderate hip OA were treated. All patients underwent bone marrow aspiration (30 cc) from the posterior iliac crest (60 cc in bilateral cases). Bone marrow aspirates were sent to the tissue engineering facility for mononuclear cell isolation and expansion by cell culture protocols. 20×10^6 ex vivo expanded BM-MSC were infused into the damaged hip joint at days 0, 7, and 14 (total infusion of 60×10^6 ex vivo expanded autologous MSC). Mean VAS improved from 4.1 to 1; Mean HHS improved from 61.9 to 85; WOMAC improved from 48.1 to 27; and VAIL score improved from 61 to 78.2. No Tönnis grade progression was seen in postoperative X-ray imaging at a mean follow-up of 27 months (range 16–40).

28.5 Surgical Treatments

28.5.1 Microfracture

Microfracture is perhaps the most widely known surgical technique used to treat osteochondral injury. As previously discussed, the outcome of microfracture treatment has been studied in more detail in the knee [26, 27], and these results are often extrapolated to the hip. Certain conditions are typically required to achieve good results using this technique, such as younger patient selection (< 40 years of age), body mass index less than 30 kg/m^2, none to mild OA (Tönnis 0–1), and treatment of focal lesions less than 4 cm^2 in size [14, 28]. Another important factor to consider is the concomitant treatment of other hip pathology (DDH, FAI, etc.) to prevent recurrence of chondral damage.

The goal of microfracture is to release bone marrow cells and growth factors into the cartilage defect [14, 28]. Reported literature shows favorable clinical outcomes and adequate fill of focal hip chondral defects using microfracture. Karthikeyan et al. [29] reported on 20 patients

with FAI and acetabular chondral defects (mean size 1.54 cm^2) treated with hip arthroscopy and microfracture, and described no OA progression at a mean follow-up of 21 months. Domb [30] reported on a series of 30 patients with a minimum follow-up of 2 years treated with arthroscopic microfracture and showed a significant clinical improvement in patient-reported outcome scores. In a classic study by Philippon et al. [31], 9 patients were treated with microfracture for acetabular chondral defects and subsequently underwent revision hip arthroscopy. Ninety-one percent of the defects were filled with good-quality cartilage in an average time of 20 months between the initial arthroscopy and the revision procedure. With regard to literature that is currently available to guide clinical decision-making, larger sample sizes, longer term outcome analysis, and studies using control groups are needed to better evaluate the effectiveness of microfracture for management of chondral injuries in the hip [28].

Microfracture and Biologic Augmentation: Biologic injection therapy used as an adjunct to microfracture could improve cartilage regeneration and may achieve better outcomes [32, 33]. Augmentation with HA, PRP, MSCs, or BMAC has been shown to enhance cartilage repair quality, with increased aggrecan content and tissue firmness [34]. Clinical results and repair tissue quality using microfracture in combination with PRP injection have demonstrated superior results compared to treatment by microfracture alone [13, 15].

28.5.2 Matrix-Induced Cell Implantation Techniques

There are several treatment options that combine biologic matrices with cell-based surgical procedures, include autologous chondrocyte implantation (ACI), matrix-induced ACI (MACI), autologous matrix-induced chondrogenesis (AMIC), and membrane seeded with expanded MSCs.

ACI has been used predominantly to treat larger chondral lesions in the knee [35]. For the treatment of hip chondral injuries using ACI, clinical studies are limited. Akimau et al. [36] described a case report of a patient with severe chondrolysis and femoral head osteonecrosis after a fracture dislocation. This injury was treated with MACI 21 months after the injury. At 1-year follow-up, the subjective hip score and range of motion improved. At 15 months follow-up, biopsy demonstrated 2 mm thick cartilage repair tissue, well populated with viable cells and integrated with the underlying bone [36]. Later, Fontana et al. described a 30 patient case-control study (15 MACI and 15 debridement alone). At 5 years follow-up, HHS was significantly better in the ACI group compared to the debridement group [37]. In another study, Mancini and Fontana [38] compared clinical outcomes of MACI and AMIC (in 26 and 32 patients, respectively), for acetabular chondral defects. Both procedures showed comparable results however AMIC had the advantage of being a low-cost, single-stage arthroscopic procedure with reduced morbidity compared to two-stage MACI [14, 38].

The use of membranes seeded with expanded MSCs has gained momentum as a treatment option, given some of the disadvantages of MACI such as donor site morbidity and insufficient coverage of the defect area due to some shrinkage effect [20]. Techniques using MSC concentrates have been extensively studied in the knee in association with many types of matrices, including type I/III collagen, hyaluronic acid-based scaffolding, and others [39–41], and have been used in a minimally invasive fashion to treat both chondral and osteochondral defects [42, 43]. Regarding techniques that utilize expanded isolates of MSCs, the matrix-associated stem cell transplantation (MAST) technique involves the culturing of stem cells for 28 days, which are then transferred to a matrix for 1 week with a non-differentiated medium, followed by a chondrogenic medium for 21 days. The seeded matrix implantation is performed arthroscopically. In the experience of our group in 15 patients treated with this technique, clinical improvement was achieved, with improved HHS at 2-year follow-up in all patients.

28.5.3 Mosaicplasty and Osteochondral Autograft Transplantation (OATS)

Mosaicplasty is an autologous osteochondral graft transplantation technique. It is widely used in the knee and, essentially, consists of the transplantation of autologous osteochondral cylindrical grafts harvested from a healthy articular area, to fill osteochondral defects in an affected joint. It begins with the measurement and preparation of the defect area, creating stable and healthy cartilage edges about the lesion and then removing cylindrical bony segments by penetrating the subchondral bone, creating holes within the defect. Osteochondral graft is then harvested from a healthy area (most commonly from the lateral trochlea of the knee, or from the lateral aspect of the involved femoral head), which is then implanted into the previously created holes. In the hip, this technique is used for femoral head osteochondral lesions, and typically requires an open surgical hip dislocation for adequate exposure [28].

Girard et al. [44] treated 10 young patients suffering from femoral head osteochondral defects with an average lesion size of 4.8 cm^2. After 30 months follow-up, the Merle d'Aubigné Postel score and the HHS improved. There was excellent graft incorporation reported and none of the patients required THA. Sotereanos et al. [45] published a case report of a young patient who underwent mosaicplasty to treat femoral head ON, after previously being treated with free fibular grafting. Mosaicplasty was performed using grafts from the inferolateral aspect of the femoral head. The pain score decreased from 90 to 9 (scale from 0 to 100), and pain-free status was maintained at the final postoperative follow-up of 5 years. In another case report, Kocadal et al. [46] described surgery in a 27-year-old male patient who had a symptomatic osteochondral defect of the femoral head treated with arthroscopic-assisted retrograde mosaicplasty, without surgical hip dislocation. At the final follow-up (26 months), the patient had painless full range of motion, and near-complete incorporation of the graft with preservation of the joint space radiographically.

Johnson and the Mayo Clinic group published their results in five patients treated with femoral head osteochondral autograft transfer from the anteroinferior medial and lateral portions of the ipsilateral femoral head [47]. Clinical outcome and radiographic progression of disease were assessed. The mean follow-up was 53.8 months. Four patients reported complete symptom relief and returned to baseline activities. HHS improved from 60.8 to 86.6 and there was no radiographic evidence of progression.

In a recent study, Viamont-Guerra et al. [48] evaluated clinical outcomes of mosaicplasty using ipsilateral femoral head autografts through a minimally invasive anterior approach in 22 cases, using the mHHS and WOMAC scores at a minimum follow-up of 1 year. The mHHS improved from 56.3 to 88.4 and WOMAC scores improved from 45.1 to 80.6. There were two patients (8%) who required subsequent hip arthroscopy to treat FAI due to cam deformity.

28.5.4 Osteochondral Allografts Transplantation

Fresh osteochondral allografts allow for treatment of large defects (>2.5 cm) with a single-stage procedure that would otherwise be difficult to treat using alternative techniques [49]. Additionally, this technique eliminates donor site morbidity, provides an immediately functioning joint surface, and provides a hyaline cartilage replacement [28]. A potential downside concerns the maintained viability of chondrocytes from the time of graft procurement to the time of implantation. This may be impacted by the duration of storage time after graft procurement. Some reports suggest that there is a substantial reduction in graft viability after 28 days of storage [28].

Khanna et al. [50] reported, in a prospective study, the outcomes of 17 patients treated with fresh osteochondral allograft. At the end of follow-up (41.6 months), 13 patients had fair to good outcomes. One patient required a repeat allograft transplantation and three patients required a total hip replacement.

In a recent study, Chen et al. [51] described their experience with fresh osteochondral allograft using the anterior (Smith Peterson) mini-open approach and anterior surgical hip dislocation. They concluded that the use of allograft avoids donor site morbidity and that the anterior approach avoids the need for a trochanteric osteotomy (and its potential complications), and reduces the potential risk of iatrogenic injury to the medial femoral circumflex artery. Furthermore, this treatment method may allow a faster rehabilitation and earlier return to function [51].

Krych et al. [52] reported on management of osteochondral defects of the acetabulum in two patients. The first patient had a superior acetabular cyst and the allograft was taken from a donor acetabulum. The second patient had fibrous dysplasia of the acetabulum and the allograft was taken from a donor tibial plateau. In the first patient, mHHS improved from 75 to 97 at 2-year follow-up. In the second patient, the mHHS improved from 79 to 100 at 3-year follow-up. MRIs (obtained after 1-year in the first patient and at 18-months in the second) showed graft incorporation and hip joint congruity in both patients.

28.5.5 Prosthetic Biocomposites for Osteochondral Defect Repair

Mardones, Mrosek, and the Mayo Clinic group [53, 54] reported the development of a biologic prosthetic composite that contains a porous tantalum (TM) or poly-e-caprolactone (PCL) scaffold combined with an "articular" surface coating of periosteum from 2-month-old rabbits (cultured under chondrogenic conditions), that forms a robust hyaline-like cartilage. Cylindrical osteochondral defects were created on the medial and lateral condyles of 10 rabbits and filled with TM/periosteum or PCL/periosteum biosynthetic composites. The regenerated osteochondral tissue was then analyzed histologically and with blinded evaluation. The mechanical properties of these prosthetic biocomposites were shown to be very similar to a normal osteochondral graft. Most of the regenerates were well integrated with the surrounding bone and showed a partial restoration of the tidemark. A hyaline-like surface was reported, although the cartilage yields were inconsistent.

In a more recent study from the same group [55], results were published using TM with autologous periosteum to reconstitute large osteochondral defects in sheep models. They divided 24 sheep into three groups: (1) trabecular metal/periosteal graft (TMPG), (2) trabecular metal (TM), and (3) empty defect (ED). At 16 weeks postoperatively, histological findings among the three groups were not statistically different. The neo-cartilage yield was lower compared with the contralateral articular cartilage controls. The authors concluded that TM enables excellent bony ingrowth and fast integration. However, combined with autologous periosteum, such a biocomposite failed to promote satisfactory neo-cartilage formation.

28.6 Summary

Management of injuries to the articular cartilage is complex and challenging, particularly in young and active patients. Pathological processes were classically thought to affect only one component of the osteochondral unit; however, due to the structural and functional interaction and biochemical crosstalk, it is now understood that alterations in a single tissue will ultimately affect all components of the unit.

The hip is a complex joint that routinely manages the distribution of substantial forces through a single weight-bearing compartment. Preserving osteochondral tissue in the injured hip is a very challenging objective, and there is no single ideal treatment option. The use of microfracture, ACI, mosaicplasty, osteochondral grafting, and other methods of articular cartilage repair have been described, with variable success rates reported. The literature related to osteochondral repair strategies in the hip is limited to small case series and case reports, and there is lack of control groups and studies with long-term follow-up. Further investigation of these treatment options as they apply to the hip is necessary to provide appropriate clinical recommendations for management of chondral injuries of the hip joint.

References

1. Lepage S, Robson N, Gilmore H, Davis O, Hooper A, St John S, Kamesan V, Gelis P, Carvajal D, Hurtig M, Koch T. Beyond cartilage repair: the role of the osteochondral unit in joint health and disease. Tissue Eng Part B Rev. 2019;25(2):114–25.
2. Pan J, Zhou X, Li W, Novotny JE, Doty SB, Wang L. In situ measurement of transport between subchondral bone and articular cartilage. J Orthop Res. 2009;27:1347.
3. Hunter DJ, Guermazi A, Roemer F, Zhang Y, Neogi T. Structural correlates of pain in joints with osteoarthritis. Osteoarthr Cartil. 2013;21:1170.
4. Bae DK, Yoon KH, Song SJ. Cartilage healing after microfracture in osteoarthritic knees. Arthroscopy. 2006;22:367–74.
5. Warrener AG. Hominin hip biomechanics: changing perspectives. Anat Rec (Hoboken). 2017;300(5):932–45.
6. Daniel M, Iglič A, Kralj-Iglič V. The shape of acetabular cartilage optimizes hip contact stress distribution. J Anat. 2005;207(1):85–91.
7. Wyler A, Bousson V, Bergot C, Polivka M, Leveque E, Vicaut E, Laredo JD. Hyaline cartilage thickness in radiographically normal cadaveric hips: comparison of spiral CT arthrographic and macroscopic measurements. Radiology. 2007;242(2):441–9.
8. Armstrong CG, Gardner DL. Thickness and distribution of human femoral head articular cartilage. Changes with age. Ann Rheum Dis. 1977;36:407–12.
9. Dallich AA, Rath E, Atzmon R, Radparvar JR, Fontana A, Sharfman Z, Amar E. Chondral lesions in the hip: a review of relevant anatomy, imaging and treatment modalities. J Hip Preserv Surg. 2019;6(1):3–15.
10. Camacho D, Mardones R. Femoroacetabular impingement: correlation between acetabular overcoverage zone and cartilage delamination área. Rev Esp Cir Ortop Traumatol. 2013;57(2):111–6.
11. Ilizaliturri V Jr, Byrd JWT, Sampson T, Guanche C, Philippon M, Kelly B, et al. A geographic zone method to describe intra- articular pathology in hip arthroscopy: cadaveric study and preliminary report arthroscopy. Arthroscopy. 2008;24:534–9.
12. Kaya M, Suzuki T, Emori M, Yamashita T. Hip morphology influences the pattern of articular cartilage damage. Knee Surg Sports Traumatol Arthrosc. 2016;24(6):2016–23.
13. Camacho D, Mardones R. Cartilage restoration and use of orthobiologics. Hip Preservation Symposium. Techniques in Orthopaedics. Article in press.

14. Chahla J, LaPrade RF, Mardones R, Huard J, Philippon MJ, Nho S, Mei-Dan O, Pascual-Garrido C. Biological therapies for cartilage lesions in the hip: a new horizon. Orthopedics. 2016;39(4):e715–23.

15. Sherman B, Chahla J, Glowney J, Frank R. The role of orthobiologics in the management of osteoarthritis and focal cartilage defects. Orthopedics. 2019;42(2):66–73.

16. Watterson JR, Esdaile JM. Viscosupplementation: therapeutic mechanisms and clinical potential in osteoarthritis of the knee. J Am Acad Orthop Surg. 2000;8(5):277–84.

17. Migliore A, Massafra U, Bizzi E, et al. Intra-articular injection of hyaluronic acid (MW 1,500-2,000 kDa; HyalOne) in symptomatic osteoarthritis of the hip: a prospective cohort study. Arch Orthop Trauma Surg. 2011;131(12):1677–85.

18. Nathani A, Gold G, Monu U, Hargreaves B, Finlay AK, Rubin EB, Safran MR. Does injection of hyaluronic acid protect against early cartilage injury seen after Marathon running? A randomized controlled trial utilizing high-field magnetic resonance imaging. Am J Sport Med. 2019;47(14):3414–22.

19. Jevsevar DS, Brown GA, Jones DL, et al. American academy of orthopaedic surgeons. The American academy of orthopaedic surgeons evidence-based guideline on: treatment of osteoarthritis of the knee, 2nd edition. J Bone Joint Surg Am. 2013;95(20):1885–6.

20. Mardones R, Camacho D, Marrugo N, Larrain C. Stem cell therapy for hip lesions: clinical applications. Hip arthroscopy and hip joint preservation surgery, 2nd edn; 2020.

21. Milano G, Sanna Passino E, Deriu L, Careddu G, Manunta L, Manunta A, Saccomanno MF, Fabbriciani C. The effect of platelet rich plasma combined with microfractures on the treatment of chondral defects: an experimental study in a sheep model. Osteoarthr Cartil. 2010;18(7):971–80.

22. Sanchez M, Guadilla J, Fiz N, Andia I. Ultrasound-guided platelet-rich plasma injections for the treatment of osteoarthritis of the hip. Rheumatology (Oxford). 2012;51(1):144–50.

23. Jo CH, Lee YG, Shin WH, et al. Intra-articular injection of mesenchymal stem cells for the treatment of osteoarthritis of the knee: a proof-of-concept clinical trial. Stem Cells. 2014;32(5):1254–66.

24. Mardones R, et al. Cell therapy for cartilage defects of the hip. Muscles Ligaments Tendons J. 2016;6(3):361–6.

25. Mardones R, Jofre C, Tobar L, Minguell J. Mesenchymal stem cell therapy in the treatment of hip osteoarthritis. J Hip Preser Surg. 2017;4(2):159–63.

26. Gobbi A, Karnatzikos G, Kumar A. Long-term results after microfracture treatment for full-thickness knee chondral lesions in athletes. Knee Surg Sport Traumatol Arthrosc. 2014 Sep;22(9):1986–96.

27. Gobbi A, Whyte GP. One-stage cartilage repair using a hyaluronic acid-based scaffold with activated bone marrow-derived mesenchymal stem cells compared with microfracture: five-year follow-up. Am J Sports Med. 2016;44(11):2846–54.

28. El Bitar YF, Lindner D, Jackson TJ, Domb BG. Joint-preserving surgical options for management of chondral injuries of the hip. J Am Acad Orthop Surg. 2014;22(1):46–56.

29. Karthikeyan S, Roberts S, Griffin D. Microfracture for acetabular chondral defects in patients with femoroacetabular impingement: results at second-look arthroscopic surgery. Am J Sports Med. 2012;40(12):2725–30.

30. Domb BG, El Bitar YF, Lindner D, Jackson TJ, Stake CE. Arthroscopic hip surgery with a microfracture procedure of the hip: clinical outcomes with two-year follow-up. Hip Int. 2014;24(5):448–56.

31. Philippon MJ, Schenker ML, Briggs KK, Maxwell RB. Can microfracture produce repair tissue in acetabular chondral defects? Arthroscopy. 2008;24(1):46–50.

32. Whyte GP, Gobbi A. Editorial commentary: acetabular cartilage repair: a critically important frontier in hip preservation. Arthroscopy. 2018;34(10):2829–31.

33. Tahoun MF, Tey M, Mas J, Abd-Elsattar Eid T, Monllau JC. Arthroscopic repair of acetabular cartilage lesions by chitosan-based scaffold: clinical evaluation at minimum 2 years follow-up. Arthroscopy. 2018;34(10):2821–8.

34. McIlwraith CW, Frisbie DD, Rodkey WG, et al. Evaluation of intra-articular mesenchymal stem cells to augment healing of microfractured chondral defects. Arthroscopy. 2011;27(11):1552–61.

35. Gobbi A, Chaurasia S, Karnatzikos G, Nakamura N. Matrix-induced autologous chondrocyte implantation versus multipotent stem cells for the treatment of large patellofemoral chondral lesions: a nonrandomized prospective trial. Cartilage. 2015;6(2):82–97.

36. Akimau P, Bhosale A, Harrison P, et al. Autologous chondrocyte implantation with bone grafting for osteochondral defect due to posttraumatic osteonecrosis of the hip – a case report. Acta Orthop. 2006;77(2):333–6.

37. Fontana A, Bistolfi A, Crova M, et al. Arthroscopic treatment of hip chondral defects: Autologous chondrocyte transplantation versus simple debridement – a pilot study. Artroscopy. 2012;28(3):322–9.

38. Mancini D, Fontana A. Five-year results of arthroscopic techniques for the treatment of acetabular chondral lesions in femoroacetabular impingement. Int Orthop. 2014;38(10):2057–64.

39. Gobbi A, Whyte GP. Long-term clinical outcomes of one-stage cartilage repair in the knee with hyaluronic acid–based scaffold embedded with mesenchymal stem cells sourced from bone marrow aspirate concentrate. Am J Sports Med. 2019;47(7):1621–8.

40. Gobbi A, Scotti C, Karnatzikos G, Mudhigere A, Castro M, Peretti GM. One-step surgery with multipotent stem cells and Hyaluronan-based scaffold for the treatment of full-thickness chondral defects of the knee in patients older than 45 years. Knee Surg Sports Traumatol Arthrosc. 2017;25(8):2494–501.

41. Gobbi A, Karnatzikos G, Sankineani SR. One-step surgery with multipotent stem cells for the treatment of large full-thickness chondral defects of the knee. Am J Sports Med. 2014;42(3):648–57.

42. Whyte GP, Gobbi A, Sadlik B. Dry arthroscopic single-stage cartilage repair of the knee using a hyaluronic acid-based scaffold with activated bone marrow-derived mesenchymal stem cells. Arthrosc Tech. 2016;5(4):e913–8.

43. Sadlik B, Gobbi A, Puszkarz M, Klon W, Whyte GP. Biologic inlay osteochondral reconstruction: arthroscopic one-step osteochondral lesion repair in the knee using morselized bone grafting and hyaluronic acid-based scaffold embedded with bone marrow aspirate concentrate. Arthrosc Tech. 2017;6(2):e383–9.

44. Girard J, Roumazeille T, Sakr M, Migaud H. Osteochondral mosaicplasty of the femoral head. Hip Int. 2011;21(5):542–8.

45. Sotereanos NG, DeMeo PJ, Hughes TB, Bargiotas K, Wohlrab D. Autogenous osteochondral transfer in the femoral head after osteonecrosis. Orthopedics. 2008;31(2):177.

46. Kocadal O, Akman B, Güven M, Şayl U. Arthroscopic-assisted retrograde mosaicplasty for an osteochondral defect of the femoral head without performing surgical hip dislocation. SICOT J. 2017;3:41.

47. Johnson JD, Desy NM, Sierra RJ. Ipsilateral femoral head osteochondral transfers for osteochondral defects of the femoral head. J Hip Preserv Surg. 2017;4(3):231–9.

48. Viamont-Guerra MR, Bonin N, May O, Le Viguelloux A, Saffarini M, Laude F. Promising outcomes of hip mosaicplasty by minimally invasive anterior approach using osteochondral autografts from the ipsilateral femoral head. Knee Surg Sports Traumatol Arthrosc. 2020;28(3):767–76.

49. Gobbi A, Scotti C, Lane JG, Peretti GM. Fresh osteochondral allografts in the knee: only a salvage procedure? Ann Transl Med. 2015;3(12):164.

50. Khanna V, Tushinski DM, Drexler M, et al. Cartilage restoration of the hip using fresh osteochondral allograft: resurfacing the potholes. Bone Joint J. 2014;96(11(suppl A)):S11–6.

51. Chen JW, Rosinsky PJ, Shapira J, Maldonado DR, Kyin C, Lall AC, Domb BG. Osteochondral allograft implantation using the smith-peterson (anterior) approach for chondral lesions of the femoral head. Arthrosc Tech. 2020;9(2):e239–45.

52. Krych AJ, Lorich DG, Kelly BT. Treatment of focal osteochondral defects of the acetabulum with osteochondral allograft transplantation. Orthopedics. 2011;34(7):e307–11.

53. Mardones RM, Reinholz GG, Fitzsimmons JS, Zobitz ME, An KN, Lewallen DG, Yaszemski MJ, O'Driscoll SW. Development of a biologic prosthetic composite for cartilage repair. Tissue Eng. 2005;11(9–10):1368–78.

54. Mrosek EH, Schagemann JC, Chung HW, Fitzsimmons JS, Yaszemski MJ, Mardones RM, O'Driscoll SW, Reinholz GG. Porous tantalum and poly-epsilon-caprolactone biocomposites for osteochondral defect repair: preliminary studies in rabbits. J Orthop Res. 2010;28(2):141–8.

55. Mrosek EH, Chung H-W, Fitzsimmons JS, O'Driscoll SW, Reinholz GG, Schagemann JC. Porous tantalum biocomposites for osteochondral defect repair: a follow-up study in a sheep model. Bone Joint J. 2016;5:403–11.

Nontraumatic Hip Osteochondral Pathologies

Graeme P. Whyte, Jordan Fried, Brian D. Giordano, and Thomas Youm

29.1 Introduction

Nontraumatic osteochondral injury to the hip resulting in joint damage and end-stage joint failure is a frequently encountered musculo-skeletal condition and can be associated with a profound reduction in quality of life. Components of the osteochondral unit are closely associated and interdependent [1]. Development of osteoarthritis is a complex pathologic process, and treatments to halt or slow the progression of osteochondral injury are advancing, recognizing that the pathologic processes that impact subchondral bone are similarly important to processes that impact overlying articular cartilage [2].

There are many pathologic conditions associated with progressive degenerative joint injury in the hip. The understanding of various forms of intra- and extra-articular impingement is advancing rapidly, as are surgical methods to treat these conditions. Regarding well-studied forms of hip osteochondral abnormality, such as Legg-Calvé-Perthes disease, slipped capital femoral epiphysis (SCFE), developmental dysplasia of the hip (DDH), and osteonecrosis, there are a number of widely available and understood treatments described, in addition to newly developed therapies that are constantly evolving. Differences in hip morphology that are associated with a number of pathologic processes result in patterns of articular cartilage injury that are specific to the hip disorder [3]. Bony realignment procedures to normalize hip joint alignment and to properly balance joint reaction forces can be combined with intra-articular procedures to repair cartilage or treat other soft tissue injuries in the hip, using cell-based methods of repair or techniques that incorporate biologic scaffolding and mesenchymal stem cells/signaling cells. Complex surgical methods to address malalignment and focal areas of osteochondral injury have been extensively studied in the knee [4], and some of these strategies can be applied to the hip joint. Targeted, individualized, and comprehensive surgical treatments are often required to preserve the native hip joint, to increase the longevity of the functional hip, and to improve quality of life.

G. P. Whyte (✉)
Weill Medical College, Cornell University, New York Presbyterian Hospital/Queens, New York, NY, USA

NYU Langone Orthopedic Hospital, New York University School of Medicine, New York, NY, USA
e-mail: info@whytemd.com

J. Fried · T. Youm
NYU Langone Orthopedic Hospital, New York University School of Medicine, New York, NY, USA
e-mail: Jordan.Fried@nyulangone.org;
Thomas.Youm@nyumc.org

B. D. Giordano
University of Rochester Medical Center, Rochester, NY, USA
e-mail: Brian_Giordano@urmc.rochester.edu

© ISAKOS 2022
A. Gobbi et al. (eds.), *Joint Function Preservation*, https://doi.org/10.1007/978-3-030-82958-2_29

29.2 Diagnostic Imaging of the Hip

29.2.1 Anteroposterior Pelvis Radiograph

Anteroposterior (AP) pelvis plain radiography is routinely used to examine a number of anatomic parameters and features that will assist with diagnosis of nontraumatic conditions affecting the hip joint. This film is taken supine or standing, with the lower extremities internally rotated 15°. Joint space narrowing and periarticular changes consistent with degenerative chondral injury are often readily identifiable. Close examination of the articulating subchondral bone may identify focal osteochondral abnormalities. There should be particular attention paid to alignment and femoral head coverage. Lateral center-edge angle (LCEA) and the acetabular index (AI) should be evaluated on each AP pelvis radiograph. The measurement of LCEA was originally described by Wiberg to evaluate lateral coverage of the femoral head [5], and this can be used to screen for and to diagnose developmental dysplasia of the hip. This measurement is made by determining the angle between a line drawn from the center of the femoral head representing the vertical axis of the pelvis and another line drawn from the femoral head to the lateral border of the sourcil (Fig. 29.1a). A normal LCEA is considered to be greater than 25°. Borderline or mild dysplasia may be considered with a measurement between 20° and 25°, depending on clinical presentation and soft tissue considerations such as excessive ligamentous laxity. The AI is another measurement used to assess coverage and was described by Tönnis [6]. This measurement is made by determining the angle between the transverse pelvic axis and a line connecting the medial and lateral extents of the sourcil. An AI between 0° and 10° is considered within the normal range.

29.2.2 False Profile Radiograph

The false profile radiograph is particularly helpful to examine for posterior degenerative joint

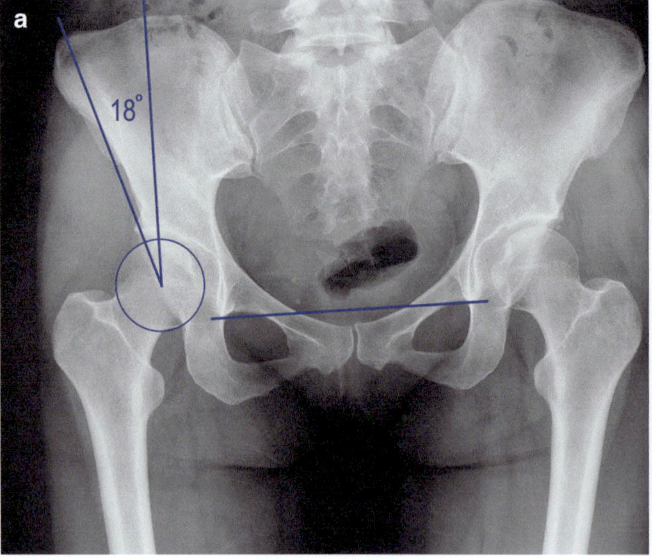
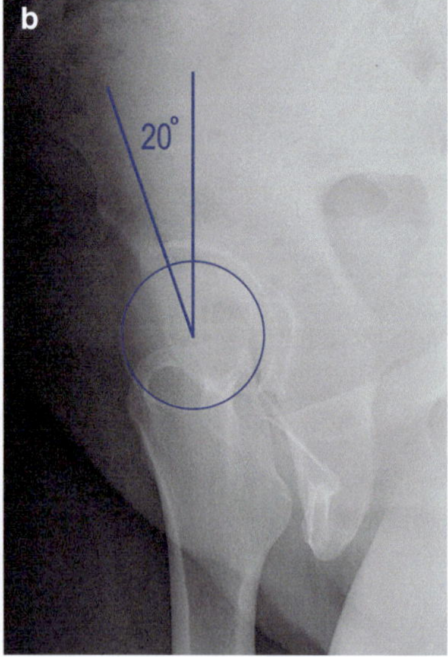

Fig. 29.1 (**a**) Lateral center-edge angle (LCEA) measured on the AP pelvis radiograph for a right hip. The LCEA corresponds to the angle between a line drawn from the center of the femoral head representing the vertical axis of the pelvis and a line drawn from the center of the head to the lateral extent of the condensed acetabular line (sourcil). (**b**) Anterior center-edge angle (ACEA) measured on the false profile radiograph for a right hip. The ACEA corresponds to the angle between a line drawn to represent the vertical axis from the center of the femoral head and a line drawn from the center of the femoral head to the anterior extent of the sourcil

changes and to assess anterior femoral head coverage by measuring the anterior center-edge angle (ACEA). This radiograph is taken with the patient standing and the pelvis rotated 65°. Measurement of the ACEA on the false profile radiograph is made by determining the angle between a vertical line drawn from the center of the femoral head and a second line drawn from the center of the head to the anterior extent of the sourcil (Fig. 29.1b) [7]. Normal anterior coverage as determined by the ACEA is considered to be an angle greater than 25°, with borderline undercoverage in the range of 20°–25°.

29.2.3 Elongated-Neck Lateral (Dunn) Radiograph

The Dunn radiograph, also known as an elongated-neck lateral view, is taken in the AP plane with the affected hip in neutral rotation, abducted 20°, and flexed to either 45° (45° Dunn view) or 90° (90° Dunn view). This radiograph is performed routinely to examine the sphericity of the femoral head-neck junction. There is variability in the

ability of different plain film views to identify femoral head-neck asphericity [8]. Asphericity of the femoral head-neck junction has been shown in some literature to be better identified on the 45° Dunn radiograph [9], as this view will profile the femoral head-neck junction closer to the 1:00–1:30 o'clock region, as opposed to the 90° Dunn view which will usually provide a profile more anteriorly at the femoral head-neck junction. Contour of the femoral head-neck junction is assessed on the Dunn view using the alpha angle. To measure this angle, a circle of best fit is created around the femoral head and a line is drawn along the femoral neck axis through and the center of the femoral head. A second line is then drawn from the center of the head to the point on the femoral head-neck junction that deviates outside the circle of best fit. The angle between these two lines is the alpha angle (Fig. 29.2). There is a wide range over which the degree of cam deformity may contribute to femoroacetabular impingement (FAI), as acetabular or femoral version abnormalities may impact the severity of impingement at any given measurement of alpha angle. There is no definitive consensus as to the specific value of

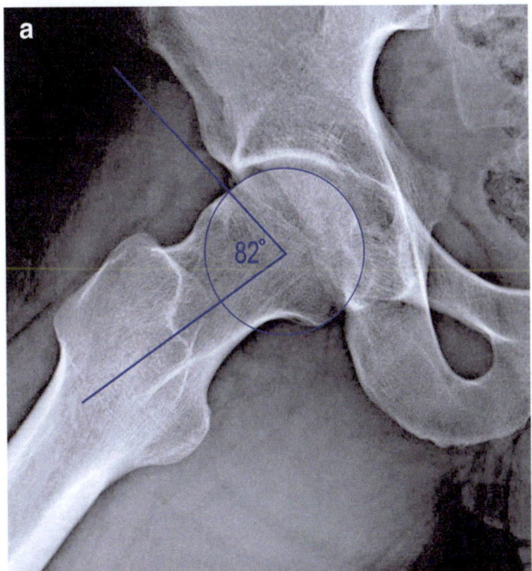
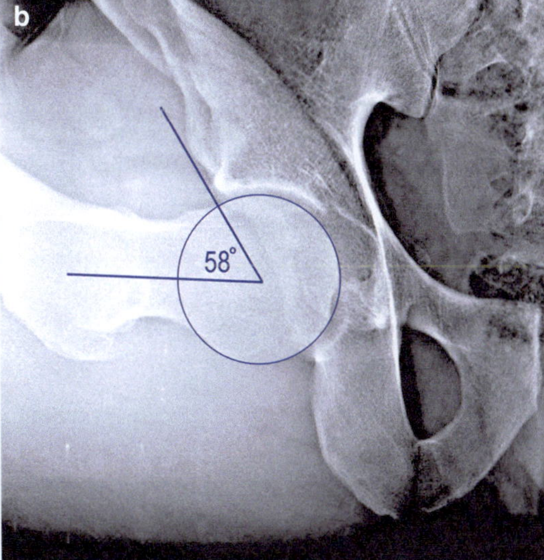
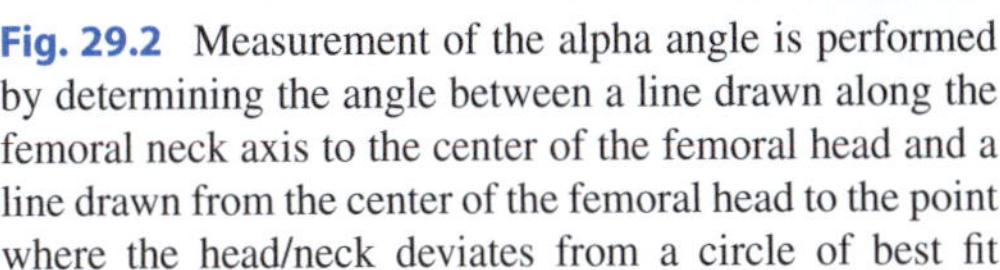

Fig. 29.2 Measurement of the alpha angle is performed by determining the angle between a line drawn along the femoral neck axis to the center of the femoral head and a line drawn from the center of the femoral head to the point where the head/neck deviates from a circle of best fit around the femoral head. This is a measure of femoral head-neck asphericity and is demonstrated on a 45° Dunn (elongated-neck lateral) view (**a**) with the hip abducted 20° and flexed 45°, and on a 90° Dunn view (**b**) with the hip abducted 20° and flexed 90°

alpha angle measurement that defines a cam lesion, although measurement of 60° or greater is often used [10]. Importantly, other anatomic parameters about the hip, such as acetabular or femoral version, may impact the clinical importance of femoral head-neck asphericity of any given degree. For instance, an alpha angle less than 60° when associated with relative femoral retroversion may be associated with more severe FAI compared to a hip with an alpha angle greater than 60° that has increased femoral anteversion.

29.2.4 Cross-Table Lateral and Frog-Leg Lateral Radiographs

The cross-table lateral view is taken with the patient supine and the contralateral limb flexed and elevated. The lower extremity is internally rotated 15° and the X-ray beam is directed at an angle of 45° to the affected hip. For the frog-leg lateral view, the hip is flexed, abducted, and externally rotated such that the heal of the affected extremity is positioned at the contralateral knee, with the patient supine and the X-ray beam directed anterior to posterior. Both the cross-table lateral and frog-leg lateral views are used to assess contour of the femoral head and the femoral head-neck junction anteriorly and posteriorly. With regard to assessing asphericity in cases of cam deformity, the Dunn lateral view is often considered superior for examining the anterolateral head-neck junction [11].

29.2.5 Computed Tomography of the Hip Joint

Computed tomography (CT) is an excellent imaging modality to assess bony alignment about the hip and to detail bony abnormalities associated with osteochondral injury. Determining the extent of subchondral bony involvement and accurately measuring the dimensions of such injury can be performed on CT examination. This is particularly useful for surgical planning when repairing osteochondral lesions. Regarding the assessment of intra- and extra-articular impinge-

ment about the hip joint, CT with three-dimensional reformatting provides many advantages over plain radiography and magnetic resonance imaging (MRI). Areas of asphericity about the femoral head-neck junction can be accurately identified and measured on CT, which is especially valuable in cases of subtle cam deformity that may not be well visualized on plain film, and in cases where the asphericity extends across several segments of the clock face [12]. Measurements of acetabular and femoral versions are most accurate on CT examination. A low-lying prominence at the anterior inferior iliac spine associated with subspine impingement is often difficult to assess on plain radiography and is easily visualized on CT with three-dimensional reformatting. Coverage abnormalities can be assessed using the coronal center-edge angle and sagittal center-edge angle on CT. For diagnosing anterior and lateral undercoverage associated with dysplasia, it must be appreciated that the numerical values of the angle measurements of LCEA and ACEA used to diagnose coverage abnormalities are based on plain X-ray imaging and do not correlate exactly with the numerical values of coverage assessed by center-edge angle measurements on coronal and sagittal CT slices.

29.2.6 Magnetic Resonance Imaging (MRI) of the Hip Joint

MRI is ideally suited to examine for nontraumatic osteochondral pathologies affecting the hip articulation and is used routinely. Gadolinium contrast may be used in conjunction with MRI in order to better visualize a number of pathologic conditions; however, with advances in high-field MRI, many pathologies are examined in detail without the use of contrast. In cases of injury associated with intra-articular impingement, labral injuries are readily assessed with MRI, in addition to focal areas of bony edema that may be identified at locations of impingement. To examine and characterize osteochondral abnormalities, proton density, T2 weighted, and fast spin echo techniques may be used. Recent advances in techniques of delayed gadolinium-enhanced MRI of cartilage (dGEMERIC) and T2 relaxation time mapping are used to evaluate abnormalities of articular cartilage.

29.3 Nontraumatic Osteochondral Pathologic Processes of the Hip Joint

29.3.1 Intra-Articular Impingement

29.3.1.1 Asphericity of the Femoral Head-Neck Junction

Asphericity of the femoral head-neck junction, otherwise known as a "cam deformity" or "cam lesion," is associated with intra-articular impingement that results from insufficient offset of the head-neck junction. This type of impingement most commonly occurs anterolaterally about the hip articulation as the hip flexes and internally rotates, resulting in FAI. This condition is associated with focal labral injury in addition to delaminating chondral injury at the acetabular periphery. As the area of cam deformity repetitively contacts the chondrolabral junction and chondral tissue about the acetabular periphery, labral tearing may occur later in the course of this condition, with chondral tissue sometimes bearing the initial brunt of injury. On the femoral side, the initial chondral injury occurs at the head-neck junction; however, later in the course of disease, there may be involvement of the weight-bearing femoral head. Larger cam lesions will have a greater alpha angle and tend to be associated with higher grade labral and chondral injury at the acetabular periphery anterolaterally.

It has been reported that femoral head-neck asphericity is an acquired deformity that develops in adolescent years prior to physical closure [13]. The development of cam morphology has been associated with particular loading patterns across the hip articulation [14]. Males tend to have greater cam deformity and higher grade intra-articular pathology related to FAI compared to females [15]. Large cam deformity and limitation of internal rotation have been associated with rapid progression of articular cartilage injury and eventual joint failure [16, 17]. Furthermore, increased alpha angles have been shown to affect gait kinematics [18]. It is important to note however that while cam morphology is associated with the development of hip pain, significant cam deformity may be identified incidentally on imaging in those without hip pain, and the true prevalence is not clear [19, 20].

Surgical treatment of cam deformity is associated with superior clinical outcomes when performed prior to the development of significant osteochondral injury. This may be accomplished arthroscopically or with the open technique of surgical hip dislocation as described by Ganz [21]. While repair of labral tear injury and treatment of associated chondral injury is often performed in conjunction with treatment of cam deformity, it is critical that the asphericity at the femoral head-neck junction is adequately recontoured during the surgical procedure in a manner that decompresses the area of impingement (Fig. 29.3). This is confirmed by intra-operative dynamic examination under arthroscopic visualization. Revision surgery to treat FAI is most commonly performed to treat residual cam deformity, regardless of the status of labral repair. Full-thickness chondral defects at the periphery of the acetabulum anterolaterally are not uncommonly identified at the time of arthroscopy, even without a long clinical history of hip pain [22]. Marrow stimulation techniques that disrupt subchondral bony architecture are not the preferable cartilage repair technique, as the goal with articular cartilage repair is to encourage the restoration of durable repair tissue [23, 24]. Although biologic scaffolds with or without augmentation using stem cells and other bioactive factors have been studied more extensively to treat injury in the knee joint [25–28], many of these strategies can be expanded for use in the hip and may be used to treat full-thickness delaminating articular cartilage injury at the acetabular periphery.

29.3.1.2 Acetabular Rim Impingement

Acetabular rim impingement involves what is termed a "pincer lesion." This impingement most often occurs anterolaterally at the acetabular rim

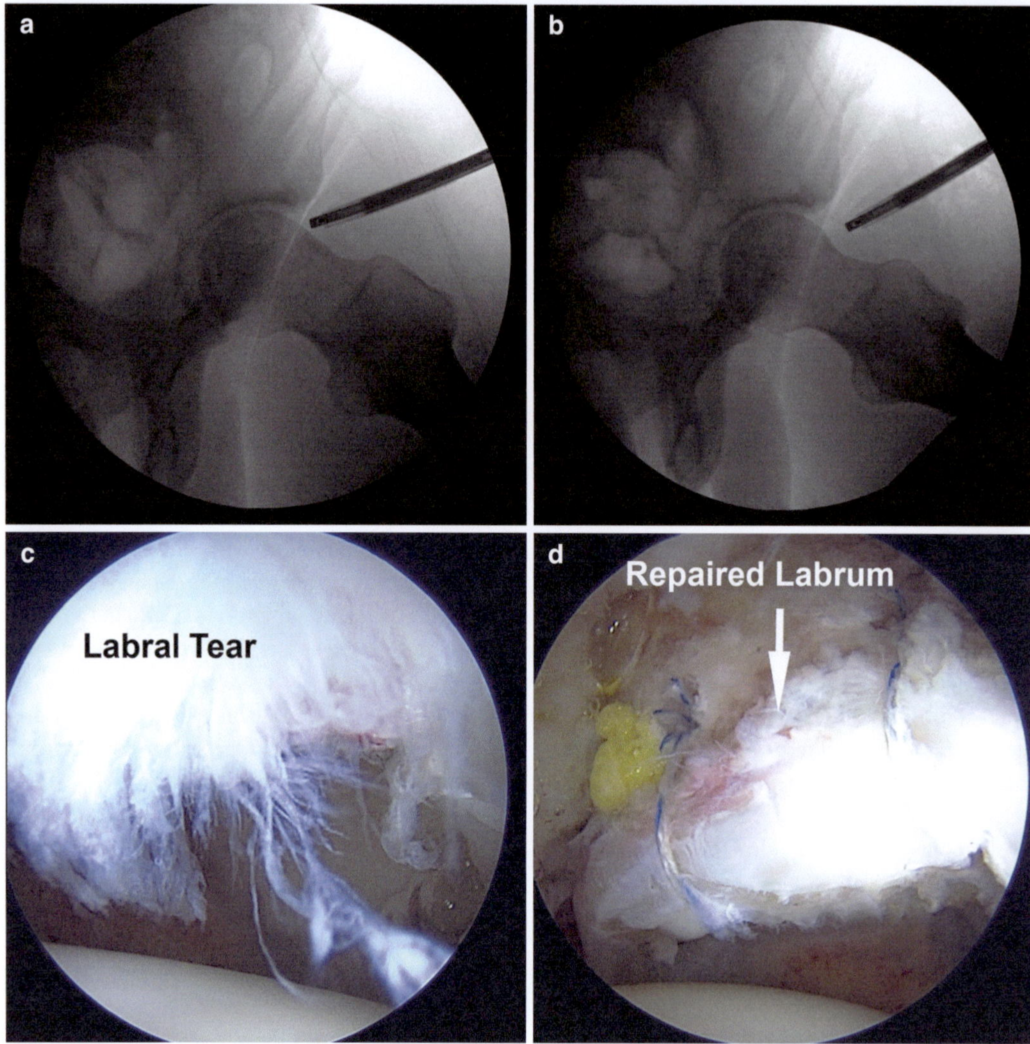

Fig. 29.3 Arthroscopic surgical treatment of femoroacetabular impingement (FAI) depicting increased alpha angle and abnormal femoral head-neck asphericity (**a**) that is corrected by osteochondroplasty to normalize contour and decompress the area of impingement (**b**). Labral tear injury (**c**) associated with the FAI and the subsequent labral repair (**d**) are depicted

and is associated with impingement of the femoral neck at a focal area of the rim. This can occur secondary to a prominent downsloping of the rim or retroversion of the acetabulum. A "cross-over sign" involving the anterior and posterior acetabular walls is indicative of this type of impingement and is a radiographic finding identified on the AP pelvis radiograph. It is important to note that when the "cross-over sign" is seen more proximally on the AP pelvis radiograph, this may not be specifically attributable to acetabular retroversion, as a prominent anterior inferior iliac spine (AIIS) can also be associated with a crossover sign and is suggestive of subspine impingement. In cases of acetabular rim impingement, the injury pattern tends to involve the labrum initially, with chondral injury occurring later in the disease course. Pathology related to both cam and pincer deformities has been reported to lead to osteoarthritis [29, 30].

Surgical treatment of acetabular rim impingement is performed using an arthroscopic or open approach in similar fashion to the treatment of cam deformity. Femoroacetabular impingement may involve "combined" lesions with both a cam and pincer deformity, and these are treated concurrently. Treatment of a pincer lesion involves recontouring the acetabular rim at the focal area of impingement and often repairing labral injury. The clinician must be keenly aware of acetabular coverage measurements, as articulating acetabulum should not be resected if there is concern of acetabular undercoverage of the femoral head. Additionally, subspine impingement can present in similar fashion to acetabular rim impingement, and in such cases there is no need to resect any of the articulating rim if sufficient decompression is accomplished by treating the subspine prominence.

29.3.1.3　Version-Related Impingement

Version abnormalities of the acetabulum and/or femur may be directly associated with FAI or may be a contributing factor to acetabular rim impingement or impingement caused by femoral head-neck asphericity. Retroversion of the femoral neck, or femoral anteversion that is lower in degree than normal, can lead to impingement of the femoral head-neck junction or femoral neck about the acetabular rim, leading to labral or chondral injury in a similar pattern seen in typical FAI when the hip is flexed and internally rotated. Likewise, a retroverted acetabulum, or an acetabulum with anteversion that is reduced compared to normal parameters, may be associated with similar intra-articular impingement. Depending on the degree of version abnormality, injury to labral and chondral tissue may develop, even in the presence of more subtle cam or pincer deformity. Cam or pincer lesions that may be considered minimally clinically significant can be associated with significant FAI and result in osteochondral injury in the setting of acetabular or femoral retroversion, or relative retroversion.

29.3.2　Extra-Articular Impingement

29.3.2.1　Subspine Impingement

The AIIS is a bony tubercle that is the site of attachment for the tendinous origin of the direct head of rectus femoris. This bony tubercle lies extra-articularly, just above the joint capsule at the anterior acetabulum. The morphology of the AIIS varies, and depending on how distal the tubercle extends and protrudes, may be associated with what is termed "subspine" impingement. On the AP pelvis radiograph, a proximal cross-over sign may be mistaken for focal acetabular retroversion in cases of a prominent AIIS. In such cases, the contour of the wall associated with the cross-over sign may be seen to extend more proximally above the rim, suggesting that this radiographic finding may represent the AIIS. Importantly, however, it has been reported that acetabular retroversion is associated with a significantly increased prevalence of subspine impingement [31]. CT imaging with three-dimensional reformatting is ideal in cases where there is suspected extra-articular impingement related to subspine impingement, as CT will clearly delineate the bony anatomy about the acetabulum and AIIS. While this is considered an extra-articular form of impingement, clinical presentation is similar to typical FAI, as the femoral head-neck junction or femoral neck will contact a prominent subspine of the AIIS with deep flexion as well as flexion and internal rotation of the hip, potentially leading to osteochondral injury that worsens over time.

Although the AIIS is an extra-articular structure, subspine impingement is treated similarly to FAI involving intra-articular anatomy. This form of extra-articular impingement is typically addressed arthroscopically by elevating the joint capsule at the anterior acetabulum to expose the impinging prominence of AIIS, identified at the 1:30–2:00 o'clock position, and then decompressing the bony prominence about the rim with a burr. Surgical treatment of subspine impingement is readily performed in conjunction with treatment for FAI with cam or pincer deformity.

29.3.2.2 Trochanteric-Pelvic Impingement

Trochanteric-pelvic impingement is a term used to describe abnormal contact between the greater trochanter and the pelvis that is usually reproduced by abduction and extension of the hip. Developmental abnormalities that affect the relative growth of the greater trochanter and femoral head can lead to a prominent and high-riding greater trochanter, often secondary to disruption of normal blood supply to the femoral head/neck, while the blood supply to the greater trochanter is maintained. Legg-Calvé-Perthes disease is frequently associated with this type of developmental malformation. Varying degrees of such deformity may also be seen in several other pathologies, such as SCFE or osteonecrosis of the femoral head related to other conditions.

Surgical treatment of trochanteric-pelvic impingement is often accomplished in conjunction with treatment of other anatomical abnormalities or joint injuries. When there is associated coxa vara, a valgus-producing osteotomy may substantially alter the position of the greater trochanter to such an extent that the impingement is relieved. When associated with a condition that is treated by the method of surgical hip dislocation, such as cam deformity or full-thickness articular cartilage injury undergoing a cartilage repair procedure, a relative neck lengthening can be performed by advancing the osteotomized portion of the greater trochanter and fixating it more distally.

29.3.2.3 Ischiofemoral Impingement

Ischiofemoral impingement occurs when there is decreased distance between the ischium and the lesser trochanter of the femur, leading to compression and inflammation of the quadratus femoris muscle. The impingement is typically most severe with hip adduction, extension, and external rotation. Pain will be located posteriorly or deep in the groin.

Surgical treatment of ischiofemoral impingement can be accomplished endoscopically or in an open manner. Commonly performed surgical technique involves recontouring of the lesser trochanter using a burr. If resection is substantial enough to impact the stability of the iliopsoas tendon insertion, the tendinous insertion may be secured using suture anchor fixation. The clinician must ensure to avoid disruption of the medial femoral circumflex artery when performing bony resection at the lesser trochanter, as this vessel is in close proximity and there is a risk of avascular necrosis of the femoral head if injured.

29.3.2.4 Iliopsoas Impingement

The iliopsoas tendon inserts at the lesser trochanter and crosses the hip articulation anteriorly at the 3 o'clock position. At the level of the acetabulum, the iliopsoas tendon is positioned directly adjacent to the capsulolabral junction and associated tissues. At this level, tendinous tension and associated focal compression at the capsulolabral complex can result in a repetitive type trauma, potentially resulting in significant labral erythema, swelling, and tear injury. A significantly greater proportion of those affected by iliopsoas impingement are female. While MRI findings may clearly indicate iliopsoas tendinitis and focal injury consistent with iliopsoas impingement, MRI is not always diagnostic. Physical examination findings that are consistent with iliopsoas impingement include pain with active hip flexion, pain with passive hip extension, and clicking anteriorly at the hip that may or may not be painful. Given the anatomic location of inflammation and injury about the hip, provocative tests consistent with a diagnosis of FAI may be positive. Image-guided corticosteroid injection into the iliopsoas tendon sheath has great diagnostic benefit.

Treatment of iliopsoas impingement will initially focus on conservative measures such as activity modification, oral analgesic medication, physical therapy, and image-guided therapeutic injection. In cases of intense pain and functional limitation related to iliopsoas impingement that is refractory to conservative treatment modalities, release or lengthening of the iliopsoas tendon is readily performed arthroscopically through a transcapsular approach. Associated pathology related to iliopsoas impingement, such as labral tear injury, is treated concurrently.

29.3.3 Dysplasia of the Adult Hip

Developmental dysplasia of the hip is most commonly associated with lateral and anterior under-coverage of the femoral head. This condition is frequently identified at the time of skeletal maturity and can lead to early end-stage osteoarthritis and total joint arthroplasty when left untreated [32]. The abnormal alignment of the hip articulation that is seen in hip dysplasia results in the inability of the joint to normally distribute stress and balance reactive forces across the articular surface. Osteochondral injury is initially more focused and then progresses over time to become diffuse, often leading to osteoarthritis and joint failure.

Treatment of hip dysplasia that is associated with joint pain and functional limitation is focused on normalization of femoral head coverage, which can be accomplished surgically in the skeletally mature hip, with minimal complication risk when performed by the experienced surgeon. Ganz et al. reported on a cohort or patients treated with a periacetabular osteotomy to correct coverage in skeletally mature patients suffering from dysplasia, and this technique has been used with increasing frequency since the first description in 1988 [33]. Due to the strong association of advanced osteoarthritis and hip dysplasia [34], early treatment is preferable to avoid or slow progression of osteochondral injury. There may be intra-articular injury such as labral tear or focal osteochondral injury in association with hip dysplasia that can be treated concurrently with periacetabular osteotomy realignment. Labral injury is common in dysplastic hips and may be treated arthroscopically [35] however the clinician must be able to recognize the presence of hip dysplasia when treating any labral injury, as repairing labral injury in a dysplastic hip without addressing the alignment abnormality is not likely to result in a long-term benefit, and may result in clinical worsening if the hip is further destabilized during an arthroscopic procedure [36]. Surgical hip dislocation may be performed in conjunction with periacetabular osteotomy to safely access the intra-articular space in order to repair labral tis-sue and to repair high-grade areas of chondral or osteochondral injury (Fig. 29.4).

29.3.4 Slipped Capital Femoral Epiphysis (SCFE)

Slipped capital femoral epiphysis is a disorder that occurs when there is separation of the femoral epiphysis from the underlying metaphysis. The slip occurs through the hypertrophic zone of the physis. The metaphysis displaces anteriorly and externally rotates, with the epiphysis remaining fixed within the acetabulum. Due to the appearance on plain film, the displacement is sometimes less correctly described as posteromedial displacement of the epiphysis. This condition leads to prominence of the metaphysis anteriorly and anterolaterally, and this alignment is maintained through skeletal maturity. Metabolic and endocrine disorders may be associated with SCFE, particularly when the condition is diagnosed outside of the usual age range. Idiopathic SCFE is the most frequent form and the disorder occurs more commonly in males. An association between SCFE and obesity has been demonstrated widely in the literature. Those aged 10–16 years old are most at risk, with females more commonly affected at ages of 10–14 years, and males more commonly 12–16 years.

SCFE can be categorized as acute (presentation less than 3 weeks), chronic, or acute on chronic. Additionally, categories of stable and unstable SCFE have been described, with unstable defined as cases where the individual is unable to weight-bear, with or without crutches [37]. The surgical treatment most commonly used for SCFE, whether stable or unstable, is single screw fixation. The screw is positioned at the center of the femoral head and perpendicular to the epiphysis. The goal is to stabilize the epiphysis on the metaphysis to prevent further displacement and to induce physeal arrest. With a greater displacement of the epiphysis relative to the metaphysis, screw placement will be more anterior at the femoral neck. Slips with greater displacement can be considered for epiphyseal reduction prior to stabilization, although there is

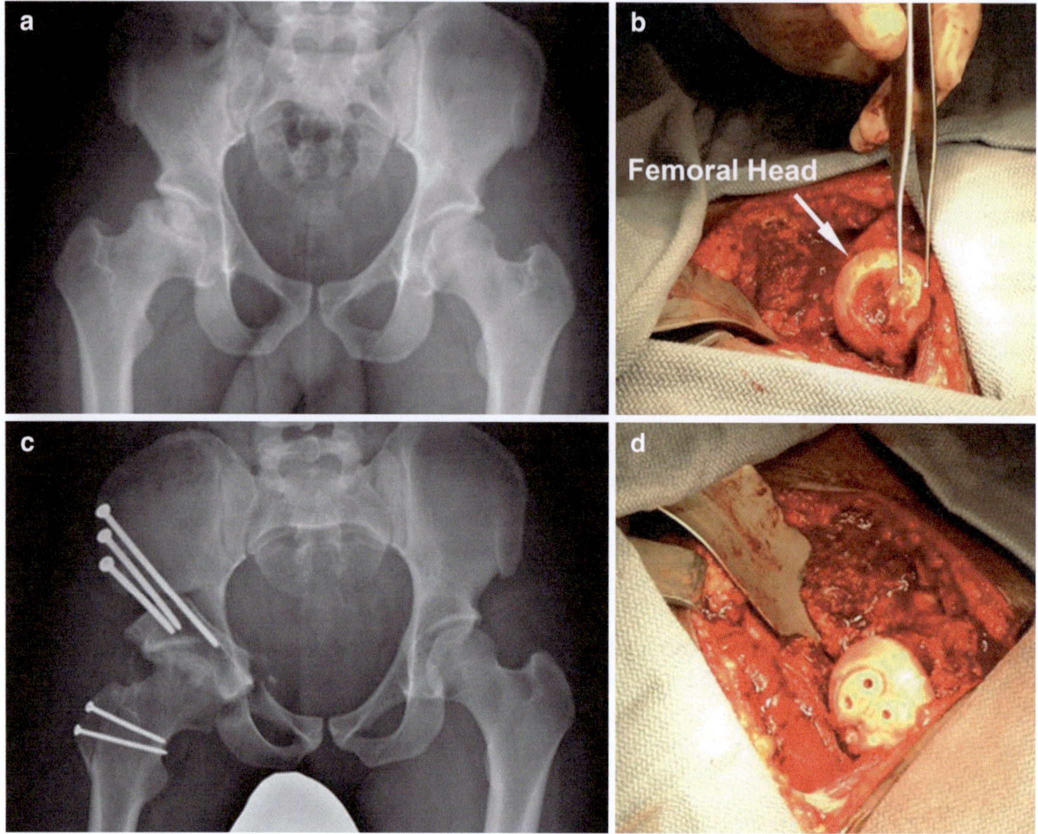

Fig. 29.4 Hip dysplasia in a skeletally mature right hip with a reduced lateral center-edge angle (**a**) in association with focal osteochondral injury to the femoral head (**b**). Normalization of femoral head coverage is accomplished by periacetabular osteotomy (**c**). Bioabsorbable headless compression screw fixation of a large osteochondral lesion of the femoral head using the approach of surgical hip dislocation in the same surgical setting is depicted (**d**)

controversy as to when this is necessary, and this type of surgical treatment must be performed with great expertise to minimize the risk of vascular disruption and avascular necrosis.

After the healed phase, those treated for SCFE are at risk of several forms of hip pathology that may lead to progressive osteochondral injury. Resultant deformity that persists into skeletally mature years can contribute to both extra-articular and intra-articular impingement. Femoroacetabular impingement is frequently encountered in those who have suffered SCFE due to the prominence anteriorly and anterolaterally of the metaphysis. The pathologic process involving SCFE has been shown to be distinct from typical idiopathic cam deformity, and it is not likely that "subclinical" SCFE is associated with a significant number of typical FAI cases [38]. The healed SCFE deformity is often associated with prominent impingement that may lead to early and rapidly progressive labral and chondral injury. Associated femoral retroversion or relative retroversion will contribute to more rapidly developing intra-articular injury. Additionally, the single screw that is used for fixation in SCFE may leave a proud screw head at the femoral neck and can contribute to impingement, particularly in cases where the screw was placed more anterior and proximal at the femoral neck in order to treat an epiphysis that was situated more posteromedially relative to the metaphysis. Arthroscopic surgery or open treatment using surgical hip dislocation can be used successfully, depending on the specific pathology and extent of injury requiring treatment. Mild SCFE deformity that is associated with a more typical FAI-type clinical picture can be readily treated with arthroscopic labral repair and osteochondroplasty. Additionally, the single screw commonly used to treat SCFE may be removed arthroscopically (Fig. 29.5). When needed for adequate exposure, open treatment using surgical hip dislocation will allow for osteochondroplasty, labral repair, treatment of articular cartilage injury, and relative neck lengthening when indicated.

29.3.5 Legg-Calvé-Perthes Disease

Idiopathic avascular necrosis of the femoral head epiphysis in children is known as Legg-Calvé-Perthes disease. This pathologic process involves the interruption of the blood supply (lateral epiphyseal vessels) to the femoral head epiphysis. This more often affects males than females and most commonly affects children of ages 4–8 years, with those less than 6 years old having the best prognosis. Later in the course of disease, blood supply to the epiphysis is reestablished; however, there are varying degrees of resultant anatomic changes affecting the femoral head and neck, such as coxa magna, depending on the age of onset and the degree of developmental deformity. Blood supply to the greater trochanter is uninterrupted during the pathologic process, resulting in varying degrees of coxa brevis and trochanteric prominence. After the healed phase, femoral head deformation and disruption of the normal articulation with the acetabulum may result in significant morbidity related to progressive osteochondral injury and eventual joint failure, in addition to trochanteric-pelvic impingement. Many of those affected by Legg-Calvé-Perthes disease are treated without surgical intervention, particularly when affected at a younger age. After skeletal maturity however surgical treatment is often considered in the young adult due to worsening hip pain and dysfunction related to deformity about the hip articulation.

Surgical treatment in young adult affected by deformity of the femoral head and abnormal articulation of the hip joint secondary to Legg-Calvé-Perthes disease is dependent on the extent of deformity. Proximal femoral osteotomy may be utilized, such as a valgus-producing osteotomy in cases of coxa brevis, and osteotomy at the pelvis may be utilized, such as periacetabular osteotomy to normalize femoral head coverage. Deformity secondary to Legg-Calvé-Perthes disease often results in both intra-articular and extra-articular impingement. Surgical dislocation of the hip can be performed safely to address femoral head-neck asphericity, labral tear, and

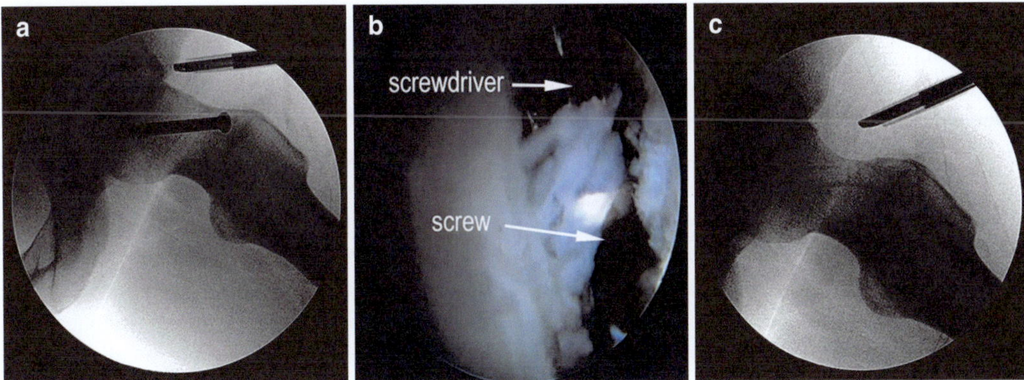

Fig. 29.5 Skeletally mature left hip in an individual suffering from femoroacetabular impingement (FAI) after previously undergoing single screw fixation at a younger age to treat a slipped femoral capital epiphysis (SCFE). (**a**) Retained screw and femoral head-neck asphericity/cam deformity are depicted on intra-operative fluoroscopy. (**b**) Arthroscopic removal of the retained single screw implanted previously to treat SCFE. (**c**) Intra-operative fluoroscopic view depicting screw removal and normalization of the femoral head-neck offset performed arthroscopically by osteochondroplasty

focal chondral or osteochondral injury. Concurrent relative neck lengthening may be used when performing surgical dislocation with trochanteric osteotomy to advance the greater trochanter distally to treat trochanteric-pelvic impingement.

29.3.6 Osteonecrosis of the Femoral Head in the Adult Hip

The primary pathogenic mechanism involved in osteonecrosis of the femoral head in the skeletally mature hip is disruption in vascular supply. Osteonecrosis of the hip is more frequently identified by the fourth decade of life, although it is not uncommon for a diagnosis to be made in the fifth and sixth decades. It is estimated that 8–12% of total hip arthroplasty procedures are performed to treat end-stage osteoarthritis secondary to osteonecrosis [39]. Osteonecrosis may occur secondary to trauma, such as proximal femoral fracture or hip dislocation, due to the disruption in blood supply to the femoral head. Osteonecrosis of the hip secondary to nontraumatic etiology has been associated with a number of predisposing factors. Corticosteroid use is a risk factor, particularly for those undergoing long-term treatment as opposed to shorter courses. Alcohol is thought to affect metabolic processes within bone marrow and is associated with osteonecrosis of the femoral head. Gaucher's disease, hemoglobinopathies such as sickle-cell disease, coagulation disorders, smoking, hyperlipidemia, and autoimmune disorders such as systemic lupus erythematosus have all been described as associated risk factors. While a detailed history is routinely taken to identify potential underlying causes, in many cases of nontraumatic osteonecrosis of the femoral head in the adult the etiology remains unidentified. The Ficat and Arlet classification of osteonecrosis of the femoral head has undergone several modifications and is frequently cited in the literature. A more recent publication by Ficat detailing the early stages, transition phase, and late stages is summarized in Table 29.1 [40].

Plain film imaging and MRI are routinely used to identify and to stage femoral head osteonecro-

Table 29.1 Stages of Osteonecrosis of the Femoral Head Summarized from Ficat (1985) [40]

Stage	Radiographic Findings	Clinical Findings
Early 0	Normal	None
I	Normal or subtle change in trabecular pattern	Pain, ↓ range of motion
II	Osteopenia, cysts, or sclerosis	Signs persist or worsen
Transition	Crescent sign, segmental flattening	
Late III	Sequestrum, change of femoral head contour, collapse, joint space maintained	Worsening pain, dysfunction, and loss of motion
IV	Advanced collapse, flattened contour, loss of joint space	Worsening pain and progressive motion loss

sis. MRI is required to accurately diagnose earlier stages, which has prognostic implications given that greater success is expected when treating the disease during earlier stages. Additionally, transient osteoporosis must be excluded from the list of potential diagnoses. Treatment of femoral head osteonecrosis leads to varying degrees of clinical success and the outcomes primarily depend on the size and location of osteochondral injury. While oral analgesic medication and restriction of weight-bearing are often recommended for symptomatic control, these are not considered therapies that will significantly alter the progression of the disease. Additionally, investigational pharmacologic therapies such as lipid-lowering medications, anticoagulants, and bisphosphonates have not demonstrated the ability to consistently and significantly alter the disease course.

Surgical treatment at the time of end-stage joint failure resulting from the progression of osteochondral pathology is typically total hip arthroplasty. Prior to end-stage disease, however, there are an array of joint-preserving surgical techniques used in earlier stages of osteonecrosis, when the disease remains more localized. Core decompression is a procedure that involves drilling holes into the area of femoral head osteonecrosis, creating pathways for vascular supply and reducing intraosseous pressure. This may be performed by drilling a single larger channel or drilling several smaller channels. Core decompression may be augmented by the application of biologic growth factors to the lesion, such as bone marrow aspirate concentrate or other cellular isolates. These types of biologic isolates provide cell signaling trophic factors and other cytokines that may be beneficial in the reparative cascade [41–43]. Core decompression with biologic augmentation is depicted in Fig. 29.6. Outcomes of core decompression are superior when performed at an earlier stage of osteonecrosis, precollapse. Vascularized or nonvascularized bone grafting procedures are commonly cited treatment options for femoral head osteonecrosis. Osteochondral grafting procedures such as osteochondral allograft transplantation may be used and can result in excellent outcomes in cases of focal osteochondral injury. More recently developed methods involving scaffolds and stem cell/

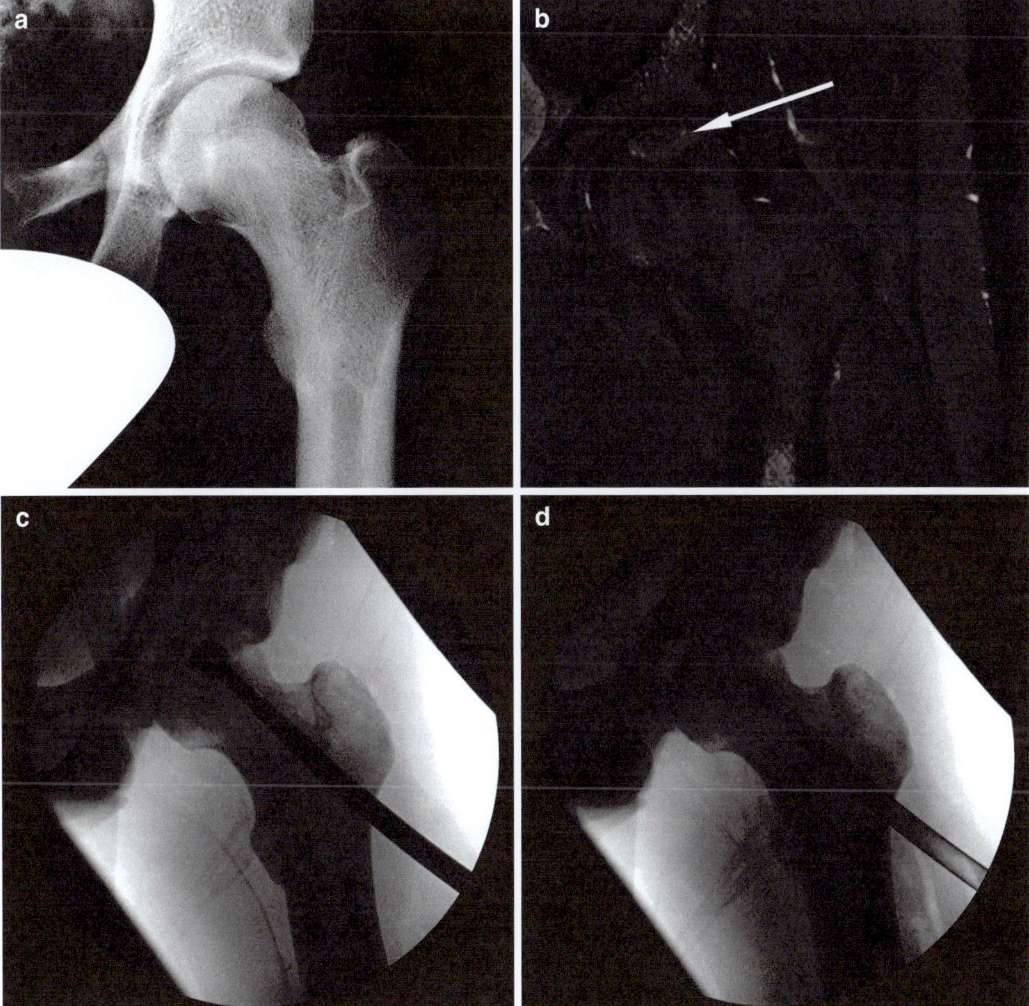

Fig. 29.6 (a) AP plain film of a skeletally mature left hip with an early-stage osteonecrosis lesion. (b) Area of femoral head osteonecrosis is visualized on a coronal slice of proton density-weighted fat-suppressed MRI (arrow). Core decompression is performed under fluoroscopic guidance by advancing an expandable reamer to the femoral head lesion over a guide pin (c). Biologic augmentation of the core decompression is depicted in (d); stem cells sourced from bone marrow aspirate concentrate are combined with demineralized bone matrix and demineralized cancellous bone allograft sponges, which are then implanted into the site of core decompression. Calcium phosphate graft is used distally to retain the biologic augmentation construct within the reamed channel and at the site of osteonecrosis

growth factor biologic augmentation have been described to repair osteochondral lesions in other joints, such as the knee and ankle [44–46], and certain cases of femoral head osteonecrosis may benefit from similar procedures. In cases where there is associated alignment abnormality about the hip that may contribute to increased focal osteochondral stresses and injury, outcomes are best when the alignment is corrected and the forces are shifted away from the affected osteochondral area. This may involve a varus- or valgus-producing proximal femoral osteotomy or periacetabular osteotomy.

29.3.7 Osteochondritis Dissecans (OCD) of the Hip Joint

Osteochondritis dissecans (OCD) is a pathology involving subchondral bone, with varying degrees of bony resorption. Progression of this condition leads to the involvement of overlying articular cartilage and chondral delamination. A fragment of subchondral bone and the associated articular cartilage separates from surrounding osteochondral tissue. OCD is usually considered to be an idiopathic condition that most commonly affects the knee joint [47–49]. OCD rarely affects the hip joint. This condition may affect the femoral head or acetabulum, with acetabular involvement being the rarest form [50]. Excluding those cases reported to be OCD that are associated with other pathologic conditions that alter the anatomy of the femoral head, isolated OCD affecting the femoral head is rare, and like acetabular OCD, is primarily discussed in case reports. OCD is not a well-understood pathology in the hip joint. There is literature to suggest that when OCD develops in the hip, there is sometimes an association with a previous diagnosis of Legg-Calvé-Perthes disease, with the diagnosis of OCD occurring years after the diagnosis of Legg-Calvé-Perthes disease [51]. It is not agreed upon whether OCD occurring subsequent to a diagnosis of Legg-Calvé-Perthes disease is a separate entity that undergoes a different pathologic process than typical idiopathic OCD.

Surgical treatment options for OCD lesions involving the hip joint include the removal of loose bodies and fixation of the lesion. Osteotomies about the hip are also considered to offload the affected area of articular surface, depending on lesion size and location, and the presence of associated malalignment. Arthroscopic and open procedures used to reconstruct or repair focal areas of injured osteochondral tissue in cases of OCD are similar to those used to treat focal osteonecrosis.

29.4 Summary

Nontraumatic osteochondral pathologies affecting the hip joint are associated with great morbidity and reduction in quality of life. These conditions frequently lead to progressive joint injury and, in more advanced cases, end-stage osteoarthritis and joint failure. There is variation in the patterns of osteochondral injury, depending on the hip morphologic characteristics associated with each particular pathologic process. To provide comprehensive hip-preserving surgical treatment of these conditions and to maximize hip joint longevity, several strategies are frequently used concurrently to treat these complex hip disorders. Treatments often involve repairing or reconstructing osteochondral tissue, potentially incorporating more recently developed biologic therapies including scaffolds, stem cells, or cell-based therapies. Additionally, realigning abnormal bony anatomy about the hip articulation to improve the distribution of stresses across the articular surface and to better manage joint reaction forces is often required to optimize outcomes. Surgical treatments can be effective for a wide range of hip disorders associated with nontraumatic osteochondral pathology, including forms of intra-articular impingement, extra-articular impingement, and other conditions that are known to lead to progressive degenerative joint injury and failure, such as osteonecrosis.

References

1. Goldring SR, Goldring MB. Changes in the osteochondral unit during osteoarthritis: structure, function and cartilage–bone crosstalk. Nat Rev Rheumatol. 2016;12(11):632–44. https://doi.org/10.1038/nrrheum.2016.148.

2. Gomoll AH, Madry H, Knutsen G, van Dijk N, Seil R, Brittberg M, et al. The subchondral bone in articular cartilage repair: current problems in the surgical management. Knee Surg Sport Traumatol Arthrosc. 2010;18(4):434–47. https://doi.org/10.1007/s00167-010-1072-x.

3. Kaya M, Suzuki T, Emori M, Yamashita T. Hip morphology influences the pattern of articular cartilage damage. Knee Surg Sports Traumatol Arthrosc. 2016;24(6):2016–23. Available from: http://www.ncbi.nlm.nih.gov/pubmed/25331654

4. Whyte GP, Gobbi A. Biologic knee arthroplasty for cartilage injury and early osteoarthritis. In: Bio-orthopaedics. Berlin: Springer; 2017. p. 517–25. https://doi.org/10.1007/978-3-662-54181-4_41.

5. Wiberg G. Studies on dysplastic acetabula and congenital subluxation of the hip joint. Acta Chir Scand. 1939;83(suppl 58):53–68.

6. Tönnis D. Normal values of the hip joint for the evaluation of X-rays in children and adults. Clin Orthop Relat Res. 1976;119:39–47.

7. Lequesne M, de Seze S. False profile of the pelvis. A new radiographic incidence for the study of the hip. Its use in dysplasias and different coxopathies. Rev Rhum Mal Osteoartic. 1961;28:643–52.

8. Meyer DC, Beck M, Ellis T, Ganz R, Leunig M. Comparison of six radiographic projections to assess femoral head/neck asphericity. Clin Orthop Relat Res. 2006;445(445):181–5. Available from: http://content.wkhealth.com/linkback/openurl?sid=WKPTLP:landingpage&an=00003086-900000000-00060

9. Hipfl C, Titz M, Chiari C, Schöpf V, Kainberger F, Windhager R, et al. Detecting cam-type deformities on plain radiographs: what is the optimal lateral view? Arch Orthop Trauma Surg. 2017;137(12):1699–705.

10. Agricola R, Waarsing JH, Thomas GE, Carr AJ, Reijman M, Bierma-Zeinstra SM, et al. Cam impingement: defining the presence of a cam deformity by the alpha angle: data from the CHECK cohort and Chingford cohort. Osteoarthr Cartil. 2014;22(2):218–25. https://doi.org/10.1016/j.joca.2013.11.007.

11. Konan S, Rayan F, Haddad FS. Is the frog lateral plain radiograph a reliable predictor of the alpha angle in femoroacetabular impingement? J Bone Joint Surg Br. 2010;92(1):47–50.

12. Khan O, Witt J. Evaluation of the magnitude and location of Cam deformity using three dimensional CT analysis. Bone Joint J. 2014;96-B(9):1167–71. Available from: http://www.ncbi.nlm.nih.gov/pubmed/25183585

13. Agricola R, Heijboer MP, Ginai AZ, Roels P, Zadpoor AA, Verhaar JAN, et al. A cam deformity is gradually acquired during skeletal maturation in adolescent and young male soccer players: a prospective study with minimum 2-year follow-up. Am J Sports Med. 2014;42(4):798–806. https://doi.org/10.1177/0363546514524364.

14. Roels P, Agricola R, Oei EH, Weinans H, Campoli G, Zadpoor AA. Mechanical factors explain development of cam-type deformity. Osteoarthr Cartil. 2014;22(12):2074–82. https://doi.org/10.1016/j.joca.2014.09.011.

15. Nepple JJ, Riggs CN, Ross JR, Clohisy JC. Clinical presentation and disease characteristics of femoroacetabular impingement Are sex-dependent. J Bone Joint Surg Am. 2014;96(20):1683–9. Available from: http://www.ncbi.nlm.nih.gov/pubmed/25320194

16. Agricola R, Heijboer MP, Bierma-Zeinstra SMA, Verhaar JAN, Weinans H, Waarsing JH. Cam impingement causes osteoarthritis of the hip: a nationwide prospective cohort study (CHECK). Ann Rheum Dis. 2013;72(6):918–23. Available from: http://www.ncbi.nlm.nih.gov/pubmed/22730371

17. Agricola R, Waarsing JH, Arden NK, Carr AJ, Bierma-Zeinstra SMA, Thomas GE, et al. Cam impingement of the hip: a risk factor for hip osteoarthritis. Nat Rev Rheumatol. 2013;9(10):630–4. https://doi.org/10.1038/nrrheum.2013.114.

18. Lahner M, von Schulze Pellengahr C, Walter PA, Lukas C, Falarzik A, Daniilidis K, et al. Biomechanical and functional indicators in male semiprofessional soccer players with increased hip alpha angles vs. amateur soccer players. BMC Musculoskelet Disord. 2014;15(1):88. Available from: http://www.biomedcentral.com/1471-2474/15/88

19. Khanna V, Caragianis A, Diprimio G, Rakhra K, Beaulé PE. Incidence of hip pain in a prospective cohort of asymptomatic volunteers: is the cam deformity a risk factor for hip pain? Am J Sports Med. 2014;42(4):793–7. Available from: http://www.ncbi.nlm.nih.gov/pubmed/24481825

20. Jung KA, Restrepo C, Hellman M, Abdel Salam H, Morrison W, Parvizi J. The prevalence of cam-type femoroacetabular deformity in asymptomatic adults. J Bone Joint Surg Br. 2011;93-B(10):1303–7.

21. Ganz R, Gill TJ, Gautier E, Ganz K, Krügel N, Berlemann U. Surgical dislocation of the adult hip. J Bone Joint Surg Br. 2001;83-B(8):1119–24. https://doi.org/10.1302/0301-620X.83B8.0831119.

22. Whyte GP, Gobbi A. Editorial commentary: acetabular cartilage repair: a critically important frontier in hip preservation. Arthroscopy. 2018;34(10):2829–31. https://doi.org/10.1016/j.arthro.2018.08.001.

23. Blasiak A, Whyte GP, Matlak A, Brzóska R, Sadlik B. Morphologic properties of cartilage lesions in the knee arthroscopically prepared by the standard curette technique are inferior to lesions prepared by specialized chondrectomy instruments. Am J Sports Med. 2018;46(4):908–14. Available from: http://www.ncbi.nlm.nih.gov/pubmed/29281796

24. Sadlik B, Matlak A, Blasiak A, Klon W, Puszkarz M, Whyte GP. Arthroscopic cartilage lesion preparation in the human cadaveric knee using a curette technique demonstrates clinically relevant histologic variation. Arthroscopy. 2018;34(7):2179–88.

25. Whyte GP, Gobbi A, Sadlik B. Dry arthroscopic single-stage cartilage repair of the knee using a hyaluronic acid-based scaffold with activated bone marrow-derived mesenchymal stem cells. Arthrosc Tech. 2016;5(4):e913–8. Available from: http://www.ncbi.nlm.nih.gov/pubmed/27709058

26. Gobbi A, Whyte GP. One-stage cartilage repair using a hyaluronic acid-based scaffold with activated bone marrow-derived mesenchymal stem cells compared with microfracture: five-year follow-up. Am J Sports Med. 2016;44(11):2846–54. Available from: http://www.ncbi.nlm.nih.gov/pubmed/27474386

27. Whyte GP, McGee A, Jazrawi L, Meislin R. Comparison of collagen graft fixation methods in the porcine knee: implications for matrix-assisted chondrocyte implantation and second-generation autologous chondrocyte implantation. Arthroscopy. 2016;32(5):820–7.

28. Sadlik B, Jaroslawski G, Puszkarz M, Blasiak A, Oldak T, Gladysz D, Whyte G. Cartilage repair in the knee using umbilical cord Wharton's jelly–derived mesenchymal stem cells embedded onto collagen scaffolding and implanted under dry arthroscopy. Arthrosc Tech. 2018;7(1):e57–63.

29. Beck M. Hip morphology influences the pattern of damage to the acetabular cartilage: femoroacetabular impingement as a cause of early osteoarthritis of the hip. J Bone Joint Surg Br. 2005;87-B(7):1012–8. https://doi.org/10.1302/0301-620X.87B7.15203.

30. Johnston TL, Schenker ML, Briggs KK, Philippon MJ. Relationship between offset angle alpha and hip chondral injury in femoroacetabular impingement. Arthroscopy. 2008;24(6):669–75. Available from: http://www.ncbi.nlm.nih.gov/pubmed/18514110

31. Lerch TD, Siegfried M, Schmaranzer F, Leibold CS, Zurmühle CA, Hanke MS, et al. Location of intra- and extra-articular hip impingement is different in patients with pincer-type and mixed-type femoroacetabular impingement due to acetabular retroversion or protrusio acetabuli on 3D CT–based impingement simulation. Am J Sports Med. 2020;48(3):661–72.

32. Murphy SB, Ganz R, Müller ME. The prognosis in untreated dysplasia of the hip. A study of radiographic factors that predict the outcome. J Bone Joint Surg Am. 1995;77(7):985–9.

33. Ganz R, Klaue K, Vinh TS, Mast JW. A new periacetabular osteotomy for the treatment of hip dysplasias. Technique and preliminary results. Clin Orthop Relat Res. 1988;232:26–36.

34. Thomas GER, Palmer AJR, Batra RN, Kiran A, Hart D, Spector T, et al. Subclinical deformities of the hip are significant predictors of radiographic osteoarthritis and joint replacement in women. A 20 year longitudinal cohort study. Osteoarthr Cartil. 2014;22(10):1504–10.

35. Ross JR, Zaltz I, Nepple JJ, Schoenecker PL, Clohisy JC. Arthroscopic disease classification and interventions as an adjunct in the treatment of acetabular dysplasia. Am J Sports Med. 2011;39(1_suppl):72–8. https://doi.org/10.1177/0363546511412320.

36. Mei-Dan O, McConkey MO, Brick M. Catastrophic failure of hip arthroscopy due to iatrogenic instability: can partial division of the ligamentum teres and iliofemoral ligament cause subluxation? Arthroscopy. 2012;28(3):440–5. https://doi.org/10.1016/j.arthro.2011.12.005.

37. Loder RT, Richards BS, Shapiro PS, Reznick LR, Aronson DD. Acute slipped capital femoral epiphysis: the importance of physeal stability. J Bone Joint Surg. 1993;75(8):1134–40. Available from: http://journals.lww.com/00004623-199308000-00002

38. Monazzam S, Bomar JD, Pennock AT. Idiopathic cam morphology is not caused by subclinical slipped capital femoral epiphysis: an MRI and CT study. Orthop J Sport Med. 2013;1(7):2325967113512467. https://doi.org/10.1177/2325967113512467.

39. Hungerford DS, Jones LC. Asymptomatic osteonecrosis: should it be treated? Clin Orthop Relat Res. 2004;429:124–30. Available from: http://journals.lww.com/00003086-200412000-00019

40. Ficat R. Idiopathic bone necrosis of the femoral head. Early diagnosis and treatment. J Bone Joint Surg Br. 1985 Jan;67-B(1):3–9. https://doi.org/10.1302/0301-620X.67B1.3155745.

41. Gobbi A, de Girolamo L, Whyte GP, Sciarretta FV. Clinical applications of adipose tissue-derived stem cells. In: Bio-orthopaedics. Berlin: Springer; 2017. p. 553–9. https://doi.org/10.1007/978-3-662-54181-4_44.

42. Gobbi A, Whyte GP. Emerging orthobiologic approaches to ligament injury. In: Bio-orthopaedics. Berlin: Springer; 2017. p. 313–24. https://doi.org/10.1007/978-3-662-54181-4_24.

43. Whyte GP, Gobbi A, Lane JG. The role of orthobiologics in return to play. In: Return to play in football. Berlin: Springer; 2018. p. 273–82. https://doi.org/10.1007/978-3-662-55713-6_23.

44. Gobbi A, Whyte GP. Long-term clinical outcomes of one-stage cartilage repair in the knee with hyaluronic acid–based scaffold embedded with mesenchymal stem cells sourced from bone marrow aspirate concentrate. Am J Sports Med. 2019;47(7):1621–8.

45. Sadlik B, Kolodziej L, Puszkarz M, Laprus H, Mojzesz M, Whyte GP. Surgical repair of osteochondral lesions of the talus using biologic inlay osteochondral reconstruction: clinical outcomes after treatment using a medial malleolar osteotomy approach compared to an arthroscopically-assisted approach. Foot Ankle Surg. 2019;25(4):449–56.

46. Sadlik B, Gobbi A, Puszkarz M, Klon W, Whyte GP. Biologic inlay osteochondral reconstruction: arthroscopic one-step osteochondral lesion repair in the knee using morselized bone grafting and

hyaluronic acid-based scaffold embedded with bone marrow aspirate concentrate. Arthrosc Tech. 2017;6(2):e383–9.

47. Edmonds EW, Polousky J. A review of knowledge in osteochondritis dissecans: 123 years of minimal evolution from König to the ROCK study group. Clin Orthop Relat Res. 2013;471(4):1118–26.

48. Gobbi A, Whyte GP. Osteochondritis dissecans: pathoanatomy, classification, and advances in biologic surgical treatment. In: Bio-orthopaedics. Berlin: Springer; 2017. p. 489–501. https://doi.org/10.1007/978-3-662-54181-4_39.

49. Gobbi A, Espregueira-Mendes J, Karahan M, Cohen M, Whyte GP. Osteochondritis dissecans of the knee in football players. In: Injuries and health problems in football. Berlin: Springer; 2017. p. 189–200. https://doi.org/10.1007/978-3-662-53924-8_17.

50. Yildirim OS, Okur A, Erman Z. Osteochondritis dissecans of the acetabulum: a case report. Joint Bone Spine. 2004;71(2):160–1. Available from: https://linkinghub.elsevier.com/retrieve/pii/S1297319X03002422

51. Kamhi E, GD ME. Osteochondritis dissecans in Legg-Calvé-Perthes disease. J Bone Jt Surg Am. 1975;57(4):506–9.

Knee Osteochondral Lesions Treatments

Ignacio Dallo and Alberto Gobbi

30.1 Introduction

Articular cartilage has limited intrinsic healing potential attributed to the presence of a few specialized and undifferentiated cells with low mitotic activity, and a lack of vessels that can promote tissue repair. Therefore, once an injury occurs, surgical intervention is necessary to maximize the chances of articular cartilage repair. A good cartilage repair will lead to good long-term functional outcomes and will avoid subsequent cartilage degeneration that could otherwise lead to the development of osteoarthritis (OA) [1]. The etiology of symptomatic knee osteochondral pathology is multifactorial. Causes include a traumatic impact, joint instability events, repetitive microtrauma, chronic overload in the setting of malalignment, obesity, and osteochondritis dissecans (OCD) lesions [2].

Osteochondral lesions are frequently found during knee arthroscopy. Curl et al. [3], in a review of 31,516 knee arthroscopies, found over 53,000 hyaline cartilage lesions in over 19,000 patients. A retrospective study of 5233 knee arthroscopies found that more than half of patients had chondral defects, with 5.2% having Outerbridge grade III or IV lesions [4]. Several surgical procedures for the repair and regeneration of the osteochondral unit have been proposed (Table 30.1). Among them, osteochondral autograft or allograft transplantation (OAT) [5], two-step procedures like autologous chondrocyte implantation (ACI), and matrix-induced autologous chondrocyte implantation (MACI), have been shown to provide good results, promoting the formation of new hyaline-like cartilage tissue [6–8]. Single-step cell-based procedures are an attractive treatment option, given the potential for cost savings and the requirement for the patient to undergo only one surgical procedure instead of two.

Multipotent stem cells sourced from bone marrow aspirate concentrate (BMAC) in combination with a biologic scaffold have demonstrated good to excellent clinical outcomes at long-term follow-up as with ACI [9–11]. Optimization of knee biomechanics through concomitant bony and/or soft tissue procedures will maximize the results when osteochondral treatment is indicated. In this chapter, we describe the current and emerging surgical procedures that can be used to treat osteochondral knee injuries in a wide range of patient age and lesion sizes in the knee joint.

I. Dallo · A. Gobbi (✉)
Orthopaedic Arthroscopic Surgery International (O.A.S.I.) Bioresearch Foundation, Milan, Italy
e-mail: gobbi@cartilagedoctor.it

© ISAKOS 2022
A. Gobbi et al. (eds.), *Joint Function Preservation*, https://doi.org/10.1007/978-3-030-82958-2_30

Table 30.1 Knee osteochondral treatment studies

Author, year	Treatment	N	Age	Follow-up	Results
Gobbi et al. (2019) [11]	HA-BMAC		48	10 years	Good to Excellent Cl
Gomol et al. (2014) [18]	ACI	110	33	4 years	Good to Excellent Cl
Levy et al. (2013) [53]	OC Allograft	129	33	13.5 years	Survivorship 66%
Berruto et al. (2014) [50]	MioRegen	1	39	2 years	Signif. Improv. Cl/MRI
de Windt et al. (2017) [47]	IMPACT	6	30.8	1.5 years	Signif. Improv. Cl/MRI
Shimomura et al. (2018) [51]	TEC	1	35	2 years	Signif. Improv. Cl/MRI

30.2 Osteochondral Autograft and Allograft Transplantation

Osteochondral autograft or allograft transplantation may be used to reconstitute the pathologic osteochondral unit in cases of osteochondral lesions. This treatment modality has the advantage of restoring the hyaline cartilage, while at the same time providing osseous support that is biomechanically strong and is capable of integrating into the native surrounding subchondral bone.

Osteochondral autograft transplantation (OAT) is typically used for lesions smaller than 2 cm^2. It is a single-stage procedure which may be performed arthroscopically or in open fashion. Cylindrical plugs are harvested from donor sites localized to non-articulating regions from within the knee joint such as the intercondylar notch or the superior medial/lateral femoral condyle. Single or multiple osteochondral plugs, of varying sizes up to 10 mm in diameter, are harvested and then transferred to the affected area. A notable disadvantage of this procedure is the technical difficulty of recreating the anatomic radius of curvature of the articular surface, as any disruption in articular congruity may lead to increased contact pressures and focused shear forces. Wu et al. demonstrated that osteochondral plugs within the knee that are 1 mm prominent result in significantly increased contact pressures, while plugs recessed 0.25 mm decrease these pressures by 50% [12]. Furthermore, treatment using the OAT technique is limited by the availability of autologous tissue, as donor site morbidity is an essential concern if multiple grafts are used.

Fresh osteochondral allograft transplantation has been studied extensively in the treatment of osteochondral injury and may be used to treat lesions greater than 2 cm^2 in surface area. This procedure is a reasonable option in both primary and revision surgery to treat osteochondral knee injuries [13]. The advantages of using allograft include the flexibility of graft sizing and the ability to use a single transplanted plug to treat the entire lesion, without the worry of donor site morbidity. Some disadvantages include reduced chondrocyte viability as a result of storage and processing, potential immunogenicity concerns, and disease transmission. Moreover, as with osteochondral autograft transfers, there may be difficulty matching the radius of curvature of the graft to the native articular surface, and prominence or recession of the graft leads to a suboptimal distribution of contact forces.

30.3 Autologous Cell-Based Scaffolds

30.3.1 Two-Stage Procedure: ACI-MACI

Autologous cell-based repair methods such as autologous chondrocyte implantation (ACI) are considered a preferred method to repair extensive osteochondral injuries. Concerning the knee joint, osteochondral lesions larger than $2–3 \text{ cm}^2$ should be considered for treatment by these methods. Second- or third-generation ACI techniques, with or without subchondral bone grafting, are currently preferentially used to treat a range of osteochondral pathology.

Histologic studies have shown that ACI results in hyaline-like Type II collagen [14]. While femoral lesions had acceptable outcomes, the patients with patellar injuries showed poor results [15].

However, when malalignment and other comorbidities are corrected concurrently or in a staged fashion, findings by the different authors were shown to be similar to femoral lesions of the knee [16–18].

Nonetheless, the apparent complexity of this technique, needing the sacrifice of periosteal tissue, the uncertain distribution of chondrocyte solution, and complications such as periosteal patch hypertrophy and arthrofibrosis prompted the scientific community to develop second-generation ACI [7, 19].

Studies of matrix-induced autologous chondrocyte implantation (MACI) for knee osteochondral lesions also show good results when appropriate concomitant procedures are performed [20]. Filardo et al. have reported the outcome's differences between patellar and trochlear injuries when treated by MACI [21]. Some studies have shown better results with MACI compared to ACI [22, 23].

At our institution, treatment of osteochondral injuries using MACI has been accomplished using a bioengineered scaffold that is entirely structured on the benzylic ester of hyaluronic acid (Hyalofast, Anika Therapeutics, Srl, Abano Terme, Italy). This material is composed of a network of fibers, 20 μm thick, with interstices of variable size. This scaffold has been shown to provide excellent physical support to enable cell-cell contact, cluster formation, and extracellular matrix deposition.

Patients with knee osteochondral lesions treated with this MACI technique have significant improvements in clinical outcome, as measured by objective and subjective assessment instruments at medium-term follow-up. Furthermore, second-look arthroscopy and MRI examination of available cases at our institution have typically demonstrated high-quality osteochondral repair tissue [7, 24].

Usually, where the depth of bony involvement is more than 8 mm, strong consideration should be given to bone grafting, which will better restore the anatomic radius of curvature of the articular surface.

However, primarily it remains a two-step procedure including an arthroscopic biopsy and subsequent implantation of the cultured chondrocytes. Apart from donor site morbidity, the risks of two surgical procedures, and the limited quantity of cartilage that could be harvested, the total cost of surgeries, scaffold, and in vitro culture still represent the major limitation of this technique [25, 26].

30.3.2 One-Stage Procedure: HA-BMAC and BIOR

The evolution of cartilage repair technique leads to the development of new scaffolds that allowed cell proliferation but did not avoid the chondrocyte harvest and cultivation.

Performing a one-step procedure avoids the two-step surgical procedures and reduces the costs of the operation by approximately five times.

Bone marrow aspirate concentrate (BMAC) contains bone marrow stem cells (BMSCs) and growth factors that are a promising option for cartilage repair and regeneration because of their differentiation potential to cartilage [27–30]. Bone marrow-derived stem cells (BMSCs) interact with a nonwoven scaffold, the HYAFF 11, that supports cellular adhesion, migration, and proliferation, promoting the synthesis of extracellular matrix components under static culture conditions [31–33]. Nejadnik et al. compared the clinical outcomes of patients treated with first-generation ACI and patients treated with autologous BMSCs and he concluded that BMSCs are as effective as chondrocytes for articular cartilage repair [34]. In our institution, we compared patients treated with matrix-induced autologous chondrocyte implantation (MACI) with patients treated with BMSCs combined with the same scaffold. We did not notice, at 3 years follow-up, any significant statistical differences between the two groups, concluding that these techniques were viable and effective [24]. It has been shown in many clinical studies that the hyaluronic acid-based scaffold with activated bone marrow aspirate concentrate (HA-BMAC) technique is a valuable method for the treatment of full-thickness cartilage lesions of the knee [35]. Different sizes of osteochondral lesions can be treated, from small injuries to significant defects up to 22 cm² showing good clinical outcomes at long-term follow-up [11, 36]. The HA-BMAC technique has proven to be effective in treatment for patients over 45 years of age [10].

30.4 The Procedure

The entire procedure is performed under general anesthesia. The patient is positioned supine for standard knee arthroscopy. The ipsilateral iliac crest is prepared and exposed for bone marrow aspiration. Examination of the knee under anesthesia is done to recognize the concomitant pathologies that will be addressed during the surgery. All cartilage lesions are then identified during diagnostic arthroscopy. At the time of the procedure, it is necessary to choose whether the procedure will be performed arthroscopically or via an arthrotomy. Arthroscopic intervention is only possible if the lesion can be fully visualized with the arthroscope and reached with instruments. If not, the procedure should be continued through an arthrotomy. Thorough debridement of the loose chondral tissue is necessary, ensuring that the border of the lesion is vertical to the subchondral plane. The calcified cartilage layer overlying the subchondral bone is removed. Care must be taken not to violate the subchondral plate. BMAC preparation is started after the lesion is prepared. Approximately 60 mL of bone marrow from the ipsilateral iliac crest is harvested, using a dedicated aspiration kit. The aspirate is centrifuged with a commercially available system to obtain the concentrated bone marrow (Angel, Arthrex, Cytomedix, Gaithersburg, MD). The dimensions of the lesion have to be measured to prepare the matching implant using a three-dimensional hyaluronic acid-based scaffold (Hyalofast, Anika Therapeutics, Bedford MA USA Srl, Abano Terme, Italy). It is also possible to prepare an aluminum foil template of the lesion, and then cut the scaffold to correspond to the contour of the aluminum foil model. When the scaffold is ready, BMAC is activated with batroxobin enzyme (Plateltex Act, Plateltex SRO, Bratislava, Slovakia). The activation process is necessary for BMAC to form a clot, which is then applied onto the prepared scaffold forming a sticky implant that is easy to apply to the lesion.

According to the chosen approach, previously prepared HA-BMAC is then implanted into the lesion. If an open technique is selected, the surgeon should apply HA-BMAC directly onto the defect. If needed, fibrin glue is added to secure the graft further. The knee is then flexed and extended to check graft stability. If the surgeon chooses an arthroscopic approach, fluid needs to be completely drained, and the lesion should be inspected arthroscopically after fluid drainage to ensure that the circumferential border is stable. The scaffold is introduced into the joint via the working portal through a valveless cannula using a grasper. The implant is placed gently filling the cartilage defect. A hook can be used to press-fit the scaffold into the lesion. The crucial part of the procedure is to check the implant stability. The joint is moved through a range of motion several times while the scaffold is observed with the arthroscope. If needed, fibrin glue is applied to improve implant stability. The working portals are sutured, but a drain should not be inserted into the joint [26, 36].

A recently described technique by Sadlik et al. [37] to repair osteochondral injury using morselized bone grafting and mesenchymal stem cells sourced from bone marrow aspirate has been termed Biologic Inlay Osteochondral Reconstruction (BIOR). This technique uses a hyaluronic acid-based scaffold embedded with BMAC in association with a malleable bone graft inlay. Although only preliminary clinical outcome data is currently available for osteochondral pathology treated with BIOR, this type of cell-based, single-stage reconstruction procedure is expected to become a preferred method of surgical treatment, given the cost-effective nature and technical versatility of the technique.

30.5 Emerging Osteochondral Treatments

30.5.1 Allograft Cell-Based Scaffolds

DeNovo NT Natural Tissue Graft (Zimmer Biomet) is an off-the-shelf human tissue allograft, consisting of juvenile hyaline cartilage pieces with viable chondrocytes with promising preliminary outcomes. The immature chondrocytes have been shown to have increased metabolic and proliferative activity when compared to adult chondrocytes and are intended for the repair of articular cartilage lesions in a one-step procedure [38]. One study showed significant improvements in MRI scores and clinical outcomes at over 2 years follow-up [39]. A more recent study showed that lesion fill at 6, 12, and 24 months was 82%, 85%, and 75%, respectively [40]. A clinical study in patellofemoral cartilage lesions showed significant improvements in KOOS scores at 8 months [41]. A prospective trial showed improvements in radiographic appearance, histology, and clinical scores at 2 years follow-up [42]. However, randomized controlled trial and long-term data are needed.

Cartiform (Osiris Therapeutics, Inc.) is a cryopreserved viable chondral allograft. It is currently available in a 2 cm^2 size to treat smaller lesions. The main advantage is that the cells remain viable, up to 70%, at 2 years [43]. This can facilitate surgical planning and elective scheduling, rather than waiting for a fresh, stored graft traditionally. There are very few clinical data of this technique; only one case report is published [44].

BioCartilage (Arthrex, Naples, FL) is a new product containing dehydrated micronized allogeneic cartilage scaffold implanted with platelet-rich plasma and fibrin glue added over a contained microfracture-treated defect can be used in the knee for small lesions. There are limited clinical studies of short- or long-term outcomes [45, 46].

IMPACT (D. Saris's Team, Utrecht) is another promising one-step, cell-based cartilage regeneration technique [47]. With this procedure, the authors prepare the cartilage defect in the standard fashion. The cartilage is then partially digested to separate the chondrons and the extracellular matrix. These are then combined with a precise ratio of allogeneic MSCs. The findings of the first study in 35 patients demonstrate good short-term clinical, MRI, and histological results. The authors conclude that allogeneic MSCs can be a safe cell source to augment or facilitate tissue regeneration through paracrine mechanisms and cellular communication in a clinical setting [48].

30.5.2 Cell-Free Scaffolds

Cell-free scaffolds have been developed with the aim of promoting and inducing tissue regeneration. To date, there are few clinical studies on knee osteochondral lesions. MaioRegen (Fin-Ceramica Faenza SpA, Faenza, Italy) is a nanostructured three-layer biomimetic scaffold with a porous composite structure. The device mimics the entire osteochondral structure with a cartilaginous type I collagen-based layer with a smooth surface, an intermediate tidemark-like layer consisting of a combination of type I collagen and hydroxyapatite, and a bottom layer composed of a mineralized blend of type I collagen and hydroxyapatite. Clinical studies using the MioRegen scaffold for osteochondral lesions have shown good clinical and MRI outcomes as reported by. Kon et al. [49] at 5 years follow-up and Berruto et al. [50] at 2 years in a cohort of 49 knees.

30.5.3 Scaffold-Free Tissue-Engineered Construct

The Tissue-Engineered Construct (TEC) technique is an autologous three-dimensional (3D) biologic structure made through simple-cell culture methods of synovial MSCs. The TEC contains an extracellular matrix, synthesized by the cells, composed of fibrillar collagen (type I–III). The construct is pliable and highly adherent to healthy cartilage because of the adhesion molecules, such as fibronectin and vitronectin, that are present in it. Shimomura et al. [51] published the first in-human, "early proof of concept," trial in five patients. They reported positive clinical and morphologic outcomes across all patients involved, without any significant adverse events at 2 years follow-up.

30.6 Concomitant Surgical Procedures

Irrespective of osteochondral repair technique, particular attention should be paid to bony malalignment, even in cases of subtle deformity. Ensuring the involved compartment is sufficiently off-loaded will provide the optimal environment for cartilage repair tissue to remodel and mature. Realignment procedures that are often required in cases of extensive cartilage injury include distal femoral osteotomy, high tibial osteotomy (HTO), and tibial tubercle osteotomy. Depending on the deformity, these procedures may be performed independently or in combination. Care should be taken to avoid shifting the loading forces to compartments with significant cartilage injury, mainly if reliable cartilage repair treatments are not used.

Combining advanced cartilage repair techniques with corrective osteotomy has demonstrated impressive improvements in clinical outcomes compared to previously published series, where little attention was paid to cartilage restoration. In cases of patellofemoral maltracking treated with tibial tubercle osteotomy alone, poor outcomes have been demonstrated in cases with associated cartilage injury [52]. In contrast to this, several centers, including ours, have demonstrated that successful outcomes may be achieved with tibial tubercle osteotomy in cases of extensive patellofemoral cartilage injury, if appropriate cartilage repair treatment is performed [6, 18, 24, 35]. This highlights the potential for substantial clinical improvement when cartilage defects are addressed with tissue repair techniques, in addition correcting bony malalignment and optimizing the loading forces across knee compartments. While corrective osteotomies are traditionally used with caution in cases of cartilage lesions affecting multiple compartments, improvements in cell-based cartilage repair techniques have the potential to enable successful treatment of articular injury in cases that may have previously been contraindicated for joint-preserving procedures, such as those considered to be early osteoarthritis.

30.7 Conclusion

Given the variety of lesion types and demographic factors that influence prognosis, no single technique is considered the preferred treatment. An individualized approach based on the patient's goals and the surgeon's preferences is crucial. Pathological background factors such as malalignment, meniscus deficiency, or ligament laxity have to be addressed to provide an optimal environment for cartilage repair.

Single-step cartilage repair eliminates the need for a two-step procedure thereby reducing the cost and morbidity to the patient.

HA-BMAC is a safe and accessible procedure that provides good to excellent clinical outcomes at long-term follow-up in small or large lesions, single or multiple injuries, and various compartments.

Recent treatment developments that employ biomaterials and cell-based therapy have demonstrated encouraging medium-term results.

We need future studies on cartilage repair based on biological and imaging biomarkers testing the inflammatory and degenerative environment of a joint to better estimate survival and success (Boxes 1 and 2).

> **Box 1 Surgical pearls of HA-BMAC procedure**
> - Complete exposure of the patellofemoral cartilage lesion is critical.
> - Use traction methods as needed to provide a comfortable working space.
> - An aluminum foil template is used to measure the prepared cartilage defect.
> - Place the HA-based scaffold against the subchondral bone on either side.
> - Fibrin glue is applied to improve implant stability.
> - All arthroscopic techniques can be performed only in cases where the entirety of the defect is appreciated.
> - Cycle the knee under arthroscopic visualization to confirm the graft seating within de defect.
> - A drain should not be inserted into the joint.

Box 2 Principal contraindications of HA-BMAC procedure

- Patients older (>60 years).
- Obese (BMI > 30).
- Severe tricompartimental OA (ICRS grade 4).
- Patients with untreated malalignment (varus/valgus >5°) or knee instability.
- Multiple intra-articular steroid injections.
- Hip disorders leading to abnormal gait.
- Rheumatic diseases, Bechterew syndrome, and Chondrocalcinosis, and gout.

References

1. Mankin HJ. The response of articular cartilage to mechanical injury. J Bone Joint Surg Am. 1982;64(3):460–6.
2. Sherman SL, Oladeji LO, Welsh J, DiPaolo ZJ. Management of patellofemoral instability in the setting of multiligament knee injury. Ann Joint 2018;3:100. https://doi.org/10.21037/aoj.2018.11.10.
3. Curl WW, et al. Cartilage injuries: a review of 31,516 knee arthroscopies. Arthroscopy. 1997;13(4):456–60.
4. Widuchowski W, Widuchowski J, Trzaska T. Articular cartilage defects: study of 25,124 knee arthroscopies. Knee. 2007;14(3):177–82.
5. Chahla J, et al. Osteochondral allograft transplantation in the patellofemoral joint: a systematic review. Am J Sports Med. 2019;47(12):3009–18.
6. Gobbi A, et al. Patellofemoral full-thickness chondral defects treated with Hyalograft-C: a clinical, arthroscopic, and histologic review. Am J Sports Med. 2006;34(11):1763–73.
7. Kon E, et al. Arthroscopic second-generation autologous chondrocyte implantation compared with microfracture for chondral lesions of the knee: prospective nonrandomized study at 5 years. Am J Sports Med. 2009;37(1):33–41.
8. Battaglia M, et al. Validity of T2 mapping in characterization of the regeneration tissue by bone marrow derived cell transplantation in osteochondral lesions of the ankle. Eur J Radiol. 2011;80(2):e132–9.
9. Gobbi A, Whyte GP. One-stage cartilage repair using a hyaluronic acid-based scaffold with activated bone marrow-derived mesenchymal stem cells compared with microfracture: five-year follow-up. Am J Sports Med. 2016;44(11):2846–54.
10. Gobbi A, et al. One-step surgery with multipotent stem cells and Hyaluronan-based scaffold for the treatment of full-thickness chondral defects of the knee in patients older than 45 years. Knee Surg Sports Traumatol Arthrosc. 2017;25(8):2494–501.
11. Gobbi A, Whyte GP. Long-term clinical outcomes of one-stage cartilage repair in the knee with hyaluronic acid-based scaffold embedded with mesenchymal stem cells sourced from bone marrow aspirate concentrate. Am J Sports Med. 2019;47(7):1621–8.
12. Wu JZ, Herzog W, Hasler EM. Inadequate placement of osteochondral plugs may induce abnormal stress-strain distributions in articular cartilage --finite element simulations. Med Eng Phys. 2002;24(2):85–97.
13. Murphy RT, Pennock AT, Bugbee WD. Osteochondral allograft transplantation of the knee in the pediatric and adolescent population. Am J Sports Med. 2014;42(3):635–40.
14. Brittberg M, et al. Rabbit articular cartilage defects treated with autologous cultured chondrocytes. Clin Orthop Relat Res. 1996;326:270–83.
15. Brittberg M, et al. Treatment of deep cartilage defects in the knee with autologous chondrocyte transplantation. N Engl J Med. 1994;331(14):889–95.
16. Vasiliadis HS, et al. Malalignment and cartilage lesions in the patellofemoral joint treated with autologous chondrocyte implantation. Knee Surg Sports Traumatol Arthrosc. 2011;19(3):452–7.
17. Gillogly SD, Arnold RM. Autologous chondrocyte implantation and anteromedialization for isolated patellar articular cartilage lesions: 5- to 11-year follow-up. Am J Sports Med. 2014;42(4):912–20.
18. Gomoll AH, et al. Autologous chondrocyte implantation in the patella: a multicenter experience. Am J Sports Med. 2014;42(5):1074–81.
19. Peterson L, et al. Autologous chondrocyte transplantation. Biomechanics and long-term durability. Am J Sports Med. 2002;30(1):2–12.
20. Ebert JR, et al. A comparison of 2-year outcomes in patients undergoing tibiofemoral or patellofemoral matrix-induced autologous chondrocyte implantation. Am J Sports Med. 2017;45(14):3243–53.
21. Filardo G, et al. Treatment of "patellofemoral" cartilage lesions with matrix-assisted autologous chondrocyte transplantation: a comparison of patellar and trochlear lesions. Am J Sports Med. 2014;42(3):626–34.
22. Macmull S, et al. The role of autologous chondrocyte implantation in the treatment of symptomatic chondromalacia patellae. Int Orthop. 2012;36(7):1371–7.
23. Nawaz SZ, et al. Autologous chondrocyte implantation in the knee: mid-term to long-term results. J Bone Joint Surg Am. 2014;96(10):824–30.
24. Gobbi A, et al. Matrix-induced autologous chondrocyte implantation versus multipotent stem cells for the treatment of large patellofemoral chondral lesions: a nonrandomized prospective trial. Cartilage. 2015;6(2):82–97.
25. Henderson I, et al. Autologous chondrocyte implantation for treatment of focal chondral defects of the knee--a clinical, arthroscopic, MRI and histologic evaluation at 2 years. Knee. 2005;12(3):209–16.
26. Minas T, Peterson L. Advanced techniques in autologous chondrocyte transplantation. Clin Sports Med. 1999;18(1):13–44. v-vi

27. Caplan AI. Review: mesenchymal stem cells: cell-based reconstructive therapy in orthopedics. Tissue Eng. 2005;11(7–8):1198–211.
28. Caplan AI. Mesenchymal stem cells: the past, the present, the future. Cartilage. 2010;1(1):6–9.
29. Dimarino AM, Caplan AI, Bonfield TL. Mesenchymal stem cells in tissue repair. Front Immunol. 2013;4:201.
30. Huselstein C, Li Y, He X. Mesenchymal stem cells for cartilage engineering. Biomed Mater Eng. 2012;22(1–3):69–80.
31. Pasquinelli G, et al. Mesenchymal stem cell interaction with a non-woven hyaluronan-based scaffold suitable for tissue repair. J Anat. 2008;213(5):520–30.
32. Lisignoli G, et al. Chondrogenic differentiation of murine and human mesenchymal stromal cells in a hyaluronic acid scaffold: differences in gene expression and cell morphology. J Biomed Mater Res A. 2006;77(3):497–506.
33. Facchini A, et al. Human chondrocytes and mesenchymal stem cells grown onto engineered scaffold. Biorheology. 2006;43(3,4):471–80.
34. Nejadnik H, et al. Autologous bone marrow-derived mesenchymal stem cells versus autologous chondrocyte implantation: an observational cohort study. Am J Sports Med. 2010;38(6):1110–6.
35. Gobbi A, Karnatzikos G, Sankineani SR. One-step surgery with multipotent stem cells for the treatment of large full-thickness chondral defects of the knee. Am J Sports Med. 2014;42(3):648–57.
36. Whyte GP, Gobbi A, Sadlik B. Dry arthroscopic single-stage cartilage repair of the knee using a hyaluronic acid-based scaffold with activated bone marrow-derived mesenchymal stem cells. Arthrosc Tech. 2016;5(4):e913–8.
37. Sadlik B, et al. Biologic inlay osteochondral reconstruction: arthroscopic one-step osteochondral lesion repair in the knee using Morselized bone grafting and hyaluronic acid-based scaffold embedded with bone marrow aspirate concentrate. Arthrosc Tech. 2017;6(2):e383–9.
38. Bonasia DE, et al. Cocultures of adult and juvenile chondrocytes compared with adult and juvenile chondral fragments: in vitro matrix production. Am J Sports Med. 2011;39(11):2355–61.
39. Tompkins M, et al. Preliminary results of a novel single-stage cartilage restoration technique: particulated juvenile articular cartilage allograft for chondral defects of the patella. Arthroscopy. 2013;29(10):1661–70.
40. Grawe B, et al. Cartilage regeneration in full-thickness patellar chondral defects treated with Particulated juvenile articular allograft cartilage: an MRI analysis. Cartilage. 2017;8(4):374–83.
41. Buckwalter JA, et al. Clinical outcomes of patellar chondral lesions treated with juvenile particulated cartilage allografts. Iowa Orthop J. 2014;34:44–9.
42. Farr J, et al. Clinical, radiographic, and histological outcomes after cartilage repair with Particulated juvenile articular cartilage: a 2-year prospective study. Am J Sports Med. 2014;42(6):1417–25.
43. Geraghty S, et al. A novel, cryopreserved, viable osteochondral allograft designed to augment marrow stimulation for articular cartilage repair. J Orthop Surg Res. 2015;10:66.
44. Hoffman JK, Geraghty S, Protzman NM. Articular cartilage repair using marrow stimulation augmented with a viable chondral allograft: 9-month postoperative histological evaluation. Case Rep Orthop. 2015;2015:617365.
45. Fortier LA, et al. BioCartilage improves cartilage repair compared with microfracture alone in an equine model of full-thickness cartilage loss. Am J Sports Med. 2016;44(9):2366–74.
46. Wang KC, et al. Arthroscopic Management of Isolated Tibial Plateau Defect with Microfracture and Micronized Allogeneic Cartilage-Platelet-Rich Plasma Adjunct. Arthrosc Tech. 2017;6(5):e1613–8.
47. de Windt TS, et al. Early health economic modelling of single-stage cartilage repair. Guiding implementation of technologies in regenerative medicine. J Tissue Eng Regen Med. 2017;11(10):2950–9.
48. de Windt TS, et al. Allogeneic MSCs and recycled autologous Chondrons mixed in a one-stage cartilage cell Transplantion: a first-in-man trial in 35 patients. Stem Cells. 2017;35(8):1984–93.
49. Kon E, et al. Clinical results and MRI evolution of a nano-composite multilayered biomaterial for osteochondral regeneration at 5 years. Am J Sports Med. 2014;42(1):158–65.
50. Berruto M, et al. Treatment of large knee osteochondral lesions with a biomimetic scaffold: results of a multicenter study of 49 patients at 2-year follow-up. Am J Sports Med. 2014;42(7):1607–17.
51. Shimomura K, et al. First-in-human pilot study of implantation of a scaffold-free tissue-engineered construct generated from autologous synovial mesenchymal stem cells for repair of knee chondral lesions. Am J Sports Med. 2018;46(10):2384–93.
52. Pidoriano AJ, et al. Correlation of patellar articular lesions with results from anteromedial tibial tubercle transfer. Am J Sports Med. 1997;25(4):533–7.
53. Levy YD, et al. Do fresh osteochondral allografts successfully treat femoral condyle lesions? Clin Orthop Relat Res. 2013;471(1):231–7. https://doi.org/10.1007/s11999-012-2556-4. PMID: 22961315; PMCID: PMC3528935.

Ankle Osteochondral Pathologies and Treatment

Gian Luigi Canata, Valentina Casale,
Valentina Rita Corbo, and Alberto Vascellari

31.1 Introduction

Osteochondral lesions of the talus (OCLT) occur in the articular cartilage and subchondral bone of the talus. They are usually associated with ankle injuries, such as sprains and fractures [1, 2]. In the absence of a history of trauma, there could be a biomechanical cause [1].Nowadays, OCLTs remain a challenge for orthopedic surgeons. These lesions are in fact associated with vague and nonspecific symptoms [3], often affecting young patients with high functional demands. There are several therapeutic options, and it is challenging to evaluate treatment results using classical diagnostic techniques, such as computed tomography (CT) and magnetic resonance imaging (MRI) [3].

During normal gait, the ankle joint supports up to five times the weight of the body, whereas during sprinting, cutting-in, or stumbling these forces increase to up to 13 times the body weight [4]. The chronic consequences, in the case of pre-vious trauma or altered biomechanics, include the development of mechanical instability, osteochondral lesions, and impingement [5].

31.2 Epidemiology

Osteochondral and transchondral lesions of the talus are relatively common injuries, occurring in up to 50% of acute ankle fractures and sprains [2, 6–8]. Considering ankle fractures, the rates of secondary chondral injury have been reported as high as 73% [9].

With respect to foot and ankle injuries in track and field athletes, more than 60% of injuries occur during training, while only 20% develop during competition [10]. Sprints and hurdling disciplines are more prone to injury than middle-distance run or jumping. Furthermore, the ankle joint is more frequently involved in jumping injuries, while the foot is more often injured during sprints [10].

Recently, the number of diagnosed OCLTs has grown, thanks to the increased use of CT and MRI scans which has allowed the diagnosis of small lesions or injuries involving only the surface of the cartilage [3].

Osteochondral lesions of the distal tibia are less frequent than the OCLT [11–13]. In spite of their low incidence, it is important to have a high level of suspicion when ankle pain is persistent and there is a history of ankle injuries [14].

G. L. Canata (✉) · V. Casale
Centre of Sports Traumatology, Koelliker Hospital,
Torino, Italy
e-mail: canata@ortosport.it; studio@ortosport.it

V. R. Corbo
Specialistic Foot and Ankle Unit, Orthopedic Institute
IRCCS Galeazzi, Milan, Italy

A. Vascellari
Medicine Center, Villorba, Italy
e-mail: info@albertovascellari.it

© ISAKOS 2022
A. Gobbi et al. (eds.), *Joint Function Preservation*, https://doi.org/10.1007/978-3-030-82958-2_31

Conventional radiographs may miss up to 50% of cases [14] therefore the diagnosis is often delayed or missed, causing a delay of surgery by on average 22 months [15].

31.3　Pathomechanics and Pathophysiology

Sports-related injuries causing inversion, forced dorsiflexion, plantarflexion, or lateral rotation of the tibia may lead to traumatic ankle lesions [16]. The traumatic insult is widely assumed as the most important etiologic factor for the development of OCLT [2]. Hintermann et al. observed in fact the presence of ankle cartilage lesions in 66% of lateral ligament injuries, as well as in 98% of deltoid ligament injuries [5].

The tibiotalar joint is highly congruent and characterized by articular cartilage with an average thickness of 1.1 mm in women and 1.35 mm in men [2]. This results in a lack of cartilage elasticity, leading the talus to a higher risk of microfractures in the underlying bone when exposed to high impact forces [3]. A possible consequence of an ankle fracture malunion is the loss of the joint congruence, with an increase of nonuniformly distributed contact pressures at the talar dome [3, 17, 18].

Therefore, varus and valgus malalignment play a role too causing the development of high pressures on the medial and the lateral border of the talus, respectively [1].

The loss of the natural tibiotalar joint congruence is also a characteristic of chronic lateral ankle instability, making it one of the most predisposing factors for developing OCLT [19–21]. Ferkel and Chams reported in fact that 95% of patients undergoing a modified Broström procedure for chronic lateral instability had intra-articular pathology [20].

The location of the OCLT may also be correlated to the mechanism of injury, when an acute trauma occurs [22, 23]. It has been reported in fact that dorsiflexion and inversion mechanisms usually contribute to the development of anterolateral talar dome lesions [6], while plantar flexion and inversion, with or without external rotation, frequently cause medial talar dome injuries [6, 24].

The damage to the articular cartilage may follow a traumatic event, such as an ankle sprain or a fracture. The thin and less elastic talar cartilage is then susceptible to cartilage lesions and microfractures, and weight-bearing worsens the damaging process by enhancing the synovial pressure and therefore the lesion size or depth as well [2]. It has been reported in fact that intermittent or continuous high local pressure may inhibit the normal bone perfusion, leading to osteonecrosis, bone resorption, and the development of lytic areas in the bone [25–28].

The pain in OCLT is most probably due to an intermittent local rise in intraosseous fluid pressure developing during weight-bearing and consequently sensitizing the highly innervated subchondral bone [2]. The fluid forms from the damaged cartilage and may be forced into the microfractured subchondral bone plate, with an inversely proportional relationship between the diameter of the defect and the fluid pressure. The consequences may be osteolysis and the eventual formation of a subchondral cyst [2].

Etiologic factors in nontraumatic OCLT may be ischemic events, subsequent necrosis, and genetics [29]. OCLTs show in fact a higher incidence among twins, suggesting a familiarity or hereditary association [30, 31]. Furthermore, a correlation has been identified between dominant familiar osteochondral lesions and a missense mutation in the aggrecan C-Type Lectin domain in Chromosome 15 [32].

31.4　Diagnosis

31.4.1　Clinical Evaluation

Symptoms of OCLT are not specific and their diagnosis remains challenging. A high suspicion is critical when even a minor trauma has been experienced, yet fails to improve after several weeks [3]. Therefore, the first step of the diagnostic process should be the collection of information on the patient's history. Nevertheless, the diagnosis of OCLT may be elusive during the early clinical stages, because it is not unusual that an ankle inversion injury resolves after few weeks, but then evolves into chronic pain, stiffness, instability, or locking [24]. For this reason, in the acute situation, the OCLT may be misdiagnosed, as long as the swelling and pain from the lateral ligament lesion prevail [2].

Patients usually report ankle pain with or after weight-bearing and sports activity with rates up to 94% of OCLT presenting with ankle pain during activity, as well as 89% of cases reporting a previous ankle injury [33]. Locking and catching are usually symptoms of a displaced fragment [2].

Lesions involving the cartilage surface may present with mild symptoms, while those involving the underlying highly innervated subchondral bone are often more symptomatic, especially in the presence of subchondral cysts [34].

During the clinical evaluation, the talar dome should be palpated with the ankle in both dorsiflexion and plantar flexion [22]. Physical examination may be relatively nonspecific in OCLT, but the most frequently detectable finding is anterior joint line tenderness [3]. Nevertheless, patients with OCLT often report diffuse, nonspecific tenderness, and its location is not a reliable method for locating the lesion [3].

Ankle joint stability may be assessed with both the anterior and posterior drawer tests, as well as with the inversion and eversion tests [22]. It is important always to compare the plantar flexion and dorsiflexion motion with the uninvolved side, both in active and passive modes [22].

Range of motion should be assessed with flexed knee, to eliminate any restriction caused by shortened gastrocnemius muscles [3].

Conditions such as soft tissue injuries, ankle fractures or stress fractures, arthritis, or syndesmotic injury should always be considered in the differential diagnosis of OCLT [3].

31.4.2 Imaging and Classification

Although it has been reported that only half of OCLTs are detected with plain films alone [33], they are a fundamental starting point to rule out other pathologies, such as fractures, arthrosis, exostoses, and neoplasm [22].

Plain radiographs should include weight-bearing anteroposterior, lateral, and mortise views [3]. Stress radiographs may show associated instability [3]. Radiographs give clear information in the case of detached osteochondral fragments and areas of compression of the subchondral bone [3]. However, it cannot identify nondisplaced lesions or cartilage defects [35] thus its use is limited [36].

The magnetic resonance imaging (MRI) allows multiplanar evaluation and the ability to visualize the articular cartilage surface and subchondral bone, as well as bone edema and other features of the surrounding tissues [29]. Furthermore, it can detect early subchondral damages [37]. MRI is also effective in evaluating the stability of a minimally or nondisplaced fragment, and this characteristic may be important when choosing or avoiding surgical fixation [3].

The gold standard for measuring the exact size and morphology of an OCLT is computed tomography (CT) [38], especially because MRI may often overestimate the size of the lesion because it considers the perilesional edema [3]. CT allows the evaluation of the integrity of the subchondral bone in multiple planes [29] and can delineate possible associated cystic changes [3]. For these reasons, it is considered invaluable in preoperative planning [38].

Several classifications for OCLT have been proposed, beginning with Berndt and Harty's classification, based on radiographic findings [6]. Afterward, Ferkel and Scaglione introduced the use of CT details [39]. Anderson suggested an MRI-based classification [40] and Cheng and Ferkel proposed the use of arthroscopic findings to better classify OCLT [41]. More recently, the ISAKOS consensus group has questioned the importance of these classification systems, introducing a therapy-based classification [42]:

1. Asymptomatic or low symptomatic lesions: conservative treatment.
2. Symptomatic lesions up to 15 mm: debridement and drilling/microfracturing/bone marrow stimulation.

3. Symptomatic lesions larger than 15 mm: fixation.
4. Cystic lesions in tibial roof or large talar cystic lesions: retrograde drilling and bone transplantation.
5. Failed primary treatment: osteochondral transplant, hemicap, or calcaneal osteotomy.

31.5 Treatment

31.5.1 Conservative Treatment

In most cases, a trial of nonoperative treatment is warranted. This consists of activity modification, protected weight-bearing, rehabilitation, bracing, and the use of nonsteroidal anti-inflammatory drugs (NSAIDs) [43, 44]. This trial is usually suggested for asymptomatic or low symptomatic OCLT [42]. The goal is not to regenerate the cartilage, in fact, but to obtain pain relief [3].

According to the International Consensus Meeting on Cartilage Repair of the Ankle, conservative treatment may be attempted for 3 months before surgery, except in the presence of an acute, displaced osteochondral fragment which can be fixed immediately and when large lesions with edema and/or ligamentous instability are detectable. In those cases, surgery should be considered at 6–8 weeks if there is no response to conservative treatment. However, conservative management should not be prolonged to 6 months, except in the presence of a demonstrable improvement in symptoms and unresolved or uncontrolled medical comorbidities [45].

It is important to consider that this approach is more successful in the pediatric population than among adults [46].

31.5.2 Surgical Treatment

A wide variety of surgical procedures have been described [47–51]; however, it has been recently concluded that no superior treatment exists for both primary and secondary OCLT [52, 53]. Each surgical technique is based on one of the following principles [42]:

- Debridement and bone marrow stimulation (BMS), using procedures such as microfracturing, abrasion arthroplasty, or drilling.
- Securing a fragment to the talar dome through retrograde drilling or fixation.
- Preservation of the hyaline cartilage using osteochondral autografts, allografts, or autologous chondrocyte implantation (ACI).

Which procedure to choose mainly depends on the duration and intensity of symptoms, the size of the lesion, and if it is a primary or secondary OCLT [42].

Assessing the presence of a malalignment or chronic ankle instability is crucial. In the presence of these conditions, the concomitant treatment of the osteochondral lesions is critical. In the case of chronic ankle instability, an additional surgical procedure should be planned, such as lateral ligament repair or reconstruction. In the case of malalignment, it may be useful planning an additional surgical procedure to correct the hindfoot axis [2].

The main surgical procedures available for the management of OCLT are:

- Arthroscopic bone marrow stimulation (BMS):
  - Debridement.
  - Drilling.
- Retrograde drilling.
- Internal fixation.
- Tissue transplantation [53]:
  - Osteo(chondral) transplantation.
    Osteochondral autograft transfer systems (OATS).
    Mosaicplasty.
    (autogenous) bone grafting.
    Autologous osteoperiosteal cylinder grafting.
    Osteochondral allograft transplantation.
  - Autologous chondrocyte implantation.
    Periosteum-covered technique (ACI).
    Matrix-associated technique (MACI).
  - Chondrogenesis-inducing techniques.
    - Autologous collagen-induced chondrogenesis (ACIC).
    - Autologous matrix-induced chondrogenesis (AMIC).

31.5.2.1 Bone Marrow Stimulation (BMS)

The BMS is the most frequently used technique for primary OCLT, because of its relatively low costs, high effectiveness, low morbidity and it allows an early return to sports [48, 53, 54]. Techniques such as arthroscopic debridement, microdrilling (MD), and microfracturing (MF) are especially suggested when the OCLT is completely detached and the lesions are <10 mm in diameter, <100 mm^2 in area, and <5 mm in depth, as the International Consensus Meeting on Cartilage Repair of the Ankle has recently established [43, 48, 55–61].

The primary objective of BMS is the removal of unstable cartilage and underlying necrotic bone, then creating perforations in the healthy bone. The result is the mobilization of progenitor cells from the bone marrow. The marrow elements come from microfractures and form a regenerative fibrous cartilage repair tissue in the osteochondral defect [60]. Even when a subchondral cystic lesion is present, several studies have reported good to excellent results following BMS alone [48, 62, 63]. In particular, large cystic lesions should be addressed by retrograde drilling, which allows lowering of the pressure within the cyst and this defect is then filled with bone graft, if necessary [64].

Retrograde drilling is also an effective procedure when an OCLT with an intact articular cartilage surface is present. To solve this problem, in fact, drill holes may be placed through a retrograde/transtalar approach, instead of the transmalleolar/transarticular one [65–67]. This method allows in fact to access the lesion from above, by drilling through the sinus tarsi, not disrupting the articular surface [68, 69]. Retrograde drilling may be performed using an anterolateral approach, but an injury of the talar attachment of the anterior talofibular ligament (ATFL) may develop, so the posterolateral approach is preferred [55].

Further aspects to be considered are the dimension and the placement of both MF and MD, because they may influence the postoperative results. The International Consensus Meeting on Cartilage Repair of the Ankle, in fact, has suggested an awl or a low-speed drill of ≤2 mm in size, with a depth of holes which results in subchondral bone bleeding or fat droplets visualization to obtain better outcomes after BMS [61].

An OCLT may be debrided to a depth of 5 mm before bone grafting is required [61].

31.5.2.2 Internal Fixation

According to the International Consensus Meeting on Cartilage Repair of the Ankle, in the presence of an intact osteochondral fragment with a diameter larger than 10 mm or a bony fragment with a thickness of at least 3 mm, internal fixation is recommended [70]. The "lift the defect, drill the bone, fill the defect with bone graft, fix the fragment with bioabsorbable or metallic screws or pins" concept is fundamental in these cases [71]. It is important to highlight that adolescents may require fixation even if the fragment is smaller than 15 mm [42], as long as the procedure does not damage the growth plate and a medial malleolar osteotomy is not required [70].

If the fragment does not fit in the original site, the osteochondral unit can be shaped by removing sclerotic bone first, then the unstable cartilage. If the fragment is still too big, the expanded cartilage can be trimmed away with a blade [70]. If the lesion is purely cartilaginous, fixation is not recommended [70].

At least one bioabsorbable compression screw is required, then additional bioabsorbable darts or pins may be useful. As an alternative, 2 mm or 2.7 mm steel screws or a bone peg may provide satisfying results [70].

31.5.2.3 Cartilage Transplantation and Chondrogenesis-Inducing Techniques

Scaffold-based procedures are an effective alternative to the reparative techniques, such as the BMS, especially in the presence of cartilage lesions greater than 1 cm, with or without the presence of cysts [72].

Thanks to the recent advancements in tissue engineering and biomaterial science, scaffolds are currently available for treating OCLT [73]. They may be used both in two-step techniques, such as

ACI and MACI, and in the more recent one-step techniques, such as the autologous matrix-induced chondrogenesis (AMIC) and the bone marrow-derived cell transplantation (BMDCT) [73, 74].

Although ACI, MACI, AMIC, and scaffolds have become popular in recent years, they have shown satisfying results only in the short- and mid-term. For this reason, and considering their high costs, they are a good option for revision surgeries or large injuries, if fixation is not possible and previous procedures have failed [64].

31.5.2.4 Osteo(Chondral) Transplantation

The osteochondral autologous transplantation surgery (OATS) consists of harvesting osteochondral cylinders from the knee to fill an ankle defect [64].

It is usually indicated for symptomatic large, cystic OCLT of >1 cm in diameter [75], including lesions that have failed previous reparative procedures, such as BMS [76–78]. The main advantage of this technique is that it can replace a talar lesion with hyaline cartilage and subchondral bone native to the host [75].

Despite a high rate of successful outcomes reported [79], several areas of controversy still exist, when considering postoperative complications such as donor site morbidity [80, 81].

Osteochondral allograft transplantation may be a valid treatment option when the size of the OCLT is >1.5 cm in diameter, as well as when knee osteoarthritis of the donor is present or there is a history of infection [82]. Furthermore, other variables that influence the postoperative outcomes are the type of OCLT, the chosen allograft and the preferred storage parameters [82]. Despite promising results which have been reported, the success rate of this procedure still ranges from 20 to 100% [53]. Further research on this issue is still needed.

31.6 Rehabilitation

A four-level activity scheme has been recently proposed for a safe return to sport after debridement and BMS of OCLT [83]. These four stages correspond to different levels of increasing activity.

Level 1: early mobilization and partial weight-bearing should be allowed from the day of surgery, by both active and passive dorsi- and plantar flexing motions. Return to normal walking is usually achieved in 4–8 weeks, according to the gravity of the lesion. At the end of this phase, proprioception exercises can be started to regain normal active stability.

Level 2: training of proprioception, endurance, and technical skills should help achieve controlled sideways movements. Pain and swelling after increased activity should subside within 24 h at the end of this phase. Return to running is usually allowed 12 weeks after surgery [84–86].

Level 3: training of speed, force, and endurance guarantees the ability to safely run or sprint on even ground. Pain may still develop after increasing activity but should be relieved within 24 h. Return to noncontact sports can be allowed 4–6 months after surgery, when muscle strength is regained [56, 87].

Level 4: training of explosive force and sports-specific skills makes it possible to run on uneven ground, sprint, twist, and turn usually within 4–6 months after surgery [58, 88].

Age <50 years [56, 67, 87], low body mass index [56], a defect size <15 mm^2 [56, 87], and early mobilization [83] can be considered as positive influencing factors.

Since MRI T2 mapping provides qualitative information on the regenerated tissue by evaluating its water and glycosaminoglycan content [89], it may be an adjunctive guiding tool in determining the progression of the rehabilitation phases.

31.7 Prevention

Strategies for ankle osteochondral pathologies are mainly focused on ankle sprains and chronic ankle instability prevention. Prevention modalities for ankle sprains may target modifiable risk factors that increase the risk of sustaining an ankle sprain, such as limited dorsiflexion and overall ankle joint ROM, reduced proprioception, strength and coordination, and deficiencies in postural control/balance [90–94].

Prophylactic interventions to minimize the risk of ankle sprains can be stratified into interventions capable of affecting mechanical function and those designed to improve proprioceptive ability and neuromuscular function about the joint [95].

Injury-prevention programs focused on musculoskeletal strengthening, stretching, neuromuscular training, and improved dynamic ankle stability have been established to prevent lower extremity musculoskeletal injuries [96]. Each component may highlight an important role in ankle sprains prevention. For example, stretching of the triceps surae can reduce the limitation of ankle dorsiflexion, which has been associated with chronic ankle instability [90, 95]. Although neuromuscular training has been shown to prevent recurrent ankle sprains [97], the evidence for their effect in reducing first-time ankle sprains is less robust. Foss et al. performed a prospective randomized control study implementing a neuromuscular training program among middle school and high school aged athletes and found reduced overall injury rates, but no significant difference in ankle injuries specifically [98].

Balancing and proprioceptive exercises may enhance both the static and dynamic postural control, necessary for athletic performance, by optimizing the body's ability to sense and correct mild deviations in joint motion [95]. Therefore, they have an important role in preventing first-time and especially recurrent ankle sprains [99–104]. Schiftan et al. in their meta-analysis found an overall significant reduction of ankle sprain incidence after proprioceptive training (relative risk = 0.65), with results supporting the intervention both for participants with a history of ankle sprain and those without a history of sprain [100].

31.8 Conclusion

Osteochondral pathologies of the talus are not uncommon in track and field athletes. Acute and chronic instability is a clear risk factor. Preventive measures and a careful prompt evaluation and management of these injuries are important issues.

When conservative treatment fails, arthroscopic debridement and drilling or microfracture are effective noninvasive techniques. When these first-line treatments cannot adequately stimulate healing, cartilage replacement surgeries like OATS, ACI, and MACI are efficacious.

A dedicated rehabilitation program respecting the biological healing processes and including neuromuscular training is a mandatory complement to ease a full functional recovery.

References

1. Bruns J, Rosenbach B. Pressure distribution at the ankle joint. Clin Biomech Bristol Avon. 1990;5:153–61.
2. van Dijk CN, Reilingh ML, Zengerink M, van Bergen CJA. Osteochondral defects in the ankle: why painful? Knee Surg Sports Traumatol Arthrosc. 2010;18:570–80.
3. Wodicka R, Ferkel E, Ferkel R. Osteochondral lesions of the ankle. Foot Ankle Int. 2016;37:1023–34.
4. Self BP, Paine D. Ankle biomechanics during four landing techniques. Med Sci Sports Exerc. 2001;33:1338–44.
5. Hintermann B, Boss A, Schäfer D. Arthroscopic findings in patients with chronic ankle instability. Am J Sports Med. 2002;30:402–9.
6. Berndt AL, Harty M. Transchondral fractures (osteochondritis dissecans) of the talus. J Bone Joint Surg Am. 1959;41-A:988–1020.
7. Ferkel RD, Dierckman BD, Phisitkul P. Arthroscopy of the foot and the ankle. Mann's surgery foot ankle e-book expert consult - Online. Amsterdam: Elsevier Health Sciences; 2013. p. 1723–828.
8. Shelton M, Pedowitz W. Injuries to the talar dome, subtalar joint, and mid-foot. In: Disorder foot ankle med surgery management. 2nd ed. Philadelphia: Saunders; 1991. p. 2274–92.
9. Leontaritis N, Hinojosa L, Panchbhavi VK. Arthroscopically detected intra-articular lesions associated with acute ankle fractures. J Bone Joint Surg Am. 2009;91:333–9.
10. Ortiz C, Wagner E, Fernandez G. Athletic injuries. In: Foot ankle sports orthopedics. New York: Springer; 2017. p. 421–6.
11. Bauer M, Jonsson K, Lindén B. Osteochondritis dissecans of the ankle. A 20-year follow-up study. J Bone Joint Surg Br. 1987;69:93–6.

12. Canosa J. Mirror image osteochondral defects of the talus and distal tibia. Int Orthop. 1994;18:395–6.
13. Chapman CB, Mann JA. Distal tibial osteochondral lesion treated with osteochondral allografting: a case report. Foot Ankle Int. 2005;26:997–1000.
14. Gianakos AL, Yasui Y, Hannon CP, Kennedy JG. Current management of talar osteochondral lesions. World J Orthop. 2017;8:12–20.
15. Mologne TS, Ferkel RD. Arthroscopic treatment of osteochondral lesions of the distal tibia. Foot Ankle Int. 2007;28:865–72.
16. Vannini F, Costa GG, Caravelli S, Pagliazzi G, Mosca M. Treatment of osteochondral lesions of the talus in athletes: what is the evidence? Joints. 2016;4:111–20.
17. Ramsey PL, Hamilton W. Changes in tibiotalar area of contact caused by lateral talar shift. J Bone Joint Surg Am. 1976;58:356–7.
18. Thordarson DB, Motamed S, Hedman T, Ebramzadeh E, Bakshian S. The effect of fibular malreduction on contact pressures in an ankle fracture malunion model. J Bone Joint Surg Am. 1997;79:1809–15.
19. DIGiovanni BF, Fraga CJ, Cohen BE, Shereff MJ. Associated injuries found in chronic lateral ankle instability. Foot Ankle Int. 2000;21:809–15.
20. Ferkel RD, Chams RN. Chronic lateral instability: arthroscopic findings and long-term results. Foot Ankle Int. 2007;28:24–31.
21. Gregush RV, Ferkel RD. Treatment of the unstable ankle with an osteochondral lesion: results and long-term follow-up. Am J Sports Med. 2010;38:782–90.
22. Talusan PG, Milewski MD, Toy JO, Wall EJ. Osteochondritis dissecans of the talus: diagnosis and treatment in athletes. Clin Sports Med. 2014;33:267–84.
23. Canale ST, Belding RH. Osteochondral lesions of the talus. J Bone Joint Surg Am. 1980;62:97–102.
24. O'Farrell TA, Costello BG. Osteochondritis dissecans of the talus. The late results of surgical treatment. J Bone Joint Surg Br. 1982;64:494–7.
25. Astrand J, Skripitz R, Skoglund B, Aspenberg P. A rat model for testing pharmacologic treatments of pressure-related bone loss. Clin Orthop Relat Res. 2003;(409):296–305.
26. Dürr HD, Martin H, Pellengahr C, Schlemmer M, Maier M, Jansson V. The cause of subchondral bone cysts in osteoarthrosis: a finite element analysis. Acta Orthop Scand. 2004;75:554–8.
27. Johansson L, Edlund U, Fahlgren A, Aspenberg P. Bone resorption induced by fluid flow. J Biomech Eng. 2009;131:094505.
28. Schmalzried TP, Akizuki KH, Fedenko AN, Mirra J. The role of access of joint fluid to bone in periarticular osteolysis. A report of four cases. J Bone Joint Surg Am. 1997;79:447–52.
29. Schachter AK, Chen AL, Reddy PD, Tejwani NC. Osteochondral lesions of the talus. J Am Acad Orthop Surg. 2005;13:152–8.
30. Erban WK, Kolberg K. [Simultaneous mirror image osteochondrosis dissecans in identical twins]. ROFO Fortschr Geb Rontgenstr Nuklearmed. 1981;135:357.
31. Woods K, Harris I. Osteochondritis dissecans of the talus in identical twins. J Bone Joint Surg Br. 1995;77:331.
32. Stattin E-L, Wiklund F, Lindblom K, Onnerfjord P, Jonsson B-A, Tegner Y, et al. A missense mutation in the aggrecan C-type lectin domain disrupts extracellular matrix interactions and causes dominant familial osteochondritis dissecans. Am J Hum Genet. 2010;86:126–37.
33. Loomer R, Fisher C, Lloyd-Smith R, Sisler J, Cooney T. Osteochondral lesions of the talus. Am J Sports Med. 1993;21:13–9.
34. van Dijk CN, Reilingh ML, Zengerink M, van Bergen CJA. The natural history of osteochondral lesions in the ankle. Instr Course Lect. 2010;59:375–86.
35. De Lee J. Fractures and dislocations of the foot. In: Surgery foot ankle. 6th ed. Maryland Heights, MO: Mosby; 1991. p. 1465–518.
36. Canale ST, Kelly FB. Fractures of the neck of the talus. Long-term evaluation of seventy-one cases. J Bone Joint Surg Am. 1978;60:143–56.
37. Dipaola JD, Nelson DW, Colville MR. Characterizing osteochondral lesions by magnetic resonance imaging. Arthroscopy. 1991;7:101–4.
38. Reis ND, Zinman C, Besser MI, Shifrin LZ, Folman Y, Torem S, et al. High-resolution computerised tomography in clinical orthopaedics. J Bone Joint Surg Br. 1982;64:20–4.
39. Ferkel R, Sgaglione N, Del Pizzo W. Arthroscopic treatment of osteochondral lesions of the talus: technique and results. Orthop Trans. 1990;14:172–3.
40. Anderson IF, Crichton KJ, Grattan-Smith T, Cooper RA, Brazier D. Osteochondral fractures of the dome of the talus. J Bone Joint Surg Am. 1989;71:1143–52.
41. Cheng JC, Ferkel RD. The role of arthroscopy in ankle and subtalar degenerative joint disease. Clin Orthop Relat Res. 1998;(349):65–72.
42. van Dijk PAD, van Dijk CN. Osteochondral lesions of the talus. In: Sports injuries of the foot and ankle: a focus on advanced surgical techniques. New York: Springer; 2009. p. 133–9.
43. Amendola A, Panarella L. Osteochondral lesions: medial versus lateral, persistent pain, cartilage restoration options and indications. Foot Ankle Clin. 2009;14:215–27.
44. Hannon CP, Smyth NA, Murawski CD, Savage-Elliott I, Deyer TW, Calder JDF, et al. Osteochondral lesions of the talus: aspects of current management. Bone Joint J. 2014;96-B:164–71.
45. Dombrowski ME, Yasui Y, Murawski CD, Fortier LA, Giza E, Haleem AM, et al. Conservative management and biological treatment strategies: proceedings of the international consensus meeting on cartilage repair of the ankle. Foot Ankle Int. 2018;39:9S–15S.

46. Looze CA, Capo J, Ryan MK, Begly JP, Chapman C, Swanson D, et al. Evaluation and management of osteochondral Lesions of the talus. Cartilage. 2017;8:19–30.

47. Tol JL, Struijs PA, Bossuyt PM, Verhagen RA, van Dijk CN. Treatment strategies in osteochondral defects of the talar dome: a systematic review. Foot Ankle Int. 2000;21:119–26.

48. Zengerink M, Struijs PAA, Tol JL, van Dijk CN. Treatment of osteochondral lesions of the talus: a systematic review. Knee Surg Sports Traumatol Arthrosc. 2010;18:238–46.

49. Loveday D, Clifton R, Robinson A. Interventions for treating osteochondral defects of the talus in adults. Cochrane Database Syst Rev. 2010;CD008104.

50. Donnenwerth MP, Roukis TS. Outcome of arthroscopic debridement and microfracture as the primary treatment for osteochondral lesions of the talar dome. Arthroscopy. 2012;28:1902–7.

51. Verhagen RAW, Struijs PAA, Bossuyt PMM, van Dijk CN. Systematic review of treatment strategies for osteochondral defects of the talar dome. Foot Ankle Clin. 2003;8:233–42. viii–ix

52. Lambers KTA, Dahmen J, Reilingh ML, van Bergen CJA, Stufkens SAS, Kerkhoffs GMMJ. No superior surgical treatment for secondary osteochondral defects of the talus. Knee Surg Sports Traumatol Arthrosc. 2018;26:2158–70.

53. Dahmen J, Lambers KTA, Reilingh ML, van Bergen CJA, Stufkens SAS, Kerkhoffs GMMJ. No superior treatment for primary osteochondral defects of the talus. Knee Surg Sports Traumatol Arthrosc. 2018;26:2142–57.

54. van Eekeren ICM, van Bergen CJA, Sierevelt IN, Reilingh ML, van Dijk CN. Return to sports after arthroscopic debridement and bone marrow stimulation of osteochondral talar defects: a 5- to 24-year follow-up study. Knee Surg Sports Traumatol Arthrosc. 2016;24:1311–5.

55. Canata GL, Casale V. Arthroscopic debridement of osteochondral lesions of the talus. In: Cartilage lesions ankle. New York: Springer; 2015. p. 27–36.

56. Chuckpaiwong B, Berkson EM, Theodore GH. Microfracture for osteochondral lesions of the ankle: outcome analysis and outcome predictors of 105 cases. Arthroscopy. 2008;24:106–12.

57. Giannini S, Vannini F. Operative treatment of osteochondral lesions of the talar dome: current concepts review. Foot Ankle Int. 2004;25:168–75.

58. van Bergen CJA, de Leeuw PAJ, van Dijk CN. Treatment of osteochondral defects of the talus. Rev Chir Orthop Reparatrice Appar Mot. 2008;94:398–408.

59. van Dijk CN, van Bergen CJA. Advancements in ankle arthroscopy. J Am Acad Orthop Surg. 2008;16:635–46.

60. Murawski CD, Foo LF, Kennedy JG. A review of arthroscopic bone marrow stimulation techniques of the talus: the good, the bad, and the causes for concern. Cartilage. 2010;1:137–44.

61. Hannon CP, Bayer S, Murawski CD, Canata GL, Clanton TO, Haverkamp D, et al. Debridement, curettage, and bone marrow stimulation: proceedings of the international consensus meeting on cartilage repair of the ankle. Foot Ankle Int. 2018;39:16S–22S.

62. Han SH, Lee JW, Lee DY, Kang ES. Radiographic changes and clinical results of osteochondral defects of the talus with and without subchondral cysts. Foot Ankle Int. 2006;27:1109–14.

63. Lee K, Park H, Cho H, Seon J. Comparison of arthroscopic microfracture for osteochondral lesions of the talus with and without subchondral cyst. Am J Sports Med. 2015;43:1951–6.

64. Pereira H, Vuurberg G, Spennacchio P, Batista J, D'Hooghe P, Hunt K, et al. Surgical treatment paradigms of ankle lateral instability, osteochondral defects and impingement. Adv Exp Med Biol. 2018;1059:85–108.

65. Taranow WS, Bisignani GA, Towers JD, Conti SF. Retrograde drilling of osteochondral lesions of the medial talar dome. Foot Ankle Int. 1999;20:474–80.

66. Kono M, Takao M, Naito K, Uchio Y, Ochi M. Retrograde drilling for osteochondral lesions of the talar dome. Am J Sports Med. 2006;34:1450–6.

67. Kumai T, Takakura Y, Higashiyama I, Tamai S. Arthroscopic drilling for the treatment of osteochondral lesions of the talus. J Bone Joint Surg Am. 1999;81:1229–35.

68. Chew KTL, Tay E, Wong YS. Osteochondral lesions of the talus. Ann Acad Med Singap. 2008;37:63–8.

69. Canata G, Casale V. Arthroscopic debridement and bone marrow stimulation for talar osteochondral lesions: current concepts. J ISAKOS. 2017;2:2–7.

70. Reilingh ML, Murawski CD, DiGiovanni CW, Dahmen J, Ferrao PNF, Lambers KTA, et al. Fixation techniques: proceedings of the international consensus meeting on cartilage repair of the ankle. Foot Ankle Int. 2018;39:23S–7S.

71. Kerkhoffs GMMJ, Reilingh ML, Gerards RM, de Leeuw PAJ. Lift, drill, fill and fix (LDFF): a new arthroscopic treatment for talar osteochondral defects. Knee Surg Sports Traumatol Arthrosc. 2016;24:1265–71.

72. Rothrauff BB, Murawski CD, Angthong C, Becher C, Nehrer S, Niemeyer P, et al. Scaffold-based therapies: proceedings of the international consensus meeting on cartilage repair of the ankle. Foot Ankle Int. 2018;39:41S–7S.

73. Gobbi A, Scotti C, Peretti GM. Scaffolding as treatment for osteochondral defects in the ankle. In: Arthroscopy: basic to advanced. New York: Springer; 2016. p. 1003–12.

74. Shimozono Y, Yasui Y, Ross AW, Kennedy JG. Osteochondral lesions of the talus in the athlete: up to date review. Curr Rev Musculoskelet Med. 2017;10:131–40.

75. Hurley ET, Murawski CD, Paul J, Marangon A, Prado MP, Xu X, et al. Osteochondral autograft:

proceedings of the international consensus meeting on cartilage repair of the ankle. Foot Ankle Int. 2018;39:28S–34S.

76. Hangody L, Vásárhelyi G, Hangody LR, Sükösd Z, Tibay G, Bartha L, et al. Autologous osteochondral grafting--technique and long-term results. Injury. 2008;39(Suppl 1):S32–9.

77. Ramponi L, Yasui Y, Murawski CD, Ferkel RD, DiGiovanni CW, Kerkhoffs GMMJ, et al. Lesion size is a predictor of clinical outcomes after bone marrow stimulation for osteochondral lesions of the talus: a systematic review. Am J Sports Med. 2017;45:1698–705.

78. Scranton PE, Frey CC, Feder KS. Outcome of osteochondral autograft transplantation for type-V cystic osteochondral lesions of the talus. J Bone Joint Surg Br. 2006;88:614–9.

79. Shimozono Y, Hurley ET, Myerson CL, Kennedy JG. Good clinical and functional outcomes at midterm following autologous osteochondral transplantation for osteochondral lesions of the talus. Knee Surg Sports Traumatol Arthrosc. 2018;26:3055–62.

80. Paul J, Sagstetter A, Kriner M, Imhoff AB, Spang J, Hinterwimmer S. Donor-site morbidity after osteochondral autologous transplantation for lesions of the talus. J Bone Joint Surg Am. 2009;91:1683–8.

81. Ferreira C, Vuurberg G, Oliveira JM, Espregueira-Mendes J, Pereira H, Reis RL, et al. Good clinical outcome after osteochondral autologous transplantation surgery for osteochondral lesions of the talus but at the cost of a high rate of complications: a systematic review. J ISAKOS Joint Disord Orthop Sports Med. 2016;1:184–91.

82. Smyth NA, Murawski CD, Adams SB, Berlet GC, Buda R, Labib SA, et al. Osteochondral allograft: proceedings of the international consensus meeting on cartilage repair of the ankle. Foot Ankle Int. 2018;39:35S–40S.

83. Gerards RM, van Eekeren ICM, van Dijk CN. Rehabilitation and return-to-sports activity after debridement and bone marrow stimulation of osteochondral talar defects. In: Cartilage lesions ankle. New York: Springer; 2015. p. 87–98.

84. Saxena A, Eakin C. Articular talar injuries in athletes: results of microfracture and autogenous bone graft. Am J Sports Med. 2007;35:1680–7.

85. Seijas R, Alvarez P, Ares O, Steinbacher G, Cuscó X, Cugat R. Osteocartilaginous lesions of the talus in soccer players. Arch Orthop Trauma Surg. 2010;130:329–33.

86. Ogilvie-Harris DJ, Sarrosa EA. Arthroscopic treatment of osteochondritis dissecans of the talus. Arthroscopy. 1999;15:805–8.

87. Lee K-B, Bai L-B, Chung J-Y, Seon J-K. Arthroscopic microfracture for osteochondral lesions of the talus. Knee Surg Sports Traumatol Arthrosc. 2010;18:247–53.

88. Zengerink M, Szerb I, Hangody L, Dopirak RM, Ferkel RD, van Dijk CN. Current concepts: treatment of osteochondral ankle defects. Foot Ankle Clin. 2006;11:331–59. vi

89. Nieminen MT, Rieppo J, Töyräs J, Hakumäki JM, Silvennoinen J, Hyttinen MM, et al. T2 relaxation reveals spatial collagen architecture in articular cartilage: a comparative quantitative MRI and polarized light microscopic study. Magn Reson Med. 2001;46:487–93.

90. de Noronha M, Refshauge KM, Herbert RD, Kilbreath SL, Hertel J. Do voluntary strength, proprioception, range of motion, or postural sway predict occurrence of lateral ankle sprain? Br J Sports Med. 2006;40:824–8. discussion 828

91. Kobayashi T, Yoshida M, Yoshida M, Gamada K. Intrinsic predictive factors of noncontact lateral ankle sprain in collegiate athletes: a case-control study. Orthop J Sports Med. 2013;1:2325967113518163.

92. de Noronha M, França LC, Haupenthal A, Nunes GS. Intrinsic predictive factors for ankle sprain in active university students: a prospective study. Scand J Med Sci Sports. 2013;23:541–7.

93. Kobayashi T, Tanaka M, Shida M. Intrinsic risk factors of lateral ankle sprain: a systematic review and meta-analysis. Sports Health. 2016;8:190–3.

94. Hagen M, Asholt J, Lemke M, Lahner M. The angle-torque-relationship of the subtalar pronators and supinators in male athletes: a comparative study of soccer and handball players. Technol Health Care. 2016;24:391–9.

95. Kaminski TW, Needle AR, Delahunt E. Prevention of lateral ankle sprains. J Athl Train. 2019;54:650–61.

96. Brunner R, Friesenbichler B, Casartelli NC, Bizzini M, Maffiuletti NA, Niedermann K. Effectiveness of multicomponent lower extremity injury prevention programmes in team-sport athletes: an umbrella review. Br J Sports Med. 2019;53:282–8.

97. Vuurberg G, Hoorntje A, Wink LM, van der Doelen BFW, van den Bekerom MP, Dekker R, et al. Diagnosis, treatment and prevention of ankle sprains: update of an evidence-based clinical guideline. Br J Sports Med. 2018;52:956.

98. Foss KDB, Thomas S, Khoury JC, Myer GD, Hewett TE. A school-based neuromuscular training program and sport-related injury incidence: a prospective randomized controlled clinical trial. J Athl Train. 2018;53:20–8.

99. Taylor JB, Ford KR, Nguyen A-D, Terry LN, Hegedus EJ. Prevention of lower extremity injuries in basketball: a systematic review and Meta-analysis. Sports Health. 2015;7:392–8.

100. Schiftan GS, Ross LA, Hahne AJ. The effectiveness of proprioceptive training in preventing ankle sprains in sporting populations: a systematic review and meta-analysis. J Sci Med Sport. 2015;18:238–44.

101. LaBella CR, Huxford MR, Grissom J, Kim K-Y, Peng J, Christoffel KK. Effect of neuromuscular warm-up on injuries in female soccer and basketball athletes in urban public high schools: cluster

randomized controlled trial. Arch Pediatr Adolesc Med. 2011;165:1033–40.

102. McGuine TA, Keene JS. The effect of a balance training program on the risk of ankle sprains in high school athletes. Am J Sports Med. 2006;34:1103–11.

103. Eils E, Schröter R, Schröder M, Gerss J, Rosenbaum D. Multistation proprioceptive exercise program prevents ankle injuries in basketball. Med Sci Sports Exerc. 2010;42:2098–105.

104. Hupperets MDW, Verhagen EALM, van Mechelen W. Effect of unsupervised home based proprioceptive training on recurrences of ankle sprain: randomised controlled trial. BMJ. 2009;339:b2684.

At-AMIC: A Reliable Solution for Talar Osteochondral Lesions

Cristian Indino, Rossella De Marco, Federico G. Usuelli, and Riccardo D'Ambrosi

32.1 Surgical Technique and Indications

Ankle sprains and fractures are common injuries that can occur in active young persons. They can develop an osteochondral lesion of the talus (OCLT) that means a defect of the talar articular cartilage and the subchondral bone. Sprains can be often recurrent and cause chronic instability of the ankle. In such type of patients, OCLTs can occur in up to 70% of cases. Debridement and microfracture therapy are the most described techniques for this condition but in lesions larger than 1.5 cm they have poorer outcomes. Autologous Matrix-Induced Chondrogenesis (AMIC®) combines microfractures with the use of Chondro-Gide® (Geistlich Pharma, Wolhusen, Switzerland), a type I-type III porcine collagen matrix, used to hold bone marrow-derived mesenchymal stem cells retained in a regeneration chamber. Usually this treatment is performed in arthrotomy with medial malleolar osteotomy. An all-arthroscopic AMIC®(AT-AMIC®) technique

has been developed, less invasive, with less comorbidities and with a faster recovery for patients.

Indications and contraindications for AT-AMIC technique are listed in Table 32.1.

Preoperative planning involves clinical examination with the evaluation of both ankles and alignment of the hindfoot, preoperative weight-bearing radiographs of the foot and ankle, and a magnetic resonance imaging (MRI) of the interested ankle to evaluate the cartilage defect.

The procedure is performed with the patient under either general or spinal anesthesia in the supine position and with feet sticking out of the surgical table, using a high tight tourniquet for all the duration of the procedure. The entire surgical procedure is performed arthroscopically with an anteromedial and anterolateral portal approaches for the ankle. The anteromedial portal is placed immediately medial to the anterior tibialis tendon,

Table 32.1 Indications and contraindications for Arthroscopic Autologous Matrix-Induced Chondrogenesis Technique

Indications	Contraindications
Osteochondral lesion	Metabolic arthropathy
	Kissing lesions
Lesion >1.0 cm^2	Major defects (i.e., defects that cannot be reconstructed)
Age 18–55 year.	
Primary or revision procedure	Non-correctable hindfoot malalignment
	Systemic disorders
	Obesity

C. Indino · R. De Marco · F. G. Usuelli (✉)
Humanitas S. Pio X, Milan, Italy

R. D'Ambrosi
IRCCS Istituto Ortopedico Galeazzi, Milan, Italy

© ISAKOS 2022
A. Gobbi et al. (eds.), *Joint Function Preservation*, https://doi.org/10.1007/978-3-030-82958-2_32

just distal to the joint line, taking care to avoid the saphenous vein. The anterolateral approach is placed lateral to the extensor digitorum tendon and medial to the lateral malleolus, being careful to avoid a superficial peroneal nerve damage (Fig. 32.1).

It is required to arthroscopically confirm the presence and the dimension of the cartilage damage. An Hintermann spreader (Integra LifeScience, Plainsboro, NJ) is percutaneously positioned to distract the ankle joint, with two 2.5 mm K-wires, placed one in the tibia and the other in the talar bone, medially or laterally, depending on the lesion side (Fig. 32.2).

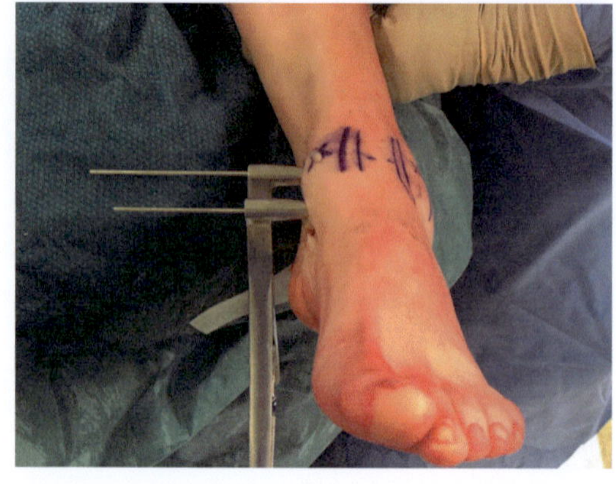

Fig. 32.2 Hintermann spreader with two 2.5 mm pins positioned on the medial side of the ankle

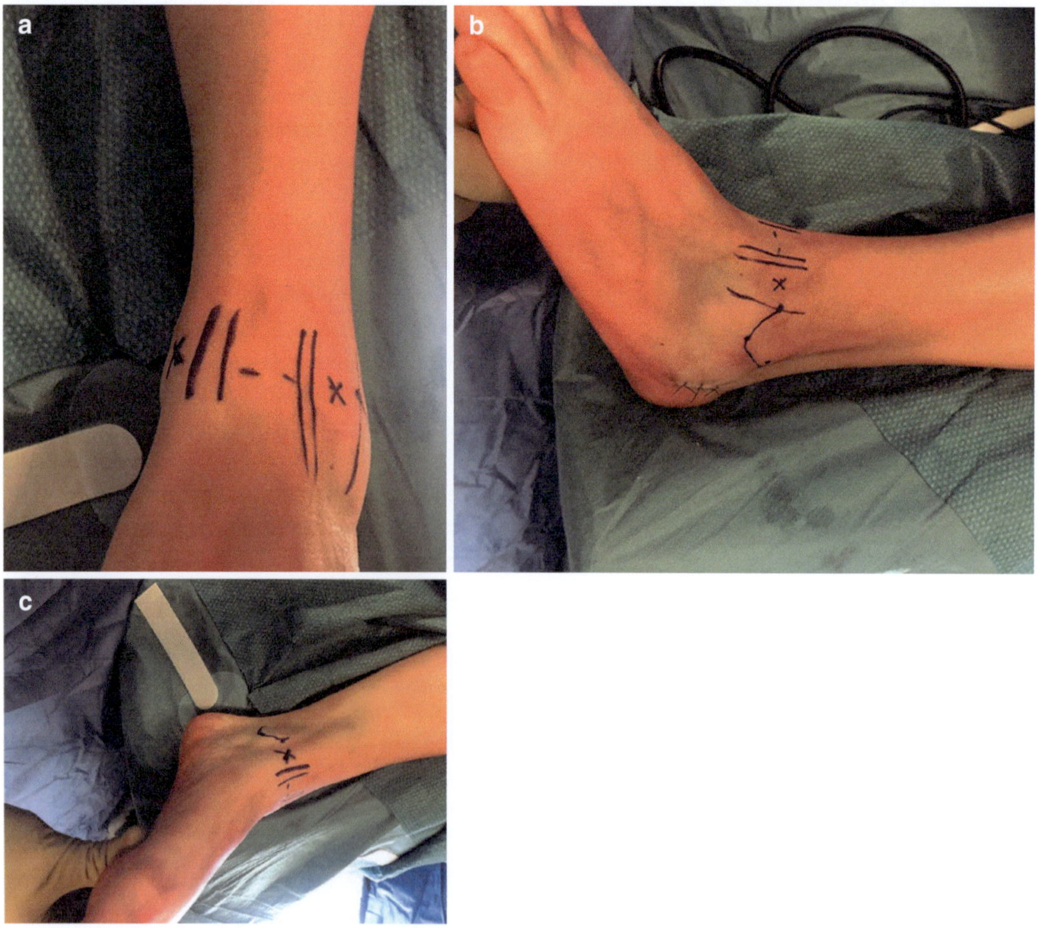

Fig. 32.1 Arthroscopic surgical planning: frontal view (**a**), anterolateral view (**b**), anteromedial view (**c**)

After identification of the lesion, the damaged cartilage with the associated necrotic and sclerotic bone is removed with a standard arthroscopic curette, in order to obtain a regular-shaped site. After that, an aluminum template provided with the matrix kit is introduced by a 5.5 mm cannula, inserted through the closest portal to the lesion, and adjusted on the lesion area.

Microfractures are performed on the subchondral bone underneath the entire size of the lesion (Fig. 32.3). If necessary to fill subchondral bone defect, cancellous bone is harvested from the ipsilateral calcaneus with an accessory lateral approach on the calcaneus cortex. The harvested bone is introduced through the cannula and impacted into the bony defect.

Now the intra-articular water is removed by dedicated suction. The membrane is prepared according to the shape of the template; it is inserted through the cannula and positioned on the lesion site. It is useful to mark the top of the matrix with the dermographic pen before the implantation, to avoid the upside-down positioning. Once the perfect coverage is obtained, the edges of the membrane are glued with a synthetic fibrin glue (Tisseel; Baxter, Deerfield, IL) introduced by a needle (Fig. 32.4). At this point, the Hintermann spreader is released and the position

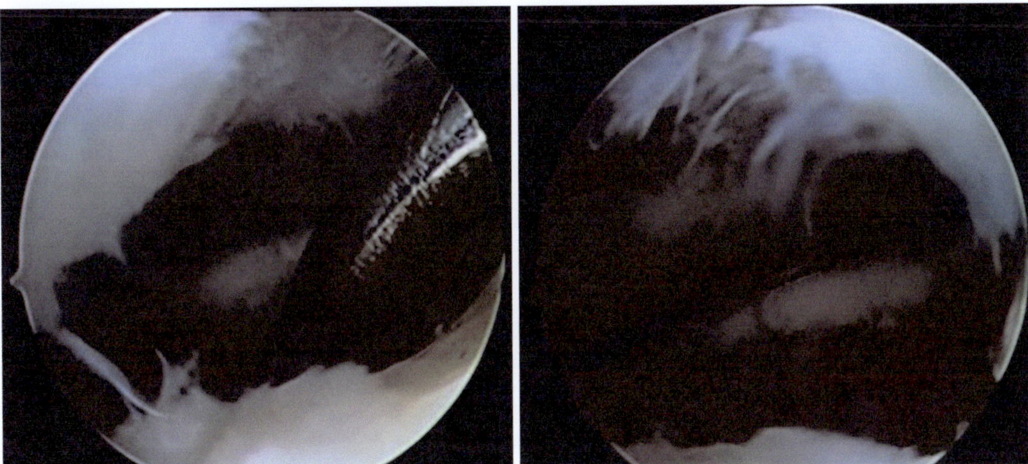

Fig. 32.3 Arthroscopic view. Microfractures made at the level of the osteochondral lesion covering the entire surface

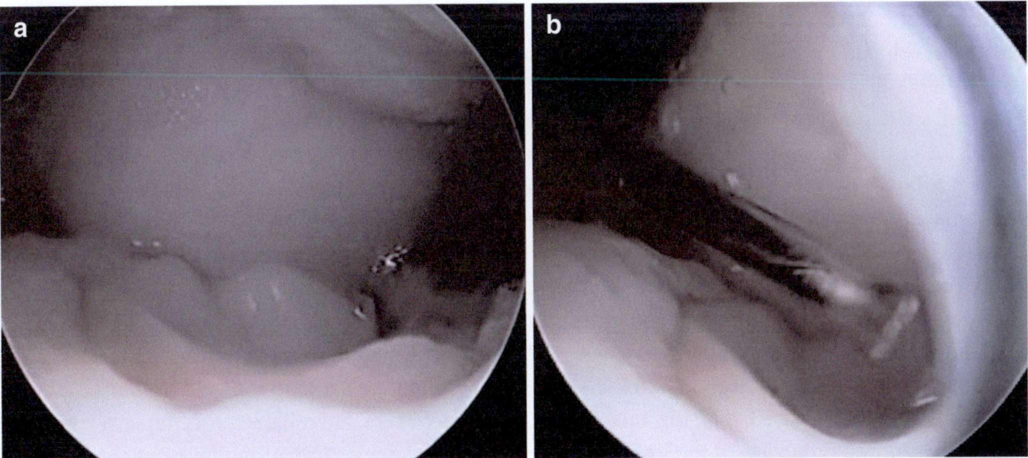

Fig. 32.4 View from anteromedial portal. (**a**) Positioning of the collagen membrane at the level of the osteochondral lesion. (**b**) The collagen matrix is fixed with a synthetic fibrin glue injected with a syringe through the 5.5 mm cannula

of the membrane is arthroscopically checked during the normal ankle range of motion. Postoperative management requires movement limitation for 15 days to avoid mobilization of the membrane and no weight bearing for 40 days. An MRI scan is required at 6 months follow-up to monitor the healing process.

Possible complications related to AT-AMIC® are matrix detachment, unknown collagen allergy, early membrane mobilization, hyperproliferative healing reaction with subsequent joint impingement, common ankle arthroscopy complications [1].

32.2   2-years Follow-up

During the follow-up period, patients need to be evaluated clinically and radiologically at 6, 12, and 24 months. Radiological evaluation includes MRI and CT scans; the lesional area is measured and defined for each patient both on CT scan and on MRI scan, according to Choi [2], using coronal length, sagittal length, depth, and area. At 12 and 24 months postoperatively, the MOCART (Magnetic Resonance Observation of Cartilage Repair Tissue) score can be performed. In 2018, Usuelli et al. [3] published a study reporting the outcome of 20 patients: at the end of the observational period, all lesions were filled with autologous bone, and no infections occurred in the postoperative period. Only in one case a single adverse event occurred: 8 months after surgery a patient developed an anterior osteophyte due to hypertrophic proliferation, causing impingement and restricted range of motion, that needed to repeat an arthroscopy in order to remove it. At 2-years follow-up, the patient was pain-free, with full range of motion and with the reduction of the lesion, as documented by CT and MRI scans.

In the first 6 months, CT examination can show an increased lesion area due to debridement and microfractures performed during the surgical procedure (Fig. 32.5) but, at the end of 2 years, the lesion area is reduced. On the other hand, MRI scans show a reduction of the lesion already at 6 months postoperatively, with progressive improvement up to 2 years (Fig. 32.6).

32.3   Population Analysis

Independent prognostic factors such as age, size of the lesion, body mass index, history of traumas, presence of osteophytes can negatively affect the outcome of OCLT treatment. In literature, it is shown that young age is related to better healing of these lesions. In a study published in 2016 [4], two groups of patients were compared: one (G_1) 33 years old or younger, and the other (G_2) older than 33 years. The aim of the study was to show that both patients older and younger than 33 years would benefit from AT-AMIC® procedure, and that younger patients would develop a greater healing rate and a greater functional gain than older patients. The 33-year old cut-off was chosen according to the existing literature on the same argument. The study showed that AT-AMIC® procedure is useful in both populations, considering that the area of the lesions measured with MRI and CT scan reduced significantly at each follow-up. There was a significant clinical improvement in both groups, in particular related to starting conditions of the ankles. We found that the functional score was higher in younger patients before and after surgery, showing a better ankle function before surgery.

There is no general consensus about the role of body mass index (BMI) in the healing process of OCLTs, considered to be a negative prognostic factor. A study has been conducted in 2017 [5], in order to assess healing and the functional outcome after AT-AMIC in 2 weight groups of patients: one with BMI $\geq$25 (OG: Overweight Group) versus the other with BMI <25 (HG: Healthy Group). In this research, it was observed that the lesion measured with MRI in the preoperative assessment in the OG was greater than HG, but not with CT scan. It is supposed that weight plays a role in the perilesional edema, increasing the size of the lesion measured with MRI. Furthermore, all patients showed a significant improvement in all clinical parameters at the final follow-up in spite of the edema around the lesion. AT-AMIC can be considered as a safe and reliable procedure even in overweight patients, with a significant improvement in quality of life. Being over-weight is not a negative predictor of outcome for this kind of procedure.

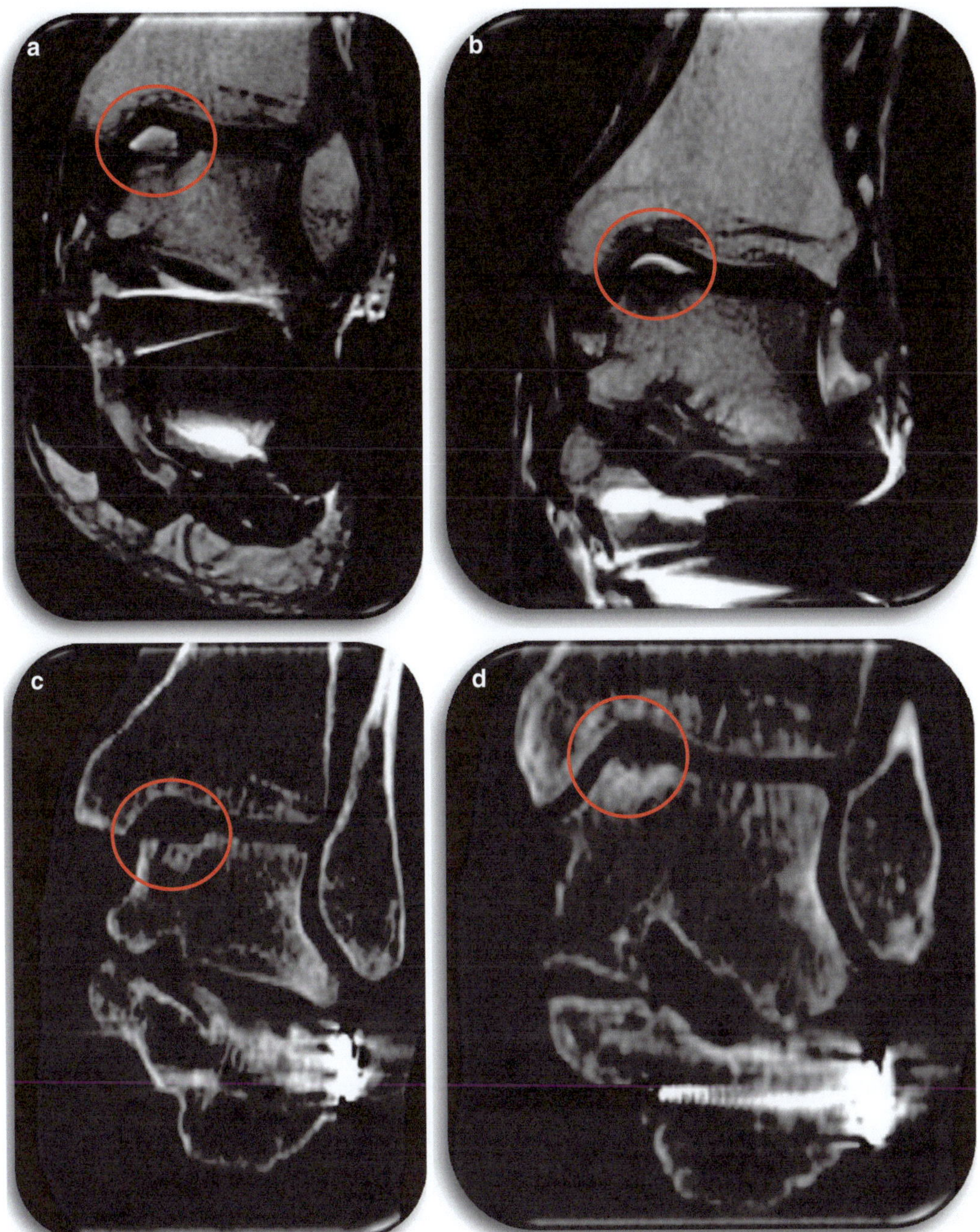

Fig. 32.5 Osteochondral lesions on the talar dome: MRI compared with CT image in coronal view. Preoperative MRI (**a**), 2-years follow-up MRI (**b**), preoperative CT scan (**c**), 2-years follow-up CT scan (**d**)

Considering that OCTLs are common entities resulting from ankle sprains or fractures in young adults who actively practice sports, return to activity seems to be a challenge. Symptomatic patients with talar chondral lesions left untreated are often unable to return to previous life and to practice sports without pain, or it results even impossible due to the high ankle functional demands. D'Ambrosi et al. in 2017 analyzed return to sports for patients after AT-AMIC pro-

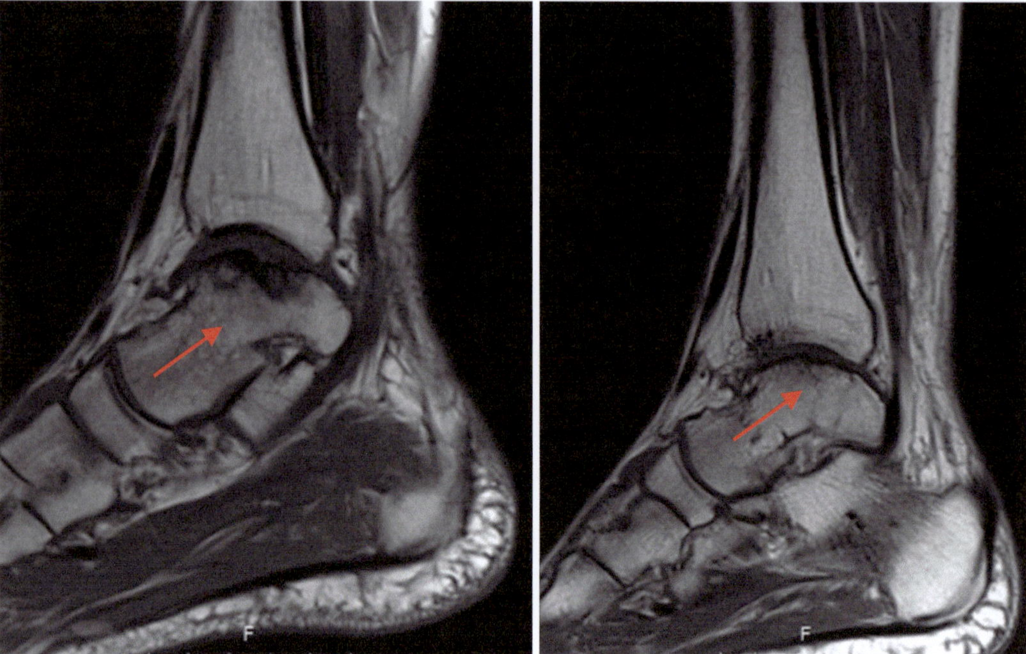

Fig. 32.6 MRI comparing a OCLTs preoperative and at 24 months follow-up in sagittal view

cedure for outcomes 2 years after surgery. Of 26 patients at 24-month follow-up analyzed, 80.8% returned to the same preinjury sport: all parameters showed a significant improvement both for physical activity and for functionality scales at a minimum follow-up of 24 months. An athletic and active population can achieve good outcomes with AT-AMIC procedure, improving ankle functionality and sports performance [6].

32.4　Take-home Message

AMIC technique is classically performed in open surgery with medial malleolar osteotomy. A new all-arthroscopic technique has been developed, with lower complication rates, less invasivity, less comorbidities and a faster recovery. Return to sports is possible in 80.8% of patients with good outcomes. To be older than 33 years and to be overweight are not absolute contraindications: good results are possible even in these groups of patients. Further studies will have to focus on the role of preoperative and postoperative bone edema in this kind of surgical procedure.

References

1. Usuelli FG, de Girolamo L, Grassi M, D'Ambrosi R, Montrasio UA, Boga M. All-arthroscopic autologous matrix-induced chondrogenesis for the treatment of osteochondral lesions of the talus. Arthrosc Tech. 2015;4:e255–9.
2. Choi WJ, Park KK, Kim BS, Lee JW. Osteochondral lesion of the talus: is there a critical defect size for poor outcome? Am J Sports Med. 2009;37:1974–80.
3. Usuelli FG, D'Ambrosi R, Maccario C, Boga M, de Girolamo L. All-arthroscopic AMIC® (AT-AMIC®) technique with autologous bone graft for talar osteochondral defects: clinical and radiological results. Knee Surg Sports Traumatol Arthrosc. 2018;26(p):875–81.
4. D'Ambrosi R, Maccario C, Serra N, Liuni F, Usuelli FG. Osteochondral lesions of the talus and autologous matrix-induced chondrogenesis: is age a negative predictor outcome?. Arthroscopy. 2016;33(2):1–8.
5. Usuelli FG, Maccario C, Ursino C, Serra N, D'Ambrosi R. The impact of weight on arthroscopic osteochondral talar reconstruction. Foot Ankle Int. 2017;38(6):612–20.
6. D'Ambrosi R, Villafane JH, Indino C, Liuni FM, Berjano P, Usuelli FG. Return to sport after arthroscopic autologous matrix-induced chondrogenesis for patients with osteochondral lesion of the talus. Clin J Sport Med, 2017;27(3):1–6.

Osteochondral Lesions of the Ankle: Talus and Distal Tibia

Edward L. Baldwin III, Sachin Allahabadi, Brian C. Lau, and Annunziato Amendola

33.1 Introduction

Osteochondral lesions of the ankle are important injuries to recognize with osteochondral lesions of the talus (OLT) specifically being the most common form of this injury [1]. The overall prevalence of OLT is <1% without an apparent increase with age [2]. Of those who do have OLT, between 50% and 70% of these lesions can be attributed to trauma, such as ankle sprains or fractures [2]. Chondral injuries can be painful and result in significant deleterious impact on patients' daily lives and function. As such, patients who have persistent pain following sprains beyond the anticipated duration of recovery should be evaluated for possible chondral lesions. Due to the limited regenerative potential of articular cartilage, it is important to recognize these injuries early and treat them appropriately.

E. L. Baldwin III · B. C. Lau · A. Amendola (✉)
Department of Orthopaedic Surgery, Duke University College of Medicine, Durham, NC, USA
e-mail: Edward.baldwin.iii@duke.edu;
brian.lau@duke.edu; ned.amendola@duke.edu

S. Allahabadi
Department of Orthopaedic Surgery, University of California San Francisco, San Francisco, CA, USA
e-mail: Sachin.Allahabadi@ucsf.edu

33.2 Ankle Anatomy

The ankle joint is comprised of the articulation between the distal tibia and talar dome, with a lateral strut provided by the fibula and its syndesmotic attachments with the tibia. This inherent bony stability is augmented by the ligamentous structures about the ankle. The tibiotalar articulation is lined with articular type 2 hyaline cartilage. However, the tibiotalar joint cartilage differs from that of the knee. Most notably, the thickness of talar cartilage is 1–2 mm [3]. In contrast, the thickness of the articular cartilage of the knee can be up to 6 mm thick depending on the location of the knee [3].

The surface of the talus can be divided into 9 equivalent surface-area zones. The zone description was originally described by Raikin and Elias to quantify incidences of OLT based on reproducible location [4]. This grid system is not universally accepted, and often lesions of the talus are described in standard anatomical terms, i.e., anteromedial.

Studies have shown that the incidence and severity of OLT vary based on location on the talus [2, 5]. The most common location as described by Dahmen et al. are medial lesions with an incidence of 77%, followed by lateral lesions at 21%. Central lesions account for 2% and 0.4% of lesions are combined medial and lateral. Medially based lesions are generally larger

© ISAKOS 2022
A. Gobbi et al. (eds.), *Joint Function Preservation*, https://doi.org/10.1007/978-3-030-82958-2_33

and deeper and occur more insidiously while lateral lesions tend to be traumatic in nature [2, 6].

The varying size of OLTs dependent on lesion location is likely related to the forces causing the lesions. When there is shear force across the joint line, the resulting damage is usually to the superficial cartilage without damage to the underlying subchondral plate. The resulting shape of these lesions will be narrow and oval shaped, as with traumatic lateral lesions [2]. On the other hand, when forces are more axial in nature or there is torsional impaction about the ankle joint, the resulting osteochondral lesion is more likely to be more medially based and deeper and broader [2].

33.3 Treatment

Once OLTs have been identified, the next step is to develop the most appropriate treatment plan individualized for the patient. An initial course of non-operative management is generally appropriate in most cases. Conservative treatment can consist of regimens that include rest, immobilization, activity modification, or NSAIDs [1]. These conservative measures can also be augmented by physical therapy that focuses on lower extremity stability, balance, and proprioception since the majority of OLT have a traumatic component and a component of ligamentous injury [1].

Other more invasive but non-surgical options can include intra-articular injections. Mei-dan et al. evaluated the use of intra-articular hyaluronic acid and platelet-rich plasma (PRP) [7]. The study found short-term (over a period of 28 weeks) improvement in pain as measured by Visual Analog Scale (VAS) scores as well as American Orthopaedic Foot & Ankle Society (AOFAS) ankle-hindfoot scale functional scores [7]. However, the lasting effectives of injections are not known as most of the data regarding these treatments is only short- to mid-term with studies of approximately 6–12 months [7]. The lack of long-term follow-up in conjunction with the knowledge that these injections do not treat the underlying structural defects makes this treatment option less fitting for those seeking longer-term relief.

Operative management is reserved for when conservative management fails.

The two general categories of surgical treatments are repair and replacement [1]. Repair treatments include, but are not limited to, microfracture and bone marrow stimulations techniques and Autologous Chondrocyte Implantation (ACI), and replacement techniques include, but are not limited to, osteochondral allograft/autograft transfer. The goal of these treatments is to attempt to reproduce as best as possible the native mechanical, structural, and chemical properties of hyaline cartilage that was lost [1]. We will examine some of these different forms that this can take in the remainder of this chapter.

33.3.1 Reparative Procedures

33.3.1.1 Microfracture (Bone Marrow Stimulation)

Historically, microfracture or bone marrow stimulation was considered the first-line treatment for osteochondral lesions of the talus. The goal for microfracture is to breach the underlying subchondral bone at the area of the defect in order to release bone marrow elements with its healing factors to restore cartilage. Bone marrow stimulation may also be achieved by other means such as curettage deep to the subchondral plate to stimulate bleeding at the base of the lesion. Lesions that are most amenable to this treatment are generally small, less than 15 mm^2 [1, 8]. A recent systematic review suggests that lesions sized 10 mm^2 should be the maximum size indication for microfracture [9].

Microfracture is generally performed arthroscopically. The first step is to perform a thorough debridement of the lesion, including removal of damaged unstable cartilage and necrotic bone as well as the calcified cartilage layer (Figs. 33.1 and 33.2). Once an appropriate amount of debridement has been completed with a stable border and base, attention is then turned to the subchondral plate. An awl or micro-drill is used to penetrate the subchondral plate to a depth of 3–4 mm or until fat droplets from underlying bone marrow are visible. Microfracture through the subchondral plate yields a bleeding base

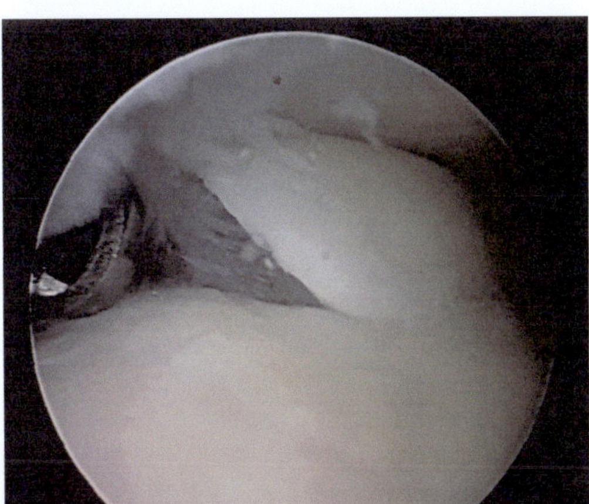

Fig. 33.1 This image is flap of cartilage from medial/shoulder talus

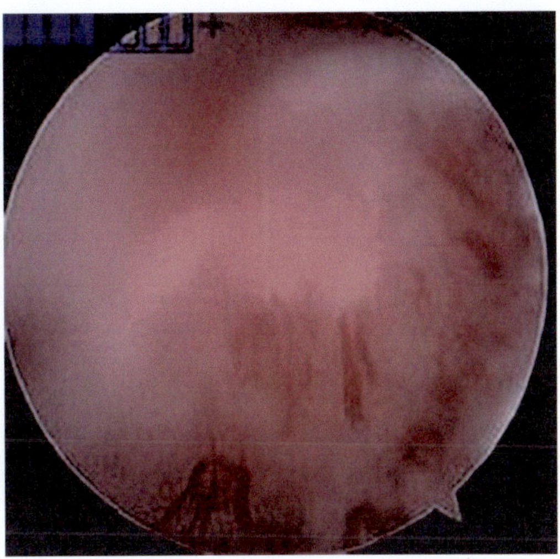

Fig. 33.3 Debrided osteochondral defect with bleeding base

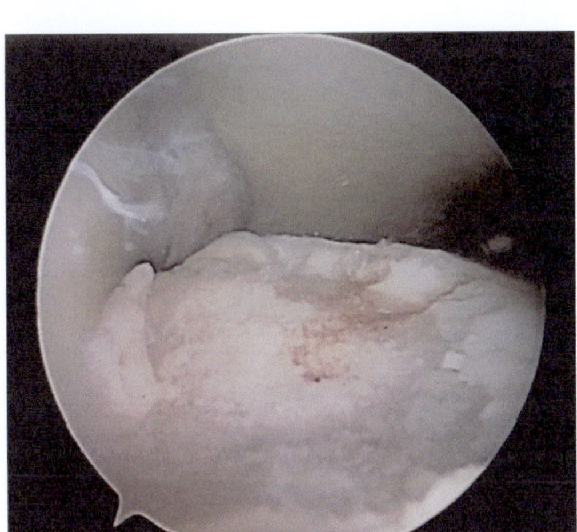

Fig. 33.2 This image is the area debrided

which (Fig. 33.3) is believed to promote healing factors into the defect to promote healing of the cartilage. Generally, the cartilage that regenerates is more fibrous in nature and has less biomechanical strength than normal hyaline cartilage [1]. Although the cartilage is not normal structurally, studies have shown that there are good short- (2 years) term outcomes of this procedure in the ankle, with up to 85% excellent clinical outcomes [1]. VanBergen et al. demonstrated average AOFAS ankle-hindfoot scores of 88 at 141 months following OLT microfracture.

Additionally, microfracture or bone marrow stimulation can be performed with biologic adjuvants. These include PRP, demineralized bone matrix (DBM) (e.g., BioCart, Arthrex Inc., Naples FL) (Fig. 33.4), or juvenile chondrocytes (e.g., DeNovo, Zimmer Inc., Warsaw, IN) (Fig. 33.5). Currently, the literature is limited in outcomes on the synergistic effects of microfracture with juvenile chondrocytes or DBM. Hogan et al., however, reviewed the use of PRP and hyaluronic acid (HA). Three of the studies they highlighted were randomized control trials with 10–24 months of follow-up. This compilation of studies demonstrated that microfracture with augmentation from HA or PRP showed improved outcomes when compared to microfracture alone (Figs. 33.1, 33.2, 33.3, 33.4, 33.5, 33.6, 33.7, 33.8, 33.9, 33.10, 33.11, and 33.12) [10]. Regarding the use of juvenile cartilage compared to microfracture, Karnovsky et al. evaluated their series of 50 patients, of which 30 were treated with microfracture and 20 augmented with juvenile chondrocytes. The average follow-up was 30.9 months and there was no significant difference demonstrated in VAS pain scores or Foot and Ankle Outcome Score (FAOS) post-operatively between the groups [11]. There is a paucity of data on the use of DBM with PRP or bone marrow aspirate and further evidence is needed.

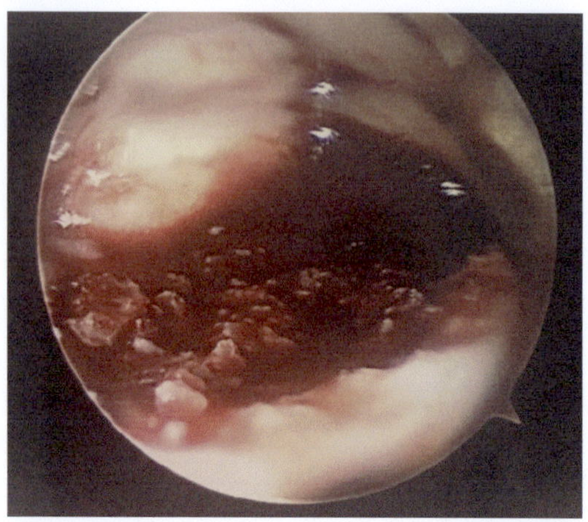

Fig. 33.4 This is after microfracture with bone grafting and BioCart placement

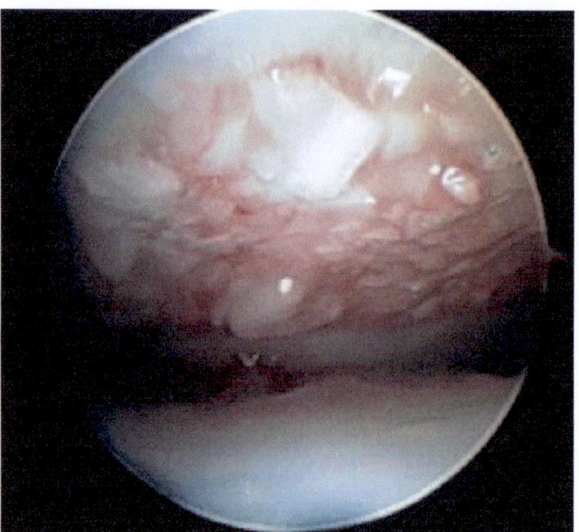

Fig. 33.5 Osteochondral defect distal tibia after microfracture and DeNovo placement

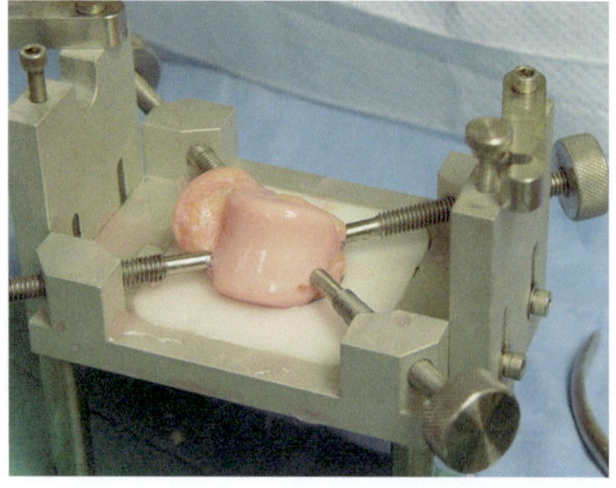

Fig. 33.6 OATS work station for harvesting allograft plug

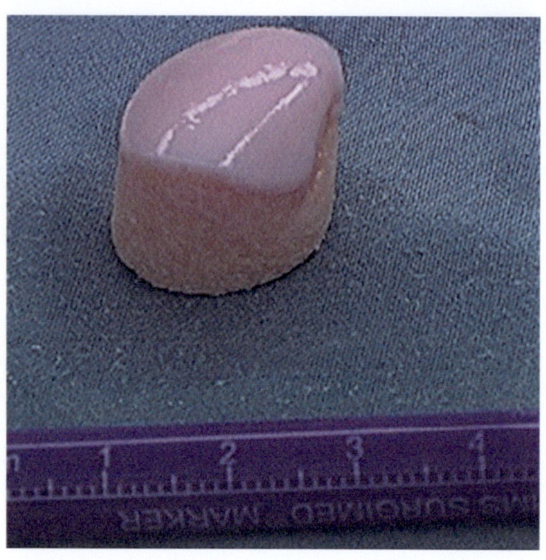

Fig. 33.7 Osteochondral Allograft plug pending implantation

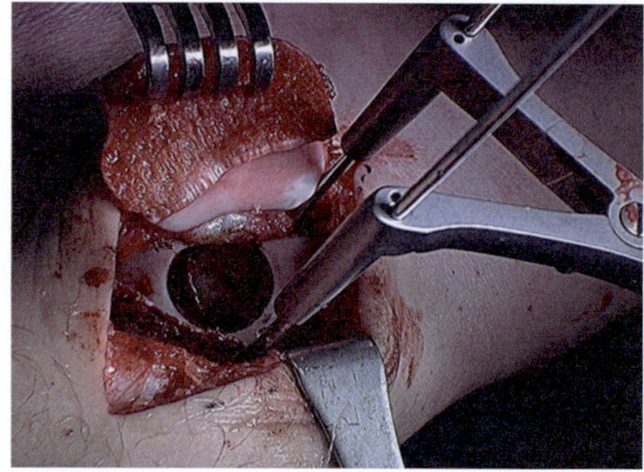

Fig. 33.8 Defect after coring, pending implantation of allograft

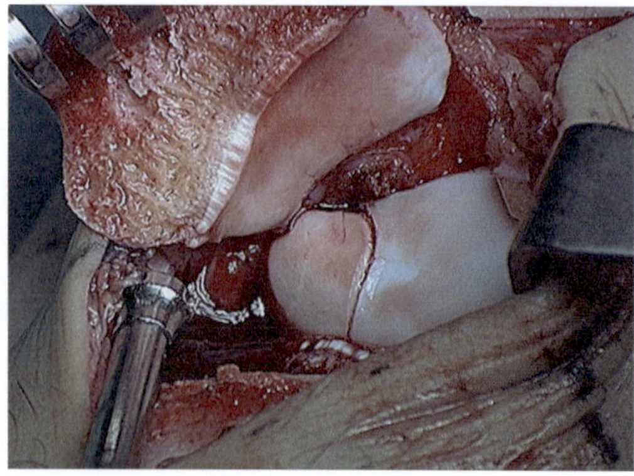

Fig. 33.9 Implanted Allograft donor plug

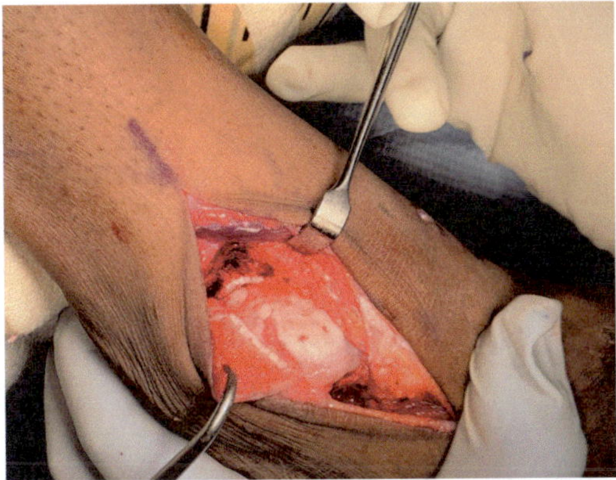

Fig. 33.10 Chaput osteotomy

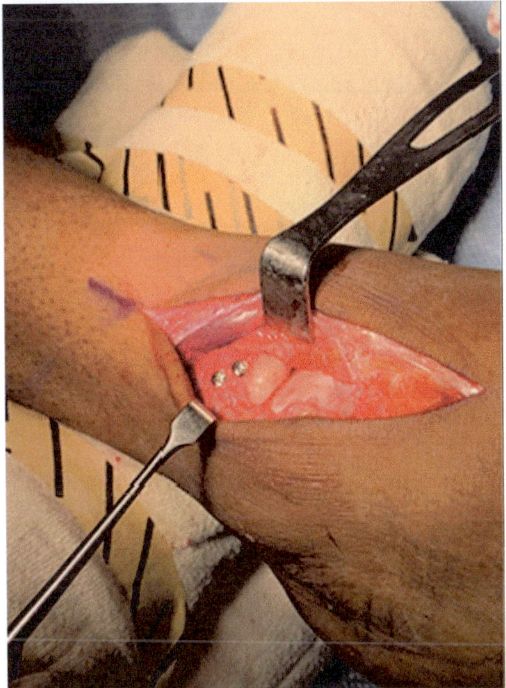

Fig. 33.11 Chaput Osteotomy follwing screw fixation

33.4 Autologous Chondrocyte Implantation (ACI)

ACI initially developed for the treatment of osteochondral lesions of the knee, but this technique may also be used for larger OLTs or when a prior bone marrow stimulation procedure has been unsuccessful. The goal of ACI is to regenerate native hyaline cartilage using one's own cells.

This is done in a two-step process, with initial procedure to harvest healthy cartilage from the patient. These chondrocytes are then cultured in vitro. After 6–8 weeks, once appropriate maturation has been achieved in the lab, the chondrocytes can be re-implanted into the talar defect where they continue to mature and fill in the void left by the defect [12]. To promote stability of the re-implanted cells, a periosteal sleeve that can be harvested from the ipsilateral distal tibia can be placed over the cells. Giannini et al. reviewed a series of 46 patients who underwent ACI for OLT with mean of 7.1 years follow-up, AOFAS scores were evaluated pre-op and at designated intervals post-op. They showed continued improvement of mean score until final follow-up (average of 87.2 months with final AOFAS score mean 92), although there were 3/46 (6.5%) treatment failures [13]. Of those three failures all demonstrated an irregular articular surface with fibro cartilaginous tissue.

33.5 Matrix-induced Autologous Chondrocyte Implantation (MACI)

MACI represents the second generation of autologous chondrocyte implantation. This again is a two-step process. The chondrocytes are harvested from the patient and placed on a scaffolding matrix. On the matrix, the chondrocytes are also grown in vitro and in a second surgery is replaced within the defect. The difference between MACI when compared to ACI is the polymer scaffolding in which the chondrocytes are cultured in with MACI [1]. This scaffolding negates the need for a periosteal component, as the scaffolding provides the stability necessary for the chondrocytes within the defect. Giza et al. retrospectively reviewed their series of 10 talar lesions treated with MACI. All patients had 2-year follow-up and they demonstrated statistically significant improvement in AOFAS hindfoot scores from 61.2 avg. to 73.3 [14]. The average size of the lesions treated was 12.9 × 10.7 mm (138.0 mm²) [14] Kreulen et al. evaluated 10 patients undergoing MACI with 7-year follow-up and similarly

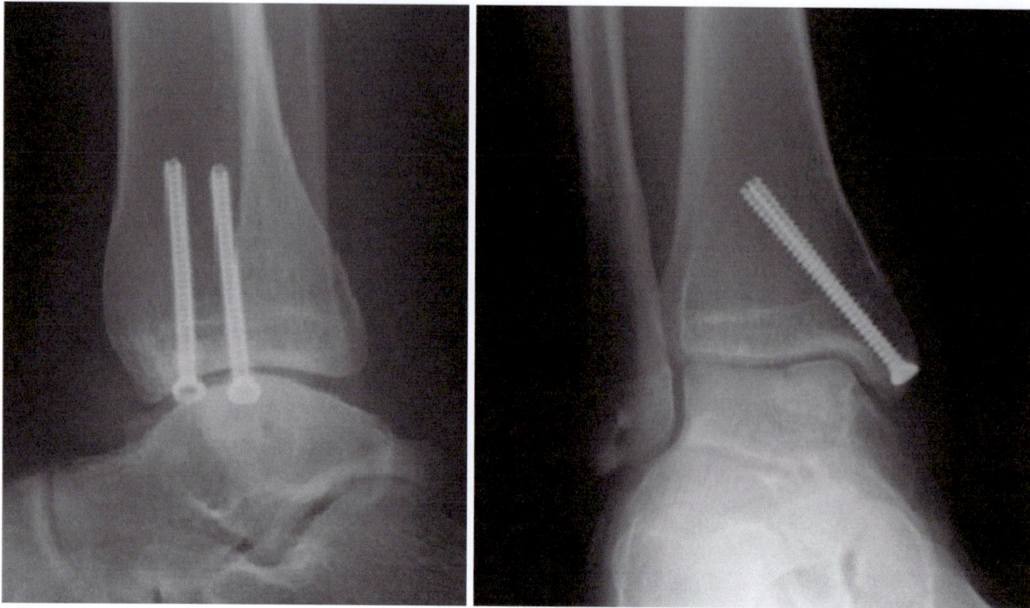

Fig. 33.12 Radiograph of Repaired medial malleolar osteotomy with fully threaded screws

found improvement in AOFAS scores of 78.3 compared to 61.8 pre-operatively. They also demonstrated improvements in physical functioning, social functioning, and lack of bodily pain [15].

33.5.1 Replacement Procedures: Osteochondral Autograft/ Allograft Transfer

Osteochondral autograft transfer in management of OLT includes re-implanting the patient's own native cartilage from a donor site into the defect of the talus. The donor site for the autologous graft is generally a non-weight bearing portion of the ipsilateral lateral femoral condyle. Once the cartilage is harvested from the donor site it is then transplanted to the lesion of the talus. This procedure may be utilized not only when defects are larger (>150 mm^2), but also if the defect has a cystic component in the subchondral bone [16]. These autologous transplantations have shown good outcomes within both the general population and the athletic population. Kennedy et al. demonstrated a mean FAOS increase from 52.67 to 86.19 in 72 patients with avg. 28 months f/u

[17]. Hangody et al. showed good outcomes in 39 high level athletes (with in a larger group of 354 patients) suffering from talar OCDs, including good-to-excellent outcomes in 92% of the talar lesions treated based on Hannover scores [18].

When defects begin to progress beyond the shoulder or larger portion of the talus then an allograft may be the preferred method of treatment of the osteochondral defect. Allografts can be obtained from a size-matched donor following pre-operative CT scans (Fig. 33.6) talar harvesting table. These allografts allow for filling of larger or irregularly shaped (Figs. 33.7, 33.8, and 33.9) defects in which trying to use an autograft may yield increased donor site morbidity [19]. Generally, use of allograft is considered a salvage procedure for larger and irregular lesions. Allografts have also been shown to be a viable treatment option in the short- and medium-term. Gaul et al. reported on 20 ankles undergoing osteochondral allograft transfer with average graft size of 3.8 cm^2. This study showed a mean 30-point improvement on the Olerud–Molander Ankle Score [20].

Frequently, osteotomies are needed to provide adequate access for autograft or allograft placement. A medial malleolar osteotomy is commonly used for medial lesions. A Chaput osteotomy may be used for laterally based lesions (Figs. 33.10, 33.11, and 33.12). (Reference Am J Sports Med. 2006 Sep;34(10):1457–63. Epub 2006 Apr 24. Talar dome access for osteochondral lesions. Muir D[1], Saltzman CL, Tochigi Y, Amendola N.)

Shimozono et al. retrospectively compared groups receiving autograft (25 patients) vs. allograft (16 patients) through a follow-up of 22–26 months. They showed that there was a significant improved FAOS and Short Form-12 (SF-12) scores with both groups [1]. However, the improvement of these scores was statistically significantly better in the autograft group as compared to the allograft group [1].

33.6 Osteochondral Lesions of the Distal Tibia

33.6.1 Epidemiology and Injury Patterns

The large majority of osteochondral lesions in the ankle are located in the talus, with early reports of distal tibial lesions beginning in the late 1980s [21, 22]. On the other hand, the proportion that occur in the distal tibia is thought to be much smaller, with sources citing a ratio of 14:1 to 20:1 of talar:tibial lesions and a rate of 2.6% in a series of 880 ankle arthroscopies [23–25]. One report exists on a case of bilateral distal tibial lesions [26].

Osteochondral lesions of the distal tibial plafond (OLTP) are thought to be rarer than OLTs in part due to differing cartilage biology and mechanics. Distal tibial articular cartilage has shown to be stiffer than talar articular cartilage, with the anteromedial tibia having the largest modulus relative to the softest locations noted in the posterolateral and posteromedial talus [27]. Furthermore, the convexity of the talus may contribute to differential loading than the concavity of the distal tibia [23, 25, 28].

OLTP may present either in isolation or coexisting with OLT. The incidence of coexisting OLT and OLTP lesions has been cited between 15.8% and 35% [25, 29, 30]. However, variances in study design may make interpretation of these numbers less reliable; some studies exclude non-operatively managed patients, some evaluate for coexisting OLTs in those with OLTPs, and some evaluate for coexisting OLTP in known OLTs.

OLTPs have been characterized by lesion location in a 3 × 3 grid creating 9 zones of equivalent surface area as described by Elias et al. [25, 31]. A zone 10 has also been described as an osteochondral lesion of the fibula [30]. "Kissing" coexisting lesions refers to lesions that occur in the same zone of injury on both the talus and tibia [29, 32].

In its original description in a study of 38 MRI scans with OLTP, 14 (37%) were medial (zones 1, 4, and 7), 11 (29%) were lateral (zones 3, 6, and 9), and 13 (34%) were central (zones 2, 5, and 8) [25]. In the AP plane, 9 (24%) were anterior (zones 1–3), 15 (39%) were posterior (zones 7–9), and 14 (37%) were central (zones 4–6) [25]. The most frequent individual zones affected were zone 4 (medial central plafond, 21%) and zone 7 (posterior medial plafond, 16%) [25].

One case series of 83 patients undergoing surgical management for OLT found 26 (31%) of patients to have coexisting lesions and 9 of those 26 (35%) were kissing lesions [29]. The most common affected zones in this study were zones 2, 4, and 5 of the tibial plafond, with 5 lesions in each of these zones (19%) [29]. Zones 2 and 4 were also shown to be the most common coexisting lesion locations (over 35% together) in another study of 297 OLTs on MRI [30]. However, lesion location may ultimately be variable [33].

Some studies have evaluated potential risk factors for or associations with coexisting lesions. Those with coexisting lesions may be of older age ($p = 0.009$) and more likely to have lateral talar OCL ($p = 0.012$) on multiple linear regression [29]. In addition, there may be an association of higher grade of talus cartilage damage, with coexisting lesions associated with ICRS grade 4 talar changes ($p = 0.034$) [29]. Furthermore, existing posterior tendon pathology or ankle ligamentous injury may be more likely in those with coexisting lesions compared to isolated OLT ($p < 0.05$) [30].

While some OLTPs may be recognized on radiographic imaging [23], MRI is the study of choice for evaluation osteochondral lesions in general.

33.7 Outcomes

Due to the relative rarity of reports on OLTP, high-quality outcome studies are limited. Most studies are retrospective in nature or are case reports/series.

33.7.1 Patient/Lesion Factors

Studies thus far have not demonstrated an association between lesion location and functional or imaging outcomes [24, 25, 34]. Furthermore, those with coexisting OLT and OLTP do not appear to have statistically significant differences in functional outcomes compared to those with isolated OLTP or compared to those with isolated OLT [24, 29, 33, 34].

33.7.2 Surgical Factors/Procedures

Similar cartilage repair and cartilage restoration procedures have been trialed in the distal tibial plafond as in the talus.

33.7.2.1 Microfracture

Arthroscopic microfracture has been trialed and reported in several studies. In general, microfracture is utilized for smaller-sized osteochondral lesions [42].

A retrospective study of 31 ankles undergoing arthroscopic microfracture with average lesion size 38 mm^2 and minimum two-year follow-up demonstrated significant improvements in FAOS from 50.5 to 74.2 ($p < 0.01$) and SF-12 scores from 38.7 to 59.5 ($p < 0.01$) [34]. AOFAS Ankle-Hindfoot scores also demonstrate significant improvements after microfracture, from 35.2 to 50.4 ($p < 0.05$) in another study in which 3/13 (23.1%) had repeat surgery and 1/13 had persistent pain and disability [33].

Tibial articular cartilage healing after microfracture has been assessed with Magnetic Resonance Observation of Cartilage Repair Tissue (MOCART) in 23 of the ankles; increasing age ($r = -0.43$, $p = 0.04$) and increasing lesion size ($r = -0.44$, $p = 0.04$) were associated with lower MOCART scores [34]. One downside to microfracture is that results demonstrate inferior repair tissue relative to normal hyaline cartilage on MRI evaluation [34].

33.7.2.2 Excision, Curettage, Abrasion Arthroplasty

Mologne and Ferkel described a cohort of 17 patients with mean follow-up of 44 months with OLTP that were treated with excision, curettage, and abrasion arthroplasty [24]. Of these, some patients had additional treatments based on surgeon choice: two underwent microfracture, five underwent transmalleolar drilling of the lesion in cases with deeper lesions, and two had autologous iliac crest bone graft to fill larger cystic cavities [24]. Significant improvements were noted in AOFAS Ankle-Hindfoot scores from pre- (52) to post-operatively (87, $p < 0.001$) [24]. Subjective questionnaires noted some improvement in 15/17 patients though five patients had restriction of ankle ROM with mean loss of 13 degrees of arc of motion [24]. Degenerative changes were noted in 8/15 patients with radiographs at final follow-up [24].

33.7.2.3 Bone Marrow-Derived Cell Transplantation (BMDCT)

BMDCT is performed under the premise of utilizing bone marrow cells (from bone marrow aspirate concentrate) with wide cell differentiation capabilities and stimulation with specific factors as may be seen in platelet-rich fibrin (PRF). The goal is for a one-step procedure to serve as a scaffold and stimulate osteochondral growth more similar to native hyaline cartilage [10]. The extent of studies on this technique and its application to the distal tibial plafond is limited. However, one study of 27 patients found improvement in mean AOFAS scores from 52.4 to 80.6 at final follow-up of up to 72 months [10]. It is important to note that the authors found a significantly higher rate of improvement in AOFAS scores in those with smaller lesions <150 mm^2 compared to those with larger lesions at 36 months follow-up (p = 0.038) and higher increase in scores for more shallow lesions <4 mm compared to deeper lesions at all time points (72 months, p = 0.003) [10]. Therefore, the effects of BMDCT in management of OLTP may be most beneficial for smaller and shallower lesions, but further studies and comparisons are needed to draw conclusions.

33.7.2.4 Osteochondral Autograft/ Allograft Transfer

Osteochondral auto- or allograft transfer is used in management of osteochondral lesions in general for larger size lesions and, for instances, when prior procedures such as bone marrow stimulation with microfracture were unsuccessful [40, 41]. A case report by Chapman and Mann describes the use of osteochondral allografting (graft from the talus) in the distal tibia in a patient who had previously failed arthroscopic microfracture [35]. That patient had improvement postoperatively with radiographic incorporation of the graft and maintenance of joint space [35]. Ueblacker et al. reported on two patients with OLTPs utilizing retrograde autografts (obtained from the femoral trochlea) with placement guided by an anterior cruciate ligament drill guide [36]. Both of these patients had post-operative healing of the plug and congruent chondral surfaces on MRI [36]. Osteochondral auto- and allograft transfers warrant more investigation for OLTP, particularly as a salvage procedure when other procedures are unsuccessful.

33.7.2.5 Other Techniques

Several other surgical techniques to manage OLTPs have been described in case reports. Each of these techniques mentioned involves cases with improvement in patient outcome, however much larger numbers and comparative studies are needed to evaluate the true effects and determine the appropriate indications [43].

Autologous chondrocyte transplantation has been utilized to treat distal tibial lesions. One report describes taking an arthroscopic cartilage biopsy from the femoral condyle, culturing the chondrocytes, and implanting the chondrocytes with a periosteal patch sealed with fibrin glue [37].

Autologous Matrix-Induced Chondrogenesis (AMIC) utilizes a collagen matrix and autologous bone graft to stimulate chondrogenesis. AMIC has been described in a case report for use in the distal tibia with use of collagen type I and III matrix, iliac crest autograft, and fibrin glue [38]. That patient returned to sport 2 months post-operatively and had maintained benefits at 3 years with intact and healthy cartilage at the three-year quantitative analysis MRI [38].

One case report describes the use of a synthetic osteochondral plug, another one-step procedure utilizing osseous and cartilaginous substitutes in attempt to induce healing of more hyaline-like cartilage [39]. In that report, while the patient had improvement in function and return to activities, follow-up MRI demonstrated more of a fibrous cartilage composition than hyaline [39].

In summary osteochondral defects of the talus and distal tibia can cause significant disability to those patients who are living with it. The treatment of these lesions is multifactorial depending on size of lesion, stability of underlying subchondral bone, as well as previous treatments. The overarching goal is to restore the cartilage surface of the talus to as close to the native condition as possible.

References

1. Shimozono Y, Hurley ET, Nguyen JT, Deyer TW, Kennedy JG. Allograft compared with autograft in osteochondral transplantation for the treatment of osteochondral lesions of the talus. J Bone Joint Surg. 2018;100(21):1838–44. https://doi.org/10.2106/JBJS.17.01508.
2. Pereira H, Cengiz IF, Vilela C, Ripoll PL, Espregueira-Mendes J, Oliveira JM, et al. Emerging concepts in treating cartilage, osteochondral defects, and osteoarthritis of the knee and ankle. Adv Exp Med Biol. 2018:25–62. https://doi.org/10.1007/978-3-319-76735-2_2.
3. Chubinskaya S, Cotter EJ, Frank RM, Hakimiyan AA, Yanke AB, Cole BJ. Biologic characteristics of shoulder articular cartilage in comparison to knee

and ankle articular cartilage from individual donors. Cartilage. 2019:194760351984774. https://doi.org/10.1177/1947603519847740.

4. Raikin SM, Elias I, Zoga AC, Morrison WB, Besser MP, Schweitzer ME. Osteochondral lesions of the talus: localization and morphologic data from 424 patients using a novel anatomical grid scheme. Foot Ankle Int. 2007;28(2):154–61. https://doi.org/10.3113/fai.2007.0154.

5. Dahmen J, Lambers KTA, Reilingh ML, Bergen CJAV, Stufkens SAS, Kerkhoffs GMMJ. No superior treatment for primary osteochondral defects of the talus. Knee Surg Sports Traumatol Arthrosc. 2017;26(7):2142–57. https://doi.org/10.1007/s00167-017-4616-5.

6. Georgiannos D, Bisbinas I, Badekas A. Osteochondral transplantation of autologous graft for the treatment of osteochondral lesions of talus: 5- to 7-year follow-up. Knee Surg Sports Traumatol Arthrosc. 2014;24(12):3722–9. https://doi.org/10.1007/s00167-014-3389-3.

7. Mei-Dan O, Carmont MR, Laver L, Mann G, Maffulli N, Nyska M. Platelet-rich plasma or hyaluronate in the management of osteochondral lesions of the talus. Am J Sports Med. 2012;40(3):534–41.

8. Murawski CD, Kennedy JG. Operative treatment of osteochondral lesions of the talus. J Bone Joint Surg. 2013;95(11):1045–54. https://doi.org/10.2106/jbjs.l.00773.

9. Ramponi L. Lesion size is a predictor of clinical outcomes after bone marrow stimulation for osteochondral lesions of the talus: a systematic review. Am J Sports Med. 2017;45(7):1698–705.

10. Hogan MV, Hicks JJ, Chambers MC, Kennedy JG. Biologic adjuvants for the Management of Osteochondral Lesions of the talus. J Am Acad Orthop Surg. 2019;27(3):e105–11. https://doi.org/10.5435/JAAOS-D-16-00840.

11. Karnovsky SC, DeSandis B, Haleem AM, Sofka CM, O'Malley M, Drakos MC. Comparison of juvenile Allogenous articular cartilage and bone marrow aspirate concentrate versus microfracture with and without bone marrow aspirate concentrate in arthroscopic treatment of Talar osteochondral lesions. Foot Ankle Int. 2018;39(4):393–405. https://doi.org/10.1177/1071100717746627.

12. Matthew M, Eric G, Martin S. Cartilage transplantation techniques for Talar cartilage lesions. J Am Acad Orthop Surg. 2009;17(7):407–14. Cited in: Journals@Ovid Full Text at http://ovidsp.ovid.com/ovidweb.cgi?T=JS&PAGE=reference&D=ovftk&NEWS=N&AN=00124635-200907000-00001. Accessed May 04, 2020

13. Giannini S, Buda R, Ruffilli A, Cavallo M, Pagliazzi G, Bulzamini MC, Desando G, Luciani D, Vannini F. Arthroscopic autologous chondrocyte implantation in the ankle joint. Knee Surg Sports Traumatol Arthrosc. 2014;22(6):1311–9.

14. Giza E, Sullivan M, Ocel D, Lundeen G, Mitchell ME, Veris L, Walton J. Matrix-induced autologous chondrocyte implantation of talus articular defects. Foot Ankle Int. 2010;31(9):747–53. https://doi.org/10.3113/fai.2010.0747.

15. Kreulen C, Giza E, Walton J, Sullivan M. Seven-year follow-up of matrix-induced autologous implantation in talus articular defects. Foot Ankle Spec. 2018;11(2):133–7. https://doi.org/10.1177/1938640017713614.

16. Ross KA, Robbins J, Easley ME, Kennedy JG. Autologous osteochondral transplantation for osteochondral lesions of the talus. Oper Tech Orthop. 2014;24(3):171–80. https://doi.org/10.1053/j.oto.2014.02.012.

17. Kennedy JG, Murawski CD. The treatment of osteochondral lesions of the talus with autologous osteochondral transplantation and bone marrow aspirate concentrate: surgical technique. Cartilage. 2011;2(4):327–36. https://doi.org/10.1177/1947603511400726.

18. Hangody L, Dobos J, Baló E, et al. Clinical experiences with autologous osteochondral mosaicplasty in an athletic population: a 17-year prospective multicenter study. Am J Sports Med. 2010;6:1125–33.

19. Bisicchia S, Rosso F, Amendola A. Osteochondral allograft of the talus. Iowa Orthop J. 2014;34:30–7.

20. Gaul F, Tírico LEP, McCauley JC, Pulido PA, Bugbee WD. Osteochondral allograft transplantation for osteochondral lesions of the talus: midterm follow-up. Foot Ankle Int. 2019;40(2):202–9. https://doi.org/10.1177/1071100718805064.

21. Bauer M, Jonsson K, Lindén B. Osteochondritis dissecans of the ankle. A 20-year follow-up study. J Bone Joint Surg Br. 1987;69:93–6.

22. Parisien JS, Vangsness T. Operative arthroscopy of the ankle. Three years' experience. Clin Orthop Relat Res. 1985;199:46–53.

23. Bui-Mansfield LT, Kline M, Chew FS, Rogers LF, Lenchik L. Osteochondritis dissecans of the tibial plafond: imaging characteristics and a review of the literature. AJR Am J Roentgenol. 2000;175:1305–8.

24. Mologne TS, Ferkel RD. Arthroscopic treatment of osteochondral lesions of the distal tibia. Foot Ankle Int. 2007;28:865–72.

25. Elias I, Raikin SM, Schweitzer ME, Besser MP, Morrison WB, Zoga AC. Osteochondral lesions of the distal tibial plafond: localization and morphologic characteristics with an anatomical grid. Foot Ankle Int. 2009;30:524–9.

26. Sopov V, Liberson A, Groshar D. Bilateral distal tibial osteochondral lesion: a case report. Foot Ankle Int. 2001;22:901–4.

27. Athanasiou KA, Niederauer GG, Schenck RC. Biomechanical topography of human ankle cartilage. Ann Biomed Eng. 1995;23:697–704.

28. Hall FM. Osteochondritis dissecans of the tibial plafond. AJR Am J Roentgenol. 2001;176:1328–9.

29. Irwin RM, Shimozono Y, Yasui Y, Megill R, Deyer TW, Kennedy JG. Incidence of coexisting Talar and Tibial osteochondral lesions correlates with patient age and lesion location. Orthop J Sports Med. 2018;6:2325967118790965.

30. You JY, Lee GY, Lee JW, Lee E, Kang HS. An osteochondral lesion of the distal tibia and fibula in patients with an osteochondral lesion of the talus on MRI: prevalence, location, and concomitant ligament and tendon injuries. AJR Am J Roentgenol. 2016;206:366–72.

31. Elias I, Zoga AC, Morrison WB, Besser MP, Schweitzer ME, Raikin SM. Osteochondral lesions of the talus: localization and morphologic data from 424 patients using a novel anatomical grid scheme. Foot Ankle Int. 2007;28:154–61.

32. Canosa J. Mirror image osteochondral defects of the talus and distal tibia. Int Orthop. 1994;18:395–6.

33. Cuttica DJ, Smith WB, Hyer CF, Philbin TM, Berlet GC. Arthroscopic treatment of osteochondral lesions of the tibial plafond. Foot Ankle Int. 2012;33:662–8.

34. Ross KA, Hannon CP, Deyer TW, Smyth NA, Hogan M, Do HT, Kennedy JG. Functional and MRI outcomes after arthroscopic microfracture for treatment of osteochondral lesions of the distal tibial plafond. J Bone Joint Surg Am. 2014;96:1708–15.

35. Chapman CB, Mann JA. Distal tibial osteochondral lesion treated with osteochondral allografting: a case report. Foot Ankle Int. 2005;26:997–1000.

36. Ueblacker P, Burkart A, Imhoff AB. Retrograde cartilage transplantation on the proximal and distal tibia. Arthroscopy. 2004;20:73–8.

37. Nakamae A, Engebretsen L, Peterson L. Autologous chondrocyte transplantation for the treatment of massive cartilage lesion of the distal tibia: a case report with 8-year follow-up. Knee Surg Sports Traumatol Arthrosc. 2007;15:1469–72.

38. Miska M, Wiewiorski M, Valderrabano V. Reconstruction of a large osteochondral lesion of the distal tibia with an iliac crest graft and autologous matrix-induced chondrogenesis (AMIC): a case report. J Foot Ankle Surg. 2012;51:680–3.

39. Pearce CJ, Lutz MJ, Mitchell A, Calder JDF. Treatment of a distal tibial osteochondral lesion with a synthetic osteochondral plug: a case report. Foot Ankle Int. 2009;30:900–3.

40. Flynn S, Ross K, Hannon C, Yasui Y, Newman H, Murawski C, Deyer T, Do H, Kennedy J. Autologous osteochondral transplantation for osteochondral lesions of the talus. Foot Ankle Int. 2015;37 https://doi.org/10.1177/1071100715620423.

41. Vantienderen RJ, Dunn JC, Kusnezov N, Orr JD. Osteochondral allograft transfer for treatment of osteochondral lesions of the talus: a systematic review. Arthroscopy. 2017;33(1):217–22. https://doi.org/10.1016/j.arthro.2016.06.011.

42. van Bergen CJ, Kox LS, Maas M, Sierevelt IN, Kerkhoffs GM, van Dijk CN. Arthroscopic treatment of osteochondral defects of the talus outcomes at eight to twenty years of follow-up. J Bone Joint Surg Am. 2013;95-A:519–25.

43. Baldassarri M, Perazzo L, Ricciarelli M, Natali S, Vannini F, Buda R. Regenerative treatment of osteochondral lesions of distal tibial plafond. Eur J Orthop Surg Traumatol. 2018;28:1199–207.

Macarena Morales and Eleonora Irlandini

34.1 Introduction

The goal of cartilage repair is to restore the articular surface and prevent the progression of the focal cartilage injury to end-stage osteoarthritis. Several clinical studies have shown the importance of restoring a smooth superficial surface, and promoting the restoration of a normal bone cartilage interface with acceptable subchondral flexibility.

Different treatment options are available to prevent and delay the progression of osteoarthritis. A broad spectrum of treatments are available, from non-pharmacological modalities to dietary supplements and pharmacological therapies. Generally, patients should be treated with the less invasive therapies before proceeding to a possible surgical treatment [1]. Prevention is crucial to delay the pathology's progression and avoid arthroplasty at an early stage (Fig. 34.1).

This chapter will focus on the non-surgical treatments, including non-pharmacological therapies, such as exercise, nutraceuticals, and the huge world of pharmacological therapies.

34.2 Non-Pharmacological Treatment

34.2.1 Exercise

Exercise, a cornerstone of the treatments for osteoarthritis, has shown to have enough evidence in order to achieve a better recovery, with less pain and cartilage protection.

One of the aims is to improve altered knee joint biomechanics, and excessive joint loading achieved mainly with strength exercises. This applies primarily to acute lesions or early osteoarthritis, but in later stages of osteoarthritis, studies report a positive but short-term effect in interventions to "unload" the medial compartment [2, 3].

There is enough evidence that physical activity and joint loading promote an increase in cartilage volume, and by itself, it does not correlate with joint narrowing [4]. However, overload and excessive training prompt joint degeneration, as well as an unload correlate with a decrease in cartilage thickness.

Range of motion, neuromuscular exercises, and stretching have benefits in pain modulation. It also reduces soft tissue inflammation, improves repair, extensibility, or stability of contractile and non-contractile tissues, facilitates movement, and improves function.

These interventions are safe, inexpensive, and should be encouraged [5, 6]. The choice

M. Morales (✉) · E. Irlandini
Orthopaedic Arthroscopic Surgery International (O.A.S.I.) Bioresearch Foundation, Gobbi N.P.O., Milan, Italy
e-mail: e.irlandini@oasiortopedia.it

© ISAKOS 2022
A. Gobbi et al. (eds.), *Joint Function Preservation*, https://doi.org/10.1007/978-3-030-82958-2_34

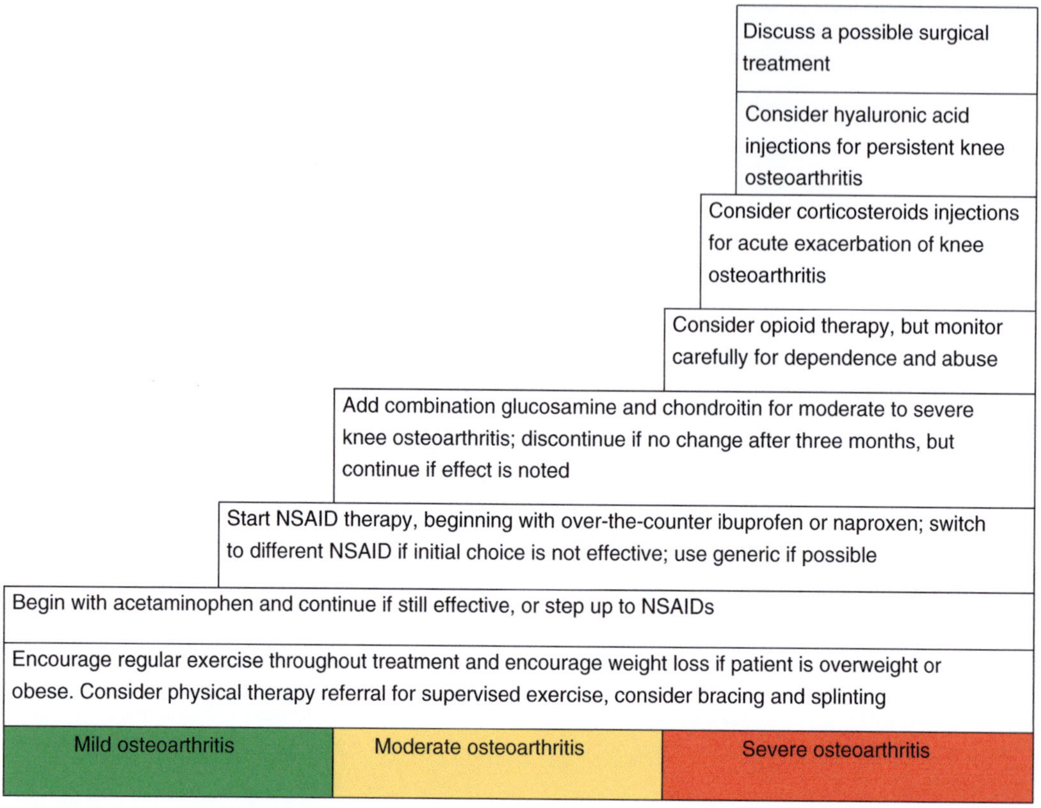

Discuss a possible surgical treatment

Consider hyaluronic acid injections for persistent knee osteoarthritis

Consider corticosteroids injections for acute exacerbation of knee osteoarthritis

Consider opioid therapy, but monitor carefully for dependence and abuse

Add combination glucosamine and chondroitin for moderate to severe knee osteoarthritis; discontinue if no change after three months, but continue if effect is noted

Start NSAID therapy, beginning with over-the-counter ibuprofen or naproxen; switch to different NSAID if initial choice is not effective; use generic if possible

Begin with acetaminophen and continue if still effective, or step up to NSAIDs

Encourage regular exercise throughout treatment and encourage weight loss if patient is overweight or obese. Consider physical therapy referral for supervised exercise, consider bracing and splinting

| Mild osteoarthritis | Moderate osteoarthritis | Severe osteoarthritis |

Fig. 34.1 Different approaches for the treatment of knee osteoarthritis

will depend upon the patient's characteristics, and they should follow an individualized program [4].

34.2.2 Physical Therapies

34.2.3 Pulsed Electromagnetic Fields (PEMF)

Stimulation, particularly low frequency (i.e., 3–50 Hz) with long durations of treatment (3–10 h a week), may exert a chondroprotective effect on articular cartilage by increasing proteoglycan synthesis and counteracting the catabolic activity of pro-inflammatory cytokines, together with positive effects also by inhibiting subchondral bone sclerosis, particularly in early OA stages [3, 7].

34.2.4 Low-Intensity Pulsed Ultrasound (LIPUS)

This may also stimulate chondrocyte proliferation and matrix production, with dose-dependent effects and greater attenuation of cartilage degeneration in the early OA phases [4]. It has promising results in vitro and has shown benefits in pain relief and knee functional recovery [8].

34.2.5 Extracorporeal Shockwave Therapy (ESW)

Has shown to downregulate in vitro the intracellular levels of TNF-α and IL-10, thus suggesting that ESW might restore TNF-α and IL-10 production by osteoarthritic chondrocytes at normal levels. It has a potential in reducing pain and improving knee function; the therapeutic effects may peak at 8 weeks after the completion of treatment. Further research is needed to arrive at a definitive conclusion [4, 9].

Other treatments such as transcutaneous electrical nerve stimulation (TENS) [3, 9], acupuncture, valgus braces, or lateral wedge insoles for pain and function in the knee appear limited [9], as well as diathermy, low-level laser therapy (LLLT), short-wave therapy, or magnetic stimulation [10]. More conclusive is the recommendation for neuromuscular electrical stimulation [11], which there is enough evidence to say is not appropriate for osteochondral treatment.

34.3 Nutraceuticals

34.3.1 Glucosamine

Glucosamine is an amino monosaccharide precursor of the synthesis of glycosaminoglycans, hyaluronic acid, and aggrecan. In vitro studies have demonstrated that glucosamine sulfate inhibits gene expression of OA. Moreover, long-term oral administration of glucosamine in guinea pigs reduces the destruction of cartilage and the inflammatory process [12]. The administration of oral glucosamine sulfate has shown to improve significantly pain and movement limitation in patients with OA over a 4-week treatment course [13].

Nevertheless, the evidence in the scientific literature is not well defined, and despite it is wide use, glucosamine showed inconsistent results between industry-sponsored and independent trials. Thus, heterogeneity among the different studies does not allow for scientific evidence in OA prevention. Whereas several clinical studies have described a symptomatic effect, independent placebo-controlled studies have not shown protective effects, alone or in combination with chondroitin sulfate, and the disease-modifying effect is controversial as well [4].

34.3.2 Chondroitin Sulfate

Chondroitin sulfate is a sulfated glycosaminoglycan, which is also an important component of the extracellular matrix. Similar to glucosamine, it presents unclear findings and recommendations for symptom relief [14]. A Cochrane review of randomized trials supports some benefits in pain relief in the short term, but most studies are of low quality. The proposed mechanism of action includes the restoration of the extracellular matrix of cartilage, the prevention of cartilage degradation, and overcoming a dietary deficiency of sulfur-containing amino acids, essential for extracellular matrix molecules [15].

Besides possible sponsor-based bias, it has been suggested that differences in origin and levels of purity may affect the clinical results of the substance.

OARSI OA recommendation guidelines state that treatment with glucosamine or chondroitin sulfate may provide symptomatic benefit in patients with knee OA. However, if no response is obtained within 6 months of treatment, it should be discontinued [16].

34.4 Pharmacological Therapies

34.4.1 Acetaminophen

Acetaminophen has been historically used as first-line therapy for pain in early osteoarthritis. It acts mainly through the downregulation of cyclooxygenase (COX) and its inhibitory action on nitric oxide [14]. Although there are studies that show that acetaminophen is superior to placebo in overall pain reduction, a recent Cochrane systematic review indicated that there is strong evidence to suggest that acetaminophen provides minimal clinically important improvement in pain and life function as a single treatment. It also evidences that pain and function do not change when using different doses of acetaminophen [17]. In the same study, they make a call to reconsider osteoarthritis treatment guidelines in their recommendation of acetaminophen as a first-line treatment [18].

Overall, when used for short periods of time it is considered to be safe, as the incidence of adverse events has shown to be similar to placebo, yet the risk of adverse events after long periods of administration is uncertain [19]. Patients could experience abnormal liver function, but the clinical importance of this finding is uncertain [14]. When used for prolonged periods, it has been related to produce greater risks such as gastrointestinal disorders or multi-organ failure [14]. Therefore, it is recommended to be administrated for short periods, with conservative dosing [14].

34.4.2 Non-Steroidal Anti-Inflammatory Drugs (NSAIDs)

NSAIDs are the most widely used pharmaceuticals in treating OA since many trials have confirmed their superiority related to acetaminophen for pain relief, but with a higher toxicity profile [14].

The increased expression of tissue-destructive enzymes and cytokines during osteochondral damage supports the rationale for their administration to control both inflammation and pain. The therapeutic effect of NSAIDs derives from their ability to reduce prostaglandin biosynthesis by competitively inhibiting cyclooxygenase 1 (COX-1), or COX-2 upregulated during the inflammatory response.

A frequent cause for concern when treating bone inflammatory lesions, such as contusions, subchondral fractures, or microfractures is the negative effect of NSAIDs on bone healing. It is thought that osteoblast behavior may be controlled by the concentration of Prostaglandin E2 (PGE2) through the relative expression of the Benjamin1 receptor activator of nuclear factor kappa-B ligand and osteoprotegerin, which is regulated through the enzymes cyclooxygenase (COX)-1 and COX-2. Even though there is enough evidence to support this, the negative effect may be dose or time- dependent because low-dose and short-duration exposure have not shown to affect union rates [20, 21].

It is proposed that PGE2 at a low dose switches osteoblast's biology in favor of bone union, inducing OPG gene expression by increasing RANKL gene expression, reducing PGEs synthesis, and increasing bone alkaline phosphatase gene expression. An opposite effect is expected when the concentration of NSAIDs exceeds certain levels [20]. Despite these findings, they should be avoided in patients with an increased risk of nonunion [21].

A common adverse reaction is gastrointestinal toxicity, caused by the inhibition of COX-1 in gastric epithelial cells, thus lowering the cytoprotective prostaglandin production. They also may expose patients to severe side effects in the renal and cardiovascular systems. COX-2 inhibitors have been developed with the aim of long-term treatment, but they create a prothrombotic state with an increased cardiovascular risk, which is contraindicated in patients with ischemic heart or cerebrovascular diseases. Non-selective NSAIDs with gastroprotective agents provide a similar analgesic effect with comparable gastrointestinal risk at less cost, making them the best option [14, 16].

Published guidelines and expert opinion are divided over the relative role of acetaminophen and NSAIDs as first-line pharmacologic therapy; their comparative safety is also important to consider [18]. Thus, the choice should be made after carefully assessing the risks according to the patient's history and characteristics and should be considered as treatment option when acetaminophen alone is ineffective for pain relief.

34.4.3 Corticosteroids

Glucocorticoids are steroid hormones usually used in early OA for pain relief. It has also been postulated that during the initial phase of post-traumatic joint damage, they may decrease the early inflammatory response and post-traumatic alterations that lead to early OA [22].

They act by binding and activating glucocorticoid receptors; this produces an upregulation of anti-inflammatory proteins that suppress the expression of local and systemic pro-inflammatory cytokines like IL-2, IL-4, IL-6, IL7, IL-17, and TNFα [23, 24].

The effects of intra-articular corticoids are dose- dependent. At low doses, it has shown beneficial animal in vivo effects stimulating an increase in cell growth and recovery. However, at high doses, corticosteroids are associated with significant gross cartilage damage and chondrocyte toxicity [22, 25]. Other studies have shown a significant decrease of cartilage turnover markers such as cartilage oligo-matrix protein and osteocalcin, suggesting that it may have a cartilage protective effect.

The main problem is that usually high doses of corticosteroids are needed due to the high clearance rate from the joint space, producing local damage such as accelerated OA progression, subchondral insufficiency fracture, complications of osteonecrosis, and rapid joint destruction, including bone loss as well as systemic effects, which are also dose-dependent [23].

Corticosteroids can produce a significant decrease in cortisol serum levels within just 4 h and up to 4 days, causing not only local damage but also systemic complications such as elevated glucose, severe impairment of bone metabolism with increased risk of fractures, and avascular necrosis of the bone. These side effects and risks of systemic absorption typically limit patients to 2–4 injections per year [23].

For this reason, other corticosteroids with low solubility, such as extended-release triamcinolone, may be more effective in pain reduction than other corticosteroids, with results higher than placebo in terms of pain reduction and global patient assessment [26]. But there is still more evidence needed to support these findings [2, 27, 28]. Corticosteroid injections in the knee are a relatively safe procedure, indicated for acute and inflammatory symptoms, they may have a good short-term effect (1–4 weeks). However, there is no evidence of long-term (6 months) clinical benefit [29]. The incidence of serious infectious complications may be as high as 1 in 3000, [30] so they should be used with caution at the lowest effective dose and considered a second- line of treatment [31].

34.4.4 Opioids

Opioids are considered second-line treatment and used when first-line treatments such as NSAIDs or acetaminophen are ineffective or contraindi-cated. Weak opioids such as tramadol are preferred because of their pain relief with no relevant abuse potential and fewer adverse effects than other opioids [32].

They inhibit pain pathways in the central nervous system through binding to mu-opioid receptors and norepinephrine and serotonin reuptake [33].

Their indication should be in combination with standard therapy in order to allow for decreased dosages. Even Cochrane's systematic review concludes that compared to placebo, tramadol alone or in combination with acetaminophen probably has no important benefit on mean pain or function in patients with osteoarthritis [34].

One of the main concerns with opioid use is the possible development of dependence; this can be avoided with a short-duration treatment.

Common adverse effects include nausea, vomiting, constipation, and sedation; more severe reactions can include respiratory depression, hypotension, paralytic ileus, urinary retention, and dehydration [35].

It is important to note that most opioids have shown to affect negatively bone remodeling and healing. In vitro studies have found a significant reduction of osteocalcin synthesis with opioid administration, thus downregulating osteoblast activity [36]. Remifentanil is an exception as it has been shown to stimulate osteoblast differentiation in vitro and inhibit differentiation and maturation of osteoclasts, thereby reducing bone resorption. Opioid antagonists such as naloxone, vasoactive intestinal peptide (VIP) and the neuropeptide Y (NPY), have been proved to promote osteogenesis [37]. These are all being studied for their possible therapeutic role in the future for osteoporosis and bone healing.

In conclusion, opioids are a second-line treatment that must be used with caution, especially in the elderly, and are not recommended for routine chronic pain treatment [4].

34.4.5 Gabapentinoids

In individuals with minor joint changes that report high levels of refractory pain to common analgesic treatments, a neuropathic pain component should be considered.

Subchondral bone is densely innervated; after destruction of subchondral structure such as in advanced phases of osteoarthritis, markers of nerve injury in sensory nerves innervating the knee are significantly increased [38]. In animal models, gradual proliferation of microglia in the dorsal horn of the spinal cord demonstrated progressive nerve injury, suggesting a gradual initiation of neuropathic pain state along with an inflammatory pain [39].

This supports the use of gabapentin, which might reduce OA pain or enhance the effect of conventionally used NSAIDs drugs [40]. In addition, there are several studies that support its effectiveness for neuropathic pain, as well its use in conjunction with other pain treatments as most of the time neuropathic and inflammatory pain coexist [41].

The most important adverse effects are the increased risk of sedation and respiratory depression, especially when combined with other central nervous system depressants such as opioids. Thus, caution must be taken when used [42].

In the future drugs targeted on neuropathic pain may be increasingly used together with other pharmaceutical agents to address symptoms. The use of gabapentin may be a valuable aid in pain management; however, future studies regarding doses and prescriptions are required [43].

34.5 Viscosupplementation

34.5.1 Hyaluronic Acid

Among the pharmacological therapies, non-steroidal anti-inflammatory drugs (NSAIDs) and intra-articular (IA) corticosteroid injections are most commonly prescribed [44]. These options have some limitations: NSAIDs show several side effects and corticosteroids injections [45], besides side effects, provide a short period of pain relief [46]. Another possible option is the injection of hyaluronic acid.

Hyaluronic Acid (HA) injections, also known as "viscosupplementation," are widely used by orthopedic surgeons for the treatment of symptomatic mild to moderate OA of the knee (Fig. 34.2) [47]. HA has been shown to reduce OA symptoms and provide a superior safety profile compared to the use of NSAIDs and corticosteroids [48]. Bannuru et al. have proved that corticosteroids had a relatively greater effect on pain in the first 4 weeks after infiltration, but HA showed a greater efficacy beyond 8 weeks [46].

Discovered by Meyer and Palmer in 1934, HA is a non-sulfated glycosaminoglycan consisting of alternately repeating D-glucuronic acid and N-acetylglucosamine units. It is a hydrophilic molecule with high solubility in aqueous environment, ensuring tissue hydration [49]. HA exists naturally in various animal tissue (especially in rooster combs) and human tissue, including umbilical cord, vitreous body, epidermis, dermis, and serum; especially, it is the major component of the synovial fluid and cartilage [50]. In these sites, HA is produced by the type B synoviocytes and fibroblast of the synovium, and

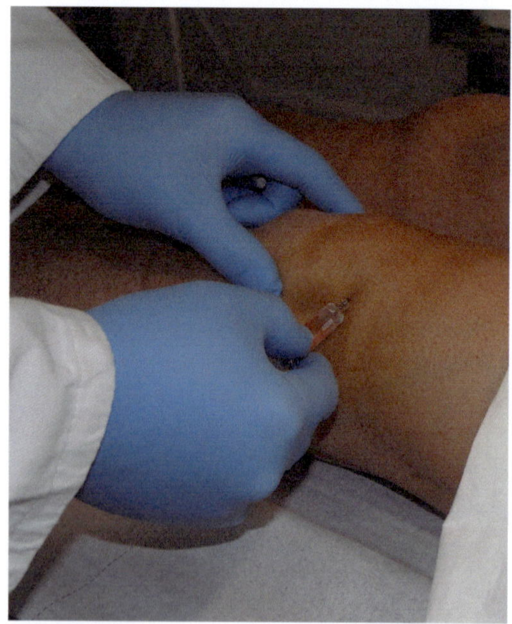

Fig. 34.2 Hyaluronic acid injection

its role is multifactorial. The primary role is to maintain the viscoelastic structural and functional characteristics of the articular matrix [51]. It provides joint lubrification and absorbs shock while also promoting chondrocytes proliferation/differentiation [52]. Moreover, HA has been shown to inhibit tissue nociceptors and the effects of the pain mediator substance P and to stimulate endogenous hyaluronan formation [53]. The healthy human knee contains approximately 2 mL of synovial fluid. The osteoarthritic knee is characterized by a considerable reduction of the concentration of molecular weight of HA, wich might generate pain and loss of function [54].

Several studies have shown that intra-articular viscosupplementation may restore the articular viscoelastic properties, exerting both chondroprotective and anti-inflammatory effects. Chondroprotection occurs through downregulation of the gene expression of OA-associated cytokines and enzymes, while the anti-inflammatory effect is due to downregulation of TNF-α, IL-8, and iNOS in synoviocytes [55]. Moreover, the therapeutic benefits of viscosupplementation occur by different actions [56]:

- Stimulation of metabolism.
- Prevention of apoptosis of chondrocytes.
- Inhibition of chondral degradation and articular inflammatory responses.
- Decrease of lymphocytes proliferation.

The chondroprotective effect also allows the use of HA in the postoperative knee; pain that persists after arthroscopy can be treated with HA injections [55].

Several HA compounds are currently used worldwide; each differs in molecular weight, composition, dosing regimens, and source [51]. The first products that have been used derived from rooster combs; this source was utilized for several years through extraction with an organic solvent like acetone and ethanol. This method allows to obtain a compound characterized by a good purity and a high molecular weight [57, 58]. The main disadvantage is the risk of viral infec-

tion between different species; moreover, hypersensitivity reactions are common.

Hence, microbial fermentation has emerged as a new alternative for HA production. The first commercially fermented HA was produced from *Streptococcus zooepidemicus,* which synthesizes HA as part of its outer capsule under the suitable culture condition [59]. This technique has become very popular because of the lower production costs and less environmental pollution compared with rooster comb HA products. On the other hand, HA produced from attenuated pathogen *Streptococci* may have the potential to be contaminated with pathogen factors [57]. Consequently, HA production through fermentation is generally recognized as safe (GRAS) microbial strains are highly appreciated. This method has emerged as an attractive alternative that could alleviate safety concern about pathogenic *S. zooepidemicus* and avian products. Host bacteria, both Gram-positive and Gram-negative, include *Bacillus* sp., *L. lactis, Agrobacterium* sp., and *E. coli* can be used.

Regarding the molecular weight, different preparations are available for intra-articular use. The low molecular weight preparation (0.5–1.5 million Da) can achieve maximum concentration into the joint and may reduce inflammation, but it presents a low elastoviscosity and consequently, it needs at least 4–5 injections [60]. The high molecular weight preparation (6–7 million Da) shows a better increase of fluid retention into the joint and a stronger anti-inflammatory effect. However, studies concerning the use of HA with different molecular weights report conflicting results but favors the high molecular weight. In order to increase the molecular weight of HA and to provide durable activity, the cross-linking technique has been commonly used. The controlled cross-linking creates a viscous gel with increased density of hyaluronan and viscoelasticity, requiring only 1 injection. The single-injection regimen is attractive because it decreases patient time expenditure and discomfort associated with the procedure [61], but clinicians report more side effects with cross-linked HA, especially synovitis.

Recently, a new type of HA has been brought to the market. It consists of a mixture of two hyaluronates, one of medium molecular weight (1200–1500 kDa), which promotes viscosupplementation, and the other of low molecular weight (200–400 kDa), which contributes to the resolution of OA articular damages. The main innovation is an excipient, the trehalose, a disaccharide that would seem to act as a protector of HA, delaying the degradation from hyaluronidase. Pre-clinical studies show that this new formulation (PROMOVIA HYDROBALANCE) lasts longer than 3 months. We are currently performing a clinical trial in order to analyze the effect in patients suffering from OA. It is a monocentric, randomized, post-marketing, and double-blinded study.

34.6 Disease-Modifying Drugs (DMDs)

34.6.1 Bisphosphonates

Bisphosphonates have been suggested as disease-modifying agents [5] as they can reduce bone turnover by inhibiting osteoclastic activity and reducing bone resorption.

They are thought to aid in subchondral bone remodeling abnormalities that increase bone turnover and result in bone loss, which may lead to degradation of overlying cartilage and joint collapse [62]. Experimental studies have shown that the microstructural impairment of subchondral bone could aggravate cartilage damage but, on the other hand, improvement of this microstructure would reduce the progression of cartilage impairments [62].

Risedronate has been shown to reduce cartilage collagen degradation marker (CTX-II). However, there is still no recommendation for its use due to the lack of evidence on the efficacy of risedronate to affect symptoms, function, and progression of knee OA [2]. Zoledronic acid has been demonstrated to provide significant symptomatic benefit and reduction on bone marrow lesions after 6 months of treatment in a double-blind control trial; other studies have found similar results but with no significant difference in WOMAC pain score after 24 months [6, 63].

Diagnostic sensitivity of bone densitometry (DXA) is low. More than 50% of patients suffering from a fragility fracture have a normal DX [64]. So it is of crucial importance to suspect osteoporosis and fragility fractures when the patient presents risk factors such as postmenopausal hypogonadism, age, prevalent osteoporotic fracture, low BMI, smoking, alcohol, rheumatoid arthritis, diabetes mellitus, glucocorticoids intake, as well as aromatase inhibitors or antiandrogens. Strontium ranelate is another antiresorptive that has been associated with reduced radiographic knee OA progression, and with meaningful clinical improvement. It was also significantly associated with decreased MRI-assessed cartilage volume loss and bone marrow lesions (BMLs) [62].

34.7 Monoclonal Antibodies

They target inflammatory mediators and can halt or delay disease progression and preserve the structure and function of damaged cartilage.

34.7.1 Interleukin-1 (IL-1) Inhibitor

IL-1 stimulates the synthesis of proteolytic enzymes, cytokines, and other inflammatory mediators, which lead to cartilage degradation and osteoarthritis [65]. The effectiveness of inhibiting IL-1 receptor for the prevention of osteoarthritis has been well documented in experimentally induced OA in animal studies [66]. IL-1 inhibitor, also known as Lutikizumab (ABT-981), has been recently developed against both IL1a and IL-1b. It is thought to be capable of decreasing inflammation and slowing OA progression. In recent randomized double-blind placebo-controlled parallel group phase IL-1 showed at 16 weeks reduced pain in WOMAC, but cartilage thickness and synovitis had similar results with the placebo group. More studies are needed in order to see if longer use correlates with better results [67].

34.7.2 Anti-Nerve growth Factor (Anti-NGF)

Also known as Tanezumab is an investigational humanized monoclonal antibody against B-NGF. It provides potential pain modulation through nociceptor sensitization. Its main effects have been evaluated in clinical trials for hip and knee OA and showed superiority in pain reduction compared to placebo [68]. However, there were cases of rapidly progressive OA with chondrolysis and subchondral bone destruction associated with increasing doses of anti-NGF antibodies and parallel therapy with NSAIDs. These safety issues have important implications for future clinical trials and need further investigation [69]. Recent data suggest that even though the changes in pain are significant, the improvements were modest and their clinical importance must be established [70].

34.8 Molecules

34.8.1 BNTA (N-(2-Bromo-4-(phenylsulfonyl) thiophen-3-yl)-2-chlorobenzamide)

Is a small molecule that targets superoxide dismutase 3 (SOD3). SOD3 is thought to have a vital role in maintaining the extracellular matrix. Normally, it is abundant in cartilage ECM, but it is markedly reduced after OA development. A recent study showed that an intra-articular injection of BNTA delays the disease progression in a trauma-induced rat model of osteoarthritis. Making it a potentially therapeutic agent [71].

34.8.2 Kartogenin

Kartogenin (KGN) is another molecule that aids in healing repair. It induces mesenchymal cells to differentiate into chondrocytes in vitro and has shown to improve OA in small animal models [72]. It has been shown to be more effective as cartilage regeneration inducer when compared with growth factors. KGN has been processed and applied in many forms, such as intra-articular injection, with growth factors, in drug delivery systems, and in combination with scaffolds [73, 74], but whether Kartogenin will be effective in patients will have to be investigated in clinical trials.

34.8.3 Vitamin D

Vitamin D deficiency is very common, especially in the elderly population. It has been associated as a progression factor for OA development, because of its effects on calcium absorption in bone metabolism [75]. The results for its treatment as a pain modulator are contradictory. Gao et al. meta-analysis concluded that vitamin D supplementation significantly decreases WOMAC pain and loss of function, but has no effect on WOMAC stiffness score or tibial cartilage volume [76]. On the other hand, the Arden et al. study shows no effectiveness of vitamin D pain reduction as well as cartilage volume loss compared to placebo [77]. Thus, vitamin D supplementation may have a positive clinical effect on pain modification, but conflicting results exist. Additional long-term clinical trials are required to further determine its clinical benefit.

It is important to highlight that before prescribing therapies such as bisphosphonates or vitamin D; a metabolic test should be ruled out (calcium, phosphorous, parathyroid hormone, alkaline phosphatase, and vitamin D levels). This is important to determine secondary and potentially treatable causes of osteoporosis, such as asymptomatic hyperparathyroidism or chronic kidney disease, [78] which can manifest with bone marrow edema; or osteoporosis-like syndrome with impaired bone quality and increased risk of fragility fractures [79].

34.9 Conclusion

Nowadays, no treatment has shown to be the gold standard. Side effects have been reported at both systemic and local level that is why the target must be the minimal therapeutic dose for the least amount of time. For achieving this the use of multimodal treatment regimens should be attempted, utilizing medications that work on multiple locations of the pain pathway via different mechanisms in order to produce analgesia.

Part of this medication regime involves the use of acetaminophen which works as a centrally acting analgesic agent by blocking the COX-3 enzyme in the thalamus, NSAIDS such as COX-1 and COX-2 inhibitors which decrease the local production of prostaglandins to decrease the local inflammatory response, as well as low dose of corticoids as initial pain and early inflammation management. Neuropathic pain can be, if suspected, targeted with gabapentinoids when subchondral bone has been damaged. Prevention of further subchondral bone collapse could be achieved with bisphosphonates, vitamin D, and other potential cartilage protectors. HA injections represent a good alternative to pharmacological therapy because they reduce the pain associated with OA with a superior safety profile.

Finally, it is important to teach patients what to expect with a pain management program and the goals with conservative treatment should be realistic. Treatment will be mostly for symptom treatment and not curative. They must be aware that other conservative measures must be taken such as exercise and/or physical therapies, as well as a possible need for future surgical treatment.

References

1. Sinusas K. Osteoarthritis: diagnosis and treatment. Am Fam Physician. 2012;86(10):893.
2. McAlindon TE, LaValley MP, Harvey WF, Price LL, Driban JB, Zhang M, et al. Effect of intra-articular triamcinolone vs saline on knee cartilage volume and pain in patients with knee osteoarthritis a randomized clinical trial. JAMA. 2017;317(19):1967–75.
3. Boopalan PRJVC, Arumugam S, Livingston A, Mohanty M, Chittaranjan S. Pulsed electromagnetic field therapy results in healing of full thickness articular cartilage defect. Int Orthop. 2011;35(1):143–8.
4. Kon E, Filardo G, Drobnic M, Madry H, Jelic M, van Dijk N, et al. Non-surgical management of early knee osteoarthritis. Knee Surg Sports Traumatol Arthrosc. 2012;20:436–49.
5. Bennell KL, Buchbinder R, Hinman RS. Physical therapies in the management of osteoarthritis: current state of the evidence. In: Current opinion in rheumatology, vol. 27. Philadelphia: Lippincott Williams and Wilkins; 2015. p. 304–11.
6. Farrokhi S, Voycheck CA, Tashman S, Fitzgerald GK. A biomechanical perspective on physical therapy management of knee osteoarthritis. J Orthop Sports Phys Ther. 2013;43:600–19.
7. Gobbi A, Lad D, Petrera M, Karnatzikos G. Symptomatic early osteoarthritis of the knee treated with pulsed electromagnetic fields: two-year follow-up. Cartilage. 2014;5(2):78–85.
8. Draper DO, Klyve D, Ortiz R, Best TM. Effect of low-intensity long-duration ultrasound on the symptomatic relief of knee osteoarthritis: a randomized, placebo-controlled double-blind study. J Orthop Surg Res. 2018;13(1):257.
9. Wang T, Xie W, Ye W, He C. Effects of electromagnetic fields on osteoarthritis, Biomedicine and pharmacotherapy, vol. 118. Amsterdam: Elsevier Masson SAS; 2019.
10. Zhou X-Y, Zhang X-X, Yu G-Y, Zhang Z-C, Wang F, Yang Y-L, et al. Effects of low-intensity pulsed ultrasound on knee osteoarthritis: a Meta-analysis of randomized clinical trials. Biomed Res Int. 2018;2018:7469197.
11. Fibel KH. State-of-the-art management of knee osteoarthritis. World J Clin Cases. 2015;3(2):89.
12. Reginster JY, Neuprez A, Lecart MP, Sarlet N, Bruyere O. Role of glucosamine in the treatment for osteoarthritis. Rheumatol Int. 2012;32:2959–67.
13. Noack W, Fischer M, Förster KK, Rovati LC, Setnikar I. Glucosamine sulfate in osteoarthritis of the knee. Osteoarthr Cartil. 1994;2(1):51–9.
14. McAlindon TE, Bannuru RR, Sullivan MC, Arden NK, Berenbaum F, Bierma-Zeinstra SM, et al. OARSI guidelines for the non-surgical management of knee osteoarthritis. Osteoarthr Cartil. 2014;22(3):363–88.
15. Singh JA, Noorbaloochi S, Macdonald R, Maxwell LJ. Chondroitin for osteoarthritis. Cochrane Database Syst Rev. 2017;2015:CD005614.
16. Zhang W, Moskowitz RW, Nuki G, Abramson S, Altman RD, Arden N, et al. OARSI recommendations for the management of hip and knee osteoarthritis, part II: OARSI evidence-based, expert consensus guidelines. Osteoarthr Cartil. 2008;16(2):137–62.
17. Leopoldino AO, MacHado GC, Ferreira PH, Pinheiro MB, Day R, McLachlan AJ, et al. Paracetamol versus placebo for knee and hip osteoarthritis [internet]. Cochrane Database of Syst Rev. 2019;25:CD013273.
18. Towheed T, Maxwell L, Judd M, Catton M, Hochberg MC, Wells GA. Acetaminophen for osteoarthritis. Cochrane Database Syst Rev. 2006;2006:Acetaminophen for osteoarthritis.
19. Bannuru RR, Osani MC, Vaysbrot EE, Arden NK, Bennell K, Bierma-Zeinstra SMA, et al. OARSI guidelines for the non-surgical management of knee, hip, and polyarticular osteoarthritis. Osteoarthr Cartil. 2019;27(11):1578–89.
20. Dodwell ER, Latorre JG, Parisini E, Zwettler E, Chandra D, Mulpuri K, et al. NSAID exposure and

risk of nonunion: a meta-analysis of case-control and cohort studies. Calcif Tissue Int. 2010;87(3):193–202.

21. Wheatley BM, Nappo KE, Christensen DL, Holman AM, Brooks DI, Potter BK. Effect of NSAIDs on bone healing rates: a Meta-analysis. J Am Acad Orthop Surg. 2019;27(7):e330–6.

22. Black R, Grodzinsky AJ. Dexamethasone: Chondroprotective corticosteroid or catabolic killer? Eur Cells Mater. 2019;38:246–63.

23. Habib GS. Systemic effects of intra-articular corticosteroids. Clin Rheumatol. 2009;28:749–56.

24. Li Y, Wang Y, Chubinskaya S, Schoeberl B, Florine E, Kopesky P, et al. Effects of insulin-like growth factor-1 and dexamethasone on cytokine-challenged cartilage: relevance to post-traumatic osteoarthritis. Osteoarthr Cartil. 2015;23(2):266–74.

25. Wernecke C, Braun HJ, Dragoo JL. The effect of intra-articular corticosteroids on articular cartilage: a systematic review. Orthop J Sports Med. 2015;3(5):1–7.

26. Conaghan PG, Cohen SB, Berenbaum F, Lufkin J, Johnson JR, Bodick N. Brief report: a phase IIb trial of a novel extended-release microsphere formulation of triamcinolone Acetonide for intraarticular injection in knee osteoarthritis. Arthritis Rheumatol. 2018;70(2):204–11.

27. Conaghan PG, Hunter DJ, Cohen SB, Kraus VB, Berenbaum F, Lieberman JR, et al. Effects of a single intra-articular injection of a microsphere formulation of triamcinolone acetonide on knee osteoarthritis pain: a double-blinded, randomized, placebo-controlled, multinational study. J Bone Joint Surg (Am Vol). 2018;100(8):666–77.

28. Paik J, Duggan ST, Keam SJ. Triamcinolone acetonide extended-release: a review in osteoarthritis pain of the knee. Drugs. 2019;79:455–62.

29. Jüni P, Hari R, Rutjes AWS, Fischer R, Silletta MG, Reichenbach S, et al. Intra-articular corticosteroid for knee osteoarthritis. Cochrane Database Syst Rev. 2015;2015:CD005328.

30. Kraus VB, Conaghan PG, Aazami HA, Mehra P, Kivitz AJ, Lufkin J, et al. Synovial and systemic pharmacokinetics (PK) of triamcinolone acetonide (TA) following intra-articular (IA) injection of an extended-release microsphere-based formulation (FX006) or standard crystalline suspension in patients with knee osteoarthritis (OA). Osteoarthr Cartil. 2018;26(1):34–42.

31. Derendorf H, Möllmann H, Grüner A, Haack D, Gyselby G. Pharmacokinetics and pharmacodynamics of glucocorticoid suspensions after intra-articular administration. Clin Pharmacol Ther. 1986;39(3):313–7.

32. Mushtaq S, Choudhary R, Scanzello CR. Non-surgical treatment of osteoarthritis-related pain in the elderly. Curr Rev Musculoskelet Med. 2011;4(3):113–22.

33. da Costa BR, Nüesch E, Kasteler R, Husni E, Welch V, Rutjes AWS, et al. Oral or transdermal opioids for osteoarthritis of the knee or hip. Cochrane Database Syst Rev. 2014;2014:CD003115.

34. April KT, Bisaillon J, Welch V, Maxwell LJ, Jüni P, Rutjes AWS, et al. Tramadol for osteoarthritis. Cochrane Database Syst Rev. 2019;2019.

35. Sun EC, Darnall BD, Baker LC, MacKey S. Incidence of and risk factors for chronic opioid use among opioid-naive patients in the postoperative period. JAMA Intern Med. 2016;176(9):1286–93.

36. Pérez-Castrillón JL, Olmos JM, Gómez JJ, Barrallo A, Riancho JA, Perera L, et al. Expression of opioid receptors in osteoblast-like MG-63 cells, and effects of different opioid agonists on alkaline phosphatase and osteocalcin secretion by these cells. Neuroendocrinology. 2000;72(3):187–94.

37. Coluzzi F, Scerpa MS, Centanni M. The effect of opiates on bone formation and bone healing. Curr Osteoporosis Rep. 2020;18:325–35.

38. Ivanavicius SP, Ball AD, Heapy CG, Westwood FR, Murray F, Read SJ. Structural pathology in a rodent model of osteoarthritis is associated with neuropathic pain: increased expression of ATF-3 and pharmacological characterisation. Pain. 2007;128(3):272–82.

39. Orita S, Ishikawa T, Miyagi M, Ochiai N, Inoue G, Eguchi Y, et al. Pain-related sensory innervation in monoiodoacetate-induced osteoarthritis in rat knees that gradually develops neuronal injury in addition to inflammatory pain. BMC Musculoskelet Disord. 2011;12:134.

40. Ohtori S, Inoue G, Orita S, Takaso M, Eguchi Y, Ochiai N, et al. Efficacy of combination of meloxicam and pregabalin for pain in knee osteoarthritis. Yonsei Med J. 2013;54(5):1253–8.

41. Enteshari-Moghaddam A, Azami A, Isazadehfar K, Mohebbi H, Habibzadeh A, Jahanpanah P. Efficacy of duloxetine and gabapentin in pain reduction in patients with knee osteoarthritis. Clin Rheumatol. 2019;38(10):2873–80.

42. Hannon CP, Fillingham YA, Browne JA, Schemitsch EH, Mullen K, Casambre F, et al. The efficacy and safety of gabapentinoids in total joint arthroplasty: a direct Meta-analysis. J Arthroplasty. 2020; https://doi.org/10.1016/j.arth.2020.05.033.

43. Li F, Ma J, Kuang M, Jiang X, Wang Y, Lu B, et al. The efficacy of pregabalin for the management of postoperative pain in primary total knee and hip arthroplasty: a meta-analysis. J Orthop Surg Res. 2017;12(1):49.

44. Altman RD, Manjoo A, Fierlinger A, Niazi F, Nicholls M. The mechanism of action for hyaluronic acid treatment in the osteoarthritic knee: a systematic review. BMC Musculoskeletal Disorders. 2015;16:321.

45. Rutjes AWS, Jüni P, da Costa BR, Trelle S, Nüesch E, Reichenbach S. Viscosupplementation for osteoarthritis of the knee: A systematic review and meta-analysis. Ann Intern Med. 2012;157:180–91.

46. Bannuru RR, Natov NS, Obadan IE, Price LL, Schmid CH, McAlindon TE. Therapeutic trajectory of hyaluronic acid versus corticosteroids in the treatment of knee osteoarthritis: a systematic review and meta-analysis. Arthritis Care Res. 2009;61(12):1704–11.

47. Colen S, van den Bekerom MPJ, Mulier M, Haverkamp D. Hyaluronic acid in the treatment of

knee osteoarthritis: A systematic review and meta-analysis with emphasis on the efficacy of different products. BioDrugs. 2012;26:257–68.

48. Bellamy N, Campbell J, Welch V, Gee TL, Bourne R, Wells GA. Viscosupplementation for the treatment of osteoarthritis of the knee. Cochrane Database Syst Rev. 2006;2006.

49. Santilli V, Paoloni M, Mangone M, Alviti F, Bernetti A. Hyaluronic acid in the management of osteoarthritis: injection therapies innovations. Clin Cases Miner Bone Metab. 2016;13(2):131.

50. Iannitti T, Lodi D, Palmieri B. Intra-articular injections for the treatment of osteoarthritis: Focus on the clinical use of hyaluronic acid. Drugs R D. 2011;11:13–27.

51. Petrella RJ, Cogliano A, Decaria J. Combining two hyaluronic acids in osteoarthritis of the knee: a randomized, double-blind, placebo-controlled trial. Clin Rheumatol. 2008;27(8):975–81.

52. Watterson JR, Esdaile JM. Viscosupplementation: therapeutic mechanisms and clinical potential in osteoarthritis of the knee. J Am Acad Orthop Surg. 2000;8(5):277–84.

53. Waddell DD, Bert JM. The use of hyaluronan after arthroscopic surgery of the knee. Arthroscopy. 2010;26:105–11.

54. Balazs EA, Denlinger JL. Viscosupplementation: a new concept in the treatment of osteoarthritis. J Rheumatol Suppl. 1993;39:3–9.

55. Zaslav K, McAdams T, Scopp J, Theosadakis J, Mahajan V, Gobbi A. New Frontiers for cartilage repair and protection. Cartilage. 2012;3(1 Suppl):77S–86S.

56. Oliveira MZ, Albano MB, Namba MM, da Cunha LAM, de Lima Gonçalves RR, Trindade ES, et al. Effect of hyaluronic acids as chondroprotective in experimental model of osteoarthrosis. Revista Brasileira de Ortopedia (English Edition). 2014 Jan;49(1):62–8.

57. Liu L, Liu Y, Li J, Du G, Chen J. Microbial production of hyaluronic acid: current state, challenges, and perspectives. Microb Cell Fact. 2011;10:99.

58. Kang DY, Kim WS, Heo IS, Park YH, Lee S. Extraction of hyaluronic acid (HA) from rooster comb and characterization using flow field-flow fractionation (FlFFF) coupled with multiangle light scattering (MALS). J Sep Sci. 2010;33(22):3530–6.

59. Chien LJ, Lee CK. Hyaluronic acid production by recombinant Lactococcus lactis. Appl Microbiol Biotechnol. 2007;77(2):339–46.

60. Bagga H, Burkhardt D, Sambrook P, March L. Longterm effects of intraarticular hyaluronan on synovial fluid in osteoarthritis of the knee. J Rheumatol. 2006;33(5):946–50.

61. Sun SF, Hsu CW, Lin HS, Liou IH, Chen YH, Hung CL. Comparison of single intra-articular injection of novel Hyaluronan (HYA-JOINT plus) with synvisc-one for knee osteoarthritis: a randomized, controlled, double-blind trial of efficacy and safety. J Bone Joint Surg (Am Vol). 2017;99(6):462–71.

62. Bellido M, Lugo L, Roman-Blas JA, Castañeda S, Caeiro JR, Dapia S, et al. Subchondral bone microstructural damage by increased remodelling aggravates experimental osteoarthritis preceded by osteoporosis. Arthritis Res Ther. 2010;12:R152.

63. Simon MJK, Barvencik F, Luttke M, Amling M, Mueller-Wohlfahrt HW, Ueblacker P. Intravenous bisphosphonates and vitamin D in the treatment of bone marrow oedema in professional athletes. Injury. 2014;45(6):981–7.

64. Bliuc D, Alarkawi D, Nguyen TV, Eisman JA, Center JR. Risk of subsequent fractures and mortality in elderly women and men with fragility fractures with and without osteoporotic bone density: the Dubbo osteoporosis epidemiology study. J Bone Mineral Res. 2015;30(4):637–46.

65. Murakami S, Lefebvre V, de Crombrugghe B. Potent inhibition of the master chondrogenic factor Sox9 gene by interleukin-1 and tumor necrosis factor-α. J Biol Chem. 2000;275(5):3687–92.

66. Evans C. Editorial: Arthritis Gene Therapy Using Interleukin-1 Receptor antagonist [Internet]. Arthritis Rheumatol. 2018;70:1699–701.

67. Fleischmann R, Bliddal H, Blanco F, Schnitzer T, Peterfy C, Chen S, et al. SAT0575 safety and efficacy of lutikizumab (ABT-981), an anti–interleukin-1 alpha/beta dual variable domain (DVD) immunoglobulin, in subjects with knee osteoarthritis: results from the randomised, double-blind, placebo-controlled, parallel-group phase 2 trial. Ann Rheum Dis. 2018:1141.2–1142.

68. Nicol GD, Vasko MR. Unraveling the story of NGF-mediated sensitization of nociceptive sensory neurons: On or off the trks? Mol Interv. 2007;7:26–41.

69. Schnitzer TJ, Marks JA. A systematic review of the efficacy and general safety of antibodies to NGF in the treatment of OA of the hip or knee. Osteoarthritis Cartil. 2015;23:S8–17.

70. Schnitzer TJ, Easton R, Pang S, Levinson DJ, Pixton G, Viktrup L, et al. Effect of Tanezumab on joint pain, physical function, and patient global assessment of osteoarthritis among patients with osteoarthritis of the hip or knee: a randomized clinical trial. JAMA. 2019;322(1):37–48.

71. Shi Y, Hu X, Cheng J, Zhang X, Zhao F, Shi W, et al. A small molecule promotes cartilage extracellular matrix generation and inhibits osteoarthritis development. Nat Commun. 2019;10(1):1–14.

72. Huang H, Xu H, Zhao J. A novel approach for meniscal regeneration using Kartogenin-treated autologous tendon graft. Am J Sports Med. 2017;45(14):3289–97.

73. Cai G, Liu W, He Y, Huang J, Duan L, Xiong J, et al. Recent advances in kartogenin for cartilage regeneration. J Drug Target. 2019;27:28–32.

74. Liu F, Xu H, Huang H. A novel kartogenin-platelet-rich plasma gel enhances chondrogenesis of bone marrow mesenchymal stem cells in vitro and promotes wounded meniscus healing in vivo. Stem Cell Res Ther. 2019;10:201.

75. Manoy P, Yuktanandana P, Tanavalee A, Anomasiri W, Ngarmukos S, Tanpowpong T, et al. Vitamin D supplementation improves quality of life and physical performance in osteoarthritis patients. Nutrients. 2017;1:9(8).
76. Gao XR, Chen YS, Deng W. The effect of vitamin D supplementation on keen osteoarthritis: a meta-analysis of randomized controlled trials. Int J Surg. 2017;46:14–20.
77. Arden NK, Cro S, Sheard S, Doré CJ, Bara A, Tebbs SA, et al. The effect of vitamin D supplementation on knee osteoarthritis, the VIDEO study: a randomised controlled trial. Osteoarthr Cartil. 2016;24(11):1858–66.
78. Wang CY, Hsu YJ, Peng YJ, Lee HS, Chang YC, Chang CS, et al. Knee subchondral bone perfusion and its relationship to marrow fat and trabeculation on multi-parametric MRI and micro-CT in experimental CKD. Sci Rep. 2017;7(1):3073.
79. Tezcan ME, Temizkan S, Ozal ST, Gul D, Aydin K, Ozderya A, et al. Evaluation of acute and chronic MRI features of sacroiliitis in asymptomatic primary hyperparathyroid patients. Clin Rheumatol. 2016;35(11):2777–82.

Principles of Rehabilitation in Cartilage and Lesions

35

Lorenzo Boldrini, Giacomo Lucenteforte, Furio Danelon, and Francesco Della Villa

35.1 Introduction

The management of different clinical cases of articular cartilage pathologies suffers the lack of a consensus for a therapeutic algorithm to date. Research about rehabilitation programs is conditioned by the fast evolution of new surgical treatments that created continuously new scenarios in the integrated management of these complex conditions, both from the point of view of the orthopedic surgeons and the rehabilitation specialists. Classically articular cartilage rehabilitation has been based mostly on a protective and maybe sometimes over-protective approach, mainly based on basic science and animal data. In the context of cartilage repair procedures as the new standard of treatment for articular cartilage pathologies, the research of the most appropriate and updated rehabilitation protocols should consider both clinical trials and clinical experience-based evidences. This approach allows to define the best rehabilitation protocols according to the latest progress in surgical and non-surgical management of cartilage and osteo-chondral lesions [1]. This book chapter will cover both the principles and the practical application of rehabilitation following chondral and osteochondral lesions, with a special reference to the knee joint.

35.2 Rehabilitation Principles

Despite the regeneration of hyaline cartilage is not yet feasible, different surgical techniques widely described in literature, as microfracture of subchondral bone, auto/allografts, cell transplantation, growth factors, and artificial matrices, can be helpful to stimulate the generation of a new articular surface. This is evidenced by clinical improvement for most of the patients, documented in different clinical trials, mostly focused on knee cartilage lesions. In this context it has been highlighted the importance of the characterization of various clinical factors besides the cartilage lesion to select the most appropriate treatment in patients evaluated for cartilage repair [2].

Considering that the reparative tissue in osteochondral lesions often does not lead to a normal restoration of the normal structure, composition, and mechanical properties of the hyaline cartilage, and keeping in mind the avascular nature of the articular cartilage, it is not surprising the relative high rate of failure of reparative tissue, mostly in the long term [2, 3].

L. Boldrini (✉) · G. Lucenteforte · F. Danelon
Isokinetic Medical Group, FIFA Medical Centre of Excellence, Milan, Italy
e-mail: l.boldrini@isokinetic.com

F. Della Villa
Isokinetic Medical Group, Education and Research Department, Bologna, Italy

A. Gobbi et al. (eds.), *Joint Function Preservation*, https://doi.org/10.1007/978-3-030-82958-2_35

That is mainly related to the biological properties of the reparative tissue that entails intermediate features between chondral and fibrous tissue. This new joint environment needs biomechanical stimulation to reach cartilage-like characteristics, avoiding the fibroblastic-like growth process. Moreover, the healing process of osteochondral lesions is conditioned by different but interdependent factors, as age, size, and location of the lesion and the anthropometric parameters of the patient. Older patients suffering from lesions in weight bearing areas have generally worst outcomes while younger patients had better results, related to presence of any vascularized areas that favors a more effective synthesis of macromolecules [2].

Among the key factors in the healing process, mechanical stimuli, like the cyclical pressure on the chondral cells, should be considered. Different from other tissues where the effect of the mechanical stimuli is better understood and applied in a clinical context, there is still room for improvement in the understanding of the mechanical loading for the osteochondral tissue. However, we do know that appropriate cyclical loading is crucial for the neo-cartilage tissue, both from a histological (macro architecture of the tissue) and cytological (synthesis of type II collagen) point of view [4–9]. This theoretical framework which is well explained in the theory of the mechanotransduction [10–13] is the crucial basic principle of rehabilitation.

To date the basic science experience suggests us that:

- Unloading and immobilization are harmful in the healing process of articular cartilage, causing proteoglycan structural/biochemical alterations [14, 15]. The negative effects of prolonged immobilization on joint homeostasis [5, 16] lead to a decrease of synthetic activities of the chondrocytes, determining less content in proteoglycan and water in the cartilage matrix [16]. Moreover, the decrease of collagen fibrils and joint lubrification affect cartilage and tendon nutrition, inducing joint stiffness and capsular contraction [17].

- Passive movement has positive effects to promote joint homeostasis and adaptations of cartilage tissue to proper exercise biomechanical requests [6, 8, 9, 16].
- Adequate weight bearing and progressive recovery of range of motion (ROM) induce enhancement in the healing process, stimulating matrix production and improving the tissue's mechanical properties of the reparative tissue, moreover preventing the degeneration [4, 10, 12, 18].
- Repetitive loading of articular cartilage, in addition to the mentioned effects, favors the maintenance of the homeostasis of articular cartilage and subchondral bone and supports the cartilage nutrition improving the diffusion of synovial fluid [2].
- Mechanical load stimulates chondrocytes to produce the specific types of collagen depending on the load degree, through still unclear mechanisms. It has been reported that grafts in a high loaded area produced a hyaline-like repair tissue, while grafts in a less loaded area produced a more fibrocartilaginous one [19].

At the same time, it has been proved the adverse effect of excessive load on reparative cartilage tissue, leading to the good practice of temporary or defined unloading of the treated area, that prevent the progressive matrix loss. In the same way, a body mass index over 30 kg/m^2 could cause complications in specific repair procedures [13], therefore reaching an optimal body weight is mandatory to manage the weight bearing. In that sense, both partial weight bearing with crutches and hydro-kinesitherapy are useful to initiate the weight bearing exercises and the gait training. Moreover, force platforms are very helpful to quantify the entity of load and to guide the therapist and the patient during the exercises focused on improving strength, proprioception, and balance since the early phases of rehabilitation [3, 20].

Rehabilitation principles largely come from afore-mentioned concepts, applied biomechanics and exercise physiology, and from few clinical trials and more consisting clinical experiences. Most of the clinical knowledge derived from research focused on knee joint. Only a few numbers of studies concern the rehabilitation in ankle, shoulder, and elbow cartilage lesions [21–24]. Nevertheless, a rehabilitation program based on clinical and functional criteria, and characterized by phases progression, allows to extent and adapts the same principles to different joints.

35.3 Rehabilitation Strategies

The rehabilitation protocol, in both operative and non-operative conditions, needs to be customized, progressive, and supervised (Table 35.1).

Customized, based on patient's and lesion's characteristics. Patient's clinical history, clinical examination, evaluation of radiological findings, and specific aspects of surgical intervention should be considered.

Progressive, based on a phase-progression model depending on clinical and functional evaluation, according to joint responses to rehabilitation-induced solicitations.

Supervised, consisting in regular clinical and functional checks from the rehabilitation and orthopedic team, including rehabilitation physician, physiotherapist, athletic trainer, surgeon.

The rehabilitation program can be resumed in a 5-phase progression:

Table 35.1 The key points of a rehabilitation protocol for cartilage condition

Customized	Type and site of lesion Surgical techniques Anthropometric data Functional expectations/Sport activity level Clinical history General health Psychological aspects
Progressive	Phase-progression rehabilitation model depending on clinical and functional evaluation, according to joint responses to rehabilitation stimuli
Supervised	Regular clinical and functional check from the rehabilitation and orthopedic team (rehabilitation physician, physiotherapist, athletic trainer, surgeon)

- Phase 1 Resolution of swelling and inflammation.
- Phase 2 Recovery of range of motion and muscle flexibility.
- Phase 3 Recovery of muscle strength and resistance.
- Phase 4 Recovery of neuromuscular control and coordination.
- Phase 5 Recovery of specific gestures.

The progression through the different rehabilitation phases is guided by the fundamental principle of the adaptation of the joint and patient's reactions to applied stimuli (Specific Adaptation to Imposed Demand) [7]. Pain and swelling reaction must always be avoided because they can worsen clinical conditions inducing a delay in recovery and a diminished trust in the rehabilitation program by the patient.

Therefore, patients are periodically supervised and controlled by the physician to assess clinical and functional parameters and to define the goal of the subsequent rehabilitation phase (Fig. 35.1). Swelling and pain must be always investigated; clinicians must distinguish among pain, soreness, and fear of pain through an accurate series of questions to the patient. Clinical and functional parameters, together with ROM and strength measurements, are registered to monitor the progression of the patient through the rehabilitation phases. Communication within the team is also relevant for a successful rehabilitation [20]; updated functional objectives and treatment indications are communicated to the rehabilitation team in order to share a common strategy for the subsequent phase.

Functional assessment of key physical measures is warranted to objectively assess the recovery and to customize the progression of load. The rehabilitation team should assess:

- ***Muscle strength*** *(e.g., Isometric or Isokinetic testing).* Strength testing is helpful during the rehabilitation of orthopedic patients in the evaluation of (1) limb symmetry index and relative deficit to the uninjured limb; (2) absolute strength (peak torque) relative to body weight and patients features (e.g., anthropometrics, sports level) [25–28], and

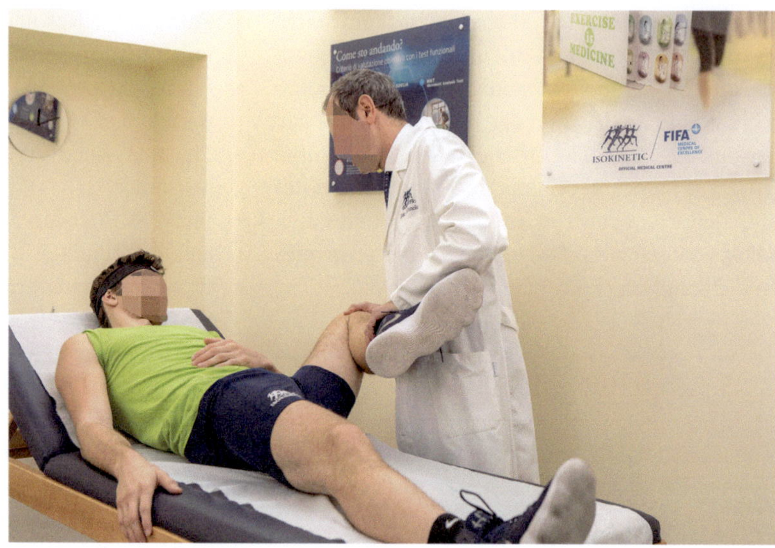

Fig. 35.1 Lower limb ROM clinical evaluation

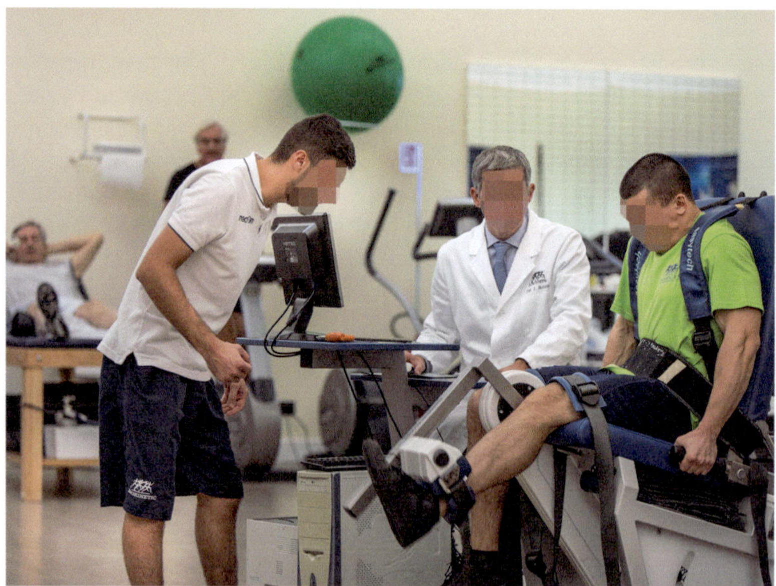

Fig. 35.2 Isokinetic knee strength test

(3) ratio between agonist and antagonist muscles (e.g., hamstring/quadriceps ratio). Strength testing (Fig. 35.2) is helpful in guiding the rehabilitation team in the decision-making processes during the progression of the rehabilitation program.

- **Cardiovascular condition** *(e.g., aerobic and anaerobic threshold test)* aerobic fitness tests may be used to identify aerobic and anaerobic thresholds and personalize the intensity of training sessions, increasing the quality of delivered sessions. An example of this kind of

tests is the threshold test, assessed during an incremental test. Aerobic threshold is identified by a capillary blood lactate concentration of 2 mmol/L [29]. Heart rate corresponding at the intensity of exercise of 2 and 4 mmol/L of blood lactate accumulation is identified and used to adjust intensity of exercises during rehabilitation [30].

- **Movement quality** *(e.g., jumping and cutting mechanics):* tests measuring the quality of movements have been published and studied widely in the knee ligament injuries domain.

Various protocols have been suggested [31, 32] aiming to clinically check various biomechanical variables (e.g., *frontal plane alignment, sagittal plane loading, asymmetries)*. This kind of tests may be useful in customizing a neuromuscular training intervention, aiming to correct an altered joint loading.

- *Performance and Sport specific measures (e.g., field testing and GPS metrics):* The clinician may incorporate additional testing in these areas to evaluate and monitor the athlete's current physical fitness. These tests are mainly focused on the evaluation of speed, as 30-m sprint running test, agility and change-of-direction ability through specific tests [33–36]. Finally, during on-field rehabilitation (OFR) external load can be monitored using global positioning system (GPS), critically helpful in optimizing the progressive increase in loading in the final phases [33].

- *Subjective recovery rating (e.g., patients reported outcomes (PROs) physical and psychological domains):* Another way to investigate clinical and functional outcomes are scales and PROs. Depending on the involved joint different scales and questionnaires can be used: for the evaluation of the knee frequently adopted scores are IKDC (International Knee Documentation Committee) and KOOS (Knee Osteoarthritis Outcome Score) [37, 38]. For the evaluation of the ankle, a useful scoring system is represented by AOFAS (American Orthopedic Foot and Ankle Society) ankle-hindfoot score [39]. Psychological PROs, such the ACL-RSI or the TSK may be also applied.

It is the author opinion that whatever are the facilities and the tests available for the evaluation, it is important to measure the progression through the rehabilitation phases and the level of recovery. Adopting clinical and functional measurement parameters allows the clinician to establish which goals of the rehabilitation program have been achieved and to schedule how to proceed in the subsequent rehabilitation phases. Define the criteria of progression and the test of evaluation, according to the surgeon and the rehabilitation team indications and experience, is therefore helpful to control the process and the results of the rehabilitation.

35.4 Rehabilitation Environments and Techniques

During the long journey of functional recovery following articular cartilage repair (up to 18 months) the patients should change frequently the rehabilitation environments (from pools to fields) to give new stimuli and progress in loading.

The *rehabilitative pool* (Fig. 35.3) is very important in cartilage pathologies where weight

Fig. 35.3 Pool assisted rehabilitation session

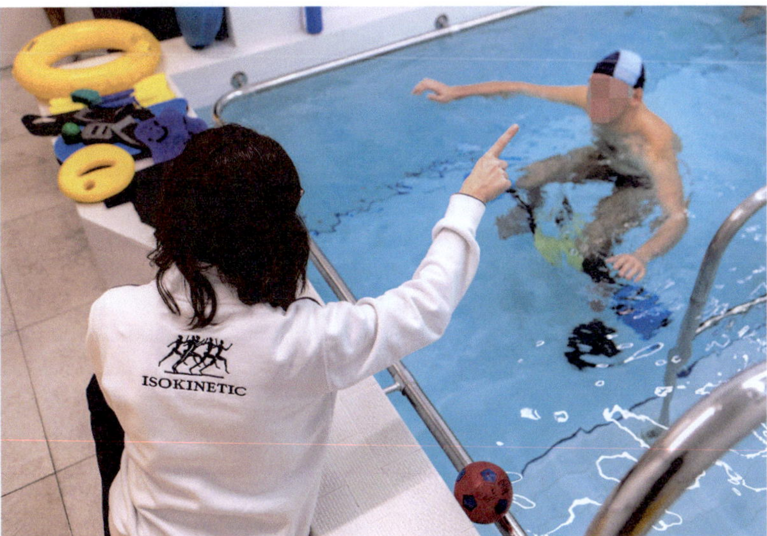

bearing is often restricted for many weeks. As well known the hydro-kinesis therapy allows early loading and joint mobilization and enhances the recovery of function [40–42]. The exercises in the water are particularly useful in the first rehabilitation phases for swelling control and recovery of range of motion, however the pool can also be used in a more advanced rehabilitation phase before the on-field rehabilitation to retrain in a safe environment sport specific gestures (jumps, changes of direction, landings, etc.) reducing impact and shear forces. In the water strength, coordination and neuromuscular performance can be improved enabling the patient to recover functional movement early on, with undoubted psychological advantages.

In the *rehabilitation gym*, patients can perform different kind of treatments in all the different phases of the rehabilitation.

In an *early stage* manual treatments and various modalities allow to enhance the recovery of passive ROM and the progressive resolution of swelling, pain, and inflammation. The control of inflammation after a trauma or surgery is important for joint homeostasis and healing. The use of ice and compression is therefore recommended in the early phases of the rehabilitation and after rehabilitation sessions. Exercises of active and assisted joint mobilization are essential to facilitate the recovery of ROM, necessary to maintain joint function. Depending on site and type of lesion or surgery restricted range of motion can be indicated for a period before allowing complete ROM exercises.

To introduce early functional movements (i.e., walking and running), commercial devices to unload the patient's body weight during treadmill ambulation can also be used [43].

In the *early and mid-stages*, the gym is also the correct environment to recover muscle strength. Neuromuscular electrical stimulation (NMES) is particularly useful in the early phases of rehabilitation when weight bearing and ROM can be restricted and complete active exercises for strengthening are not yet allowed.

Moreover, NMES can be applied to prevent and treat a delay of muscle activation caused by arthrogenic inhibition in case of persistent pain or swelling [44, 45]. In the mid stage, the patient will start progressive load in strengthening exercises. Depending on the type of injury or surgery the patient will be allowed to progress to open and closed kinetic chain (OKC and CKC) exercises. For example, in osteochondral knee pathologies, according to the site of the lesions, OKC exercises can be introduced early in the tibio-femoral joint surface lesions, while for patellofemoral lesions OKC and CKC strengthening need to be limited in terms of range of motion relatively to the exact position of the lesions. In author's experience, the progression of load in strength recovery is based on the following sequence:

(1) *isometrics,* (2) *free weights,* (3) *elastic resistance* (4) *manual resistance exercises* in the early stage, (5) *body weight,* (6) *isotonics and* (7) *isokinetics* in the mid stage, (8) *plyometrics and* (9) *functional exercises* in the late stage of the rehabilitation program.

Regardless of the type of exercise, loads can be gradually increased every session or maintained for a week before increasing the intensity of exercises. In general the principles of specificity, overload, and adaptation of the exercise must be considered to promote an effective recovery of muscle strength and function, keeping always under control joint reactions to avoid pain and effusion at any time [46].

In the mid stage, proprioceptive exercises are introduced to enhance the function of the sensorimotor system and the integration of information from peripheral mechanoreceptors [47, 48]. Balance exercises progress from *bilateral to unilateral, open to closed eyes, stable to unstable surfaces* and *simple to complex activities* up to *faster speeds, perturbations, and multidirectional stimuli.*

Functional full weight bearing exercises are gradually performed to recover the correct movement pattern for daily life activities, like walking and stair climbing.

Conditioning exercises for recovery of physical fitness are also part of the rehabilitation program. Positive effects of aerobic training are well known for general health and can promote vascularization of tissues and the processes of healing [49–51]. For athletes, the recovery of fitness conditioning is mandatory before the return to the team and useful to prevent negative effects of the fatigue in sport activity. Therefore, aerobic and anaerobic training are pursued during the entire rehabilitation process starting with exercises not stressing the involved joint, moving to exercises at low impact (i.e., arm ergometer, stationary bike, elliptical machine) up to exercises in complete weight bearing and sport specifics (flat treadmill, hill treadmill and on-field activities) [52].

In dedicated spaces it is also possible to restore neuromuscular control and movement quality using exercises with an external focus of attention [53, 54]. Special settings provided with cameras and force platform can be used for kinetic and kinematic analysis as well as for correction of movement patterns during rehabilitation sessions (Fig. 35.4). Real time and delayed video showing of specific movements in frontal and sagittal planes can be used to make the patient aware of quality of movement, developing a better consciousness of the correct pattern execution. Moreover, the use of a force platform allows the patient to understand how to manage ground reaction forces during functional and sport specific movements (i.e., jumping and cutting activities). Particular attention is placed at the capacity of the patient to absorb ground reaction forces during the eccentric phase of the movement, which is fundamental for joint protection after cartilage injury or surgery [55, 56].

The final stage of the rehabilitation is performed *on the field*. OFR is important to allow athletes to return to sport, however the on-field sessions can be useful also for non-athletes with the goal to restore general functional movements according to individual target and expectations. Therefore, the intensity and type of exercises will be adapted for each case.

A progression in five stages during the on-field rehabilitation can be adopted as previously described by our group for rehabilitation and return to sport after knee articular cartilage repair [52]. In athletes after cartilage repair procedure the on-field rehabilitation last about 8 weeks and includes aerobic and anaerobic conditioning, running, jumping, accelerations, decelerations, pivoting, cutting maneuvers, and sport specific gestures. The intensity and quality of on-field sessions can be monitored using heart rate monitors and GPS systems to prevent overloads and to distribute correct load during training sessions.

35.5 Psychological Aspects

Injuries can affect patient's well-being, especially in case of chondral and osteochondral injuries, that require long recovery times. Monitoring psychological aspects is strongly suggested to facilitate the rehabilitation process, preventing and eventually managing negative psychological responses that can affect the quality of rehabilitation and patient's outcomes. Different studies have shown that positive affective responses and high compliance with a rehabilitation plan are predictors of a successful recovery and thereby return to sport in athletes [57, 58]. For these reasons, it is advisable to work with stress management techniques, starting with learning to take responsibility and to view an injury as a challenge instead of as a failure. Moreover, it is important to create an environment that decreases negative affective responses, maximizing the rehabilitation adherence. These techniques must be shared among all the rehabilitation professionals: doctors, physiotherapists, coaches, and other professionals working with injured subjects [59]. A useful and often successful technique is represented by the goal setting, based on what the patient is trying to achieve, taking in account the psychological variables such as self-efficacy and self-satisfaction. It has been observed that it can enhance performance providing direction to the individual's effort, enhancing persistence and

Fig. 35.4 Neuromotor training session with visual feedback

facilitating the development of new strategies for improving performance [60]. Evaluation scales for fear of reinjury [61] and psychological readiness [62, 63] can be adopted to allow the patient come back to previous activities, particularly for sport active and professional athletes.

35.6 Rehabilitation Protocols

Protocols are guidelines for clinicians created from science and clinical experience. The use of protocols is particularly helpful for daily practice, where timing, goals, interventions, and criteria of progression can be shared between the rehabilitation team and with the patient. The protocol can vary depending on surgical techniques and type and site of lesions, making it difficult to resume or propose a common protocol for osteochondral lesions.

Most of the published rehabilitation protocols deal with knee joint. As an example of functional rehabilitation progression, a rehabilitation protocol for knee osteochondral cartilage surgery in sport subjects [64] is proposed in Table 35.2.

35.7 Conclusions

In conclusion, rehabilitation is a very complex phenomenon that involves the patient beyond the pathology being treated. It embraces different spheres of the "person" with the aim of improving his/her state of health. Considering the complexity of every individual person, the rehabilitation team must work carefully to avoid reducing treatment path to simple standardized rehabilitation schemes or restricting its focus to basic scientific evidence. Embedding these aspects with the clinical experience and the ability to manage the patient are currently the fundamental determinants to obtain the best possible outcomes in the rehabilitation of chondral and osteochondral injuries.

Table 35.2 Example of rehabilitation protocol of functional progression following osteochondral cartilage surgery in sport subjects [64]. Timing is indicative and should be calculated in the aftermath (a posteriori)

Phase	Goals	Interventions	Criteria to progress
Phase 1: From surgery to the end of the 3rd month	• Protect the site of surgery • Reduce pain and inflammation • Start recovery of ROM • Recovery of walk	• No weight bearing (WB) for 6 weeks • Progressive WB weeks 6–8 • Passive ROM on selected degrees • Electrical stimulation, isometrics • Active mobilization of the ankle • Pool exercises form week 3 • Stationary bike from week 6 • Stretching exercises	• Full active knee extension • Knee flexion > 120° • No or minimal pain/swelling • Correct walk pattern
Phase 2: 4th and 5th months	• Recovery of full range of motion • Increase of muscular strength • Recovery of daily life activities	• Full WB (lessen daily physical activity in presence of swelling) • Mobilization of the patella • Extension as contralateral knee • Aerobic activity low impact • Proprioceptive exercises • Eccentric strengthening of the triceps • Closed kinetic chain exercises • Pool advanced exercises	• Full range of motion • No pain and swelling. • Able to walk on a treadmill at 6 km/h for 10 min without pain and effusion
Phase 3: 6th and 7th months	• Progressive strength recovery • Return to running on treadmill	• Full WB • Maintenance of full range of motion • Stretching exercises • High speed isokinetic training • Advanced proprioceptive exercises • Open kinetic chain exercises • Running on a treadmill	• No pain and effusion • Running without pain/swelling at 8 km/h for 10 min • Recovery of strength >80% contralateral limb
Phase 4: 8th and 9th months	• Recovery of coordination • Recovery of full muscular strength and endurance	• Recovery of strength in the gym • Targeted neuromotor training • Aerobic conditioning	• No pain and effusion • No difference between the two limbs in the isokinetic strength • Proper quality and control of movement (Movement analysis test or similar functional evaluation) • Proper aerobic and anaerobic threshold (depending on type and level of sport activity)
Phase 5:from 10th to 12th month	• Recovery of sport specific skills • Return to sport • Prevent risk of reinjury	• Exercises on the field • Complete neuromotor training • Maintenance of strength and aerobic conditioning	• No pain and effusion • ROM, strength and neuromotor control as contralateral limb • Proper execution of sport specific skills

References

1. Howard JS, Ebert JR, Hambly K. Current concepts in cartilage management and rehabilitation. J Sport Rehabil. 2014;23(3):169–70.
2. Brittberg M, Imhoff AB, Henning M, Mandelbaum BR. Cartilage repair: current concepts. Guilford: DJO Publications; 2010. 206 p.
3. Wilk KE, Macrina LC, Reinold MM. Rehabilitation following microfracture of the knee. Cartilage. 2010;1(2):96–107.
4. Zhao Z, Li Y, Wang M, Zhao S, Zhao Z, Fang J. Mechanotransduction pathways in the regulation of cartilage chondrocyte homoeostasis. J Cell Mol Med. 2020;24(10):5408–19.
5. Vanwanseele B, Lucchinetti E, Stüssi E. The effects of immobilization on the characteristics of articular cartilage: current concepts and future directions. Osteoarthr Cartil. 2002;10(5):408–19.
6. Salter RB, Simmonds DF, Malcolm BW, Rumble EJ, MacMichael D, Clements ND. The biological effect of continuous passive motion on the healing of full-thickness defects in articular cartilage. An experi-

mental investigation in the rabbit. J Bone Joint Surg Am. 1980;62(8):1232–51.

7. Eckstein F, Faber S, Mühlbauer R, Hohe J, Englmeier K-H, Reiser M, et al. Functional adaptation of human joints to mechanical stimuli. Osteoarthr Cartil. 2002;10(1):44–50.

8. Eckstein F, Hudelmaier M, Putz R. The effects of exercise on human articular cartilage. J Anat. 2006;208(4):491–512.

9. Bader DL, Salter DM, Chowdhury TT. Biomechanical influence of cartilage homeostasis in health and disease. Arthritis. 2011;2011:979032.

10. Shioji S, Imai S, Ando K, Kumagai K, Matsusue Y. Extracellular and intracellular mechanisms of mechanotransduction in three-dimensionally embedded rat chondrocytes. PLoS One. 2014;9(12):e114327.

11. Khan KM, Scott A. Mechanotherapy: how physical therapists' prescription of exercise promotes tissue repair. Br J Sports Med. 2009;43(4):247–52.

12. Leong DJ, Hardin JA, Cobelli NJ, Sun HB. Mechanotransduction and cartilage integrity. Ann N Y Acad Sci. 2011;1240:32–7.

13. Buschmann MD, Gluzband YA, Grodzinsky AJ, Hunziker EB. Mechanical compression modulates matrix biosynthesis in chondrocyte/agarose culture. J Cell Sci. 1995;108(Pt 4):1497–508.

14. Säämänen AM, Tammi M, Jurvelin J, Kiviranta I, Helminen HJ. Proteoglycan alterations following immobilization and remobilization in the articular cartilage of young canine knee (stifle) joint. J Orthop Res. 1990;8(6):863–73.

15. Behrens F, Kraft EL, Oegema TR. Biochemical changes in articular cartilage after joint immobilization by casting or external fixation. J Orthop Res. 1989;7(3):335–43.

16. Hambly K, Bobic V, Wondrasch B, Van Assche D, Marlovits S. Autologous chondrocyte implantation postoperative care and rehabilitation: science and practice. Am J Sports Med. 2006;34(6):1020–38.

17. CSCS RCMPDSMeA. Postsurgical orthopedic sports rehabilitation: knee & shoulder. 1st ed. St. Louis, Mo: Mosby; 2006. 640 p.

18. Sanchez-Adams J, Leddy HA, McNulty AL, O'Conor CJ, Guilak F. The mechanobiology of articular cartilage: bearing the burden of osteoarthritis. Curr Rheumatol Rep. 2014;16(10):451.

19. Peterson L, Brittberg M, Kiviranta I, Akerlund EL, Lindahl A. Autologous chondrocyte transplantation. Biomechanics and long-term durability. Am J Sports Med. 2002;30(1):2–12.

20. Wilk KE, Briem K, Reinold MM, Devine KM, Dugas J, Andrews JR. Rehabilitation of articular lesions in the athlete's knee. J Orthop Sports Phys Ther. 2006;36(10):815–27.

21. DePalma AA, Gruson KI. Management of cartilage defects in the shoulder. Curr Rev Musculoskelet Med. 2012;5(3):254–62.

22. Hensley CP, Sum J. Physical therapy intervention for a former power lifter after arthroscopic microfrac-ture procedure for grade IV glenohumeral chondral defects. Int J Sports Phys Ther. 2011;6(1):10–26.

23. Wilk KE, Macrina LC, Cain EL, Dugas JR, Andrews JR. Rehabilitation of the overhead Athlete's elbow. Sports Health. 2012;4(5):404–14.

24. Domayer SE, Welsch GH, Stelzeneder D, Hirschfeld C, Quirbach S, Nehrer S, et al. Microfracture in the ankle. Cartilage. 2011;2(1):73–80.

25. Nunes RFH, Dellagrana RA, Nakamura FY, Buzzachera CF, Almeida FAM, Flores LJF, et al. Isokinetic assessment of muscular strength and balance in Brazilian elite futsal players. Int J Sports Phys Ther. 2018;13(1):94–103.

26. Kannus P. Isokinetic evaluation of muscular performance: implications for muscle testing and rehabilitation. Int J Sports Med. 1994;15(Suppl 1):S11–8.

27. Gaines JM, Talbot LA. Isokinetic strength testing in research and practice. Biol Res Nurs. 1999;1(1):57–64.

28. Cvjetkovic DD, Bijeljac S, Palija S, Talic G, Radulovic TN, Kosanovic MG, et al. Isokinetic testing in evaluation rehabilitation outcome after ACL reconstruction. Med Arch. 2015;69(1):21–3.

29. Foster C, Fitzgerald DJ, Spatz P. Stability of the blood lactate-heart rate relationship in competitive athletes. Med Sci Sports Exerc. 1999;31(4):578–82.

30. Jacobs I. Blood lactate. Implications for training and sports performance. Sports Med. 1986;3(1):10–25.

31. Paterno MV, Schmitt LC, Ford KR, Rauh MJ, Myer GD, Huang B, et al. Biomechanical measures during landing and postural stability predict second anterior cruciate ligament injury after anterior cruciate ligament reconstruction and return to sport. Am J Sports Med. 2010;38(10):1968–78.

32. Myer GD, Ford KR, Hewett TE. Tuck jump assessment for reducing anterior cruciate ligament injury risk. Athl Ther Today. 2008;13(5):39–44.

33. Buckthorpe M, Della Villa F, Della Villa S, Roi GS. On-field rehabilitation part 1: 4 pillars of high-quality on-field rehabilitation are restoring movement quality, physical conditioning, restoring sport-specific skills, and progressively developing chronic training load. J Orthop Sports Phys Ther. 2019;49(8):565–9.

34. Buckthorpe M, Della Villa F, Della Villa S, Roi GS. On-field rehabilitation part 2: a 5-stage program for the soccer player focused on linear movements, multidirectional movements, soccer-specific skills, soccer-specific movements, and modified practice. J Orthop Sports Phys Ther. 2019;49(8):570–5.

35. Young W, Russell A, Burge P, Clarke A, Cormack S, Stewart G. The use of sprint tests for assessment of speed qualities of elite Australian rules footballers. Int J Sports Physiol Perform. 2008;3(2):199–206.

36. Sayers MGL. Influence of test distance on change of direction speed test results. J Strength Cond Res. 2015;29(9):2412–6.

37. Higgins LD, Taylor MK, Park D, Ghodadra N, Marchant M, Pietrobon R, et al. Reliability and validity of the international knee documentation commit-

tee (IKDC) subjective knee form. Joint Bone Spine. 2007;74(6):594–9.

38. Collins NJ, Prinsen CAC, Christensen R, Bartels EM, Terwee CB, Roos EM. Knee injury and osteoarthritis outcome score (KOOS): systematic review and meta-analysis of measurement properties. Osteoarthr Cartil. 2016;24(8):1317–29.

39. Christensen BB, Foldager CB, Jensen J, Jensen NC, Lind M. Poor osteochondral repair by a biomimetic collagen scaffold: 1- to 3-year clinical and radiological follow-up. Knee Surg Sports Traumatol Arthrosc. 2016;24(7):2380–7.

40. Buckthorpe M, Pirotti E, Villa FD. Benefits and use of aquatic therapy during rehabilitation after ACL reconstruction - a clinical commentary. Int J Sports Phys Ther. 2019;14(6):978–93.

41. Stuart AR, Doble J, Presson AP, Kubiak EN. Anatomic landmarks facilitate predictable partial lower limb loading during aquatic weight bearing. Curr Orthop Pract. 2015;26(4):414–9.

42. Kutzner I, Richter A, Gordt K, Dymke J, Damm P, Duda GN, et al. Does aquatic exercise reduce hip and knee joint loading? In vivo load measurements with instrumented implants. PLoS One. 2017;12(3). https://doi.org/10.1371/journal.pone.0171972.g006.

43. Liang J, Lang S, Zheng Y, Wang Y, Chen H, Yang J, et al. The effect of anti-gravity treadmill training for knee osteoarthritis rehabilitation on joint pain, gait, and EMG. Medicine (Baltimore). 2019;98(18):e15386.

44. Doucet BM, Lam A, Griffin L. Neuromuscular electrical stimulation for skeletal muscle function. Yale J Biol Med. 2012;85(2):201–15.

45. Paternostro-Sluga T, Fialka C, Alacamliogliu Y, Saradeth T, Fialka-Moser V. Neuromuscular electrical stimulation after anterior cruciate ligament surgery. Clin Orthop Relat Res. 1999;368:166–75.

46. McArdle WD, Katch FI, Katch VL. Exercise physiology: nutrition, energy, and human performance. Philadelphia: Wolters Kluwer Health; 2014.

47. Aman JE, Elangovan N, Yeh I-L, Konczak J. The effectiveness of proprioceptive training for improving motor function: a systematic review. Front Hum Neurosci. 2015;8:1075.

48. Duman I, Taskaynatan MA, Mohur H, Tan AK. Assessment of the impact of proprioceptive exercises on balance and proprioception in patients with advanced knee osteoarthritis. Rheumatol Int. 2012;32(12):3793–8.

49. Alghadir AH, Gabr SA, Al-Eisa ES, Alghadir MH. Correlation between bone mineral density and serum trace elements in response to supervised aerobic training in older adults. Clin Interv Aging. 2016;11:265–73.

50. Warburton DER, Nicol CW, Bredin SSD. Health benefits of physical activity: the evidence. CMAJ. 2006;174(6):801–9.

51. Alghadir AH, Aly FA, Gabr SA. Effect of moderate aerobic training on bone metabolism indices among adult humans. Pak J Med Sci. 2014;30(4):840–4.

52. Mithoefer K, Hambly K, Logerstedt D, Ricci M, Silvers H, Della VS. Current concepts for rehabilitation and return to sport after knee articular cartilage repair in the athlete. J Orthop Sports Phys Ther. 2012;42(3):254–73.

53. Park SH, Yi CW, Shin JY, Ryu YU. Effects of external focus of attention on balance: a short review. J Phys Ther Sci. 2015;27(12):3929–31.

54. Wulf G, Chiviacowsky S, Schiller E, Ávila LTG. Frequent external-focus feedback enhances motor learning. Front Psychol. 2010;1:190.

55. LaStayo PC, Woolf JM, Lewek MD, Snyder-Mackler L, Reich T, Lindstedt SL. Eccentric muscle contractions: their contribution to injury, prevention, rehabilitation, and sport. J Orthop Sports Phys Ther. 2003;33(10):557–71.

56. Lepley LK, Wojtys EM, Palmieri-Smith RM. Combination of eccentric exercise and neuromuscular electrical stimulation to improve quadriceps function post-ACL reconstruction. Knee. 2015;22(3):270–7.

57. Daly JM, Brewer BW, Raalte JLV, Petitpas AJ, Sklar JH. Cognitive appraisal, emotional adjustment, and adherence to rehabilitation following knee surgery. J Sport Rehabil. 1995;4(1):23–30.

58. Ninedek A, Kelt GS. Sport physiotherapists' perceptions of psychological strategies in sport injury rehabilitation. J Sport Rehabil. 2000;9(3):191–206.

59. Ivarsson A, Tranaeus U, Johnson U, Stenling A. Negative psychological responses of injury and rehabilitation adherence effects on return to play in competitive athletes: a systematic review and meta-analysis. Open Access J Sports Med. 2017;8:27–32.

60. Theodorakis Y, Beneca A, Malliou P, Goudas M. Examining psychological factors during injury rehabilitation. J Sport Rehabil. 1997;6(4):355–63.

61. Hsu C-J, Meierbachtol A, George SZ, Chmielewski TL. Fear of reinjury in athletes. Sports Health. 2017;9(2):162–7.

62. Conti C, di Fronso S, Robazza C, Bertollo M. The injury-psychological readiness to return to sport (I-PRRS) scale and the sport confidence inventory (SCI): a cross-cultural validation. Phys Ther Sport. 2019;40:218–24.

63. Glazer DD. Development and preliminary validation of the injury-psychological readiness to return to sport (I-PRRS) scale. J Athl Train. 2009;44(2):185–9.

64. Marcacci M, Zaffagnini S, Kon E, Delcogliano M, Di Martino A, Filardo G, et al. Maioregen. Arch Ortop Reumatol. 2009;120(3):35–7.

MIX
Papier aus verantwortungsvollen Quellen
Paper from responsible sources
FSC® C105338

If you have any concerns about our products,
you can contact us on
ProductSafety@springernature.com

In case Publisher is established outside the EU,
the EU authorized representative is:
Springer Nature Customer Service Center GmbH
Europaplatz 3, 69115 Heidelberg, Germany

Printed by Libri Plureos GmbH
in Hamburg, Germany